The Engines of Pratt & Whitney
A Technical History

The Engines of Pratt & Whitney
A Technical History

Jack Connors

LIBRARY OF FLIGHT

Ned Allen, Editor-in-Chief
Lockheed Martin Corporation
Bethesda, Maryland

Published by
American Institute of Aeronautics and Astronautics, Inc.
1801 Alexander Bell Drive, Reston, VA 20191-4344

American Institute of Aeronautics and Astronautics, Inc., Reston, Virginia

1 2 3 4 5

Library of Congress Cataloging-in-Publication Data

Connors, Jack.
　The engines of Pratt & Whitney : a technical history / Jack Connors ; Ned Allen, editor-in-chief.
　　p. cm.
　ISBN 978-1-60086-711-8
　1. Airplanes--Motors--History. 2. Aircraft gas-turbines--History. 3. Jet engines--History. 4. Pratt & Whitney Aircraft Group--History. I. Allen, Ned. II. American Institute of Aeronautics and Astronautics. III. Title.
　TL703.P7C58 2009
　629.134′35—dc22

2009036431

Cover design by Jim Killian.
Cover images from left to right: Pratt & Whitney logo, the R-2800 engine, George Mead, Frederick Rentschler, Leonard Hobbs with the J57; courtesy of Pratt & Whitney, A United Technologies company.

Copyright © 2010 by the American Institute of Aeronautics and Astronautics, Inc. All rights reserved. Printed in the United States of America. No part of this publication may be reproduced, distributed, or transmitted, in any form or by any means, or stored in a database or retrieval system, without the prior written permission of the publisher.

Data and information appearing in this book are for informational purposes only. AIAA is not responsible for any injury or damage resulting from use or reliance, nor does AIAA warrant that use or reliance will be free from privately owned rights.

Dedication

This book is dedicated to my wife Evelyn who endured the unbelievable chaos in my den while I tried to work on this book over a ten-year period.

It is also dedicated to the Pratt & Whitney engineers who, during World War II, developed the engines to give the Army Air Corps/Army Air Forces and Navy airmen control of the air, and who have worked since then to make further gains in aircraft propulsion for both military and commercial applications.

I also acknowledge the kind appreciation from those pilots who made it back to base while operating Pratt & Whitney engines beyond the specification ratings. Such appreciation softens the sharp, taunting remarks by overly optimistic airframe salesmen and our competitors when they referred to us as "the Iron Works" because, in their estimation, our engines were too heavy—a criticism seldom made by pilots returning from combat missions.

FOREWORD

Taking on any book project is a daunting task, and taking on the history of an aircraft engine company is even more so. Although aircraft engines are thought of as inanimate objects, the people behind them—engineers, designers, test engineers, production people, plus the myriad of other skills necessary to bring an aircraft engine into production—puts humanity into the equation. Jack Connors is to be applauded for taking on this monumental task and tackling it with the professionalism in all that he does, no doubt fortified on occasion by his favorite Irish adult beverage.

Jack's book is long overdue. With the advantage of searching through the Pratt & Whitney archives, Jack has unearthed a treasure trove of information. Being a long-time employee, he was in the fortunate position of rubbing shoulders with every mover and shaker in the company from the late 1940s to his retirement in the 1980s. His unique sense of humor and wit allowed him to gain the confidence and friendship of these folks, everyone from shop floor personnel to top-level executives. But do not be fooled by Jack's amiable, easy-going personality. He is one very sharp and dedicated engineer with a profound sense of history.

The above mentioned qualifications make Jack the ideal person to write this book. I know it has been a long and difficult road for him but as they say, if it was easy, everyone would be doing it. Knowing all the top people has made for an ideal mixture of technical descriptions interspersed with the personalities involved, plus many anecdotes thrown in for good measure—and this is straight from the horse's mouth. Jack was there experiencing the successes as well as the heartaches of failure. Intermingling brief autobiographies with the technical issues gives the reader a unique insight into what it takes to develop an engine. At times it would appear that a prime requisite to survive on the job was to have a good sense of humor.

Starting with the piston engine, Jack walks the reader through the early engines such as the Wasp and Wasp Jr., and ends up with the monumental R-4360, the absolute pinnacle of aircraft piston engine development. It has been said on many previous occasions but is worth repeating here, it is a good thing jets came along when they did; the complexity of engines such as the

R-4360 was simply stupendous. Just as the R-4360 became a viable engine, the majority of engineers were siphoned off into the brand new business of gas turbines. Starting from nothing, Pratt & Whitney picked up the reins and learned the ropes in no time. Cutting their teeth on license-built centrifugal flow Rolls-Royce Nenes, Pratt & Whitney quickly realized that the future was in axial flow turbines. Amazingly, their first effort, the immortal J57, was a world beater. Understanding the difficulty of designing an axial flow compressor, it is all the more remarkable that this fledgling effort turned out be so successful. Indeed, its success resulted in Luke Hobbs being awarded the prestigious Collier Trophy in 1952 for his work on it.

The camaraderie that existed at Pratt & Whitney during Jack's tenure comes through loud and clear. Morale, even under difficult circumstances, was always of the highest order. Another observation that comes through unmistakably is the respect that was displayed for everyone at the company. Co-workers could joke and kid around, yet display the utmost respect toward each other. The employees actually enjoyed their jobs and looked forward to going to work. This contrasts sharply with today's modern corporations, run by faceless and humorless MBAs, and seemly staffed by equally faceless and humorless employees sitting in front of CAD screens where quarterly earnings command the utmost attention. It is with the preceding in mind that Pratt & Whitney's glory days have been so well-documented with all the personalities, foibles, and yes, failings of human beings. We owe Jack Connors a debt of gratitude.

Graham White
Lake City, Florida

Contents

Preface xv

Acknowledgments xvii

Chapter 1   Pre-Flight Briefing 1

Introduction 1
How the Marketplace Changed Since 1925 1
Pratt & Whitney's Five Defining Moments 8
Pratt & Whitney's Founders 9
The Founding of Pratt & Whitney Aircraft Company 12
The R-1340 Wasp Joins the Navy and Air Mail Service 16
The Formation of United Aircraft & Transport Corporation .. 18
The Break Up of UA&TC 19
Pratt & Whitney's Great Contribution in World War II 22
The Aircraft Gas Turbine Business after World War II 23
Rentschler's Accomplishments 27
Rentschler's Final Years 28
References 28

Chapter 2   The Early Years 31

George Mead and His First Pratt & Whitney Engine 31
Liquid-Cooled Engines 32
Air-Cooled Engines 33
Summary of Aircraft Engine Growth 47
George Mead and the T-Engine 48
Mead and Willgoos Design the Wasp in 1925 50
Mead Flies with Wasp on Mail Routes 57
Epilogue for George J. Mead 58
References 60

ix

CHAPTER 3   THE LATER PISTON ENGINE YEARS ... 61

Introduction ... 61
Hobbs Reveals the Secret of Success ... 63
Succession of Pratt & Whitney Engines ... 64
The Wasp and the Hornet ... 67
Roles of Wasp and Hornet in International Commercial Aviation ... 79
Wasp and Hornet Engines Are Refined ... 82
The Contributions of Two Engineers ... 120
References ... 123

CHAPTER 4   THE PISTON ENGINE EXPERIENCE ... 125

Introduction ... 125
Pratt & Whitney Flight Test Organization ... 125
R-4360 Test Flight ... 127
R-4360 Flies from Hawaii to Chicago ... 129
R-1830 Brings the Crew in Safely ... 130
R-1830 and Superchargers ... 132
Life as a Test Engineer ... 134
R-2800C Program ... 136
Controllable-Pitch Propeller ... 137
R-2800-32W's Rough Idle ... 138
R-1830s and the Luftwaffe ... 138
R-1830 Bearing Problem ... 139
R-1535 and Howard Hughes ... 141
Boeing 247's Wasp Engine Whine ... 142
Pratt & Whitney Engines with Fuel Injection ... 143
Liquid-Cooled Engines—Side Trip to Nowhere ... 145
Looking Back on Piston Engine Development ... 149
Final Observations on Piston Engines ... 157
References ... 159

CHAPTER 5   TRANSITION TO GAS TURBINES ... 161

Looking Back ... 161
Beginnings of PT1 ... 162
PT1 Program ... 167
Hobbs Moves to Gas Turbines ... 172
PT2 Program Begins ... 176
PT2 in the Lockheed Constellation ... 182
Two Engineers Rise to Prominence at Pratt & Whitney ... 182
References ... 186

Chapter 6 WWII Ends and Turbojet Development Begins	189
Pratt & Whitney Service Organization	189
Army and Navy Engine Training School	192
Turbojet Development Begins	196
American Companies Enter the Jet Age	197
European Development	199
Pratt & Whitney's First Turbojet Success	203
The J48 (JT7)	208
References	214

Chapter 7 Birth of the Two-Spool Turbojet	215
Introduction	215
Genesis of the J57	216
Meeting the Challenge of Creating Enough Propulsion	219
Concerns Regarding High-Pressure-Ratio Compressors	221
Technical Concerns of a Twin-Spool Turbojet	221
Air Force Approval for PT4 to JT3 Switch	223
Description of JT3-8	224
Description of the JT3-10	226
Soderberg Recommends a New Start	227
JT3A with a Wasp Waist Design	229
The J57 in the B-52	238
JT3 Commercial Derivative	252
Hobbs' Reflections on the Dawn of the Jet Age	254
References	255

Chapter 8 Four More Turbojets	257
Introduction	257
J75 (JT4) Twin-Spool Turbojet	257
J52 (JT8)	270
J91 (JT9)	279
J60 (JT12) Turbojet	285
References	290

Chapter 9 Transition to Turbofans	291
Introduction	291
Birth of the Turbofan	291
T57/PT5—Pratt & Whitney's Most Powerful Turboprop	293

JT10 Afterburning Turbofan	294
JT3D/TF33—Turbofan Fever Fuels Innovation	296
Three Programs Herald the Future	310
Suntan Project	310
Conclusion	317
References	317

CHAPTER 10   HIGHER AND FASTER 319

Introduction	319
RL10 Program	319
J58—Growing Mach Capability	321
Liquid Air Condensing Engine (LACE)	333
Florida Research & Development Center	334
References	339

CHAPTER 11   GOING COMMERCIAL 341

Introduction	341
Origin of the TF30 (JTF10A)	341
JT8D—Feet First into Commercial Service	348
40 Years in Service	364
References	366

CHAPTER 12   CHALLENGES AND NEW TURBOFANS 367

Introduction	367
Supporting Activities for Changing Times	367
Developing Space Technology	373
Developing New Engine Technology	377
Supersonic Transport Engine Program	379
F100—The Ultimate Military Engine	382
The Great Engine War	393
The Battle Won	395
References	395

CHAPTER 13   HIGH-BYPASS FANS 397

Introduction	397
Light Weight Gas Generator (LWGG) Program	397
C-5 Engine Program	399
Path to the JT9D Program	403
JT9D Program for Boeing's 747 Begins	405
JT9D Engine Description	409

Synopsis of the First 16 Months in the Air	414
Evolution of the JT9D Family	415
Adventures in Marketing	417
References	424

CHAPTER 14 THE MODERN ERA — 427

Introduction	427
JT10D Path to the PW-2037	427
Evolution of the B-757 and PW2037	434
International Aero Engines (IAE) and the V2500	439
PW4000 Series of High-Bypass Engines	442
Engines Following the PW4000	447
PW6000	448
What Became of the F100?	451
F119—Advances in Engine Development	452
Integrated Product Development (IPD)	458
Thrust Vectoring	463
F135—How Times Have Changed	464
PW1000G	469
Advances in Engine Technology	472
References	472

CHAPTER 15 LOOKING BACK 80 YEARS — 473

Aviation in History	473
Collier Awards as Overview of Aviation History	474
Piston Engine Era	474
Gas Turbine Engine Era	476
Progress in Airframe Technology	478
Airframe's Demand for Power	481
Progress in Engine Technology	483
Progress in Engine Controls	492
Then and Now	496
Pratt & Whitney's Finest Moments	496
Rentschler's Legacy	500
References	502

APPENDIX PRATT & WHITNEY MEDALLION — 503

AFTERWORD — 505

Early Influences	505
Early Days at Pratt & Whitney	506
Engine Reliability—The Key to Commercial Success	507

Propulsion Paradigm Shift *507*
Meeting the Inventors *508*
Golden Eagles *510*
Closing Comments *511*

INDEX **513**

SUPPORTING MATERIALS **529**

PREFACE

Pratt & Whitney neither authorized me to write this history nor made any comments on the manuscript. Pratt & Whitney's only connection with my effort is its generous permission for me to use material from their archives. This book is a labor of love for me and about 25 other retired engineers who have helped me with the history as seen through their eyes.

In 1950 Pratt & Whitney published a fine history of its first 25 years in *The Pratt & Whitney Aircraft Story*. My purpose is to provide an engineering history of the company's piston and gas turbine engines that includes engine descriptions, stories about the people who made the history, and how the engine programs came into being. I have tried to use as many as possible personal stories from the participants to put a human face on the engineering details.

There were two events in 1999 that coalesced to launch my effort. The first was a suggestion by retired senior Pratt & Whitney executive Don Jordan and the second was an invitation from Dick Wellman, General Manager of the Customer Training Center (CTC) at the time. Don Jordan asked me at a luncheon with about 50 retired engineers to look around at all the aircraft engine experience in the room. He said, "When these fellows go, nobody will ever know what they did. Somebody ought to write the history of what they worked on." There was only one man in that room who was so naive that he committed to writing a history of Pratt & Whitney. Little did I know what I was getting into!

Dick Wellman, General Manager of the CTC in 1998 was an aviation enthusiast and had jurisdiction over the remnants of the Pratt & Whitney Archives. After the original archives building was demolished, the contents were bundled up for storage in the basement of one of the older buildings. The contents consisted of about 1000 boxes and 50 file cabinets. There was no card index file, let alone a computer-generated database or spreadsheet.

In 1998 Jesse Hendershot, a retired CTC instructor, volunteered to go through each box and file cabinet to create a computer-generated card index file for the archives. I joined Jesse in this effort the following year. Later Dick Wellman provided an attractive home for the archives in one of his hangars and opened the archives for use by Pratt & Whitney employees as well as aviation buffs.

After resurrection of the archives, at least seven years of research in the archives, more than 20 interviews with the history makers, a couple of crashed computers, several printers, bales of discarded rough drafts, and countless hours of frustrated labor the book is complete. I am greatly indebted to Pratt & Whitney for the privilege of unrestricted access to the archives. Without such access, this book would not have been possible.

Jack Connors
December 2009

ACKNOWLEDGMENTS

This effort is not a solo performance. I have tried to include input from others to make the history more complete and meaningful to those who are interested in Pratt & Whitney's piston and gas turbine history. Initially, I planned to cover only gas turbines. However, the piston engine history is just as exciting (and in many instances more so). Those engineers who made the company number one in piston engine technology also used their talents to make the company successful in the jet era. My treatment of the piston engine history, however, does not repeat what is in the *Pratt & Whitney Aircraft Story*. I have focused on material not previously presented about people and engines.

CONTRIBUTORS

The following people have been very generous in giving material appropriate for the history. If there are any errors or inconsistencies, the blame is mine.

Andy Anderson	Mick Jefferies
Bill Andersen	Don Jordan
Bob Abernethy	Joel Kuhlberg
Gordon Beckwith	Ron Lindlauf
Don Brendal	Bill Martens
Niles Brook	Bob McAvoy
Jim Brown	Frank McAbee
Roger Bursey	Jack McDermott
Larry Carlson	Bill McGaw
Nils Carlson	Captain Al Mitchell
Walt Doll	Gene Montany
Bob Fitzgerald	Tom Pelland
Frank Gillette	Don Pascal
Jesse Hendershot	Captain Bob Polson
Phil Hopper	Don Rudolph
Cliff Horne	Bill Russell
Dean Iosbaker	Harry Schmidt

Ed Schneider	Hans Stargardter
Anne Schneider	Stan Taylor
Bill Sens	Bob Toft
Ted Slaiby	Tom Tumicki
Dick Smith	Dana Waring
Bob Stanwood	Jack Wells

The following two non-Pratt & Whitney gentlemen have helped me with piston engine fundamentals and prevented me from making wildly misleading statements on engines. Once, again, they are held blameless in the event of my error:

Graham White, aviation historian and author of *Allied Aircraft Piston Engines of World War II*; *R-2800—Pratt & Whitney's Dependable Masterpiece*; and *R-4360—Pratt & Whitney's Major Miracle*.

Kimble D. McCutcheon, President of the Aircraft Engine Historical Society and author of *Tornado—Wright Aero's Last Liquid-Cooled Piston Engine*.

For access to the Pratt & Whitney Archives, special thanks to Richard C. Wellman, retired General Manager of Customer Training. Together with Jesse Hendershot and Dean Iosbaker, we resurrected the archives. Thanks also to Gary E. Minor, Director of Government Affairs at United Technologies Corporation.

I would also like to mention in appreciation the two professional archivists employed by Pratt & Whitney who originally organized the material in the archives, Harvey Lippincott and Anne Millbrooke.

And finally, much thanks to Gene Montany and Jesse Hendershot for their complete review of the book and suggestions to improve the text.

Chapter 1

PRE-FLIGHT BRIEFING

INTRODUCTION

To understand the story of Pratt & Whitney and put its role in aviation in perspective, this "pre-flight briefing" explains what was going on in government and the private sector during an 80-year time line. The pre-flight briefing consists of how the aviation marketplace has changed since 1925; how the military and commercial market for air-cooled piston engines developed; how Pratt & Whitney's principal founder, Frederick Rentschler, guided the company through five challenging events that occurred during his tenure; how Pratt & Whitney worked with the automobile industry to produce its engines for World War II (WWII); and how Pratt & Whitney entered the gas turbine era. To review basic concepts of aviation, see the last chapter.

HOW THE MARKETPLACE CHANGED SINCE 1925

The aviation marketplace over the past 80 years has been affected by a series of military actions overseas, advances in propulsion technology, and increasing involvement of more people in the engine procurement process. Changes in the marketplace have occurred because military and commercial aviation requirements diverged, financial risks in new engine ventures became greater, and foreign and domestic governments became more involved in aviation.

MILITARY AND COMMERCIAL AVIATION DIRECTIONS DIVERGED

In the piston engine era, it made sense to take a military engine that had accumulated about a quarter of a million engine-hours in service and offer that as a commercial engine. In the gas turbine era, this practice continued up through the introduction of turbofans, after which time the design of military and commercial engines moved in different directions. The J57, J75, and TF33 were the last of the Pratt & Whitney gas turbines to start out with military applications and then move into commercial service. Even in this case, commercial engines were piling up more hours in service than their corresponding models in the military. As a consequence, problems were discovered in commercial applications before the military engines encountered them. Subsequently,

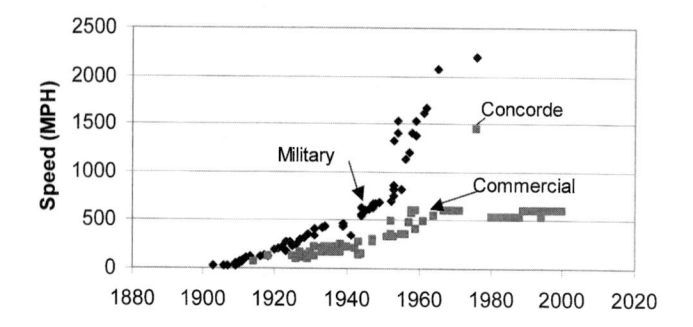

Fig. 1 Commercial aviation speeds leveled off below Mach 1 while military speeds climbed into the supersonic regime. The highest speeds were from the military YF-12 and the SR-71 reconnaissance aircraft.

commercial fixes to problems found their way back into the military engines. In the 1970s, the divergence of the military and commercial applications was so pronounced that Pratt & Whitney formed two separate divisions, the Government Products Division and the Commercial Products Division, to deal with the two widely different customer bases (see Figs. 1 and 2).

Greater Financial Risk of New Engine Ventures

The airplane grew in complexity over an 80-year time span, resulting in better performance but with fewer applications and a higher price tag. Pratt & Whitney's first engine, the 450-hp (horsepower) Wasp piston engine, powered almost 100 different aircraft models. In fact, the average number of applications for all Pratt & Whitney piston engines was over 40, a number strongly influenced by WWII needs. It was a marketplace in which an engine company said to the airframers, "Here's the engine. Use it."

In the gas turbine era, the situation changed after the introduction of the turbofan in the early 1960s. Then the airframers came to the engine companies and announced, "Here are our requirements for an engine. What can you do for us?"

Not only were there fewer applications for gas turbines, compared to piston engines, but the cost to bring an engine into commercial service became horrendous. For example, in the later period of the 80-year time span, it cost about $2 billion to bring an engine into commercial service. In addition, the engine manufacturer would not get to the break-even point until 10 to 12 years after program go-ahead! This high-risk situation created the need for partners to make the financial risk more manageable.

Another aspect of the risk was the requirement for the engine manufacturer to launch its engine program about a year before the airplane company launched its program. The engine company was out on a limb for millions of dollars until the airplane company made the commitment to go ahead.

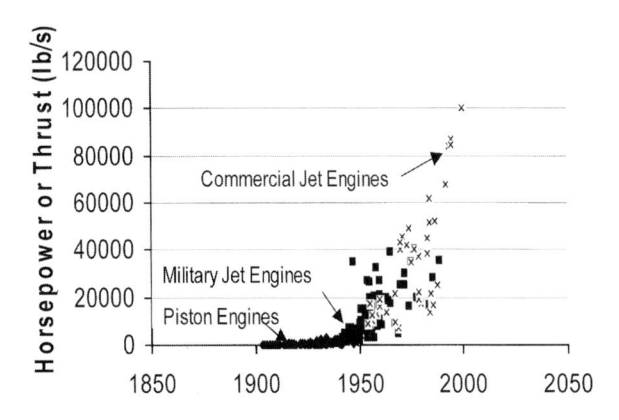

Fig. 2 The power output of commercial aircraft kept climbing while that of the military engines leveled off around 40,000 lb of thrust.

Sometimes, the airframer would discover (after trying to line up a couple of major domestic and foreign airlines) that there really was not sufficient airline interest in a particular airplane to warrant a full go-ahead. In the meantime, of course, the engine company suffered a multi-million dollar hemorrhage.

In the piston engine era, the purchasing decisions were made at the top of military and commercial organizations because those executives and generals knew their engines. However, the gas turbine era brought on such a paradigm shift in technology that those at the top no longer knew engines. So they created organizations of experts. In the piston engine era, General Arnold, head of the Army Air Corps, might have consulted with a couple of his associates and then launched a new engine program. However, a gas turbine engine program such as the F100 program involved about a hundred people evaluating proposals from the airframers and engine companies for many hours (see Fig. 3).

The modern procurement of an engine for a commercial application would not involve as many man-hours as a military application because the commercial application was not as complex as that of an Air Force fighter engine. Nevertheless, commercial applications could involve about 10 to 15 percent of the proposal evaluation effort in a military engine program. What this meant in practical terms was that the marketing function in both military and commercial markets became more complicated and the proposal preparation effort grew considerably.

FOREIGN COMPANY AND GOVERNMENT INVOLVEMENT

In the late 1920s, Germany's BMW manufactured Pratt & Whitney engines under license. In the early 1930s, Japan's Mitsubishi also built Pratt & Whitney engines under license. However, around the 1970s, foreign engine companies were not content with building engines under license. They wanted

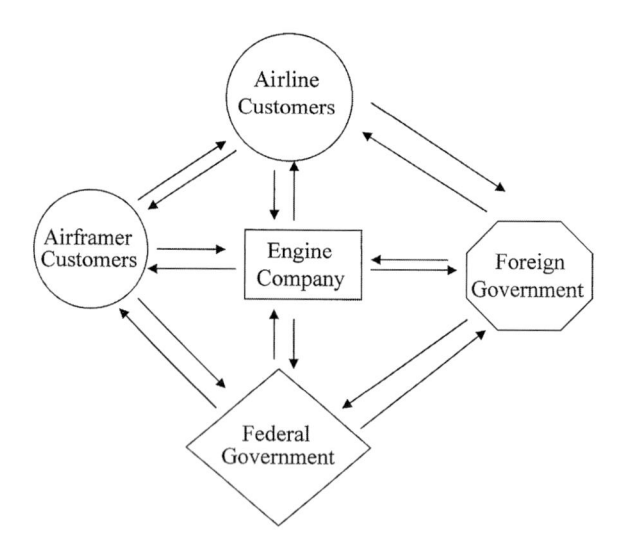

Fig. 3 The aviation business became more complex for engine companies who had to communicate with more agencies.

to participate in all phases of design, development, production, sales, and product support. In some instances foreign partners were beneficial, breaking into areas where American engine companies did not have a chance on their own (governments often have high-ranking politicians act as pitchmen for products made in their countries).

In both the military and commercial ventures abroad it has become necessary to share the production of the equipment in order to make the sale. When foreign governments own the airlines in their countries, one might get the order for engines from a foreign airline, but that contract award is subject to review by government officials. Many times the sale will not go through unless there are some local jobs created by sharing the production of some hardware. Even in domestic government markets, local politicians sometimes weigh in with the Department of Defense (DoD) as to why the engine company in their district should get the contract because of employment needs in their particular states.

How the Military and Commercial Engine Market Developed

We take commercial aviation for granted today because we can go to just about any place in the world in less than two days. We forget that at the beginning of our Pratt & Whitney time line, there was no commercial aviation in the sense that a traveler could buy a ticket in New York and fly to Los Angeles. Let us review how commercial aviation got started and where U.S. aviation stood compared to the rest of the world in the first decade of Pratt & Whitney's existence (1925–1935).

There was no question that the United States was the world leader in aviation prior to 1910 because of the pioneering work of the Wright Brothers and Glenn Curtiss. However, by the end of World War One (WWI), the United Kingdom and continental Europe took over the leadership in military aviation. Shortly thereafter, Europe also took over the leadership in commercial aviation because of direct subsidies from its governments, which saw a real need for transportation by air.

In the mid-1920s and early 1930s, the United States, which had stimulated aviation indirectly with air mail contracts, changed the picture, as was vividly demonstrated in the MacRobertson air race in 1934 from London to Melbourne, Australia—a 12,722 mile journey. The British designed a special one-of-a-kind airplane (the D.H.88 Comet) and were the winners with a time of 52 hours and 38 minutes. However, the second and third place winners were off-the-shelf American commercial aircraft—the Dutch KLM Airlines' Douglas DC-2 (flown by KLM pilots at 90-hr elapsed time) and the Boeing 247 (flown by Roscoe Turner at 92-hr elapsed time). The race's sponsor, Sir Macpherson Robertson, was overjoyed with the results that proved that a transport plane could reach Australia from England in four days [1].

Two significant factors affected the growth of aviation in the 1920s: air mail contracts in the mid-1920s and Daniel Guggenheim's investment in aircraft safety in the late 1920s.

AIR MAIL CONTRACTS STIMULATED AVIATION

The U.S. government paid airlines to carry the mail with an economic incentive that allowed them to use aircraft large enough to also carry passengers. This helped establish the United States as a world leader in commercial aviation.

President Coolidge, in his Third Annual Address to Congress on December 8, 1925, expressed his thoughts on aviation [2]:

> Aviation is of great importance both for national defense and commercial development. We ought to proceed in its improvement by the necessary experiment and investigation. Our country ... has made records for speed and for the excellence of its planes. It ought to go on maintaining its manufacturing plants capable of rapid production, giving national assistance to the laying out of airways, equipping itself with a moderate number of planes, and keeping an air force trained to the highest efficiency.

Between 1918 and 1925, airmail delivery was the job of the Post Office. It contracted with pilots to fly the routes in WWI surplus aircraft, de Havilland DH-4s and Curtiss R-4LMs. Because of 89 airplane crashes, by 1925, about 50 percent of the original 40 contracted pilots had died. By this time, most of the WWI surplus aircraft and the approximately 20,000 Liberty engines produced for WWI had been used up. The time was right for the purchase of new aircraft and engines.

During the Coolidge administration, Congress passed the Air Mail Act of 1925 (also known as the Kelly Bill). Direct subsidies, done in Europe, were not the American way. Therefore, aid to struggling airlines had to be more subtle, such as generous compensation for carrying the mail. Sir Winston Churchill had a pragmatic view toward commercial aviation development [3]:

> Our business is first to facilitate the development of civil aviation, to develop the routes and the key aerodromes.... But the effort which is to sustain it must be a spontaneous effort arising from the country and the trade, and the last thing we can do in regard to that is to make sure that we do not get in the way of it.

In 1925, the American public did not see commercial aviation as a safe way to travel: there were too many crashes of barnstorming stunt pilots to give the traveling public a sense of confidence. Therefore, in 1925, President Coolidge appointed the Morrow Committee to establish an overall policy for American aviation. The committee came up with the following recommendations:

1) Aviation is critical to the national defense. Therefore, aircraft manufacturers should get some government help.

2) Private flying serves the national interest. The government should help stimulate private flying.

3) The government needs to establish regulations for pilots and aircraft manufacturers and give the public confidence that commercial flying is a safe form of transportation.

4) The government should facilitate organized flying around the country by establishing airways—similar to waterways and highways.

5) The government should expand its air mail contracts.

Thanks to these recommendations, Congress enacted the Air Commerce Act of 1926. The Aeronautics Board function under the Department of Commerce was to promote aviation, improve navigation aids, provide more lighted airways, investigate accidents, and develop charting. The rest of the recommendations were covered in amendments to the Kelly Bill in 1926 and 1927.

The general public paid little attention to the Kelly Bill. Instead, the great stimulus to air travel came right after the Lindbergh Atlantic crossing in 1927. Daniel Guggenheim recognized the public relations power of Lindbergh's flight and financed Lindbergh's trip around the United States and into Mexico and South America. It was then that Colonel Lindbergh met Ambassador Morrow's daughter, Anne, who would subsequently become Mrs. Lindbergh.

President Hoover's Postmaster General was Walter F. Brown, who was not at all satisfied with the way airlines carried the mail. Their schedules were not tied together. In other words, there was no air transportation network, but a collection of independent airlines, some of which were frequently operating with expenses exceeding income. Brown took a vigorous approach towards creating a real airline system. He worked with former Assistant Secretary for Commerce, Mr. William P. McCracken, to launch the McNary-Watres Act of 1930. This new bill amended the Kelly Bill to spell out the following activities:

1) Carriers were to be paid for mail hauling according to the space available in their aircraft. In other words, Brown wanted to see bigger airplanes so that the carriers could make money carrying passengers in addition to the mail.

2) The postmaster could arrange air mail routes to suit the best interest of the public. However, his was the judgment deciding what was in the best interest of the public. Judging from the rapid growth of the U.S. airline industry as a result of his stimulation, he seems to have been acting in the public's best interest.

3) The air mail contracts would be awarded to the lowest bidder who owned an airline that operated daily over six months on a schedule with routes of more than 250 miles.

Walter Brown, under the McNary-Watres Act, had considerable authority over the airlines who participated in flying the mail. They were given a financial incentive to operate larger multi-engine airplanes that could carry passengers in addition to the mail. The idea was to develop a substantial passenger business that could become more profitable than just carrying the mail so that the airmail subsidy could be gradually phased out.

The net result was an air transportation system provided by responsible airlines. United Air Lines took the northern route across the United States, as well as the West Coast. TWA had the middle transcontinental route. American Airways had the southern transcontinental route. In addition, Northwest flew from Minneapolis to Seattle, and Eastern Air Transport covered the Atlantic coast.

GUGGENHEIM ENCOURAGED COMMERCIAL AVIATION

Daniel Guggenheim's son, Frank, served as a naval aviator in WWI, which was where he developed his love of aviation. After the war, Frank was disappointed that the United States was so far behind the Europeans in both military and commercial aviation. Frank convinced his father, a man of considerable wealth, that something should be done to get the country going with commercial aviation. Daniel Guggenheim provided the seed money to stimulate the development of commercial aviation. He provided millions of dollars from 1926 to 1930 in the following areas:

1) *Aeronautical meteorology*—Guggenheim helped develop a working system of weather reporting along the San Francisco to Los Angeles route. A network of weather reporting stations and teletype communications let a pilot flying that route know ahead of time what the weather would be. Eventually the Weather Bureau took over the operation.

2) *Instrument flying*—Guggenheim sponsored the development of instrument flying with the help of Lieutenant Jimmy Doolittle, Professor William G. Brown from the Massachusetts Institute of Technology (MIT), Sperry's directional gyro, and Kollsman's precision altimeter. This made all-weather flying feasible and safe.

3) *Development of navigational aids*—This included beacon lights, radio beacons, and two-way radio communications between the pilot and ground stations.

4) *The safe aircraft competition*—The point of this competition was to lower the landing speed and improve the take-off capability of aircraft.

5) *Specified single-engine operation*—Until this time, twin-engine aircraft were considered risky compared to the trimotors, which could fly even if one engine failed. Guggenheim wanted this capability even on twin-engine aircraft to make the passengers feel more secure in their air travels. When TWA was negotiating the specifications for the Douglas DC-1, Charles Lindbergh, a consultant to TWA, insisted that Douglas make a fully loaded aircraft capable of taking off on one engine from any TWA station and fly to the next station on that one engine.

6) *Established engineering courses*—Development of aeronautical engineering courses in colleges gave the fledging aircraft companies the latest technology.

7) *Popularizing aviation*—The Guggenheims sponsored Charles Lindbergh's travels around the country. "Lucky Lindy," as he was named at his ticker-tape parade in New York City, won American hearts and became a national hero. Dancers were soon doing a jitterbug-type of dance called "the Lindy."

Pratt & Whitney's Five Defining Moments

The story of Pratt & Whitney's early days is not one smooth time line of events but rather one with five major events, one of which was a major shift in aircraft engine types from piston engines to gas turbines. Those five "defining moments" occurred during the leadership of Frederick B. Rentschler, the visionary who, with George J. Mead, founded the company in 1925:

1) *Founding of Pratt & Whitney*—Rentschler concluded from WWI that the best airplane could only be designed around the best engine. Furthermore,

he believed that "it should follow that a superlative engine could not be denied in production and that some degree of success must necessarily follow" [4]. He was correct. By 1945, he supplied about half the American aircraft engine horsepower in WWII and dominated the commercial engine market.

2) *Forming United Aircraft & Transport Corporation (UA&TC)*—William Boeing and Frederick Rentschler, both aviation visionaries, saw a great market for commercial aviation. They put together a complete aviation empire of engines, airplanes, airlines, and airports.

3) *Breaking up UA&TC in 1934*—After President Roosevelt's victory in 1932, Senator Hugo Black started an investigation of how the previous administration had awarded air mail contracts, strongly implying improper actions. In 1934, Roosevelt annulled all mail contracts. The subsequent UA&TC breakup was collateral damage.

4) *Contributing to the WWII effort*—In 1940, the U.S. government urged Pratt & Whitney to create a huge output of aircraft engines by getting automobile companies to participate in production.

5) *Getting into the aircraft gas turbine business*—Pratt & Whitney had dominated the aircraft piston engine business during WWII, but in 1945 found itself an outsider in the new field of aircraft gas turbines. How Pratt & Whitney came to once again dominate the market has to be the company's finest moment (see Chapter 7).

Pratt & Whitney's Founders

There can be no discussion of Pratt & Whitney's jet engine history without some background on the founders, Frederick B. Rentschler and George J. Mead. Their combination reminds me of statements by a prominent entrepreneur in Boston when I was a college student in the 1940s. He was discussing the need for different talents in a successful enterprise and used the building of a bridge across the Charles River as an example. He could get all the technical know-how to build a bridge from professors at MIT. But he added that one also needs the type of visionary who knows where to build the bridge! In Pratt & Whitney's history, Rentschler was the visionary and Mead was the man with the technical know-how, a perfect combination for success.

Frederick B. Rentschler

George Adam Rentschler was a German immigrant who established the family foundry and manufacturing business in Hamilton, Ohio, and sent his sons to Princeton. It was Frederick (class of 1909) who became president of Wright Aeronautical Corporation after WWI and later cofounder of Pratt & Whitney Aircraft (Fig. 4).

When WWI started, Frederick Rentschler joined the Army Air Service (predecessor of the Army Air Corps) and was assigned Government

Fig. 4 Frederick B. Rentschler (courtesy of Pratt & Whitney Archives).

Representative for Production at the Wright-Martin Company. Wright-Martin was formed in 1916 to produce Hispano-Suiza aircraft, one of best engines of WWI, under license for the French government. The company's publication, *Wright-Martin Aircraftings*, described Rentschler in November 1918 as "six feet of tireless energy and unusual ability in the service of Uncle Sam." The article went on to say [5]:

> Personally, Captain Rentschler is a bully fellow, with just the right mixture of temper and temperament, of perfectly normal habits, with a leaning towards the movies. The only thing he takes seriously, other than his work here, is eating, and he is classified as a 'good feeder.' He is inclined to be a bit of a hermit, much to the regret of many of his friends here, and he has no idiosyncracies other than those expressed in this salute.

This is fairly consistent with other peoples' assessment of Mr. Rentschler. *Time Magazine*,* for example, 33 years later in 1951 described Rentschler as one who thought, talked, breathed, and dreamt engines. The article used an engine analogy in describing his ability to handle a normal workload easily (implying "cruise power setting"), saying that Rentschler could turn on extra power (take-off power) when needed. The article also pointed out that Rentschler was a private man who used memos to notify his top people of raises rather than direct contact—even those who knew him for more than 20 years called him "Mr. Rentschler."

At the end of WWI, Wright-Martin's board of directors put aside $3 million to be used to form a small aviation engine company, while the rest of the assets were sold to the Mack Truck Company. George Houston, president of

---

*You can read the original article, "Mr. Horsepower," at http://www.time.com/time/magazine/article/0,9171,890116,00.html.

Wright-Martin, approached Rentschler with a proposal for Rentschler to become the vice-president and general manager of the new aviation company until Wright-Martin was liquidated. After that, Rentschler would become president of the new company. Rentschler agreed.

Right away, the new company, Wright Aeronautical Corporation in Paterson, New Jersey, began selling Hispano-type engines, a 180-hp type to the Navy and a 300-hp version to the Army.

GEORGE J. MEAD

George Mead (Fig. 5) trained at MIT and was working directly for Henry Crane as an experimental engineer at Wright-Martin. Rentschler was very impressed with Mead and wanted to take Mead to Wright Aeronautical as a member of his core team [4]:

> Through the war period George Mead was working directly under Henry Crane as experimental engineer. I knew that Crane had great regard for Mead's ability. I had Mead in mind to take with me initially in Wright Aeronautical, but, in talking with him, I found he was very anxious to accept temporarily a job at Army's McCook Field as Chief of the Power Plant Laboratory. He believed and I agreed with him, that it would give him an exceptional opportunity to test and know almost all types of current aviation engines. I continued to keep in touch with Mead from time to time while he was at McCook Field.
>
> In time Henry Crane resigned to accept the position ... as special advisor to Alfred Sloan in General Motors. When it came time to pick his successor, my thoughts immediately turned to George Mead and I brought him back from McCook Field and placed him in charge of engineering. This occurred late in 1919.

Henry Crane filled in as Vice President of Engineering at Wright Aeronautical until Mead completed his work at the government laboratory at McCook Field.

Fig. 5 George J. Mead (courtesy of Pratt & Whitney Archives).

The Founding of Pratt & Whitney Aircraft Company

Rentschler and Mead at Wright Aeronautical

Wright Aeronautical was working on both liquid-cooled and air-cooled engines. There was a great engineering controversy between the two concepts. It was felt that the liquid-cooled engine would result in a slimmer installation within a sleek fuselage, while the greater diameter of the air-cooled engine would create more drag. However, Rentschler and Mead were putting their bets on the air-cooled engine configuration.

Admiral Moffett, head of the Navy Bureau of Aeronautics, told Rentschler of Charles Lawrance's great progress with a nine-cylinder air-cooled engine and that the Navy was very much interested in buying a production quantity of this engine. (In 1917 Lawrance had concentrated on the development of a small air-cooled engine at his company, the Lawrance Aero-Engine Corporation.) The problem was that Lawrance had practically no production facilities and his backers were anxious to merge the company with a larger organization. Rentschler correctly read the Navy's interest in air-cooled engines. Mead took over the development of the Lawrance nine-cylinder engine and began shipping production engines.

The situation in Frederick Rentschler's own words [4] at that time was:

> By 1924 Wright was a going, vigorous factor in the powerplant field. I believe it was the only company which consistently was earning reasonable profits and continuing to expand its horizon and business. I think it would be fair to rate it as the outstanding engine company of the period from an all-around viewpoint. This does not mean that it did not have a very worthy competitor in Curtiss, which stuck pretty closely to its single D-12 type, an excellent engine. Packard also was developing a liquid-cooled engine of approximately this same power.
>
> However, I personally was not at all satisfied in my position at Wright and all of this dissatisfaction had to do with those above me, our Board of Directors. These men were almost all investment bankers, none of whom had any real appreciation of what we were trying to do, none of whom was a substantial stockholder in the company, and none of whom took any real interest in its affairs or in its management.
>
> Finally it reached a point where I was discouraged enough to feel that I should give it up and return to Hamilton. This latter idea came to focus following the death of my father and because, too, my older brother had gone to New York as executive-vice president of National City Bank.
>
> However, an illness prevented any immediate action, except for my resignation from Wright, which I believe was in June or July of 1924. Toward the first of the year, I had completely recovered and began charting my future course. At that time I found it was virtually impossible to give up the thought of not going on with aviation, at any rate until I had made a determined effort to find some way of resuming what had then become a consuming passion.

Rentschler and Mead Form Pratt & Whitney Aircraft Company

Rentschler Leaves Wright Aeronautical. Planning to resign, Rentschler shared his intentions with his core team at Wright Aeronautical—Andrew Willgoos (Chief of Design), John Borrup (Superintendent of the Experimental Shop), Don Brown (Materials & Purchasing Department) and Charles Marks (Production Engineer). When they expressed their preference to go with him to the new company he was founding, he told them to "sit tight" until he could think the whole thing out. Again, in his own words [4]:

> In considering what I thought I had learned about aviation to that date one thing stood out forcibly in my mind. The French, through their Hispano engine, were dominant in the air throughout the critical war period. It seemed very definite that the best airplane could only be designed around the best engine; also, that the best airplanes only could make up the finest air force. It seemed conclusive to me that the final objective of national defense so far as the air was concerned was to have the world's dominant powerplants. I knew the entire field of powerplants very intimately and began to feel more and more as I went along that given the opportunity we might be able to establish real leadership in that field. Obviously if my conclusions were correct, it should follow that a superlative engine could not be denied in production and that some degree of success must necessarily follow.
>
> There was definitely a place for an air-cooled radial in the 400- to 500-hp field but it had to have characteristics well beyond anything yet accomplished. Nevertheless, it appeared possible from an engineering point of view. I was not only interested in something which would outperform the largest air-cooled radial then in existence, but I sincerely believed we could displace the liquid-cooled types, which meant we would have an engine of equal interest to both Navy and Army. The latter at this time were vigorous proponents of liquid-cooled engines for excellent reasons, as they were then superior.
>
> With this gleam in my eye, I first went to Washington and had a confidential talk with Admiral Moffett, head of the Bureau of Aeronautics, whom I had known intimately for a number of years. He was extremely cordial and listened attentively to my belief that given the right circumstances I could produce quickly a 400- to 500-hp air-cooled radial of unmatched efficiency. He then asked the pertinent question of how I proposed to do this, and I explained that was partly the reason for coming down to see him—to find out if there was any possibility of any kind of assistance from the Bureau. He then meticulously pointed out that any financial help in the first instance was completely impossible as they had no such authority. However, he said that he did have confidence in my judgment and experience based upon the past and if I ever found a way of getting this type of an experimental engine in his hands, after being proved through testing, he was sure the Navy would be overwhelmingly interested. I realized before I went down that I

could not look to them for any financial assistance to begin a company. I primarily wanted to know where I personally stood with him, and this I found out.

RENTSCHLER GETS THE GREEN LIGHT FOR PRATT & WHITNEY. Next, Rentschler took up the subject with his brother Gordon of National City Bank of New York. Gordon's initial reaction was to ask about going back to Hamilton, Ohio, and starting a new company there. The problem with that was that Rentschler would have to go to an investment banker and then get back into the same hassle he had gone through at Wright Aeronautical. Investors would be looking for more immediate profits rather than investing for the long haul. Then Gordon suggested Niles-Bement-Pond Company, which had more than enough working capital to explore new opportunities. In addition, they both knew the Niles Tool Works in Hamilton, Ohio, which was part of Niles-Bement-Pond.

Rentschler spoke to Mr. Cullen, president of Niles-Bement-Pond, who had lived in Hamilton, Ohio, and was one time president of the Niles Tool Works. Mr. Cullen also knew Rentschler's father as well as young Frederick Rentschler—not a bad set of circumstances with which to start out!

In the middle of April 1925, Rentschler went to Hartford to discuss his business plans with Clayton Burt, the general manager of the Pratt & Whitney Company owned by Niles-Bement-Pond, which had unused factory space in Hartford, Connecticut. The net result of the discussions was a contract dated July 14, 1925 between Pratt & Whitney and Frederick Rentschler and George Mead [4]:

> I had persuaded Mr. Cullen and Pratt & Whitney Company that the aviation project must be a separate company and that I was to have complete control and take complete responsibility for its management.... Now, immediately following my trip to Hartford and the realization that I might be given a green light, I first contacted George Mead and explained what I had done to date. I explained that I felt I was putting him in a rather difficult position but, nevertheless, I also felt obligated to conform to the promise he had exacted from me almost a year before. I insisted that he think it over for a couple of weeks and then let me know this conclusion. At the end of that time Mead told me that he definitely wanted to come along and felt that under the circumstances he should resign forthwith; then I explained to him the details of the contract, which was then in course of preparation and shortly thereafter he was available.

PRATT & WHITNEY'S FOUNDING CORE TEAM. Mead and Rentschler then talked to Don Brown and Andrew Willgoos about the future, and shortly thereafter they brought in John Borrup and Charlie Marks—all members of their core team at Wright Aeronautical. Earle Ryder (from Aeromarine) was hired to complete the core team. The founding core team of Pratt & Whitney were

- Frederick Rentschler, President
- George Mead, Vice President for Engineering
- Andrew Willgoos, Chief of Design
- Don Brown, Factory Manager
- John Borrup, Superintendent of the Experimental Department
- Charles Marks, Production Engineer
- Earle Ryder, Consulting Engineer

THE MARKETPLACE IN 1925

Rentschler got the required financial backing from the Niles-Bement-Pond Company in Hartford, Connecticut, by making a convincing sales pitch to their management. One could imagine what was going through the minds of senior management as they listened to this young aircraft engine executive. What kind of a product would he bring to the marketplace? Could he be successful in the marketplace over his competitors?

First of all, he came with credibility. He and his family had been known by some Niles-Bement-Pond executives. Rentschler managed Wright Aeronautical and the company made money while other companies were financially marginal. He would bring with him a team of the most knowledgeable engineers in the aircraft engine business.

The marketplace, however, was somewhat "iffy." H. Mansfield Horner's (former United Aircraft CEO) testimony before the Mead Committee on October 10, 1945 described Pratt & Whitney's plight in the 1925 time period [6]:

> At this time, in the Spring of 1925, the idea that was to become Pratt & Whitney Aircraft was conceived, and a few months later, Mr. Rentschler discussed the possibility of the development and manufacture of aviation engines with the Bureau of Aeronautics of the Navy Department. The Navy advised him that there were no experimental funds to finance this project, but *if* a dependable engine of approximately 400 hp could be developed to meet their requirements and *if* its performance could be successfully demonstrated on a naval endurance test, they would then be prepared to order more experimental engines. *If* these were successful in flight demonstration, quantity orders *might* follow, and a worthwhile enterprise *might* be developed.

Notice the three *if*s and two *mights* in this description of the marketplace. The potential customer, the Navy, was showing some interest in air-cooled engines while the other potential customer (the Army) did not feel that air-cooled engines were the preferred approach to aircraft propulsion! That situation must not have been encouraging to potential investors, but the fact that Frederick B. Rentschler said he could do it with his team gave Niles-Bement-Pond the confidence to go ahead with the deal. One can understand why Niles-Bemont-Pond decided to accept the proposal: Rentschler had a proven track record of success at Wright Aeronautical, a superb engineering team to bring the engine to the marketplace, and a thorough knowledge of the needs of the marketplace.

NAME CONFUSION

There were two Pratt & Whitneys, and this has caused some confusion locally in Connecticut, as well as throughout the world, since both companies became world famous. When Rentschler founded the Pratt & Whitney Aircraft Company in 1925, he got the name from Niles-Bement-Pond because they owned the Pratt & Whitney Company, which was known throughout the world as the producer of precision manufacturing machinery. This company was founded in 1860 by two young machinists, Francis Pratt and Amos Whitney. These two men established the standard inch in the United States and created "real mass production" with interchangeable parts made with precision machinery. Prior to their time, what was called "mass production" was really a collection of parts that required men called "fitters" to lightly file a little here and there to make it possible for all of the parts to go together to form an assembly. Francis Pratt and Amos Whitney eliminated the need for fitters. The adjective that best describes the Pratt & Whitney Company is accuracy, and their story is well told in *Accuracy for Seventy Years* (available at http://www.prattandwhitney.com/pratt-whitney-history.htm). They had nothing to do with aviation, however, and they sold their company in 1901 to Niles-Bement-Pond, who retained the name.

Mr. Rentschler wisely took the name of Pratt & Whitney with the addition of the word *Aircraft* to distinguish the two companies. From 1925 to 1930, both companies shared the same address in Hartford, Connecticut. In 1930, Rentschler moved his operations to East Hartford into an area big enough to include an airport. At one time or another just about all of the aviation celebrities flew into Rentschler Field. When I came to work for Pratt & Whitney Aircraft in 1948, the local people used the term *Pratt and Whitney* for the machine tool company in Hartford and the single word *Aircraft* to refer to Rentschler's company in East Hartford. Gradually the name of Pratt & Whitney Aircraft was shortened to Pratt & Whitney. During conversations in the aviation industry you often hear a further contraction to "Pratt."

THE R-1340 WASP JOINS THE NAVY AND AIR MAIL SERVICE
PRATT & WHITNEY'S FIRST DESIGN

In May 1925, Mead and Willgoos (Fig. 6) roughed out the design of the company's first engine, which was in the 400–500 hp class, for the Navy. The

Fig. 6 Andrew Willgoos (courtesy of Pratt & Whitney Archives).

initial design work for this engine, soon to be known as the Wasp R-1340, was designed in Willgoos' garage in Upper Montclair, New Jersey (see Fig. 7). Chance Vought, who was proposing a new observation plane (the Corsair) for the Navy at the time, provided valuable input. Rentschler credits Vought with worthwhile advice on the engine weight target and the installation details. By the end of the year, this engine was running in a test stand in Hartford and was passing its type test in early March 1926.

The designation, R-1340, is government terminology in which the letter R stands for "radial" and the 1340 is the total volume swept out by the cylinders in cubic inches (also called cubic inches of displacement). The term *Wasp* was coined by Faye Rentschler, who thought the engine sounded like a wasp when it was running; obviously, Mr. Rentschler was sensitive to the suggestions of his spouse.

Rentschler understood the fundamentals of effective marketing. Ever since his meeting with Admiral Moffett, with whom he discussed Pratt's commitment to a new air-cooled engine, he made sure, on a regular basis, that he and

Fig. 7 Birthplace of the Wasp R-1340 (courtesy of Marjorie Betts, daughter of Willgoos).

Mead kept the "downstairs desks in the Bureau" up to date on what was going on in the engine design and development process.

THE WASP JOINS THE MILITARY

The rate of progress was so swift that Rentschler on October 5, 1926, was negotiating with the Navy for a 200-engine production order. Two years later, the Army was buying Wasp engines to replace its liquid-cooled Packard and Curtiss engines. The R-1340 Wasp engine installation was about 1000 lb lighter than that of the Liberty engine and about 650–700 lb lighter than the Curtis D-12. The final proof that an air-cooled radial engine could beat a liquid-cooled engine was a fly-off between two Navy Curtiss fighters—one powered by the Curtiss D-12 liquid-cooled engine and the other powered by the Wasp air-cooled engine. The Wasp-powered fighter had a slight margin in speed, had somewhat better climb capability, and could turn inside the D-12-powered aircraft.

THE POST OFFICE PICKS UP THE WASP

Another comparison of air-cooled radial engines vs liquid-cooled engines came up in the late 1920s when the U.S. Post Office asked for bids from airlines for transcontinental mail delivery. William Boeing found that his mail plane could carry about a 1200-lb payload with the Wasp, but only about 300–400-lb payload with the Liberty engine. Boeing won the competition, and the Boeing Air Transport Company made money with the Wasp-powered aircraft. Another factor that made the Boeing 40A mailplane profitable was the reliability of the Wasp engine. It could run well over 250 hr without needing any attention. The Liberty engine, in contrast, caused frequent unscheduled landings because of reliability problems. What surprised Boeing's competitors was the fact that Boeing's winning bid on the mail contract was about one third of what the competitors bid. The greater surprise, of course, was that Boeing was actually making money.

THE FORMATION OF UNITED AIRCRAFT & TRANSPORT CORPORATION

It is fair to say that by the end of the 1920s and into the early 1930s the radial air-cooled engines made by Wright Aeronautical and Pratt & Whitney were establishing new records in both military and commercial applications. William Boeing, head of Boeing Air Transport (which built airplanes), had been working closely over the years with Rentschler. Boeing approached Pratt & Whitney in the late 1920s to talk about forming a new larger company. Rentschler recalled [4], "now out of the close cooperation between Boeing and Pratt & Whitney came the establishment of the first sound and successful air transport operation."

The net result of those discussions was the United Aircraft & Transport Corporation (UA&TC), which included Pratt & Whitney, Boeing, Standard Steel Propeller of Pittsburgh, the Hamilton Company, Sikorsky, Chance Vought, Stearman Aircraft Company, and a combination of smaller airlines comprising United Air Lines. Mr. Boeing was the new chairman and Mr. Rentschler became the president of the corporation. Rentschler summarized the situation [4]:

> Now, in brief review ... we came into the early '30s with Pratt & Whitney Aircraft not only well established in the industry but having made, I believe, an important contribution to aviation. It had revolutionized the Army and Navy Air Forces to the point of world supremacy; it was instrumental in establishing the first successful air transport operation; its leadership in engines was beyond question, all of which in retrospect was a pretty good three or four years' job.

Air travel was very slow to develop because it was expensive for passengers. Boeing introduced its Monomail, an all metal, low-wing aircraft. This aircraft was followed next by the first U.S. luxury commercial-transport aircraft, the twin-engine Boeing 247. Mead was pushing for a Hornet-powered aircraft in the 16,000-lb gross weight class. Pratt & Whitney's Hornet engine (the engine after the Wasp) was rated at about 600 hp vs its Wasp engine at about 450 hp. However, Rentschler and Boeing were more influenced by the United Air Lines pilots who claimed that they could not fly or land a 16,000-lb gross-weight aircraft.

It turned out that Mead was right! The subsequent chain of events a year later resulted in the Douglas DC-2 in the 16,000-lb gross weight class, and later the fabulous DC-3 powered initially by the Wright Cyclone engines. This put Wright Aeronautical into the driver's seat in commercial engine applications. Around 1935 or 1936, Mead got together with Igor Sikorsky to rough out the design of a four-engine aircraft for the airlines. United Air Lines became interested in this aircraft, and approached the airframers for bids to build such an aircraft. Douglas was the winner with a design that would be known as the DC-4. During WWII, it was better known as the C-54 Army Air Corps military transport (R5D, Navy) used all over the world.

THE BREAK UP OF UA&TC

In 1934, Senator Hugo Black started an investigation of what happened under Postmaster Brown. After all the dust settled there was nothing to investigate. What Brown did, as he interpreted his job, was to act in the best interest of the public. It was true; however, that the small operators were not given the same status as the big operators who were carrying passengers in addition to the mail. What Brown did accomplish was to begin the building of a great interlocking air-travel system. Senator Black gave up on this phase of his investigation.

Next Senator Black focused his attention on William Boeing and Frederick Rentschler. Little did he realize or prehaps care that he was about to punish the very aviation innovators whose efforts created transcontinental air travel and would be essential to winning the forthcoming air war in Europe. The senator was disturbed that Pratt & Whitney Aircraft had made so much money from a relatively small investment up front. He called George Mead to testify before the committee. The newspaper *Hartford Times* reported on Mead's testimony before a senatorial committee [7]:

> Mr. Mead's spirited defense against the questionings and inferences of Chairman Black constituted the first such reply to critics and was a feature of the hearing.
>
> Mr. Mead contended that the organizers of Pratt & Whitney had put in much work, contributed to the community welfare, produced standard engines for the Army and Navy and, altogether, had produced much out of practically nothing.
>
> "We have gone through a depression and have been able, by good management and a good product, to keep on the right side of the ledger and that has taken a tremendous amount of work. The reason our business in Hartford has been successful was that we had experience over a number of years in the aircraft industry. We realized there was a change in the business—we were going from one type of powerplant to another, from water-cooled to an air-cooled engine—and we felt we would like to develop a product strictly on our own, instead of working for someone else.
>
> We took a tremendous gamble. We left a business that was running, in which we had played a part for a number of years, and in order to make this successful we had to bring out a product that was exceptional. The engine that we originally built in Hartford is a standard engine for the Army and Navy, and for commercial operators here and abroad."

In 1934, President Roosevelt canceled all current air-mail contracts, and portrayed the current carriers as having benefited from a "spoils system" in which the smaller operators were squeezed out. Then he directed the Army Air Corps to fly the mail until new contracts could be awarded.

Twelve Army pilots died in crashes while flying the mail in the first month. Eddie Rickenbacker, who received a Medal of Honor for his service as a WWI pilot, called the President's action "legalized murder." Unfavorable publicity forced the President to have his Postmaster General, Jim Farley, announce that the airlines would soon be back in the business of flying the mail. However, Farley had his own ideas on handing out contracts and did not appear to have the primary goal of nurturing a great air transportation system.

What Presidents Coolidge and Hoover (who immediately preceded Roosevelt) attempted was to provide seed money to get airlines on a sound financial basis to carry passengers. People were not flying long distances because there was no real interlocking air system convenient to potential air travelers. For example, one could fly from New York to some Midwestern city, but then not necessarily find an air connection to keep going west towards California. It was airplanes during the daylight hours and trains by night. What these presidents did was to pay more for air mail if flown in a big airplane that could also carry passengers. Gradually the airlines flying the mail built up regular passenger service and these revenues put the airlines into the black on their balance sheets.

Boeing and Rentschler provided coast-to-coast service via UA&TC. Passengers could leave Newark Airport at 9:15 a.m. in a Ford Trimotor to Chicago. They would leave Chicago at 4:40 p.m. heading to Cheyenne, Wyoming, and then onto Oakland, California, arriving at 1:45 the next afternoon. Total elapsed time: 31 hours. UA&TC had 125 airplanes in 1931, and was flying about a million miles a month. They carried 16,000 passengers in 1929, 37,000 passengers in 1930, and 50,000 in 1931.

Postmaster Farley set the ground rules that no airline that had been at the 1933 meeting with Walter Brown could bid on the new air mail contracts and contracts would only by awarded to airlines that operated multi-engine aircraft. This restriction was his way of keeping out the same small airline operators that Brown wanted to keep out. Farley's ground rules generated new names in the airlines and forced some executives to find new jobs. American Airways became American Airlines. Eastern Air Transport became Eastern Airlines. Transcontinental & Western Airlines added "Inc." after their names and became TWA. United Air Lines did not need to change its name because its contract was under Boeing Air Transport. Braniff and Delta joined the new system.

Senator Black and his committee wrote the Air Mail Act of 1934 (the Black-McKellar Bill), which repealed all previous laws relating to the air mail business. The bill created new agencies such as the Interstate Commerce Commission to oversee the airlines and fix the airfare prices and air mail payments.

Another part of the bill, which greatly affected Boeing and Rentschler, required airlines to disassociate themselves from their aircraft and engine manufacturing companies. As a result of his treatment by Senator Black's committee and the need to separate his airline operations, William Boeing subsequently decided to get out of aviation and this led to the breakup of UA&TC. Boeing took the manufacturing properties of UA&TC west of the Mississippi and Rentschler took those companies east of the river. Rentschler's new company was called United Aircraft Corporation, comprising Pratt & Whitney, Chance Vought, Sikorsky, and Hamilton Standard (a combination of Pittsburgh Standard Propeller and Hamilton Propeller).

Rentschler commented on the break up [4]:

> I am touching upon this particularly because I know it had some real bearing on Pratt & Whitney during the early '30s. We were then approaching the period of the so-called air-mail investigation, or 'castigation' as some have called it....
>
> Naturally the success of Pratt & Whitney and of United [UA&TC] made us a perfect target for the New Dealers. When they found no air mail scandal they just had to find something else to their liking. There was no use picking to pieces the companies which had gone bankrupt and had contributed nothing to aviation, so it was inevitable that they turned their main fire on United, whose only crime was that we had earned a reasonable profit in a field where most others lost their shirts. It was of no interest to the investigators the contribution which United, its subsidiaries, and its personnel as a whole had made to both military and transport aviation.

Paul Fisher, editor of United Aircraft's in-house publications, spoke with William Boeing in 1953 about the breakup of UA&TC nearly 20 years earlier. Fisher commented (in a letter to Jim Douglas, Boeing Airplane Company, Oct. 29, 1956):

> Boeing declared that the castigation that Rentschler and himself, in particular, had suffered at the hands of the Black Committee had led him to attempt to erase every thought of aviation and its many (airline) routes from his mind.
>
> Midway in the conversation, it suddenly occurred to him ... how he had regarded Mr. Rentschler, in the early days of the relationship between Boeing and Pratt & Whitney Aircraft, as a man of intense vigor. He said that he had marked it almost at once despite Rentschler's aeronautical reputation for aloofness and modesty. When the time came to form United Aircraft & Transport Corporation, he recalled, it seemed to him that he should take the role of the elder statesman or, as he said, 'coach' and encourage Rentschler to carry through. Rentschler was then a man of 41 and, as I gathered it, Boeing felt that the fact that he was about seven years older meant to both of them that the roles that he had laid out were wise choices.

Boeing never went back to his aviation company after the Black Committee experience, but continued other pursuits in lumber and sailing. Rentschler, on the other hand, did not have any other pursuits. He stayed with aviation and guided United Aircraft through WWII.

PRATT & WHITNEY'S GREAT CONTRIBUTION IN WORLD WAR II

In 1940, President Roosevelt, with great foresight, appointed William Knudsen of General Motors to head up war production activities. George Mead, who had resigned from Pratt & Whitney in 1939, became Knudsen's

Fig. 8 Leonard "Luke" Hobbs, who joined in 1927 and became engineering manager in 1935 and corporate vice president in 1944 (courtesy of Pratt & Whitney Archives).

assistant for aviation. This was a fortunate combination of managerial experience and technical capability as the United States prepared for its forthcoming role in WWII.

Mead worked tirelessly with his former engineering protégé, Leonard Hobbs (see Fig. 8), in getting the automobile companies ready for piston engine production to augment the output of Pratt & Whitney in East Hartford. In addition, Pratt & Whitney set up a plant in Kansas City to provide additional manufacturing capability. It was almost a two-year lead time to go from go-ahead to the first production engine.

In addition to the plants in East Hartford and Kansas City, there were the following manufacturing facilities for Pratt & Whitney engines: Ford; Buick; Chevrolet; Continental; Jacobs; Nash Kelvinator; Canadian Pratt & Whitney; and Commonwealth Aircraft Corporation of Australia.

The total output of Pratt & Whitney engines in WWII period was 363,619 (see Fig. 9), which amounted to about 50 percent of the engine horsepower of the American manufacturers for the Army and Navy Air Forces. When one recalls that Rentschler was originally president of Wright Aeronautical Corporation, which provided about 35 percent of the engine horsepower in WWII, one could credit his contribution to the war effort as 85 percent of the engines horsepower. Quite a contribution!

THE AIRCRAFT GAS TURBINE BUSINESS AFTER WORLD WAR II

From 1940 to 1945, Pratt & Whitney and Wright Aeronautical were not allowed to get into jet propulsion because the U.S. military wanted those companies to focus exclusively on piston engine development and production. The British were making progress in jet propulsion and were concerned about an invasion by the Germans. They asked the United States to push the development of aircraft gas turbines, and made their technology available to General "Hap" Arnold, who looked to General Electric and Westinghouse to

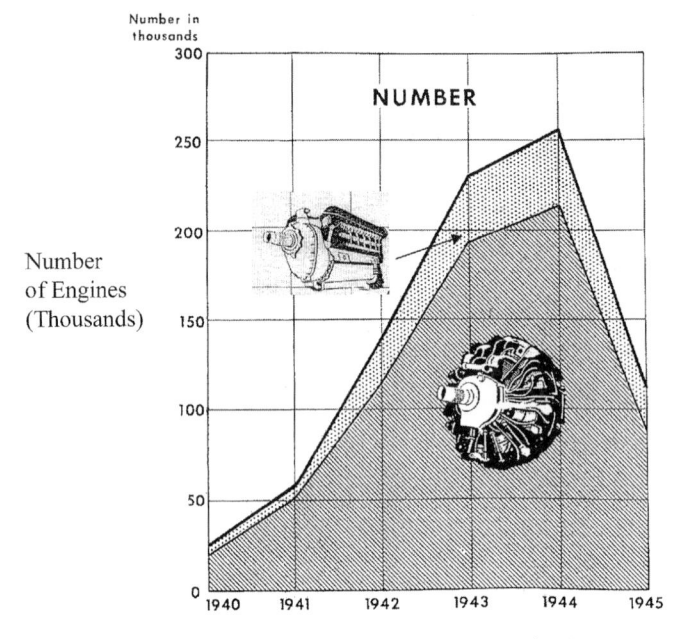

*Excludes 122 jet engines shipped in 1944 and 1208 in 1945.

Fig. 9 Air-cooled engines were predominant in World War II. In 1944, the total engines were a little over 250,000, of which about 210,000 were radial air-cooled engines and about 40,000 were liquid-cooled engines (source: Department of Commerce, Civil Aeronautics Administration, 1945).

become major manufacturers of aircraft gas turbines because of their extensive experience in turbomachinery.

The marketing outlook for Pratt & Whitney in 1945 was not anywhere near as promising as what Rentschler found in 1925. The company was not in the aircraft gas turbine market and did not even have the technology to be in the market. The new competitors were exceptionally strong because they got into the business fortuitously at government expense during the war, and now were supplying the propulsion needs of the military aviation industry. The commercial aviation market was not yet ready for the jet age.

The two old competitors in the piston engine era, Pratt & Whitney and Wright Aeronautical, approached the jet age completely differently. Wright decided to build British engines under license as a means of getting into the marketplace very quickly while minimizing its capital investments. Pratt & Whitney decided to continue in the gas turbine era as it had done in the piston engine era by heavily investing in research and facilities to design, develop, and manufacture aircraft gas turbines that could successfully compete in the world marketplace. This approach meant, to investors, making a huge immediate investment for a payoff way in the future compared to other investment opportunities.

GENERAL ELECTRIC LANDS THE U.S. AIR FORCE MARKET

The British sent one of Frank Whittle's experimental engines to General Electric with a complete set of drawings for a more advanced model. (Whittle's concepts are discussed in detail in Chapter 9.) In addition, Whittle visited them in early 1942 to give their engineers whatever help needed to build the engine. General Electric did an amazing engineering job to get engines out to California for flight test in the Bell XP-59A in the latter part of 1942. General Electric at Schenectady, which had years of experience with steam turbines and some practical experience with gas turbines for industrial use, started in on axial-flow gas turbines for military aircraft. At first they designed and built a turboprop (T31) and then later an axial-flow turbojet (J35). Subsequent to the Bell Aircraft's successful flights, the Lockheed Company built the XP-80, whose first flight was with a British engine, the de Havilland Goblin. Later in 1944 General Electric came through with the I-40 for the Lockheed P-80.

Stanley Hooker of Rolls-Royce has pointed out that both General Electric and Pratt & Whitney started their first jet engines with British-designed engines—the Whittle engine to General Electric and the R-R Nene to Pratt & Whitney [8].

WESTINGHOUSE GRABS THE U.S. NAVY MARKET

Westinghouse started off right away on an axial-flow gas turbine known as the 19A (Westinghouse nomenclature), which was the first American axial-flow turbojet. The axial-flow design was a natural selection for Westinghouse because it had extensive experience in designing axial blowers for the Navy, and felt that axial-flow engines would result in a lower frontal area to minimize aircraft drag compared to centrifugal flow engines. Axial-flow compressors pump air up to higher pressure by a series of airfoils acting on the incoming airstream at essentially a constant outer diameter. On the other hand, centrifugal-flow compressors compress air by moving from a lower radial position to a higher radial position relative to the centerline of the engine. As a result, the centrifugal compressor has a higher overall diameter than an axial-flow compressor at the same pressure ratio. Centrifugal-flow engines were mostly installed in airplane fuselages while axial engines could operate in either the fuselage or an external nacelle mounted on the wing.

The 19A was developed into the 19B, which had an annular combustion chamber. The McDonnell Phantom was powered by two of the 19B (J30 in Navy terminology) engines and was the Navy's first jet aircraft. Westinghouse developed a larger engine, the 24C, which became the J34. This engine powered the McDonnell Banshee Navy aircraft. Westinghouse formed an Aircraft Gas Turbine Division in early 1945 in preparation for steady business from the Navy.

Rentschler Breaks into the Market in 1945

Rentschler's business objective in 1945 was to have Pratt & Whitney design and build its own aircraft gas turbines and dominate the gas turbine market as it had done in the piston engine market. This required an organization that had the technology and facilities to design gas turbines, a development group with proper facilities to perform the necessary testing to ensure reliability and fuel efficiency, and the manufacturing facilities to produce high-quality aircraft gas turbines. This could not be acquired overnight.

Even so, Rentschler, with the substantial help of Hobbs, put together a plan to catch up to the competition in five years. Their plan consisted of three efforts to do the following:

1) *Develop the capability to design high-performance aircraft gas turbines.* Rentschler invested about $15 million for construction of an altitude test facility to run turbojet engines under temperatures and pressures that would simulate altitude flight conditions. He also provided other test facilities and capital for research in the components of jet engines. Hobbs assigned Perry Pratt, developer of the R-2800C Double Wasp, to form the Technical & Research Group to design aircraft gas turbines.

2) *Involve the Development Group immediately in development of the jet engine.* Rolls-Royce sent two of its Nene engines to the U.S. Navy for testing in late 1946. The 150-hr test was successful and the Navy was impressed. The Nene, at 5000-lb thrust, was the most powerful engine in the free world at that time; however, the Navy wanted to have the engine manufactured in the United States. Aircraft engine consultant Philip Taylor had the rights from Rolls-Royce to market the engine in the United States but he had no facilities to "americanize" the engine to Navy specifications. The Navy wanted Pratt & Whitney to take over the job. Pratt & Whitney negotiated with both Rolls-Royce and Taylor to produce the engine for the Navy. This became a gigantic challenge because all new manufacturing drawings had to be made and design and development had to be done to permit the engine to operate on aviation gasoline in addition to jet fuel as it had in the United Kingdom.

3) *Develop the capability to manufacture gas turbines.* Pratt & Whitney had a superb production facility for making high-quality aircraft engine parts. Now Rentschler had to get this production team to switch to making jet engine parts. An opportunity occurred when the Navy made an initial order for 500 J30 engines because Westinghouse did not yet have its production facility ready. The order later was cut to 130 engines, but it was still enough to train the manufacturing department.

Within ten years after the 1945 plan to catch up and dominate the market, Pratt & Whitney achieved Rentschler's objectives. It designed its first axial-flow turboprop (PT2/T34) and turbojet (JT3/J57). The J42 (Nene derivative) in the Navy Panther and the J48 (a more powerful development) in the Navy

Cougar and Air Force F-94 were in service. The T34 (the company's first turboprop engine) found applications in the Douglas Cargomaster C-133 and the Super Guppy. The J57 found a huge market in the B-52, KC-135, F-100, F-101, and F102 for the Air Force, as well as in the Douglas F4D Skyray and the A3D Skywarrior for the Navy. The commercial version of the J57 (JT3C in company terminology) was moving into commercial aviation in the Boeing 707 and the Douglas DC-8.

Westinghouse got out of the business even though they did excellent work and achieved great results in axial-flow turbojet engines initially. They were victims of the phenomenon of "the tragedy of early success"—a term Dick Foley, a Pratt & Whitney compressor project engineer, mentioned while we discussed experimental work. You design something that works great the first time, and you begin to feel you know what you are doing. Soon reality sets in and all kinds of trouble happens with the device that initially worked fine, which is what happened with Westinghouse. Their initial efforts were great achievements, but development problems started to appear once the engine got into service. They soon realized that aircraft gas turbines required a much more substantial investment than stationary gas turbines.

After 1945, General Electric approached the company to have Pratt & Whitney act as General Electric's production facility. Pratt & Whitney gracefully declined the offer. Later Allison took over that role. General Electric was able to pass off its J33 and J35 projects to Allison, who did some further development and a great deal of production for them.

RENTSCHLER'S ACCOMPLISHMENTS

When Frederick Rentschler died in 1956 he left an inspiring wake of aviation achievements:

1) *Correctly selected air-cooled radial engines as the technological advance.* Rentschler was an avid believer in air-cooled, radial engines when popular technical opinion was in favor of liquid-cooled engines. He was right. All commercial and military transports in the United States used air-cooled, radial engines—mostly Pratt & Whitney's.

2) *Pushed development of air-cooled engines.* After becoming dissatisfied that Wright's board of directors would not invest sufficiently for future developments in air-cooled engines, he founded Pratt & Whitney Aircraft in East Hartford. The two major engine companies, Pratt & Whitney and Wright Aeronautical, supplied about 85 percent of the aircraft engine horsepower for military aircraft in WWII.

3) *Made commercial aviation possible by developing and manufacturing dependable engines.* With William Boeing, he managed to pull together a number of independent airlines to establish the country's first viable coast-to-coast air transportation system, operated by UA&TC.

4) *Co-launched, with Boeing, jet air travel with the JT3 engine.* These engines were in the Boeing B-707 in 1958 and in the Douglas DC-8 in 1959.

In 1961, about five years after his death, his company had these business statistics for engine installations: 80 percent of piston engines flown by airlines; 40 percent of aircraft in the U.S. military; and 60 percent of all jet transports.

RENTSCHLER'S FINAL YEARS

Soon after WWII, Rentschler retired to Florida. In a 1956 commemorative Pratt & Whitney article, Paul Fisher wrote that "Mr. Rentschler's strength and vitality seemed unbreakable. Then when Mrs. Rentschler died in March, 1953, he showed his first signs of tiring. He had always been a gentle, quiet-spoken man. Now he grew even gentler." He spent time with his daughter Ann and his five-year-old grandson.

Rentschler died on April 25, 1956. In June 1955, in his report [9] to the board of directors, he had this to say:

> On the whole I think we can be reasonably optimistic about our present position in the industry from virtually every standpoint. I would call attention particularly to my earlier remarks that, while these next two or three years seem definite and assured, immediately thereafter we will be face to face with new, important, and efficient engineering competition. We certainly could not expect to continue to spread-eagle the field and it is a foregone conclusion that our advantage will dwindle in two or three years from now. However, I am unwilling to believe that, based on our past performance and the fact that we are all fully aware of the situation ahead, we cannot do something more than just hold our own in this forward competitive effort. To do this fully and from both the military and commercial transport point of view will require some extraordinary effort.

REFERENCES

[1] Gwynn-Jones, T., *Farther and Faster*, Smithsonian Institution Press, Washington, DC, 1991, p. 257.

[2] President Calvin Coolidge, Third Annual Address to Congress on December 8, 1925; http://www.presidency.ucsb.edu/ws/index.php?pid=29566 ["National Defense" section, retrieved April 2009].

[3] Sir Winston Churchill, *House of Commons Debates, 123 (col. 138)*, December 15, 1919, as quoted in Miller, R. E., and Sawers, D., *The Technical Development of Modern Aviation*, Praeger Publishers, New York, 1970.

[4] Rentschler, F. B., "An Account of Pratt & Whitney Aircraft, Company 1925–1950"; Rentschler wrote his reflections in 1950 and distributed copies; under the auspices of Pratt & Whitney President Arthur E. Wegner, it was published by the company in Oct. 1986 for internal distribution.

[5] "Close-Ups: Captain Fred Brandt Rentschler," *Wright-Martin Aircraftings*, Wright-Martin Aircraft Corporation, New Brunswick, NJ, Nov. 1918, p. 4.
[6] Horner, H. M., "Testimony Before Mead Committee," Pratt & Whitney Archives, Oct. 10, 1945.
[7] *The Hartford Times*, "Mead Defends P&W Motor as Our Own," Hartford, CT, Jan. 19, 1934.
[8] Hooker, S., *Not Much of An Engineer*, Society of Automotive Engineers, Warrendale, PA, 1984.
[9] Rentscheler, F. B., "Report to the Board of Directors of United Aircraft," June 1955.

Chapter 2

THE EARLY YEARS

GEORGE MEAD AND HIS FIRST PRATT & WHITNEY ENGINE

Young George Mead, as an aspiring engineer-to-be, kept his father's automobile in top running condition. That was necessary because Dr. George N. P. Mead needed reliable transportation to complete his rounds in Massachusetts. Young Mead's schooling included St. George's School in Newport Rhode Island, the Choate School in Wallingford, Connecticut, and MIT (1911–1915).

Mead started his engineering career in 1915 at the Sterling Engine Company in Buffalo, New York, where he worked on the development of a new internal combustion engine. In 1916, he joined the Crane Simplex Automobile Company, in New Brunswick, New Jersey, as a test engineer. Then in 1916 the Wright-Martin Company took over the Simplex Company. Wright-Martin was working on the production and development of the Hispano-Suiza aircraft engine for the French government, and Mead reported to Henry Crane. Mr. Rentschler put it this way [1, p. 7]:

> Wright-Martin took over the Simplex plant in New Brunswick, and its chief asset, Henry Crane, who did a masterful job in Americanizing this design (Hispano-Suiza) in his engineering department and getting the engine into large scale production. Towards the end of the war this plant was turning out approximately 1000 engines a month.

When Rentschler became the president of Wright Aeronautical Corporation, he approached Mead with the job of chief engineer in the new organization. Mead was definitely interested, but wanted to get more firsthand experience with engines by working at McCook Field in the Army Air Service Power Plant Laboratory. Late in 1919, Mead joined Rentschler at Wright Aeronautical and then followed Rentschler to Pratt & Whitney Aircraft Company in 1925 as Vice-President of Engineering.

George Mead started with the idea of the ideal cylinder, which in his view should be "square," the bore and stroke to be the same. At this point the cylinder could be either liquid-cooled or air-cooled. Then one could look at different groupings of cylinders to produce an engine. Early aircraft engines took the same form as automobile engines: water-cooled in-line or in a V-configuration. After examining all the possible arrangements, Mead was convinced that the radial air-cooled engine would be the lightest configuration. It had uniform cooling because each cylinder was cooled in the same way with ambient air at the same velocity and temperature. His first Pratt & Whitney engine was the radial, air-cooled R-1340 Wasp.

LIQUID-COOLED ENGINES

Figure 1 shows where the Wasp fits in with weight per horsepower compared to other air-cooled and liquid-cooled* engines in the 1925 time period [2]. The lines in the figure connect the points for air-cooled radial engines as well as for the liquid-cooled engines. Thus, the installation of liquid-cooled engines (water-cooled in this era), which include the radiator and associated plumbing, were heavier than the air-cooled engines.

The Wright E-4 is the Wright-Martin's version of the Hispano-Suiza, which continued to be manufactured by Wright Aeronautical up to about 1923. It was used in the U.S. Navy Vought VE-7 airplanes. Mead had much to do with that engine when he was working for Wright Aeronautical. This engine set an endurance record in the 1920s when it ran at full throttle for 325 hr. Other water-cooled engines were the Liberty, the Curtiss V-1400, and the Packard 1A-1500. The V-1400 was a further development of the excellent Curtiss D-12 engine. The Wright Aeronautical T-3A was a further development of the earlier

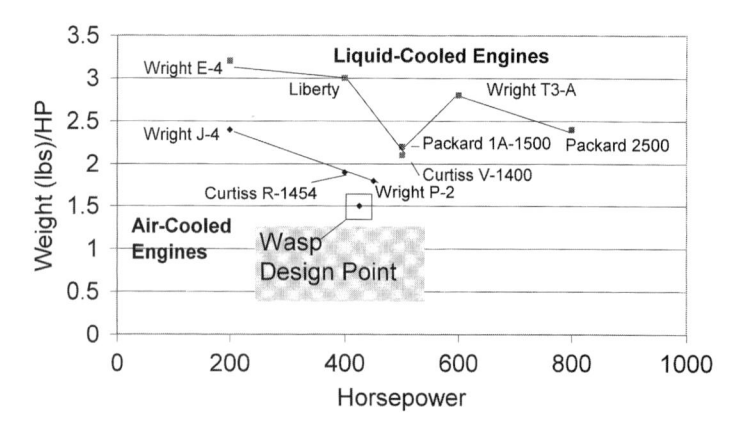

Fig. 1 Goal for the Wasp's weight/horsepower of 1.5. Upper line is for liquid-cooled engines (square symbols); lower line for air-cooled engines (diamond symbols) [data from 2].

---

*The emphasis in this book is on air-cooled engines. When nonair-cooled engines are discussed, the term *liquid-cooled* applies to both water-cooled and ethylene glycol-cooled engines.

Wright Aeronautical T-2. The Packard 2500 was the largest liquid-cooled engine in service in 1925 and had a relatively low weight for liquid-cooled engines at that time.

AIR-COOLED ENGINES

The Wright Aeronautical E-4 liquid-cooled engine was replaced in service by the Wright Aeronautical J-4 air-cooled engine in the Vought UO-1 Navy aircraft. The J-4 had been developed from the nine cylinder Lawrance J-1, which had come into the Engineering Department of Wright Aeronautical when Charles Lawrance's company merged with Wright Aeronautical. Mead was Chief Engineer at Wright Aeronautical at that time and had much to do with the growth of the J-1 into the J-4. He improved the way cylinders were connected to the crankcase. He designed the attaching flange as part of the steel cylinder, instead of putting a flange on the cooling cylinder surrounding the steel cylinder.

The R-1454 was originally a Wright Aeronautical engine that had been built in 1920, and then taken over by Curtiss as an Army Curtiss radial engine program. Sam Heron of McCook Field (Army Air Service) designed the cylinders. The Wright Aeronautical P-2 was the largest air-cooled radial engine in service in 1925 (under Navy contract).

THE R-1340 WASP

By this time, Mead and his Chief Designer Andrew Willgoos, working for Pratt & Whitney, targeted the Wasp at 1.5 pounds per horsepower (Fig. 2).

Mead gave a great deal of credit to Sam Heron for his enthusiastic support of air-cooled radial engines. Heron published articles on air-cooled engines

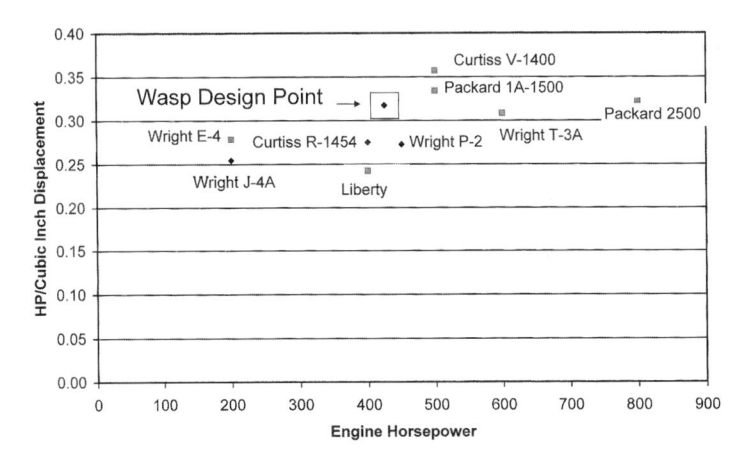

Fig. 2 The goal of a third of a horsepower per cubic inch cylinder displacement was ambitious for air-cooled engines of that time [data from 2].

in the *Journal of the Society of Automotive Engineers* (SAE) from 1922–1925. The air-cooled radial engine made sense to Mead because he felt that air sweeps by each finned cylinder head would provide the same amount of cooling for each cylinder. Mead pointed this out in his 1925 SAE paper [2]:

> Strange as it may seem, an air-cooled cylinder is simply a logically designed water-cooled cylinder with the water jacket replaced with suitable fins properly disposed. This whole subject has been ably covered by S. D. Heron on numerous occasions, and great credit is due him for his able advocacy of the air-cooled engine.

Later in the 1920s, Mead wrote an article for *Aero Digest* with the following concluding comments [3]:

> Therefore, it is evident that air-cooling has not only survived all possible service tests, but has, as a result of its superior performance, superseded water-cooling for all commercial work, and for practically every military purpose. The future is sure to bring even more decided improvements in air-cooled powerplants, and in the performance of planes equipped with them.

OVERVIEW OF THE AIR-COOLED RADIAL ENGINE

Before going into the details of specific Pratt & Whitney piston engine models, an overview of American and European engines is beneficial.* The word *piston* comes from a Latin root, *pinsare*, which means *to pound*. This is descriptive because the chamber pressure pushes a piston down the cylinder to produce work and in the process pounds on the crankpin through the connecting rod. In the Otto cycle of internal combustion engines, the piston in the cylinder works only 25 percent of the time. The remaining time is taken up with intake, compression, and exhaust. The nine-cylinder Wasp's crankshaft gets nine poundings during two revolutions. The 28-cylinder Wasp Major's crankshaft gets 28 poundings throughout two revolutions and the 42-cylinder Wright Tornado gets 42 poundings over two revolutions. A gas turbine's shaft gets so many more poundings per revolution that the shaft "thinks" the power transmission is smooth.†

A heat engine (either piston or gas turbine) converts only a fraction of the energy from fuel to useful work. The efficiency of a heat engine is equal to

---

*Due to differences in source material, there is bound to be some inconsistencies; therefore, it is important to squint a bit in order to see the general trends in the figures throught this book while not using a magnifying glass to look at differences between specific data points.

†I am greatly indebted at this point to Herschel Smith for his painstaking compilation of piston engine data on air-cooled rotaries, radials, and in-lines; liquid-cooled engines; and single-cylinder and cast block configurations [4].

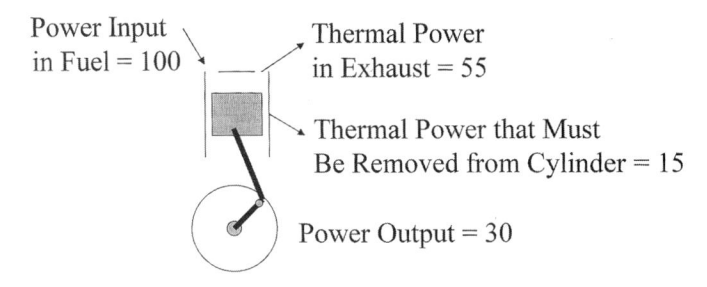

```
                Power Input              Thermal Power
                in Fuel = 100            in Exhaust = 55

                                         Thermal Power that Must
                                         Be Removed from Cylinder = 15

                                         Power Output = 30
```

Fig. 3 It is surprising that a 30-hp engine releases more than twice that power into the atmosphere through the exhaust, oil system, and cylinder head.

the work done divided by the heat energy added. The energy in the fuel is represented by the 100 in Fig. 3. Only about 30 percent of this energy is converted to work. The rest of the energy (70 percent) is discharged into the atmosphere through the exhaust valve, the cylinder head, and the oil system. Therefore, as we pull more power out of the cylinder, we must have a cylinder head designed to transfer more heat efficiently through it, whether it is liquid-cooled or air-cooled.

The critical cooling areas are the exhaust valve and the cylinder heads. In liquid-cooled engines, hollow exhaust valves and liquid sodium were effective in transferring the heat from the valve head to the cooler stem. Air-cooling the cylinder head required several developments. The rate at which heat is transferred from the cylinder head to air can be expressed as the product of three factors:

$$Q = h \times [A_{Fin}] \times [T_{Fin} - T_{Air}]$$

Q = Heat transfer in BTUs per hour
h = Heat transfer coefficient in (BTUs per hour per square foot per degree F)
A_{Fin} = Surface area of fins in square feet
T_{Fin} = Cylinder fin temperature in degrees F
T_{Air} = Air temperature at the cylinder head in degrees F

The heat transfer coefficient is proportional to air density and the velocity of the air over the cylinder head cooling surface. Therefore, the air velocity must be high as it passes by the head surface. That is why baffles were used to duct the cooling air as efficiently as possible. The other significant factor in cooling the head is to have as much surface area as possible. One can easily notice that the later model engines had much more cooling fin area per head than the earlier models. The "slices of fins" kept getting thinner and deeper with time. The trend from cast aluminum heads to forged aluminum heads permitted finer and deeper "slices" for the fins (see Fig. 4).

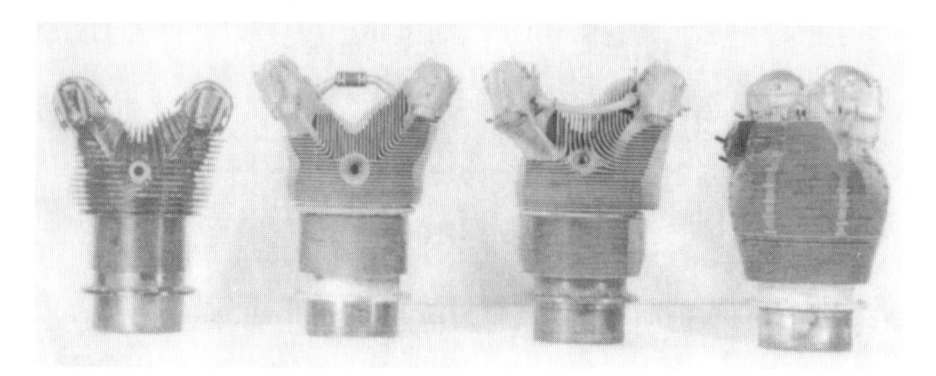

Fig. 4 There has been a constant improvement in cylinder fin area from the early days of cast aluminum heads (Wasp) up to forged aluminum heads (Wasp Major).

In addition to the cooling requirements, a great deal of attention must be paid to power production. Figure 3 is a schematic representation of the familiar Otto cycle. The power stroke occurs only 25 percent of the cycle in which the total of the four steps occur in only 40–60 milliseconds corresponding to 3000 and 2000 rpm respectively! The power stroke in this example occurs in one-fourth of the total time or 10 to15 milliseconds (Fig. 5). There are two spark plugs at opposite sides of the cylinder. At ignition, the two flame fronts approach each other as well as moving toward the piston, which is receding during its power stroke. Taking the 6-in. bore and stroke example, one sees that the time it takes the two flame fronts to meet is close to the time it takes for the piston to complete its power stroke (Fig. 6).

The point of this "back of the envelope" analysis is to show that the power stroke depends upon a number of things happening in an extremely short time. It means that higher rotational speeds and larger pistons soon run into a real performance barrier.

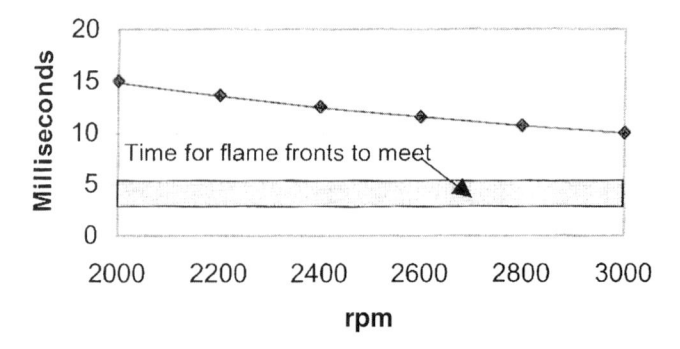

Fig. 5 Shown here is the time it takes for a 6-in. piston to complete the power stroke and the time it takes for two flame fronts from opposite sides of the cylinder to meet in the middle of the cylinder (flame speed is shown as a band).

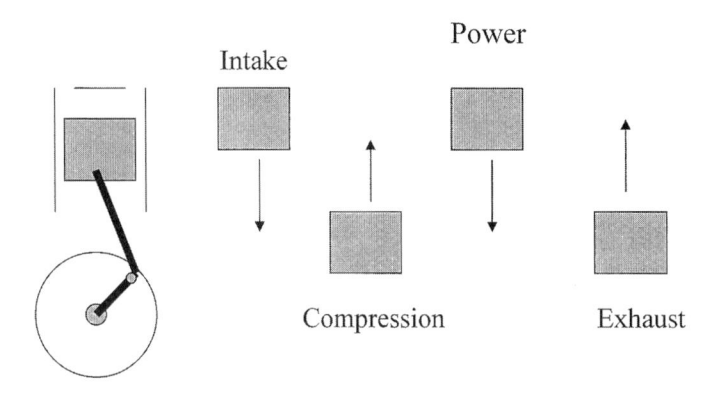

Fig. 6 Over a range of rotational speeds from 2000 to 3000 rpm, the time interval for two complete revolutions is in the 40–60 millisecond range.

The mean effective pressure (MEP) is the average pressure acting on the piston in its movement during stroke. One can calculate a MEP for a given engine if one measures the horsepower of the engine and the rpm:

$$\text{Horsepower} = \{MEP \times L \times A \times n\} \div 33{,}000$$

MEP = mean effective pressure in pounds per square inch
L = Length of stroke in feet
A = Area of piston in square inches
n = Number of power strokes per minute, which can be calculated from revolutions per minute (rpm) multiplied by the number of cylinders (C) divided by 2

This equation can be expressed also in the following terms.

$$\text{Horsepower} = \{MEP \times D \times C \times N\} \div 792{,}000$$

MEP = mean effective pressure in pounds per square inch
D = Cylinder displacement in cubic inches
C = Number of cylinders
N = Rotational speed in revolutions per minute

The resulting MEP is called BMEP (brake mean effective pressure) because it is based upon measured engine horsepower output. The product of D and C is the engine displacement in cubic inches.

Let us look at each of these terms, starting with the dimensions of the cylinder. Even though there are a variety of combinations of bore and stroke to produce a certain cubic inch displacement, in practice there is rather a

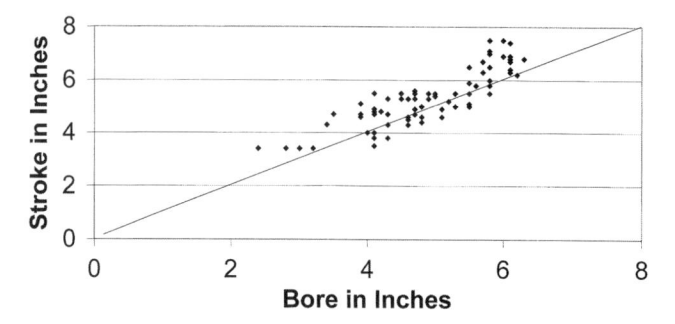

Fig. 7 For 100+ air-cooled radial engines, the stroke and bore are usually within plus or minus 25 percent (data from [4]).

limited range of variables. Figure 7 shows the relationship between bore and stroke for a large number of air-cooled radial engines.

Mead said he preferred a "square cylinder"—one where the stroke and bore were the same. A typical classroom problem in calculus for students is to design a can so that the area of metal is a minimum to contain a certain volume. The answer comes out to be length and diameter are equal. In other words, when the bore and stroke are the same, the cylinder surface area is a minimum for a given value of displacement.

At the beginning of the power stroke in the four-cycle engine a number of events have to happen sequentially, starting approximately 25 degrees before top dead center. First of all, a spark is struck from two opposite sources near the top of the cylinder. The combustible mixture now starts to burn from two opposite sources, which creates moving flame fronts heading away from the spark plugs. Then at the same time the piston is moving down on its power stroke—in the order of 10 milliseconds at 3000 rpm, for example. One can visualize that at flame speeds in the order of 1 inch per millisecond, it would take a minimum of 3 milliseconds for two opposite flame sources to meet in a 6-in. cylinder bore. In the meantime, the piston has moved an appreciable fraction of the stroke away from the flame front. Between the flame speed limitation and the cooling requirements, it is obvious that there is a practical maximum cylinder size. This practical maximum bore size seems to be about 6.1–6.3 in.

The cylinder wall surface increases as the square of the piston diameter. The volume increases faster as the cube of the piston diameter. This means the bigger the piston diameter, the more difficult it is to get rid of the cylinder heat. This surface to volume ratio is one of the factors setting the maximum piston diameter and displacement per cylinder. The following figure shows the trend in displacement per cylinder with time. Most of the data lie between 100–200 cubic inches per cylinder (see Fig. 8). As the bore increases, the ratio of surface area to volume decreases. The practical limit seems to be around 2/3 when the bore reaches 6.3 in. The range of bores for radial engines is contained within bore sizes from 2.4 in. to 6.3 in. This range was set before 1930!

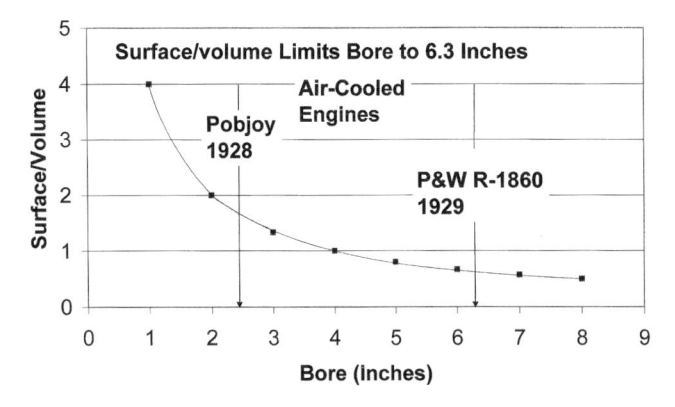

Fig. 8 Surface/volume to cylinder (data from [4]).

BRAKE MEAN EFFECTIVE PRESSURE (BMEP). There are a number of factors affecting BMEP, such as the pressure losses into and out of the cylinder, the effectiveness of the constant-volume combustion (a function also of valve timing), turbulence losses in the cylinder, back pressure in the exhaust system, and the manifold pressure. The constant volume combustion process must produce the required maximum pressure rise before the piston moves too far down on its stroke. The ideal situation would be instant combustion, but that does not occur. The flame front must travel across the cylinder volume from two spark plugs on opposite sides of the cylinder in such a short time that it is necessary to advance the spark before top dead center. A typical number for spark advance for an R-2800A series engine is 20 degrees before top dead center.

The higher the manifold pressure, the better because there can be more molecules in the fuel–air mixture reacting and converting the chemical energy in the fuel–heat energy. That is true as long as the temperature is not too high. What happens at a certain limiting higher temperature is detonation, which is the result of an abnormally rapid fuel combustion process causing a high-pressure wave. This wave causes not only mechanical damage to the cylinder in a way similar to the *water hammer* effect in hydraulics, but also causes areas of the cylinder and piston to overheat.

The early air-cooled radial engines in production were the rotary type. The rotation of the finned cylinders was the method of cooling. This may sound good to the cooling expert, but there are limitations to this design. WWI pilots can tell you about gyroscopic moment effects. Technically these engines were air-cooled radial engines that rotated for cooling. In this chapter I use the term *rotary engines* for them and *radial engines* for air-cooled radial engines whose cylinders do not rotate. Another name for the conventional radial engines is *stationary radials*, in contrast to rotating radial engines or rotaries (see Fig. 9).

Fig. 9 Nine-cylinder Gnome rotary engine with propeller mounting cage fastened to engine case (photo by the author at the New England Air Museum).

In the rotary engine, the propeller and the engine rotate as one unit while the crankshaft is fixed to the aircraft mounting structure. In the 110-hp class this was a popular engine for such aircraft as the Sopwiths and Nieuports of WWI. The advantages of this arrangement are the cooling of the cylinders and a flywheel effect from the rotating engine, which smoothes out the combustion pulses.

Bill Blaufuss, a retired Pratt & Whitney representative in California, flew aircraft powered by rotary engines in WWI. He told me the reason why the pilots wore scarves was to wipe the castor oil off their flying goggles. Castor

Fig. 10 Nine-cylinder Lawrance J-1, created about six years after Gnome, has conventional radial engine characteristics with the cylinders and crankcase in a fixed position while the crankshaft rotates (as in a conventional reciprocating engine) (photo by the author at the New England Air Museum).

The Early Years

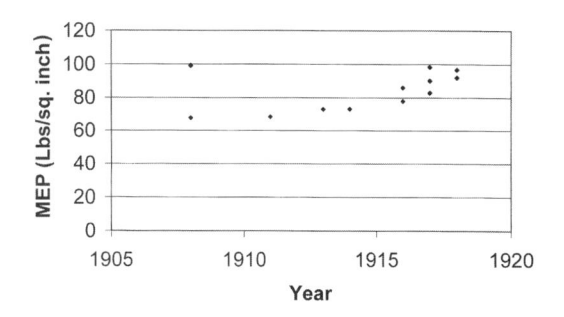

Fig. 11 BMEP for rotary engines was in the range of 70–100 psi (data from [4]).

oil was the lubricant mixed in with the fuel. Whatever oil was not lubricating the pistons and cylinder walls or burned in the cylinder came out the exhaust and annoyed the pilot. According to Bill, castor oil in the exhaust had certain unpleasant medical consequences not related to aeronautical engineering.

The engine in Fig. 10 is the first of a line of engines (J-1) at Wright Aeronautical that became the famous Whirlwinds. The J-5 powered Charles Lindbergh's Spirit of St. Louis to Paris in 1927. As mentioned in Chapter 1, Mead and Willgoos refined the J-1 and put the J-4 into production at Wright Aeronautical.

BMEP FOR AIR-COOLED ENGINES. BMEP is a measure of the output of the cylinder. When higher output per cylinder was required, the rotary engines (Fig. 11) gave way to air-cooled radial engines. Figure 12 shows the higher BMEPs for radial engines. [In the following discussion MEP is used in the same sense as BMEP. The following "MEPs" are calculated from measured (or spec) horsepower, rpm, and displacement, and therefore represent BMEP.] Compare Figs. 11 and 12 to see how much better the radial engines than the rotary engines were in terms of MEP.

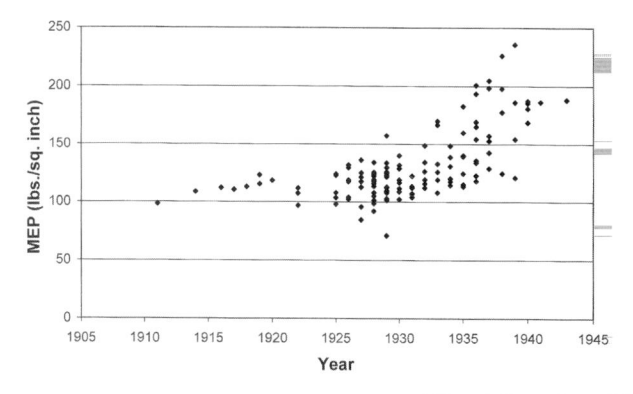

Fig. 12 MEPs for radials climbed toward 250 psi (data from [4]).

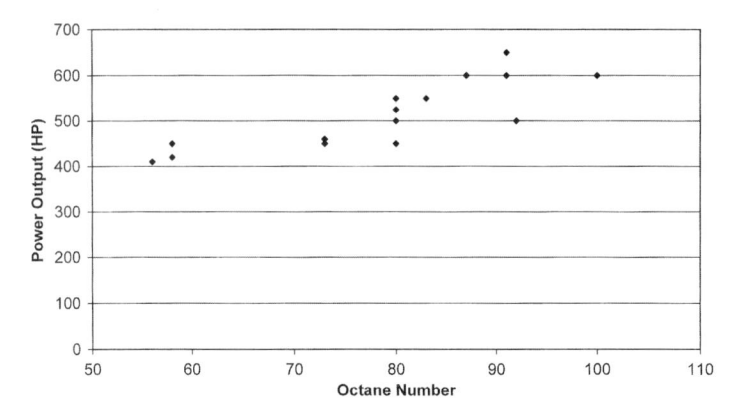

Fig. 13 Increase in horsepower with fuel octane number.

EFFECT OF FUEL OCTANE RATING ON ENGINE HORSEPOWER. Engine power can be increased by increasing the octane number of the fuel (see Fig. 13). In this example, a Wasp engine was run on the standard fuel and then the manifold pressure was increased until detonation was about to occur. Such low octane ratings seem obsolete to us now, but when the Wasp was shipped, it could run on 58 octane fuel, aviation gas at that time. Air density is proportional to the manifold pressure divided by the absolute temperature of the air. The higher the density of the air, the higher the engine's horsepower. In this example, detonation was avoided because the fuel contained detonation inhibitors, either by altering the chemical composition of the fuel or by using detonation-suppression additives. In this particular demonstration, the engine power was more than doubled without encountering detonation.

There is a rating scale for aviation gasoline that ranges from Octane Numbers 70 to 100 and Performance Numbers (PN) from 100 to 165. Until and throughout WWII, Octane Numbers up to 100 seemed to be adequate. However, more power could be pulled out of engines that used fuels with more resistance to detonation, and after WWII some commercial piston engines required PNs of about 145 to avoid detonation. When you see PNs such as 115/145, the lower number is with the mixture control set at lean and the higher number is with the mixture control set at rich.

Detonation, sometimes referred to as "knock," is associated with that part of the fuel–air mixture that burns last in the cylinder. The condition of detonation and its characteristic sound is described by MIT professors C. F. Taylor and E. S. Taylor as follows [5]:

> Laboratory research has shown that this characteristic sound is always accompanied by the following phenomena:
> 1) A large increase in apparent flame speed, near the end of the flame travel.

2) An increase in the rate of pressure-rise in the last part of the charge to burn, accompanied by an increase in the maximum pressure of this part of the charge.

3) High-frequency pressure waves within the cylinder gases, starting near the end of the combustion process.

4) A distinct change in radiation from the flame near the end of the combustion process.

The power output of an engine is sensitive to the temperature of the mixture entering the cylinder. For example, some fuels are satisfactory when the mixture temperature is "normal." However, if the mixture temperature increases to 100° F, for example, then the engine power would have to be decreased by 20 to 40 percent in some cases to avoid detonation.

Intercoolers before the engine intake manifold or between supercharger stages help to avoid the level of inlet temperatures that cause detonation. The supercharger discharge temperature rises for a given pressure ratio as the efficiency of the supercharger is lowered. A supercharger is an air compressor, usually in the form of a centrifugal compressor, that pumps up the air from ambient pressure level to a higher level to get more pounds of air into the cylinder (with a corresponding amount of fuel) for greater power from the cylinder. As a result of its experimental programs, during the 1930s Pratt & Whitney made great progress in supercharger efficiency by reducing temperature rise.

Another way to increase the engine power is to add a water/alcohol injection at take-off. The evaporation of the liquid mixture when it is exposed to air lowers the temperature of the air-fuel mixture entering the cylinder. This increases the density of the air entering the cylinder and results in higher power output. At the same time, the lower mixture temperature that enters the cylinder means an increased safety margin from reaching the detonation temperature. One could also increase manifold pressure to get more power out of the engine while still keeping the cylinder inlet temperature slightly below the detonation temperature. Another way to increase power while avoiding detonation is to increase the fuel–air mixture into the rich area. This power increase is accompanied by a substantial increase in fuel consumption, making it not a method that can be used for very long. The selection of fuel for the engine as well as the temperature of the fuel–air mixture entering the cylinder can have a pronounced effect on whether or not detonation occurs.

DISPLACEMENT PER CYLINDER. In Fig. 14, a square cylinder with a bore and stroke of 6.3 in. would correspond to a displacement of about 200 cubic inches. Most of the data is between 50 and 200 cubic inches per cylinder.

NUMBER OF CYLINDERS. Figure 14 shows the maximum displacement per cylinder is about 200 cubic inches. Nine cylinders seem to be the practical

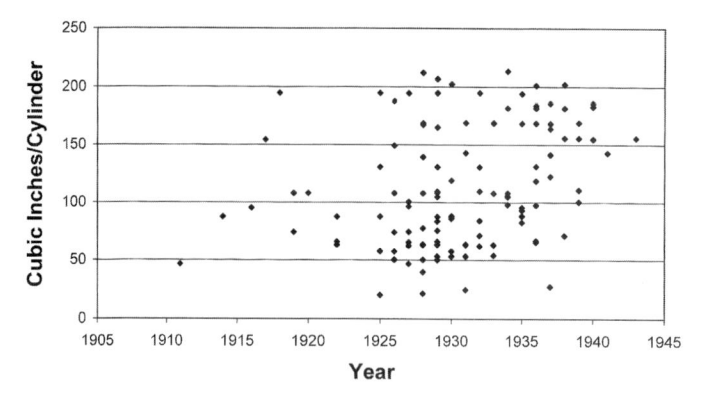

Fig. 14 Displacement per cylinder vs time (data from [4]).

maximum number of cylinders that can be put together in a single row. Therefore, 1800 cubic inches is about the practical maximum displacement for a single row. We have just seen that displacement per cylinder has a limiting value and that there is a practical limit to the number of cylinders in a single radial row. Starting in the 1930s, engine designers began to add a second row to increase the power by a factor of two. The world record for the number of cylinders in an aircraft engine goes to Wright Aeronautical for its 42-cylinder, liquid-cooled *Tornado* engine. Pratt & Whitney built five two-row engines (14 to 18 cylinders) and a four-row engine (28 cylinders).

ROTATIONAL SPEED. The engine power increases linearly with rotational speed (rpm). The higher the rotational speed, the higher the power until one runs into the practical limitations of stress on the rotating parts: the piston running faster than the flame fronts propagate from the spark plugs; the fuel–air mixture entering the cylinder; and the discharge of the products of combustion from the cylinder. Pitson speed is related to rotational speed and piston stroke. This is a more fundamental variable because it is related to linear speed and can be compared to flame front propagation. Table 1 lists some numbers for pistons speed for a 6-in. piston with a 6-in. stroke. Flame propagation speeds are in the 50 to 90 ft/sec ballpark.

TABLE 1 PISTON SPEEDS FOR 1000 TO 3000 rpm

rpm	Piston speed (ft/sec)
1000	17
2000	33
2500	42
3000	50

THE EARLY YEARS

HORSEPOWER INCREASES SHARPLY IN THE 1930s. There was slow progress up until the mid-1930s, but then a dramatic increase that continued throughout WWII. Figure 15 shows the trend in specific power (horsepower/cubic inch displacement) with year. This data takes the effect of displacement out of the picture; so, we have eliminated adding cylinders to increase the horsepower. Still, there is a marked increase in specific power starting in the 1930s. This power increase was due to 1) the development of higher PN fuels, which permitted higher manifold pressures without detonation (a great advantage the Allies had over the Germans in WWII); 2) development of more efficient superchargers, which produced the desired pressure increase without an abnormal increase in temperature; 3) forged aluminum cylinder heads with much greater cooling-fin surfaces; 4) higher rpm with reduction gears; and 5) intercoolers keeping the fuel–air mixture below the detonation temperature.

In 1919 Sanford Moss of General Electric recognized that there was considerable energy in the exhaust, some of which could be expanded through a turbine to drive a supercharger. This is essentially a mini gas turbine used to increase manifold pressure. Another variation of the gas turbine cycle would be to extract as much power as possible from the piston engine exhaust and put this power into the piston engine power shaft. Wright Aeronautical did that with the Turbo-Compound engine R-3350. As will be discussed in Chapter 4, Pratt & Whitney did look into turbo-compounding the R-2800 and the R-4360. However, the advent of the gas turbine cut that pursuit out of the picture.

SPECIFIC WEIGHT HISTORY OF AIRCRAFT ENGINES. *Specific weight* is the engine weight in pounds divided by the engine horsepower. We have seen the progress in the air-cooled engine family from the rotary to the radial engines. Another important measure of performance is the weight per horsepower.

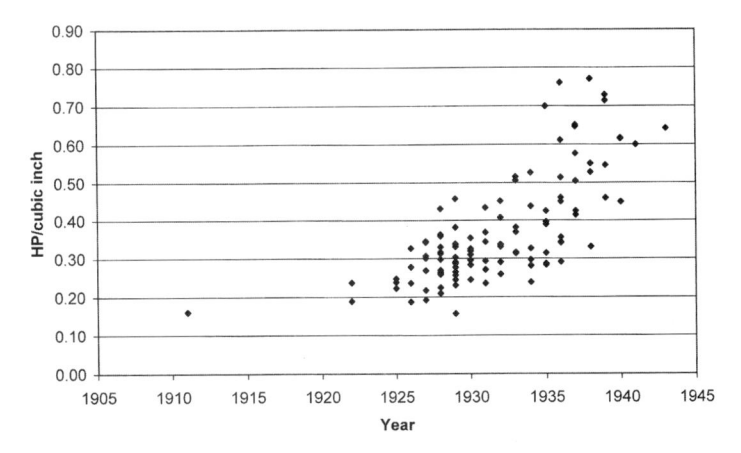

Fig. 15 Power increases in air-cooled radial engines (data from [4]).

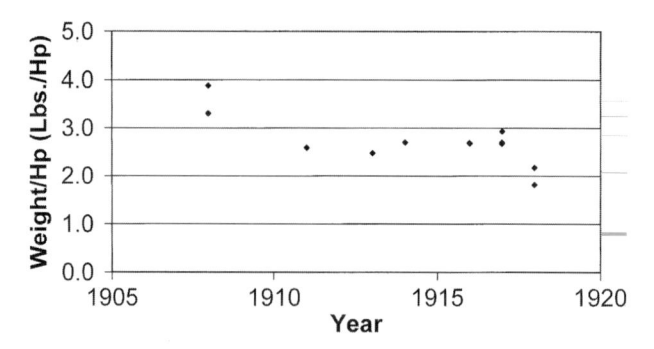

Fig. 16 Weight for rotary engines.

Figure 16 shows the progress in rotary engines (the simplest of all air-cooled engines); however, they were limited in total output and gave way to the stationary radial engines.

Rotary engines came about because of the cooling effect on the cylinders that accompanied the whirling action. The cooling effectiveness technology at that time was poor, and whirling the row of cylinders around seemed like a good idea. The limitation was reached when the cylinders pulled themselves off the rotating crankcase. On the instrument panel of the WWI Nieuports were the ominous words, "maximum speed of revolution must not exceed 1400 rpm."

The significance of engine weight was demonstrated when the Wasp replaced the Liberty engine in a pursuit-type aircraft. The reduction in aircraft weight permitted the aircraft to climb faster. The landing speed was lower as well. Because lift is proportional to the square of the velocity, the lower lift meant lower velocity. The weight per horsepower for the radial engines improved down to one pound of engine weight per horsepower (see Fig. 17).

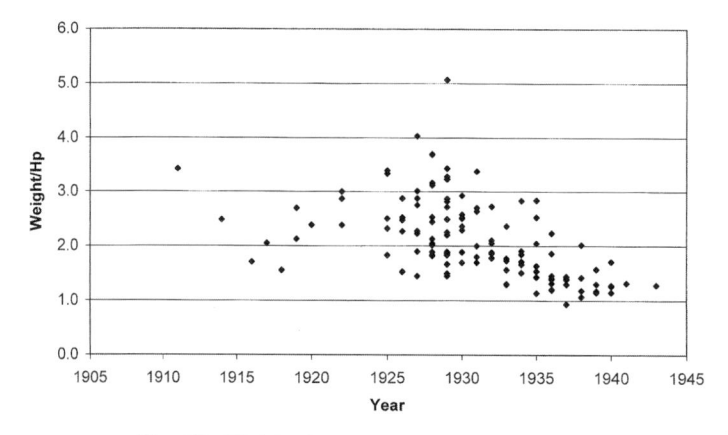

Fig. 17 Weight for radial air-cooled engines.

THE EARLY YEARS

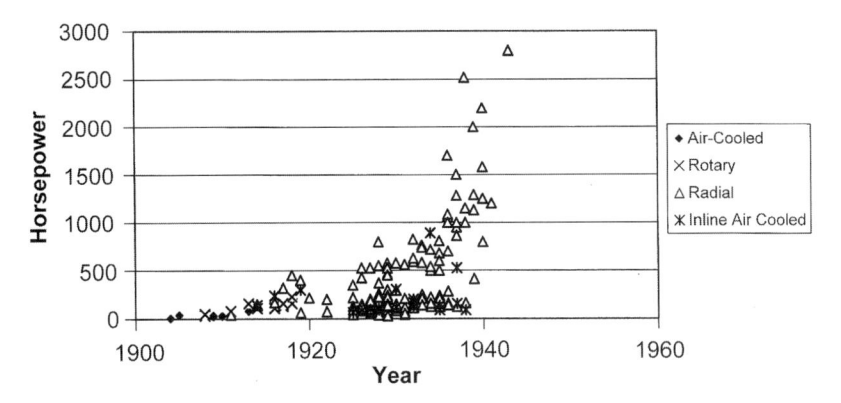

Fig. 18   Growth in horsepower for air-cooled engines.

SUMMARY OF AIRCRAFT ENGINE GROWTH

Figure 18 covers the complete spectrum of air-cooled engines from the first engines through the rotaries and the radials. (The inline air-cooled engines are included for reference.) Note the explosive growth in horsepower from the mid-1930s through WWII. Liquid-cooled engines, from the earliest to separate cylinder engines and cast block engines, show the same exponential growth in horsepower as for the air-cooled engines.

The 2800-hp point in Fig. 19 is the Fiat A.S.6, a 24-cylinder engine that powered the Macchi-Castoldi seaplane to 440 mph in 1934. This remains a world's record for seaplanes. According to Smith, the engine had an 8:1 compression ratio and ran on "dope" racing fuel—mixed by Rod Banks (popular "dopes" included tetra-ethyl-lead, amyl nitrite, ethyl nitrate and ethyl nitrite) [4]. The earliest water-cooled engines were copies of

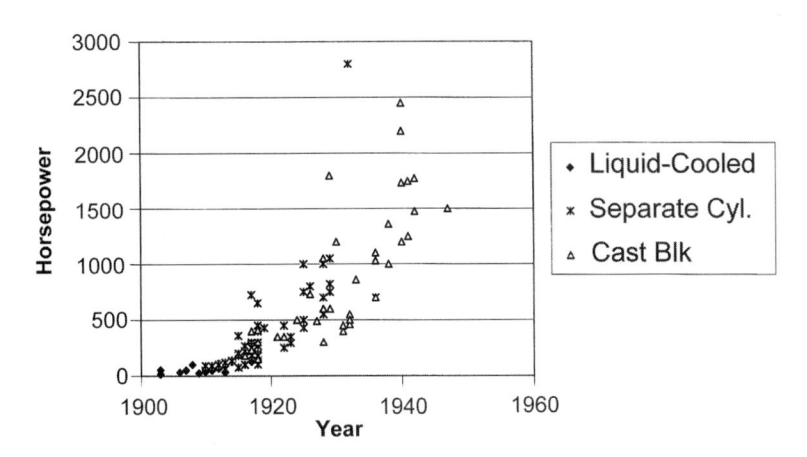

Fig. 19   Growth in horsepower for liquid-cooled engines.

automotive types. Then the trend was towards separate cylinders. The final development was for engines with a fully cast monoblock. Both the liquid-cooled engines and the air-cooled engines reached their peak with the Rolls-Royce Merlin and the P&W R-4360 Wasp Major. Which approach was the better one: liquid- or air-cooled? Some fine fighter aircraft were built around both types. When it came to commercial aviation or military transports; however, there is no question that air-cooled engines dominated the applications.

GEORGE MEAD AND THE T-ENGINE

The Navy-Wright NW-1 racing plane was built for the 1923 Pulitzer races (see Fig. 20). The aircraft was designed by the Navy Bureau of Aeronautics and built in the Vought plant. Rex Beisel, of the Navy Bureau, did the general design of the aircraft and Charles Chatfield did the stress analysis. Mead, Borrup, Willgoos, and Chatfield were part of the Wright Aeronautical team that would follow Frederick Rentschler to Pratt & Whitney Aircraft. Chatfield ended up as Secretary to United Aircraft years later and Chance Vought Aircraft also became part of United Aircraft.

Fig. 20 Aviation pioneers at the Pulitzer Races at Selfridge Field in Michigan; standing from left to right are Chance Vought, Andrew Willgoos, Jack Borrup, Rex Beisel, George Mead, L. W. Sanderson (pilot), C. Ross (Navy Inspector), C. G. Peterson (Project Manager), and C. H. Chatfield with mechanics sitting (courtesy of Pratt & Whitney Archive).

The term *pursuit engine* as used in the 1920s became *fighter engine* in the 1930s. Mead and Willgoos at Wright Aeronautical continued the development of liquid-cooled engines because at that time both the Army and the Navy saw liquid-cooled engines as the path toward higher and higher horsepower. Also at the end of WWI, the Hispano-Suiza was one of the best engines available. Then, after Wright-Martin dissolved into Wright Aeronautical, Mead continued further development of his own engine designs known as the T (Tornado) engines. The V-12 Tornado had 1947 cubic inches of displacement at a weight of about 1000 lb.

There was an interesting article in *Aviation* in 1921 that describes Mead's work ethic and reputation at that time [6]:

> George J. Mead, the present chief engineer of the Wright Co., had been associated with the development of the aeronautical engines since the time of the first work of the Simplex Co. A careful study was made by him of the different types of engines developed during the war, in an effort to determine which type was the most efficient and embodied the greatest possibilities for future development. This study included such engines as 6-cylinder verticals, 8-cylinder Vees, 12-cylinder Vees, radials, and rotaries. Each type was measured by the requirements of pursuit aviation, such as weight per horsepower, compactness, which involved not only length but head resistance, ease of overhaul, durability, and manufacturing possibilities. The result of this work was the decision to continue the development of a water-cooled, 90-degree Vee engine, as this type for pursuit work had no rival, considered both from design possibilities and the actual performance of engines of this type.
>
> Mr. Mead had been particularly identified with the Model "H" development and, therefore, was the logical man to undertake the evolution of the old Model "H" into the Model "H-3," which is now the standard type of pursuit engine for the American Air Service.

A Navy requirement was that the engine be smaller than the Liberty engine. One of Mead's innovations was to simplify the installation of the engine in the airframe: there were only three connections to be made for the engine cooling system to the aircraft—one to the pump and one to each of the cylinder blocks. Also, the externals of the engine were positioned to facilitate flight-line maintenance. This became the T-series engines, designed for the Navy Torpedo aircraft (Fig. 21). The T-engine was a 60-degree V-12 liquid-cooled engine with a power rating of 600 hp at a weight of 1150 lb. In 1923 this type of engine was in two Navy Douglas Torpedo DT-4 airplanes, which performed well in the air races for heavy aircraft at St. Louis. The same engines powered the Wright Fighters the next day and set a new world's record of 230 mph. The first 50-hr test was

Fig. 21   The 1923 Mead-Wright 12-cylinder, 575-hp T-3 engine (courtesy of Pratt & Whitney).

completed in March 1922. These water-cooled engines demonstrate that Mead and Willgoos had considerable experience before they began working on the Wasp engine in August 1925.

MEAD AND WILLGOOS DESIGN THE WASP IN 1925

The following comment by Mead explains how the design for the Wasp engine was chosen [7]:

> This is an era of specialization, and aviation is no exception to the rule. As in the case of other industries, the greatest progress in aviation is possible by specialization. Our policy from the start has been to specialize, not only in powerplants but in powerplants of a single type. After a careful and comprehensive engineering analysis of practically all types, coupled with extensive experience during the last 10 years in the design, construction, and operation of both water-cooled and air-cooled engines, we decided that the fixed air-cooled radial type offered the most possibilities. This type makes available the maximum dependability with the lightest powerplant. The low powerplant weight per horsepower give the best possible airplane performance, and airplane performance is the most important measure of an engine.

With the specifications pretty much defined by the Navy, Mead described how he arrived at the Wasp's configuration [7]:

> High crank speeds are necessary for light weight per horsepower. To date, no radial engine has had a service operating-speed of more than 1800 rpm. The mean effective pressures have been relatively low; 120 lb

was regarded as good. We decided that the required guaranteed power could be obtained with 1344-cu. in. displacement on the basis of 125-lb mean effective pressure and 1900 rpm. This meant that the average engine must be capable of at least 420 hp or 130-lb mean effective pressure to assure a safe margin above the guarantee.

Another factor ... is the kind of fuel.... With proper cylinder and intake conditions, the detonating property of the fuel definitely limits the brake mean effective pressure. Provision also had to be made for an over-speed of 2400 rpm, which would be attained in a dive by a fighter.

The proper proportions of bore and stroke were next given serious consideration. In a radial engine a number of particular limitations must be considered, such as cylinder and valve cooling, link-rod angularity and overall diameter, in addition to the usual considerations, such as proper valve and port sizes, weight of reciprocating parts and the like. We prefer what is known as a square engine, that is, with bore and stroke equal. This proportion provides the best cylinder-cooling, moderate connecting-rod angularity and the minimum diameter as well as the requisite room for large valves and ports. With these specifications determined, studies were made of the detail-design problems involved.

The following description of key components for the Wasp engine (1340 cubic inch displacement) is very similar to the Hornet (1690 cubic inch displacement). In fact about 80 percent of the parts of the Wasp are common to the Hornet. These two engines were designed within a year of each other in the mid-1920s. The Wasp consists of two major parts, the nose and power section and the accessory and blower section. The engine is mounted in the aircraft on the accessory and blower section, which is a significant advantage because the nose and power section can be removed from the accessory and blower section while the rest of aircraft is not disturbed.

ONE-PIECE CONNECTING ROD

Mead's starting point in the design was a one-piece master rod. Figure 22 shows the master rod with one link rod, a design he came up with in 1919 during the design of the first large engine for the Army Air Service. High rotational speeds are necessary for low weight per horsepower. Experience has shown that under excessive loads, fatigue cracks can occur between the split and the knuckle-pin hole for the conventional split-master rod. Also, the one-piece configuration does not have to add extra counterbalance weight to compensate for the weight of four clamping bolts in the conventional split master rod design. The master rod is subjected to considerable bending forces from the link rods. Hence there is the I-section configuration for the master-rod shank and the link rods.

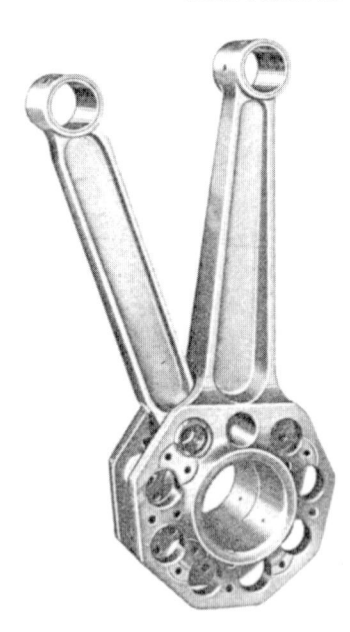

Fig. 22 The one-piece master rod provides the strongest configuration for this highly stressed part (courtesy of Pratt & Whitney Archives).

Two-Piece Crankshaft

Mead commented that on first glance the solid master rod approach seemed to move the problem from the master rod to the split-crankshaft. Obviously the crankshaft must be in two pieces when a solid master rod is used. The challenge is to create a crankshaft in two pieces that is just as stiff and strong as a solid crankshaft. The initial development of the Wasp was with a split master rod while the two-piece crankshaft was being developed (see Fig. 23). The shaft is in three parts. The front and rear sections are held together by a long through-bolt. The proper angularity of the two parts is secured by a spline. The shaft is carried by three anti-friction bearings. Two roller bearings box in the crank cheeks and a deep-row ball bearing close to the propeller takes the thrust and radial loads of the propeller. Torsion vibration problems are eliminated by using a short, wide diameter crankshaft.

Fig. 23 Mead pointed out that power is transmitted from the master rod to the propeller through a crankpin integral with the crankshaft (courtesy of Pratt & Whitney Archives).

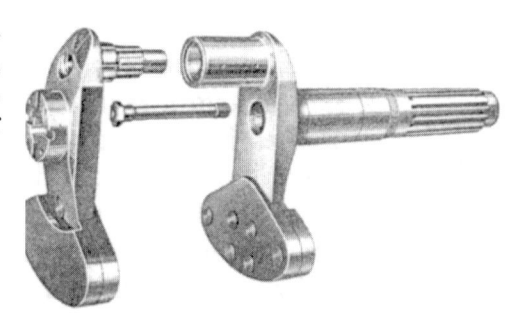

THE EARLY YEARS

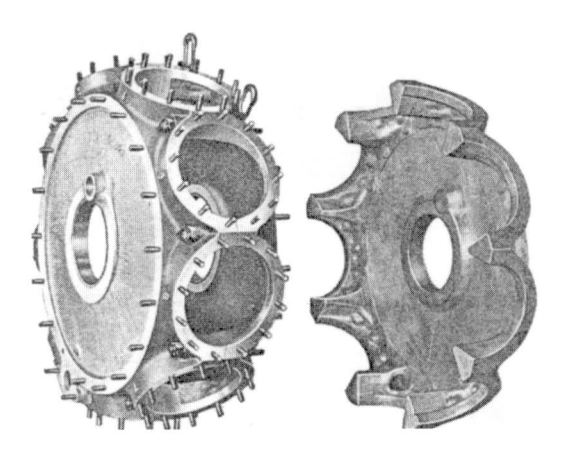

Fig. 24 The crankcase consists of two identical halves bolted together by nine through-bolts. This type of design reduces manufacturing costs (courtesy of Pratt & Whitney Archives).

FORGED ALUMINUM CRANKCASE

The crankcase is made up of two identical halves held together by long through-bolts between the cylinders (see Fig. 24). This case configuration was designed to guarantee uniformly strong parts in production. The crankcase material is forged aluminum. In Mead's design the bearing load was equal on both halves so that "no working occurs between the crankcase sections" [7].

NOSE SECTION

The hemispherical nose section bolts onto the front of the crankcase (see Fig. 25). This section carries the valve tappets and encloses the cams that drive the push rods.

Fig. 25 The nose section provides the structural support for the propeller thrust and load ball bearing as well as for the tappets (courtesy of Pratt & Whitney Archives).

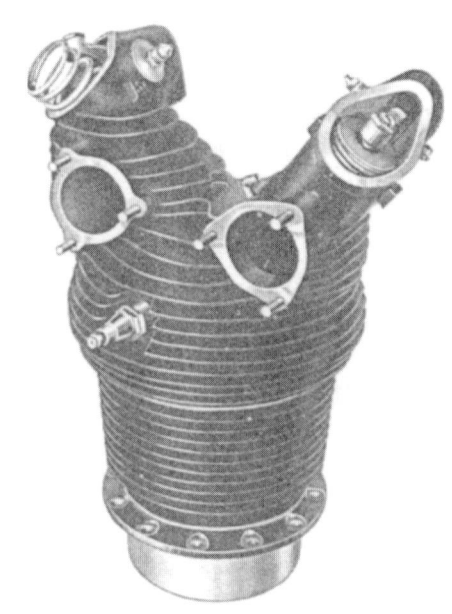

Fig. 26 Rear view of the cylinder head, with the exhaust on the left and the intake on the right (courtesy of Pratt & Whitney Archives).

Fig. 27 Power section (courtesy of Pratt & Whitney Archives).

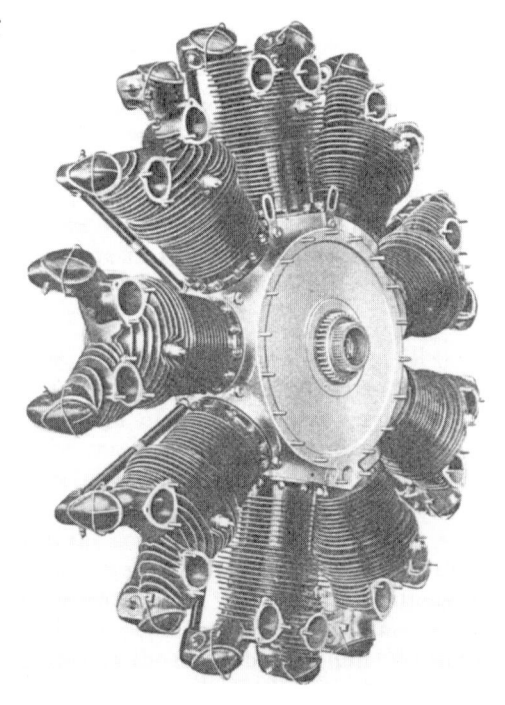

CYLINDER HEAD

Figure 26 shows the cylinder head. The assembly is made up of an aluminum head screwed and shrunk onto a steel barrel. The valves are seated on aluminum bronze rings shrunken by heat onto the cylinder head. The rocker arms and push rods (on front of the engine in the nose section) actuate the valves. Each rocker housing has a removable cover.

POWER SECTION

Figure 27 shows the power section of the engine. It is possible to separate this power section from the accessory section while the engine is on the aircraft. The accessory section stays in the aircraft and the power section can be changed if necessary. Because the engine mounting system is on the blower section, which also carries the accessory section, the power section can be removed without disturbing the connections to the aircraft. This means that one can pull off the power section and replace it with another one with minimum disturbance to the aircraft. This feature makes the maintenance effort of this radial engine easier compared to a liquid-cooled engine.

ACCESSORY AND BLOWER SECTION

There are ten mechanical drives in the accessory and blower section (see Fig. 28). These are powered by three main spur gears, about 120 degrees apart, which are driven from the crankshaft. The accessories are two magnetos, a starter, a tachometer, a generator, scavenge and pressure oil pumps, a fuel pump, and a step-up gear for the supercharger (see Fig. 29). The gears are spur

Fig. 28 Accessory and blower section, which mounts on the airplane (courtesy of Pratt & Whitney Archives).

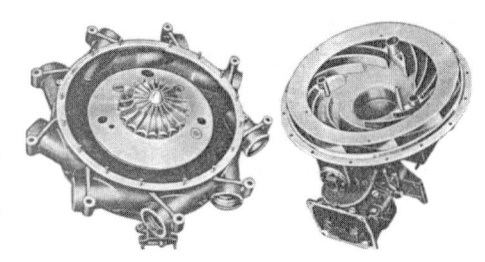

Fig. 29 The supercharger; the high-speed impeller directs the fuel–air mixture to each of the nine cylinders (courtesy of Pratt & Whitney Archives).

gears for easy removal of the accessory and blower sections from the power section. Figure 30 shows the installation of the Wasp in the Corsair O2U. Note how easy it is to get at the accessories from either side of the aircraft.

QUALITY IN MANUFACTURING

Mead's design of the Wasp paid attention to performance, reliability, service, maintenance, and production. Mead speaks to the quality of the aircraft engine production [8]:

> Briefly, quality depends on a proper design as well as the selection, manufacture, and inspection of the materials used in the fabrication of

Fig. 30 The installation of the Wasp in the Corsair O2U (courtesy of Pratt & Whitney Archives).

The Early Years

Fig. 31 Mead (left) flies on a Boeing Mail Plane to experience air travel with the Wasp engine (courtesy of Pratt & Whitney Archives).

the powerplant.... Proper material is of utmost importance. There can be only one standard of quality and that is the best. It will be a serious matter if any aircraft engines of inferior quality reach the hands of the public, since not only the reputation of the manufacturer but the welfare of the industry as a whole will suffer.

Mead Flies with Wasp on Mail Routes

Boeing Air Transport began its transcontinental mail service from Chicago to San Francisco on July 1, 1927 with twenty-four new Boeing mail planes, which were equipped with Pratt & Whitney Wasp engines. A round trip occurred daily, flying between 3,800 and 4,000 miles.

One hot July evening, Mead hopped on board a Boeing mail plane out of Omaha, Nebraska (see Fig. 31). A plane had just arrived from its 500-mile trip from Chicago with about 900 lb of mail, and Mead was going on the westbound plane piloted by Jack Knight. The first stop was North Platte, and as soon as the plane came to a stop on the runway, the night crew attended to the fueling and filling of the oil tank. The mail was taken off right away, including the mail for President Coolidge, who was at Rapid City, South Dakota.

The next stop was Cheyenne, Wyoming (altitude 6400 ft). The 500-mile flight from Omaha was so smooth that Mead dozed off feeling as though he was on a train. The following stop was Rock Springs, Wyoming (altitude 6400 ft). Mead recalled the take-off [9]: "As we taxied down the field to take off, I wondered how quickly the Wasp could lift the eight or nine hundred pounds of mail, together with the ship and its occupants. Barker [the pilot] turned the plane around and headed directly for the hangar, and up we went, without any apparent effort." Mead arrived in Salt Lake City in nine hours "without any sleep for two nights and no breakfast" where he dined with Mr. Hubbbard, the Vice President in Charge of Operations at the Boeing Air Transport who flew in from San Francisco. "A mere six hundred and twenty-five mile journey in the air over the Sierra Nevada mountains for a dinner engagement is nothing to the officials of the Air Mail," Mead recalled proudly [9]. From Salt Lake City, Mead flew back to Omaha to spend a few days studying the Boeing operations.

Then he headed to San Francisco for a meeting with Mr. Johnson, president of Boeing Transport. His next trip, later the same day, was a 40-hr train journey from San Francisco to Seattle, and then another two-night-and-one-day journey from Seattle to Salt Lake City. Mead remembered [9]:

> I had begun to feel that there wasn't any thrill in flying with the mail, and that it was rather a slow and monotonous way to get about.... My two train journeys entirely cured me of this feeling. Of all the hot, dusty, uncomfortable ways of traveling, the railroad certainly takes the prize. For this reason, I was very glad to take off again headed east behind a faithful old Wasp with Boonstra [Boeing pilot] at the stick.

George Mead arrived in Omaha, where a young Stromberg carburetor engineer named Leonard Hobbs was one of the people interested in Mead's briefing about his experiences, hearing about the Wasp throughout the Boeing Air Transport System. He would soon join Mead at Pratt & Whitney and make a world-famous name for himself in aircraft engines.

EPILOGUE FOR GEORGE J. MEAD

After a day's visit with Mead and Rentschler, Louis M. Lyons, a reporter from the *Boston Globe*, wrote an informative article about Mead (published September 8, 1935). Lyons wrote that Mead believed aviation was so complicated that no one man knew everything but had to work with a team to produce successful aircraft engines. Mead was quick to give credit to others such as Frank Caldwell, inventor of the variable-pitch propeller. Professor Hunsaker of MIT, designer of the U.S. Navy's NC4 that crossed the Atlantic in 1919, made sure that Lyons understood that Mead was an outstanding engineer. Lyons included Rentchler's comments in his article: Rentschler claimed that all aeronautical development came in the aftermath of a better engine. He could speak with authority as an airplane and engine manufacturer as Sikorsky and Chance Vought aircraft manufacturers were part of Rentschler's United Aircraft along with Hamilton Standard and Pratt & Whitney.

Paul Fisher offers following picture of Mead from an interview with Leonard Hobbs in 1950 [10]:

> There is no question of his courage, and points out that physically he stood up under illnesses that were very real and not imagined. He carried his problems with him, was a bad sleeper, a perfectionist, and an explorer of problems, the last two chemically bad for an unrested man with such slender resources of physical strength.

At UA&TC Mead insisted on a bigger aircraft size for the Boeing 247, but accepted defeat cheerfully. Rentschler, years later, admitted that Mead was right. Hobbs said that Mead would "explore every possible data and every forward thought.... In certain respects ... he came near to being a genius if he wasn't a genius. Certainly he was if you take the old definition that genius is nothing but an infinite capacity for pains."

George Mead retired from United Aircraft in June 1939. About four months later, President Roosevelt appointed him Vice-Chairman of the National Advisory Committee for Aeronautics (NACA). It was in this capacity that Mead was able to get NACA involved in engine research. The following year, Mead went to Washington, where he created the aeronautical section of the National Defense Commission, which later became the War Production Board. He was the sparkplug and the planner behind the huge aircraft engine production by the automotive companies for WWII aircraft.

The details of Mead's appointment were explained by S. Paul Johnson, Director of the Institute of Aeronautical Sciences in a letter to H. M. Horner, President of United Aircraft (July 28, 1950):

> A short time after Roosevelt's "Fifty thousand airplane speech," he turned over the problem of coordinating the entire aircraft procurement program to Henry Morgenthau ... [who] summoned two groups of aviation people.... The group consisted of E. P. Warner, J. C. Hunsaker, George Mead, Robert Hinkley, and myself [and we] agreed among themselves that a Special Assistant to the Secretary of the Treasury should be established and that George Mead was the proper candidate for the job. We then returned to Mr. Morgenthau's office with that report.... George Mead served for about a month as Special Assistant to the Treasury before the whole outfit had been transferred over to the Knudsen organization.... George Mead, however, had the unique distinction of being the only Assistant Secretary of the Treasury for Aeronautics in the history of the country.

WWII production of aircraft and engines was an almost unbelievably monumental task (see Fig. 32). The United States was fortunate to have

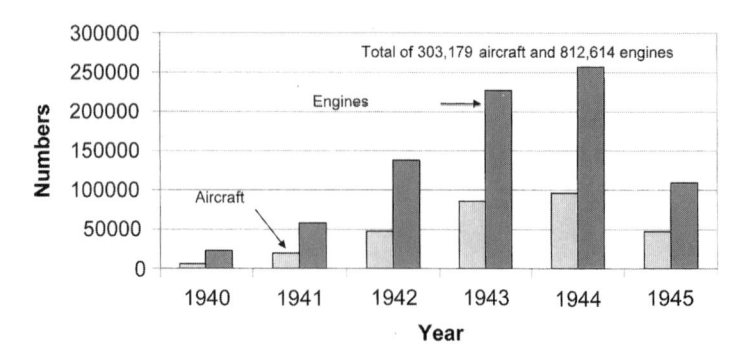

Fig. 32 Wartime production (data from the Dept. of Aviation Statistics/Dept. of Commerce).

George Mead pulling this effort together. Mead is remembered for his many capabilities [10]:

> His acumen as a good businessman, not a common thing with his kind of brilliance.... The fact that he bore in Luke's words 'the whole engineering load' (i.e. forward planning as well as critical decisions) in the years from 1925 until 1933 and 1934.... Implied throughout was the fact that Mead was a good team man and his particular kind of fine ability matched the equally fine talents of Rentschler.

In the next chapter, we'll see how Mead and his team laid the groundwork for the twin-row, radial, air-cooled engine we know as Pratt & Whitney's Wasp.

REFERENCES

[1] Rentschler, F. B., "An Account of Pratt & Whitney Aircraft Company 1925–1950"; Rentschler wrote his reflections in 1950 and distributed copies; under the auspices of Pratt & Whitney President Arthur E. Wegner, it was published by the company in Oct. 1986 for internel distribution.
[2] Mead, G. J., *Some Aspects of Aircraft Engine Development*, Society of Automotive Engineers, Vol. XVII, Warrendale, PA, Nov. 1925.
[3] Mead, G. J., *Aero Digest*, Aeronautical Digest Publishing, New York.
[4] Smith, H., *A History of Aircraft Piston Engines*, Sunflower University Press, Manhattan, KS, 1986.
[5] Taylor, C. F., and Taylor, E. S., *The Internal Combustion Engine*, International Textbook Company, Scranton, PA, 1938.
[6] "The Development of an American Pursuit Engine," *Aviation*, Dec. 26, 1921.
[7] Mead, G. J., "Wasp and Hornet Radial Air-Cooled Aeronautic Engines," *Journal of the Society of Automotive Engineers*, Dec. 1926.
[8] Mead, G. J., "Quality Problems in Aircraft Engine Production," *The Bee Hive*, UA&TC, Pratt & Whitney Aircraft Division, May 1929.
[9] Mead, G. J., "Journeyings with the Boeing Airmail," *The Bee Hive*, UA&TC, Pratt & Whitney Aircraft Division, Oct. 1927.
[10] "Conversations with Luke Hobbs," interview, Paul Fisher files, Pratt & Whitney Archives, Feb. 8, 1950.

Chapter 3

THE LATER PISTON ENGINE YEARS

INTRODUCTION

Leonard "Luke" Hobbs is a critical figure in the success of Pratt & Whitney during the 1930s. Hobbs was born in Carbon, Wyoming in 1896. His father was a civil engineer whose jobs took the family throughout the west from Texas to the Canadian border. Perhaps it was all the moves that discouraged young Hobbs from such a peripatetic profession, but one thing was clear: "he knew and admired the mobility of the steam engine, the horse, and the mule but once he had seen a motor car, he never doubted his ultimate vocation—automotive engineering"[1] (Fig. 1).

Hobbs obtained his engineering degree from Texas A&M in 1916 and started a Masters degree at Kansas State College. He left college, however, to join the Army Corps of Engineers during WWI, and served in the 42nd (Rainbow) Division in France. After military duty, he returned to Kansas State in 1919 and earned his Masters degree in 1920. For next three years, he worked on aircraft engines at McCook Field (predecessor of the Wright Field Powerplant Laboratory). He was in the outstanding company of C. F. Taylor, head of the Powerplant Laboratory, later to become professor of internal combustion engines at MIT; Sam Heron, pioneer designer of internally cooled valves and air-cooled cylinders; George Mead, later Vice President of Engineering at Pratt & Whitney; T. E. Tillinghast, later to be sales manager at Pratt & Whitney; and Lieutenant Chuck McKinney, who would be another Hobbs team member in Field Engineering at Pratt & Whitney. Frequent visitors from Wright Aeronautical Corporation to the McCook Field Powerplant Laboratory included Frank Mock of Stromberg Motor Devices Company, who would become a mentor for Hobbs; Phil Taylor, who would later cross paths with Hobbs regarding the Rolls-Royce Nene engine after WWII; Andrew Willgoos, who would work for Hobbs in the development of piston and gas turbine engines; and Earl Ryder, who would also be one of the Mead-Hobbs team at Pratt & Whitney.

While at McCook Field, Hobbs earned his nickname that was [2]:

> ... bestowed haphazardly and without reason. 'Leonard' had a formal ring. Luke Mc Glook was a reigning comic-strip character. Hobbs'

Fig. 1 A 13-year-old Hobbs enjoys the thrill of an early flying machine (courtesy of Pratt & Whitney Archives).

absorption in his work was the antithesis of comedy. The 'Luke' must have been as appropriate as it was capricious, for it has clung to him for three decades now, and aside from his wife, his mother, a few men of cloth, and a Mr. Rentschler, no one today addresses him as Leonard.

Extensive interviews with Luke Hobbs were conducted by Paul Fisher, editor of *The Bee-Hive* Pratt & Whitney publication. The following excerpt tells us Hobbs' frame of mind before joining Pratt & Whitney [2]:

> In 1925, at the time of the founding of Pratt & Whitney Aircraft, Hobbs learned about the fact within a week, possibly less.... He was with Stromberg and knew well the engineering section in Wright, Curtiss, and Packard; he would estimate that there were less than 25 engineers of sound reputation in those days specializing in the aviation powerplant field. At any event, the appearance of a company whose objective was air-cooled engines was something of an occasion in this mind, for the liquid-cooled engines, particularly the Liberty, still dominated the field. An important point that he makes is that although Lawrance had brought the air-cooled engine to this country, and Wright had manufactured a few, it was still in an experimental stage. Surplus Liberties (an engine that further refinement had made into a pretty good one of its day after enduring a sad WWI reputation) still were plentiful. There could be no successful commercial or air mail application of Liberty-powered aircraft

The Later Piston Engine Years

because of the prohibitive weight. What airmail there was, too, until 1925, when contracts were let to private competition, has been carried in military DHs [WWI-era de Havilland aircraft].

Hobbs Reveals the Secret of Success

Hobbs explained Pratt & Whitney's success in the air-cooled engine field in the same interview with Fisher [2]:

> The Wasp that emerged was light, durable, of high power and of revolutionary application in the whole aviation field. But Hobbs points out that there is no reason for straining to emphasize features of new invention. What the Wasp did have was an 'excellence of design'; what Pratt & Whitney had, too, was a first-rate intelligence in Rentschler's operation in the financial, policy, and management field, and an equally astute intelligence, both in forward engineering and general organization, in George Mead.
>
> Nor would Hobbs minimize the kind of competition Pratt & Whitney had. Both Wright and Packard were, for the time, relatively strong, but what gave P&WA [Pratt & Whitney Aircraft] its edge, beside its excellence of its engines, was the fact that collectively its men were "hungry"— in the prize fight meaning. "They were not only able people," Luke said, "but they literally worked like hell." The standard work week was 50 hours, but virtually everyone worked far longer and *worked*. Engineering paid almost no attention, as a group, to the clock. Moreover, they were young men. "There was," Luke said at another part of the talk, "a hell of a lot to be done, for you see they were a little outfit that had blossomed, very early in their life, into a big one, and luckily for them they didn't realize they were big. It was then that Vaughn, whose Wright people had been tinkering with as many as six engines, cleared the decks and concentrated solely on what ultimately became the 1820. Pratt & Whitney's engineering, in turn, was spread over five or six projects. The R-985C, the 1535, 1830, etc. The two-row stuff. The 2270 earlier, which the Navy had said was too big. The 1535 was a good engine from financial returns, but in an engineering sense, Hobbs thinks it was not right and should not have been built; a bigger engine should be been brought along. P&WA began to slip and he now thinks it was a turning point in Pratt & Whitney's history. The Douglas DC-2 had been our first real commercial loss. Wright [Aeronautical] was "beginning to run us off our feet."
>
> For it was at this time that George Mead decided that the 150-hr qualification tests and all standard tests were simply not enough. He laid down the rule that above and beyond the specifications in performance that were stated there must be a margin of power in Pratt & Whitney engines that carried beyond the usual operations. The 1830s on the PBYs were satisfactory, but no good in commercial service…. Mead's belief

that performance always should be geared to the more rigorous use, especially if the use was commercial. The development of many new tests ... 100 hours at maximum horsepower.... The way the engineering staff began to get in "condition" again.

That was how George Mead created Pratt's philosophy of conservatism. Hobbs again elaborates with Fisher [2]:

So P&WA reached 1935 with its tremendous lead in the field dissipated. Luke would judge that we were then about even with Wright [Aeronautical]. His rough diagram shows an ascent with the beginning of 1935 for us, but immediately the bearing trouble plunged P&WA [and all powerplant companies] precipitously down. [The master rod bearings were beginning to fail as the military engines reached higher power outputs and rpms—particularly in a dive where the rotational speeds could reach 3000 rpm.] The ultimate solution in the lead-silver-indium bearing he characterizes as both a credit and a criticism of the company. In the first place, he says, we wasted far too much time in telling ourselves that in bearings, go to the bearing experts; the same sort of inaction could have been equally critical three years later in the case of the supercharger and General Electric. But Hobbs also thinks that from 1935 on we were on the upbeat. We were a stride ahead of Wright [Aeronautical] when war came, and the swing in the war years was steadily and pleasantly upward.

SUCCESSION OF PRATT & WHITNEY ENGINES

Not many people appreciated the great technical requirements of aircraft engines. Hobbs used to illustrate the challenge of aviation engines with the statistics shown in Fig. 2. The aircraft engine weighs a little over one pound

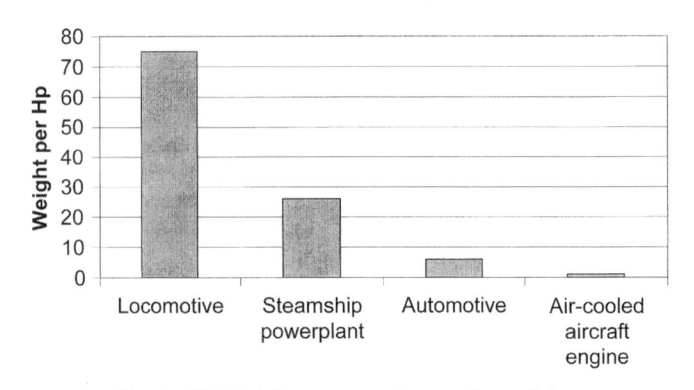

Fig. 2 Hobbs' data comparing engine weights.

Fig. 3 Wright Parkins (courtesy of Pratt & Whitney Archives).

per horsepower, a value so small that it is almost unnoticeable on this chart. The next lightest is the automobile engine at about 6 lb/hp. Steamboat and locomotive engines weigh 20 to 75 times as much as aircraft engines.

Wright Parkins (Fig. 3) joined the team in 1928 as a development engineer and served with Hobbs and Willgoos from the piston engine era through the successful launching of Pratt & Whitney into the jet era. Parkins was born in 1897 in Dakota Territory before its statehood. His parents died before he entered high school, and he lived with his aunt and uncle in Seattle. As soon as he graduated from high school, he enlisted as a private in the Field Artillery and served in France in WWI. Fisher writes about his days at the University of Washington [3]:

> Parkins' studies, drafting work, and rowing all became eclipsed when a sociology major, Annabel McLeod, was introduced to him on the school campus. The slim, merry-eyed girl was contained where Parkins was exuberant. When Parkins got his engineering degree, she gave up any dreams she had of re-making the masses into a lovelier group and since 1923 has concentrated all her sociological theories upon Parkins alone.

Parkins was a part-time worker as a draughtsman at the Almen Barrel Engine Company while in college, then worked for that company for another two years after graduation. The company had an engine program sponsored by the U.S. Army Air Corps at McCook Field and often sent Parkins to visit that facility. He was attracted to the activities at McCook, and before long he joined the group there in 1925. One of his assignments was to plan and install dynamometers for their test cells. In 1928, he came to Pratt & Whitney as a development engineer.

In 1931, Parkins spent a few months in California, reducing drag for Pratt & Whitney engines in a Fokker F-32. A contemporaneous issue of *The Bee-Hive* described the experience [4]:

> Just recently Wright's name has appeared in many newspapers and practically every aeronautical publication in the country, due to his excellent work in developing a new type ring cowl designed especially for pusher type engines where tandem installation is required. He has just recently returned to Hartford, after spending several months in California, where he has been at work on this project. Through the development of what has since become known as the Parkins Ring Cowl and its installation on the pusher type motors in the Fokker F-32, the largest American land plane, cylinder temperatures have been greatly reduced and the plane's speed increased approximately fifteen miles per hour.

Hobbs outlined aviation's achievements leading up to WWII as follows [5]:

> American aircraft engines, generally, have led the world since aviation started its upward surge following the World War, which provided the first powerful stimulus to aeronautical design and development. That development has continued at a constantly accelerating pace, and the engines have been improved and perfected in performance, reliability and low cost to a point where they constitute one of the outstanding accomplishments of a mechanical age.
>
> Aviation's achievements have been spectacular and have captured the public's imagination. The actual accomplishment, however, has been by means of the slide rule and the drafting board, the wind tunnel and the metallurgical laboratory. It is a story of painstaking research and methodical engineering progress—yet paradoxically carried on at a dizzy pace. It is a story, also, of precision manufacture and skilled craftsmen, for the manufacturing standards are almost unbelievably high.

Fisher sums up the succession of the Pratt & Whitney piston engines [2]:

> Under Mead's direction, Hobbs, Willgoos, Ryder, Wright Parkins, and a handful of Pratt & Whitney Aircraft's original team of engineers took the radial air-cooled engine from its original horsepower of 400 to a point in its development where it and its successors yielded twice that output and laid the groundwork for the twin-row, radial, air-cooled engine which evolved as the R-1830, and the R-2800 Double Wasp. Hobbs specifically developed the R-2000 for the Douglas DC-4 transport which did such an immeasurably important job throughout WWII.

When Pratt & Whitney lost the engine installation for the Douglas DC-2 in 1932, it was a sobering turning point in Pratt & Whitney's history. Douglas selected the Wright Aeronautical Cyclone engine because it used less oil.

THE LATER PISTON ENGINE YEARS 67

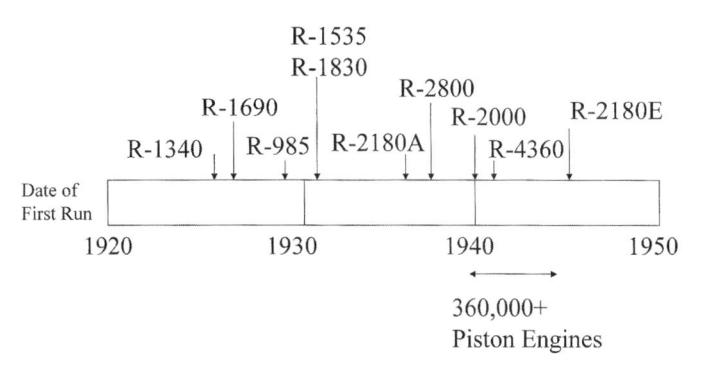

```
                    R-1535
                    R-1830
                           R-2800
           R-1690          |  R-2000   R-2180E
   R-1340  |  R-985  R-2180A  |  R-4360
Date of
First Run
   1920       1930        1940       1950
                          ←——→
                          360,000+
                          Piston Engines
```

Fig. 4 Time line of engines that reached production status.

Figure 4 shows a time line of piston engines developed by Pratt & Whitney. From 1925 to 1930, Pratt & Whitney produced the Wasp, Hornet, and Wasp Jr. In the 1930s, twin-row engines of increasing horsepower were developed. The most powerful engine was the Wasp Major, which came into use in the early 1940s. Also indicated on the chart is the huge number of engines produced during WWII. The last of the piston engine developments was the R-2180E for the Saab-90 aircraft right after WWII.

THE WASP AND THE HORNET

The Wasp, Hornet, and Wasp, Jr. used very similar technology. The next step in the evolution of Pratt & Whitney engines was to get more out of a cylinder than was possible in the first three production engines. The Hornet pushed the cylinder bore from the Wasp's 5.75 in. to 6.125 in. The larger Hornet (R-1860) increased the cylinder bore from the Hornet's 6.125 in. to 6.25 in. The conclusion from that excursion was to stay below a bore diameter of 6.25 in. The R-1690 cylinder displacement was about 188 cubic inches and the R-1860 cylinder displacement was about 207 cubic inches. The company concluded that 200 cubic inches was a maximum value for cylinder displacement, corresponding to a bore of 6.2 in. and a stroke of 6.6 in.

The way to increase engine power is to put as many cylinders in a single radial row as possible. Experience has shown that nine seems to be the maximum for a single row. Thus a single row could provide 200×9 or 1800 cubic inches of displacement. This would be a very ambitious way to achieve the 1800 cubic inches of displacement. A more conservative approach would be to make two rows of seven cylinders to get 14×130 cubic inches of displacement per cylinder to reach about 1800 cubic inches total. The example in Fig. 5 shows the relationship between cylinder bore and displacement per cylinder in cubic inches, illustrating the different approaches Wright Aeronautical (R-1820 Cyclone) and Pratt &Whitney (R-1830 Twin Wasp)

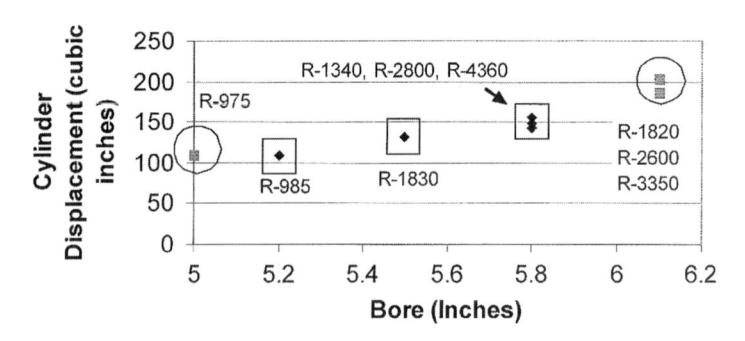

Fig. 5 Comparison cylinder displacement in Pratt & Whitney (diamond) and Wright Aeronautical (square) engines.

took to designing their engines. Pratt & Whitney made a mechanically more complicated engine with two rows to be conservative about taking power out of cylinders (130–150 cubic inches of cylinder displacement). Wright Aeronautical's three big engines (R-1820, R-2600, and R-3350) were designed close to the maximum practical limit of displacement per cylinder (close to 200 cubic inches per cylinder). Pratt & Whitney's R-2800 and R-4360 engines were designed with displacement per cylinder in the Wasp class. Even more conservative, of course, was the R-1830, an engine with the highest production run in the history of aviation. Perhaps there is something to be said for conservatism in aircraft engines. Pratt & Whitney's R-985 and Wright Aeronautical's R-975 smaller engines had much lower cylinder displacement than the bigger engines and enjoyed a greater surface to volume ratio, which helped in cylinder cooling.

R-1340 WASP

The major characteristic of this engine and its successors was what Hobbs referred to as "excellence of design." Mead took into account the needs of the maintenance crews and the cost of operation for the customers. The Wasp started out at 410 hp in 1925 and, with further development and improvement in fuels, eventually reached 600 hp (see Fig. 6 and Table 1). While the experimental R-1340 Wasp never flew, its progeny has wide-ranging applications, as shown in Table 2.

R-1690 HORNET A

The Hornet is about an inch longer and about 4 in. in diameter larger than the Wasp. It is not easy at a quick glance to tell a Hornet from an early Wasp when they are not side by side. The Hornet has a bore of 6.125 in. compared to the Wasp bore of 5.75 in. (see Table 3). The design philosophy is identical as they are only a year apart.

Fig. 6 The first R-1340 Wasp was a development engine that never got to fly (courtesy of Pratt & Whitney Archives).

The Hornet was an important engine for its time period and quickly took its place in the transport market. The German company BMW obtained a license in 1929 to manufacture the engine in Europe. It was used extensively in Junkers transports. According to the *World Encyclopedia of Aero Engines*, the fine German engine BMW132 was "distantly descended from Pratt & Whitney's Hornet ... the BMW 114 had been developed and run with cylinders almost identical with the Hornet" [6]. BMW's later development, the BMW 801, had a bore and stroke of 6.14 in. (compared to the Hornet's bore of 6.125 in. and a stroke of 6.375 in.). This 14-cylinder engine (2550 cubic inches displacement) was rated initially at 1600 hp and made the Focke-Wulf 190 a first-rate aircraft for Germany in WWII.

The Hornet, like the Wasp, was a significant force in establishing commercial aviation (Fig. 7). The Hornet, in addition to dependability, had greater

TABLE 1 SPECIFICATIONS FOR THE R-1340 WASP

Horsepower	410–600
rpm	1900–2250
Number of cylinders	9
Weight (lb.)	650–938
Bore (in.)	5.75
Stroke (in.)	5.75
Design start	1925
First run	1925
First flight	1926
Production quantity	34,966

TABLE 2   APPLICATIONS FOR THE R-1340 WASP

Model	hp/rpm take-off or military	Normal hp	Weight (lb)	Diameter (in.)	Length (in.)	Aircraft installations
Wasp A	410/1900	410/1900	745	51.44	42.63	Boeing F2B, F3B, 40A
Wasp B	450/2100	420/2000	670	50.63	43.38	Atlantic C-5; Boeing 4B1 & 4B4; Curtiss XO-12; Douglas O-32A & BT-2; Fokker F-10A; Ford C-4; Thomas Morse XO-19; Vought XF2U-1
Wasp C1	420/2000	420/2000	750	51.44	42.60	Amphibions N-2-C; Bellanca CH-400 & Model D; Boeing 100 & 204; Buhl CA-6W; Curtiss 6000A & A6A; Fairchild 71, 71A; Ford 5AT-B, C & CS; Hamilton H-45; Lockheed Air Express 3; Altair 8D & 8G; Vega 5 & 5A; Metal G2W; N. American Super Universal (Fokker A52); Ryan B-7; Stinson SM-6B; Zenith Z-6-A
Wasp SC1	450/2100	450/2100	745			Bellanca F-2, CH-400; Boeing Alpha 4-A, 4E; Curtiss 6000A; Detroit DL1; Douglas Dolphin 8-114; Fairchild C-96; FC-2-W2; Laird CL-RW-450; Lockheed 5C, 10C, & C101; Sikorsky S-36BS, S-39B & 38C
Wasp D	450/2100	450/2100	705	51.44	42.57	Boeing P-12C, XP-12G; Ford C-4A; Northrop ZC-19; Thomas Morse ZO-22
Wasp T1D1	525/2100	525/2100	763	51.44	42.63	Fokker F-22

The Later Piston Engine Years

Wasp S1D1	550/2200	550/2200	763	51.44	42.63	Bellanca F; Boeing F4B4; Lockheed UC-85; Orion 9D & 9D2
Wasp SE	500/2200	500/2200	750	51.44	42.59	Boeing P-12D, 12K, XP-12L, XP-12H; Detroit C-23 & C-25; Lockheed Vega Y1C-17
Wasp S1H1	600/2250	550/2200	865	51.80	43.01	Grumman Mallard G-73
Wasp S1H1-G	600/2250	550/2200	930	51.81	47.80	Boeing 247D & C-73; Australia's Whirraway; de Havilland Otter; Fokker S-13
Wasp S3H1	600/2250	550/2200	865	51.80	43.01	Bellanca 31-50; Grumman G-73; Lockheed 10E; N. American NA-16-1, NA-15-3; Canada Car & Foundry Norseman; Fiat G49-A; Macchi MB323; Piaggio P-150
Wasp S3H1-G	600/2250	550/2200	953	51.80	47.81	Douglas EJ-2 & 25-2; Junkers JU-52
Wasp S6H1	600/2250	500/2200	864	51.44	44.08	North American N-16-2
Wasp S1H2	600/2250	550/2200	868	51.80	45.90	Sikorsky S-55; Westland-Sikorsky S-55
R-1340-1	410/1900	410/1900	670	50.67	43.37	Curtiss XA-4; Douglas RD-1; Fairchild C-8 & XF-1; Ford C-4A, B (engine sold as Wasp A)
R-1340-3	450/2100	450/2100	670	50.67	43.39	Curtiss P-3A; Douglas O-32.A; Thomas Morse XO-19 (engine sold as Wasp B)
R-1340-4	450/2100	450/2100	705	51.44	42.57	Boeing F4B-4A; Douglas RD-3; Vought OSU-1, O3U-1 (engine sold as Wasp D)
R-1340-5	450/2200	450/2000	672	50.63	43.38	Thomas Morse XO-19; Douglas O-32 (engine sold as Wasp B)
R-1340-6	500/2200	500/2200	705	51.44	42.57	N. American NJ-1, SNJ-1; Vought O3U-6 (engine sold as Wasp D)

(Continued)

TABLE 2 APPLICATIONS FOR THE R-1340 WASP (CONT)

Model	hp/rpm take-off or military	Normal hp	Weight (lb)	Diameter (in.)	Length (in.)	Aircraft installations
R-1340-7	450/2100	450/2100	700	51.43	42.59	Boeing P-12; Douglas BT-2, 2A, 2B, 2B1, 2BG, 2BR, 2C1, 2C,G, O-32A; Lockheed C-12; Thomas Morse O-19A, B, C (engine sold as Wasp C)
R-1340-8	500/2200	500/2200	730	51.50	43.00	Boeing F4B-3, -4, -4A (engine sold as Wasp SD)
R-1340-9	450/2000	450/2000	700	51.43	42.59	Boeing XP-12A, B; Douglas YO-22; Fairchild F-1A; Thomas Morse O-19 (engine sold as Wasp SC)
R-1340-10	500/2200	500/2200	730	51.50	43.00	Boeing F4B-4; Douglas RD-2 (engine sold as Wasp SD)
R-1340-11	450/2100	450/2100	743	51.50	42.62	Boeing P-12C, G; Ford C4A, B
R-1340-12	550/2100	550/2100				Thomas Morse O-19D (engine sold as Wasp D)
R-1340-13	375/1850	375/1850		51.44		Boeing F4B-4; Curtiss XSOC-1; Vought O3U-3 & O3U-6
R-1340-14	550/2100	550/2100				Thomas Morse O-19D, E; Boeing P-12C, G (engine sold as Wasp TDG)
R-1340-15	575/2200	450/2200		51.50	47.62	N. American NJ-1; Vought O3U-6 (engine sold as Wasp D1)
R-1340-16	550/2200	550/2200	763	51.43	42.63	Boeing P-12C, G; Thomas Morse O-19D, E (engine sold as Wasp SC)
R-1340-17	500/2200	500/2200		51.50	43.25	N. American SNJ-1; Boeing F4B-4 (engine sold as Wasp S1D1)
						Boeing P-12D, E, K & XP-12L (engine sold as Wasp SD)

The Later Piston Engine Years

Model				Notes
R-1340-18	550/2100	550/2100		Curtiss SOC-1,2,3; Boeing P-12F; N. American NJ-1, SNJ-1; N.A.F. SON-1
R-1340-19	500/2000	500/2200	715	51.50	43.25	Boeing P-12F (engine sold as Wasp SE)
R-1340-21	500/2000	600/2200	715	51.44		Boeing Y1P-26 (engine sold as Wasp S2E)
R-1340-22	550/2100	550/2100	798	51.50	42.63	Curtiss SOC-1A, B, SOC-2A, B, SOC-3A, B; N.A.F. SON-1A
R-1340-23	540/2100	575/2200	715	51.44		Boeing P-12J (engine sold as Wasp SD)
R-1340-25	600/2200	600/2200	715	51.44		Sold as Wasp C2 with fuel injection
R-1340-27	500/2000	570/2200	715	51.50	43.25	Boeing P-26 & P-26C, P-29A
R-1340-29	550/2100	450/2200	715	51.50	43.25	Douglas C-29
R-1340-31	550/2100	500/1920	742			Boeing XP-29
R-1340-32	575/2200	575/2200	882			Boeing YP-29, A, B
R-1340-33	600/2120	550/2200	792	51.50	46.75	Boeing P-26B (similar to -27 except redesigned for Marvel fuel injector)
R-1340-36	600/2250	550/2200	877	51.50	44.20	Boeing YOSS-1; Curtiss XO2C-1; SOC-4; N.A.F. XOSN-1; N. American SNJ-2, -3;
R-1340-38	500/2000	570/2200				Bellanca JE-1 (engine sold as Wasp SE)
R-1340-39		550/2000				Boeing P-29
R-1340-40						Sikorsky HO4S1 (S-55)
R-1340-41	600/2250	550/2200	842			Bellanca L-11; Detroit ZC-23; N. American Y1BT-10; Northrop A-17AS (engine sold as S3H1)
R-1340-42						N. American T-6G (engine-converted R-1340 AN-1 by Navy for Hydromatic prop)
R-1340-43	550/2200	550/2200	864	51.50	42.25	Lockheed XC-35
R-1340-45	600/2250	550/2200	930	51.44	47.94	Convair YBC-3; N. American BC-2;
R-1345-46						Goodyear ZP4K (engine converted AN-2)

(Continued)

TABLE 2 APPLICATIONS FOR THE R-1340 WASP (CONT)

Model	hp/rpm take-off or military	Normal hp	Weight (lb)	Diameter (in.)	Length (in.)	Aircraft installations
R-1340-47	600/2250	550/2200	864	51.44	42.94	N. American AT-6, BC-1, BC-1A, BC-11
R-1340-48						Kaman HOK-2 (engine converted to AN-1)
R-1340-49	600/2250	550/2200	864	51.44	42.94	Comm. Australia Wirraway; Lockheed UC-36B; N. American BC-1A, -1B, AT-6A.
R-1340-51	600/2250	550/2200	863	51.44	42.94	Curtiss O-52
R-1340-53	600/2250	550/2200	930	51.70	47.70	Boeing C-73 (engine sold as S1H1-G)
R-1340-55	600/2250	550/2200	865			Bell YH-12B, XH-12 & Model 48
R-1340-57	600/2250	550/2200	878	52.00	45.50	Sikorsky H-19A, B, C, HRS-1, -2, YH-19.
R-1340-96	450/2100					Douglas RD-3 & -4 (engine sold as Wasp D)
R-1340 AN-1	600/2250	550/2200	865 mag 878 alum	51.81	43.00	Boeing AT-15BO, XAT-15; Bellanca AT-15BL; Cessna C-106A; Chase XPG-4; Fairchild XAT-13, AT-13; McDonnell AT-15MC; Noorduyn YC-64. C-64A, C-64, UC-64B, AS, AT-16; Harvard II; N. American SNJ-2, -3, -4, -5, -6, AT-6B, C, D, F Piasecki HRP-1, -2
R-1340 AN-2	550/2200	550/2200	938	51.81	47.80	Goodyear Navy ZNPK, ZNPM

THE LATER PISTON ENGINE YEARS 75

TABLE 3 SPECIFICATIONS FOR THE
R-1690 HORNET A

Horsepower	525–875
rpm	1900–2300
No. of cylinders	9
Weight (lb)	795–1075
Bore (in.)	6.125
Stroke (in.)	6.375
Design start	1926
First run	1926
First flight	May 1927
Production quantity	2,944

power than the Wasp so that the airlines could carry more payload than would have been feasible with the Wasp. The Hornet engine played an important role in mapping out routes across the Atlantic and Pacific Oceans for the Sikorsky S-42 (Fig. 8). In these early engines, the rocker mechanism was lubricated by grease. In the later model Hornet shown in Fig. 9, oil under pressure was piped into the rocker boxes. Once again, applications for a Pratt & Whitney engine were impressive (see Table 4).

Fig. 7 The Junkers W34 was used extensively by Union Airways Ltd. in South Africa (courtesy of Pratt & Whitney Archives).

Fig. 8 The Sikorsky S-42 came into service in 1934 with Pan Am; the aircraft established 10 world records and had retractable flaps and controllable-pitch propellers (courtesy of Pratt & Whitney Archives).

Fig. 9 A late-model Hornet (courtesy of Pratt & Whitney Archives).

TABLE 4 APPLICATIONS FOR THE R-1690 HORNET A

Model	hp/rpm take-off or military	Normal hp	Weight (lb)	Diameter (in.)	Length (in.)	Aircraft installations
Hornet A		525/1900	795	55.44	44.75	Boeing 95; Keystone C2H, B-3A; Martin T-4-M, 74
Hornet A2	525/1900	525/1900	795	55.44	44.75	Boeing 40B, 40B-4, 80-A1
Hornet C	600/2000	600/2000	840	55.44	45.66	Convair 17; Douglas O-38C, ZO-38B; Vought O3U-2
Hornet T1C1	675/2000	675/2000	871	55.44	45.66	Vought V-80P
Hornet T2D1	660/2000	660/2000	880	55.44	45.75	Convair 16; General GA-43; Sikorsky S-40A
Hornet T2E	650/2000	650/2000	975	54.69	44.23	Junkers (New Guinea) JU-52
Hornet S5E	700/2050	700/2050	975	54.69	44.23	Vought (Siam) V92S
Hornet S1E-G	875/2300	750/2250	1064	54.44	49.38	Junkers JU-86 (S. Africa); Sikorsky S-42, A, S-43, B
Hornet S1E2-G	875/2300	750/2250	1070	54.44	49.67	Focke Wulf Condor FW-200; Lockheed Electra 14-H2; Lockheed 18-H, C-56A (18H), C-56C
Hornet S1E3-G	875/2300	750/2250	1087	54.69	50.52	Faucett F-19
R-1690-1	525/1900	525/1900	800	55.44		Keystone XLB3A; Thomas Morse YO-20
R-1690-3	525/1900	525/1900	800	55.44		Keystone LB7; Douglas O-38, A
R-1690-5	525/1900	525/1900	841	54.43	44.78	Curtiss YA-10; Douglas O-38B; General Aviation C14B; Hamilton UC-89
R-1690-7	525/1900	525/1900	850	54.43	45.65	Douglas O-38C

(Continued)

TABLE 4 APPLICATIONS FOR THE R-1690 HORNET A (CONT)

Model	hp/rpm take-off or military	Normal hp	Weight (lb)	Diameter (in.)	Length (in.)	Aircraft installations
R-1690-9	625/2000	625/2000	850	54.43	45.75	Curtiss YA-10; Douglas O-38E
R-1690-11	775/2200	700/2150		50.14	48.50	Martin RB-12A, YB-12 (smaller heads)
R-1690-13	625/2000	625/2000	850	55.20	46.25	Douglas O-38E, O-38F
R-1690-21	760/2200	675/2150	850	54.56	50.75	Martin B-12, RB-12A
R-1690-23	800/2300	750/2250	1068	54.56	50.00	Sikorsky OA-8, Y10A-8
R-1690-25	875/2300	750/2250	1086	54.75	50.52	Lockheed C-59
R-1690-28	525/1900	525/1900				Great Lakes TG-1; Martin T4M-1 (engine sold as A1)
R-1690-32	525/1900	525/1900	795			Martin P3M-2, T4M-1 (engine sold as Hornet A2)
R-1690-38	600/2000	600/2000	840			NAF-T4 N-1; Vought SU-2, SU-3 (engine sold as Hornet C)
R-1690-40	600/2000	600/2000	840			Vought SU-1, -2, -3 (engine sold as Hornet C)
R-1690-42	600/2000	600/2000	861			Vought SU-1, -2, -3, -4
R-1690-52	850/2500	750/2250	1064	54.44	49.38	Lockheed XR40-1; Sikorsky JRS-1, S-43
R-1690-54	875/2300	750/2250	1087	54.69	50.52	Lockheed XR40-1, R50-2; Sikorsky JR-1, S-43 (engine sold as S1E3-G)

R-1860 Hornet B

Because the progression in cylinder bore from the Wasp to the Hornet gave good results, the temptation was to continue to larger bores and strokes. The R-1860 had a 6.25-in. bore and 6.75-in. stroke (see Fig. 10 and Table 5). The results made it look as though the Hornet B had as large a total displacement as one could use in a single row. It was too close to the upper limits of bore size and the challenge of cooling a cylinder with the minimum surface area to volume. Also, the potential power increase that might be derived from higher rpm was cut down by the time the flame fronts could meet while the piston was moving away from top dead center.

The lesson learned from the Hornet B was that higher potential horsepower could be achieved more effectively by adding another row of cylinders instead of increasing the cylinder bore size. The lesson was repeated in other engines, as shown in Table 6. It is interesting to note that the Wright Cyclone was just about 2 percent smaller than the Hornet B in displacement per cylinder.

Roles of Wasp and Hornet in International Commercial Aviation

In 1927, a fledgling airline under the guidance of the visionary Juan Trippe came into being as Pan American Airways, Inc. (Pan Am) when, on October 19, a Wasp-powered Fairchild FFC-2 seaplane delivered 30,000

Fig. 10 Dubbed the "Big Hornet," the R-1860 defined the limits to cylinder bore in inches (courtesy of Pratt & Whitney Archives).

TABLE 5 SPECIFICATIONS FOR THE
R-1860 HORNET B

Horsepower	575–650
rpm	1950–2000
Number of cylinders	9
Weight (lb)	985
Bore (in.)	6.25
Stroke (in.)	6.75
Design start	1928
First run	1928
First flight	
Production	446

letters (251 lb) to Havana from Key West. The next 15 years would see the development of international commercial passenger travel throughout South America, and across the Pacific and Atlantic Oceans. The Wasp and Hornet engines contributed substantially to that development.

In the very early 1930s, aircraft did not have room for enough passengers to make them economically viable in domestic commercial aviation and did not have the range for international travel. Economic viability was not reached in domestic commercial airlines until the DC-3, when 24 to 28 passenger capacities were reached. Even at that, the DC-3 range was only about 1200 miles.

In Pan Am's early days, the challenge was to acquire long-range aircraft capable of profitably carrying enough passengers and mail thousands of miles over water. Juan Trippe, his consultant Charles Lindbergh, Igor Sikorsky, Glenn L. Martin, and Wallwood Beal (Chief Engineer at Boeing) collaborated with Pratt & Whitney on a series of aviation advancements. Trippe's clever business arrangements with foreign governments and companies transformed a vision into reality over the subsequent 15 years.

The flying boat made sense in the early 1930s because the aircraft had to be big, have a long range compared to land-based aircraft, and fly huge distances over water. The task of establishing bases was easier for the flying boats compared to clearing land and building airports for land-based aircraft. Table 7 contains a tabulation of aircraft and engines in the Pan Am fleet during the glamorous era of the Clippers.

Pan Am made a gigantic contribution to the conduct of WWII due to their already established airport facilities throughout the Far East, South America, and Europe. Pratt & Whitney's contribution was in providing dependable engines for more than 80 percent of the flying boats. The last 12 aircraft were built by Boeing and powered by Wright Aeronautical's R-2600 Double Cyclones. Pratt's R-2800 was not ready in time for the flying boat era.

TABLE 6 APPLICATIONS FOR THE R-1860 HORNET B

Model	hp/rpm take-off or military	Normal hp	Weight (lb)	Diameter (in.)	Length (in.)	Aircraft installations
Hornet B	575/1800	55/1900	830			Bellanca C-27; Boeing Y1C-18, 221; Convair 16, Commodore, Fleetster 20A; Junkers JU-52; Keystone B-4A
Hornet B1	575/1950	575/1950	860	56.94	44.78	American Airplane & Engine Pilgrim 100A; Convair Fleetster; Sikorsky S-41B
Hornet B1-G	575/1950	575/1950	952	56.94	50.17	Boeing YB-9
Hornet S1B1-G	575/2000	575/2000	960	56.94	50.42	Martin YB-13
Hornet S2B1-G	575/2000	575/2000	960	56.94	50.42	Boeing Y1B-9A
R-1860-1	575/1950	575/1950	985	56.93	44.76	Convair C-11; General Aviation C-16, YC-20; Keystone LB-12 (engine sold as B)
R-1860-3	550/1950	550/1950	985			Keystone LB-8
R-1860-5	550/1950	550/1950				Bellanca C-27 (engine sold as B–G)
R-1860-7	575/1950	575/1950		56.93	44.78	Boeing Y1C-18
R-1860-11	630/1900	600/2000				Boeing Y1B-9A
R-1860-13	575/1950	575/1950				Boeing YB-9
R-1860-17	630/1950	600/2000				Martin YB-13
R-1860-19	650/2000	600/2000				Bellanca C-27A

TABLE 7   FLYING BOAT ENGINES IN PAN AM'S FLEET

Aircraft	Number	Engine model	Horsepower	Number per a/c	Total engines
S-38	38	P&W R-1340	450	2	76
S-40	3	P&W R-1690	660	4	12
S-41	3	P&W R-1860	575	4	12
S-42	10	P&W R-1690	875	4	40
S-43	12	P&W R-1690	850–875	2	24
Consolidated	14	P&W R-1860	575	2	28
Martin M-130	3	P&W R-1830	950	4	12
Boeing 314	12	Wright R-2600	1500–1600	4	48
VS-44[a]	3	P&W R-1830	1200	4	12
Installed engines				Total	264[b]

[a] These aircraft were bought for American Export Airlines, which eventually were absorbed by Pan Am after WWII.
[b] There probably was about another 20 percent of engines as spares.

Immediately after WWII, the long-range capability of aircraft such as the DC-4 and land-based subsequent aircraft sounded the final knell for the flying boats.

WASP AND HORNET ENGINES ARE REFINED

R-985 WASP JR.

This engine started out at 300 hp because there was a significant market for a dependable engine smaller than the Wasp. The design philosophy was the same as its bigger brothers, Wasp and Hornet. Many young military

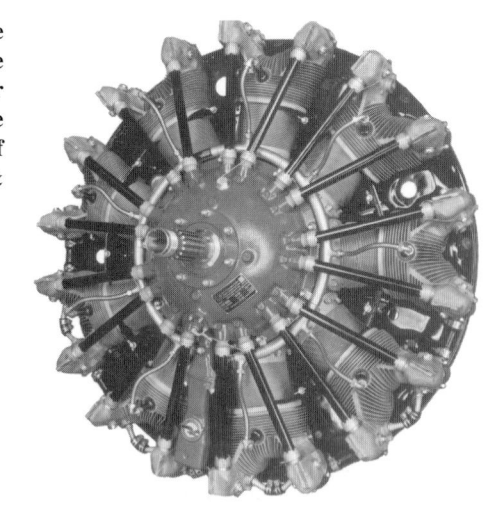

Fig. 11   This is a later model of the R-985, evident from the fine cuts in the cooling fins and the oil lubricated rocker mechanisms. The oil connection to the rocker boxes is not in place for six of the nine cylinders (courtesy of Pratt & Whitney Archives).

THE LATER PISTON ENGINE YEARS

TABLE 8   SPECIFICATIONS FOR THE
R-985 WASP JR.

Horsepower	300–450
rpm	2000–2300
Number of cylinders	9
Weight (lb)	565–684
Bore (in.)	5.1875
Stroke (in.)	5.1875
Design start	August 1929
First run	November 1929
First flight	
Production quantity	39,037

officers learned to fly on training aircraft powered by the R-985 or the R-1340 and that gave them a good feeling about Pratt & Whitney products (Fig. 11 and Tables 8 and 9).

R-2270 — THE FIRST TWIN ROW

Even while designing the Big Hornet (R-1860), Pratt & Whitney designed its first twin-row engine using the Wasp-size cylinders. This R-2270 engine had two rows of seven for a total of 14 cylinders (see Fig. 12 and Table 10). Its primary use was to explore ways of putting two rows together mechanically rather than seeking how to get the maximum power out of twin rows. The figure gives away its time period by the valve covers that indicate grease-packed lubrication. This experimental one-of-a-kind engine is number X-24, the first and only twin-row engine of this particular design.

Once the maximum horsepower per cylinder and the maximum number of cylinders per row were established, the next step was to add another similar row. This approach provided an increase in power without an increase in the size. There were questions about the cooling effectiveness of the second row in a twin-row radial engine configuration. Enough information and experience was gained in the design, building, and testing of this engine to permit the launching of the high-production R-1830 twin-row engine and the Twin Wasp Jr., R-1535.

R-1830 TWIN WASP

This is the company's most successful engine, if one measures success by the number of engines produced. It was a major commercial transport and military bomber engine. The first two production twin-row engines, R-1830 (Fig. 13, Tables 11 and 12) and R-1535, departed from the one-piece master rod philosophy of the Wasp, Hornet, and Wasp Jr. Mead, Hobbs, and Willgoos

TABLE 9 APPLICATIONS FOR THE R-985 WASP JR.

Model	hp/rpm take-off or military	Normal hp	Weight (lb)	Diameter (in.)	Length (in.)	Aircraft installations
Wasp Jr. A	300/2000	300/2000	565	45.75	41.06	Bellanca 300W; Boeing 4D, 4DM, 4DM-1; Convair BT-7, Y1PT-12, 21C; Douglas C-26; Dolphin 3; Fairchild 51A; Lockheed Orion 2D; Sikorsky C-28, S-39B; Spartan C5-301; Stinson L-12A; Stinson W
Wasp Jr. S1A	400/2300	400/2300	565	45.75	41.06	Boeing 4DX; Sikorsky S-39C
Wasp Jr. T3A	420/2200	420/2200	575	45.75	41.09	Bellanca CH-400W; Boeing 81; Sikorsky S-39C-SP
Wasp Jr. TB	440/2300	420/2200	645	45.75	41.59	Koolhoven FK-51; Waco S3HD
Wasp Jr. T1B3	450/2300	450/2300	682	46.10	42.43	Boeing A75N1
Wasp Jr. SB	450/2300	400/2200	645	45.75	41.59	Barkley-Grow T8P-1A; Beech SNB-1, AT-11, D17S, 18SA, B18S, C18SA; Fairchild Sekani; Grumman G21A, G21AA; Howard DGA-11; Waco UC-72D; Koolhoven FK-50, -51; Lockheed 10A, 10A-1, 12, 12A-1A; Spartan 7WA; Stinson UC-81E, XC-81D, FA
Wasp Jr. SB2	450/2300	400/2200	665	46.06	42.38	Grumman JRF-2, -3, G21A, G21; Sikorsky XSS-1, S-39; Waco SRE
Wasp Jr. SB3	450/2300	400/2200	682	46.10	42.43	Beech D18S; Fairchild F-11X; Grumman G-21A; Howard DGA-15P
Wasp Jr. B4	450/2300	450/2300	684	46.10	47.69	Bell 42; Sikorsky H-5F, G, H, S-51, A; Bendix Model J

Wasp Jr. B5	450/2300	450/2300	682	46.10	42.43	Beech D18S; Spartan 12A
R-985-1	300/2200	300/2000	565	45.75	41.09	Boeing (Stearman) BT-5; Douglas OA-4, B, C; Convair BT-7, Y1PT-12; Curtiss XP-21; Douglas C-26, OA-4; Seversky BT-8; Sikorsky C-28; Stinson L-12A, UC-81, XC-81D
R-985-3	420/2200	420/2200	656			Douglas OA-4, B, C, YOA-5
R-985-9	400/2200	400/2100	645	45.75	41.59	Douglas OA-4A, B, C, YOA-5, C-26B
R-985-11	440/2300	440/2200				Seversky BT-8
R-985-11A						N. American BT-14A (-11 engines were modified by Army to get -25 features to complete BR-14 contract)
R-985-13	450/2300	400/2200				Lockheed Y1C-36, UC-36, Y1C-37, UC-37 (similar to -11)
R-985-17	450/2300	400/2200	648	45.75	42.38	Beech C-45, JRB-1, YC-43, UC-43B; Grumman OA-9; Lockheed UC-40A, B, C, D; N. American BT-14.
R-985-19	450/2300	400/2200	648	45.75	42.38	Beech F-2
R-985-21	440/2300	420/2200	648	45.75	42.38	Platt-LePage XR-1, -1A; Ryan YO-51; Vultee BT-13.
R-985-23	450/2300	400/2200	650	45.75	42.38	Beech JRB-1, UC-45; Howard UC-70, -70B, C, D; Spartan UC-71
R-985-25	450/2300	450/2300	685	45.75	42.38	Air Research XBT-11; Beech AT-7, -7B; Fleetwing XBT-12; N. American BT-14, -14A; Vidal Research XBT-16; Vultee BT-13.
R-985-27	450/2300	450/2300	685	45.75	42.38	N. American BT-14, -14A; Vulltee BT-13
R-985-33						Howard UC-70; Spartan UC-71; Waco UC-72 (sold as Wasp Jr. SB)

(Continued)

TABLE 9 APPLICATIONS FOR THE R-985 WASP JR. (CONT)

Model	hp/rpm take-off or military	Normal hp	Weight (lb)	Diameter (in.)	Length (in.)	Aircraft installations
R-985-38	400/2200	400/2200				B/J Aircraft OJ-2
R-985-39						Beech C-45G, H (converted AN-1 for 100-amp generator)
R-985-39A						Beech C-45G, H (converted AN-3 for 100-amp generator)
R-985-46	400/2200	400/2200				B/J Aircraft OJ-2
R-985-48	450/2300	400/2200		45.75	41.59	Beech GB-1, JRB-2; Grumman JRF-1, -1A, XJ3F-1, OA-13A; Howard GH-1, -2, -3; Lockheed XR20-1, JO-1, -2, -3; Vought OS2U-1, XOS2U-1
R-985-50	450/2300	400/2200	664	45.88	41.65	Beech GB-1, -2A; Grumman JFR-4C; Howard GH-1, -2, -3; Lockheed JO-2B, -1, -3, R30-2, UC-36A; Vought OS2U-2B
R-985 AN-1,	450/2300	450/2300	674 mag 682 alum	46.25	43.05	Avions Max Holste MH-1521; Piasecki HUP-2, H-25; Beech F2A, B, SNB-1, -2, -2C, JRB-4, -5, -6, AT-7, -7A, -7B, AT-11, -11A, G-17S, UC-43, -43B, C-45A-H; Boeing XBT-17;

						Convair CQ-3, YCQ-1, Convair Valiant SNV-1, -2, BT-13A, B; Fleetwings BT-12; Fletcher PQ-11, YCQ-11A; Grumman OA-13; Lockheed UC-40D; Platt Le Page XR-1A; SikorskyYR-5, XR-5
R-985 AN-3	450/2300	450/2300	682	46.25	43.05	Beech SNB-1, -2, JRB-4, -5, -6, AT-7C, G17S, C-45G-H; Convair SNV-2, BT-13A, B
R-985 AN-2	450/2300	400/2200	680 mag	46.25	43.06	N.A.F. OS2N; Vought OS2U-2, -3
R-985 AN-8	450/2300	400/2200	688 alum	46.25	43.06	N.A.F. OS2N; Vought OS2U-3
R-985 AN-4, AN-10	450/2300	400/2200	674 mag 682 alum	46.25	43.06	Beech JRB-1 (AN-4), JRB-4, C-45F, JRB-3, G17S (AN-4) (AN-10 like AN-4 except aluminum cases)
R-985 AN-5	450/2300	450/2300	684	46.25	48.00	Bristol MK171; Sikorsky H5G, H, S-51, HO2S-1, HO3S-1, YR-5A, R-5D, XR-5A, R-5A; Westland-Sikorsky S-51
R-985 AN-6, AN-12	450/2300	400/2200	680 mag 688 alum	46.25	43.06	Avro Anson V; Grumman JRF-5; Howard NH-1, DGA-15P
R-985 AN-6B, AN-12B	450/2300	400/2200	680 mag 688 alum	46.25	43.06	Airspeed Oxford V (AN-6B); de Havilland DHC-2 (AN-6B); Grumman JRF-6B. (AN-12 is like AN-6B except it has aluminum cases.)
R-985 AN-14B	450/2300	400/2200	682 mag 690 alum	46.25	43.06	Avro Anson V; McDonnell XHJH-1, XHJD (H-1)

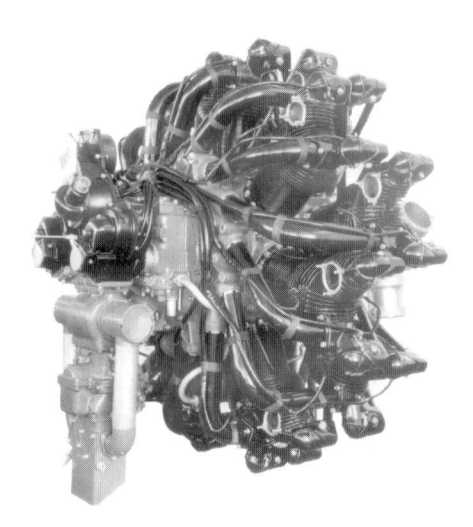

Fig. 12 This experimental R-2270 twin-row engine had a modest power output of 870 hp.

apparently had decided to make sure that the longer shaft of the two-row configuration would be of a more conservative design. This, of course, required two split master rods and a one-piece crankshaft. The horsepower per cylinder was similar to the Wasp R-1340 but with a smaller bore diameter (which helped make cylinder cooling more effective).

EXPERIMENTING WITH THE COMPRESSION RATIO

Pratt & Whitney engine cylinders have a compression ratio (maximum volume divided by the minimum volume) in the range of 5 to 7. The earlier engines started off with the lower compression ratios and gradually increased to about 7:1. Increasing compression ratio makes the engine more susceptible to detonation, because the increasing compression results in higher temperatures of the fuel–air mixture. During WWII, the company experimented with increasing the R-1830 compression ratio from 6.7 to 8.0

TABLE 10   SPECIFICATIONS FOR THE R-2270

Horsepower	870
rpm	
No. of cylinders	14
Weight (lb)	
Bore (in.)	5.75
Stroke (in.)	6.24
Design start	1928
First run	1930

Fig. 13 This R-1830 with its 14 cylinders in two rows of seven shows its up-to-date technology in finely finned cylinder heads and pressure lubricated rocker mechanisms (courtesy of Pratt & Whitney Archives).

to improve the specific fuel consumption for engines in the C-47. The requirement came from a concern about ferrying the C-47s across the Atlantic. Two experimental engines were built with special pistons to increase the compression ratio.

Bob Fitzgerald, a test engineer, who came to Pratt & Whitney right out of college in 1939, ran one of the experimental engines (X-38), in the Twin Wasp Group, for 1000 hr in a simulated C-47 endurance demonstration test (take-off, climb, cruise, and landings) with special high-performance fuel. X-39 was the other engine in the program. The program was cancelled after achieving the objective, perhaps because of the need for aviation gasoline with a PN greater than 100 octane fuel or when the initial concern was alleviated by some change in the ferrying process.

On both Pratt & Whitney's first two production twin-row engines (R-1830 and R-1535), the pistons were smaller than on single-row engines of

TABLE 11 SPECIFICATIONS FOR THE R-1830 TWIN WASP

Horsepower	750–1350
rpm	2400–2700
Number of cylinders	14
Weight (lb)	1162–1467
Bore (in.)	5.5
Stroke (in.)	5.5
Design start	December 1929
First run	April 1931
First flight	June 1931
Production quantity	173,610

TABLE 12 APPLICATIONS FOR THE R-1830 TWIN WASP

Model	hp/rpm take-off or military	hp/rpm normal	Weight (lb)	Diameter (in.)	Length (in.)	Aircraft installations
S2A5-G	950/2550	830/2400	1235	47.88	56.75	Martin Clipper
SB-G	1000/2350		1135	48.00	56.66	Junkers JU-90
SB3-G	1000/2600	900/2450	1310	48.00	55.49	Douglas DC-3A; Douglas DSTA
SC-G	1050/2700	900/2550	1423	48.00	59.90	Block 17B (French); Curtiss H-75-C1; Douglas DC-3A, DSTA; Seversky S2; Block 153-C1 (French)
SC3-G	1050/2700	900/2550	1438	48.13	60.94	Sovoia Marchetti SM.87 Seaplane
SC3-G	1050/2700	900/2550	1473	48.13	61.50	Douglas DC-3A, DSTA
S1C3G	1200/2700	1050/2550	1467	48.19	61.16	Douglas C-48B, C
S1C3-G	1200/2700	1050/2550	1473	48.19	61.67	Douglas DC-3C; Lockheed 18.08, S56D, C-66; Model 18.10, C57, A, B; Sikorsky VS44A
S3C4-G	1200/2700	1100/2550	1492	48.19	63.44	Bristol Beaufort II; Comm. of Australia Boomerang; Curtiss H-75C; Douglas DC-3C; Vought V-167
S4C4-G	1200/2700	1050/2550	1495	63.48	62.92	Douglas DC-3C; Lockheed 18-14
SSC7-G	1200/2700	1100/2550	1572	48.19	67.44	Curtiss H-81A
R-1830-1		800/2400		48.00	57.00	Martin XB14;
R-1830-7	950/2450	850/2450				Northrop XA-16
R-1830-9	950/2450	850/2450	1292	48.00	55.88	Northrop XA-16; Seversky P-35;
R-1830-11	1000/2600	850/2450	1320			Boeing XB-15; XC-105
R-1830-13	1050/2700	900/2550	1370	48.06	59.25	Curtiss P-36A, D, E, F, RP-36
R-1830-17	1200/2700	1000/2600	1403	48.00	59.25	Convair YA-19; Curtiss P-36A, C
R-1830-21	1200/2700	1050/2550	1433	48.00	59.25	Douglas C-41 (DC-2), C-41A (DC-3)
R-1830-23	1100/2700	950/2700	1436	48.50	63.00	Curtiss P-36B
R-1830-31	1050/2550	1000/2300				Curtiss XP-42; Martin YB-10A; Seversky XP-41
R-1830-33	1200/2700	1100/2550	1480	48.06	63.48	Convair P-66, RB-24A, B, XB-24, YB-24, PB3Y-3; Martin RB-10B

R-1830-35	1200/2700	1100/2550	1450	48.06	61.59	Republic YP-43, P-43B, RP-43

Model	Col2	Col3	Col4	Col5	Col6	Aircraft
R-1830-35	1200/2700	1100/2550	1450	48.06	61.59	Republic YP-43, P-43B, RP-43
R-1830-37	1200/2700	1100/2550	1433	48.00	59.25	Martin XA-22; Republic RB-43
R-1830-41	1200/2700	1100/2550	1490	48.06	62.42	Convair RB-24C
R-1830-43	1200/2700	1040/2550	1500	48.56	62.59	Convair P4Y-1, XC-109, C-109; Liberator A-22, B-24D, B-24E, B-24G, B-24H, C-87A, C-87B, XB-41; Lockheed C-57C
R-1830-43A	1200/2700	1040/2550	1500	48.56	62.59	Ford B-24M, B-24E, C-87B, F-7
R-1830-45	1050/2700	900/2700	1438	48.13	61.50	Republic AT-12; Seversky P-35A
R-1830-47	1200/2700	1050/2550	1473	48.19	61.16	Republic RP-43, P-43D
R-1830-49	1200/2700	1050/2550	1473	48.13	61.50	Lockheed A-28; Republic RP-43A, B, C
R-1830-51	1200/2700	1050/2550	1473	48.19	61.16	Convair XA-19C; Douglas C-52, A, B, C
R-1830-57	1200/2700	1050/2550	1465	48.19	61.78	Republic P-43A1, E
R-1830-59	1200/2700	1100/2550	1575	48.40	60.74	Convair B-24D
R-1830-61	1200/2700	1100/2550	1495	48.19	63.48	Convair LB-30, Convair Liberator III
R-1830-62	640/1890	710/2300				Grumman JF-1
R-1830-63	1200/2700	1100/2550	1495	48.19	63.48	Convair P-66
R-1830-64	900/2500	850/2450	1295	48.00	55.50	Convair PBY-1, 2, XPBY-1; Douglas TBD-1, -1A, XTBD-1
R-1830-65	1200/2700	1100/2550	1500	48.56	62.59	Convair B-24E, F, XB-24G, C-87, A, XC-109, C-109, F-7, A
R-1830-65A	1200/2700	1100/2550	1500	48.56	62.59	B-24B, J, L, C-87A, B
R-1830-66	1050/2700	900/2250	1370	48.00	59.25	Convair PBY-3
R-1830-67	1200/2700	1040/2550	1500	48.56	62.59	Lockheed PBO-1, RA-28A
R-1830-68	1050/2700	900/2550		48.10	59.25	Sikorsky XPBS-1
R-1830-70						Douglas XTBD-1
R-1830-72	1050/2700	900/2550	1405	48.00	59.25	Convair XPB2Y-1, PBY-4, XPBY-5A
R-1830-74	1200/2700	1050/2550	1475	48.19	57.94	For N.A,F, testing in PBY-5s
R-1830-75	1350/2800	1100/2600	1555	48.40	59.63	Ford YB-24K, B-24 N, XB-24N
R-1830-76	1200/2700	1100/2550	1550	48.06	71.31	Grumman F4F-3, XF4F-3, XF4F-4
R-1830-78	1200/2700	1100/2550	1575	48.06	73.39	Convair PB2Y-2
R-1830-80	1200/2700	1050/2550	1473	48.13	61.50	Convair XPB2Y-1, PBY-4

(Continued)

TABLE 12 APPLICATIONS FOR THE R-1830 TWIN WASP (CONT)

Model	hp/rpm take-off or military	hp/rpm normal	Weight (lb)	Diameter (in.)	Length (in.)	Aircraft installations
R-1830-82	1200/2700	1050/2550	1465	48.06	60.28	Comm. of Australia Boomerang, Convair PBY-5A; Douglas R4D-1 (DC-3)
R-1830-82A	1299/2700	1050/2550	1465	48.06	60.28	Convair PBY-5
R-1830-84	1050/2700	900/2500	1423	48.06	61.59	Convair PBY-3
R-1830-84A	1200/2700	1050/2550	1438	48.06	61.59	Lockheed R50-3
R-1830-86	1200/2700	1100/2550	1560	48.19	67.44	Eastern FM-1; Grumman F4F-3, F4F-4, F4F-7
R-1830-88	1200/2700	1100/2550	1595	48.19	67.44	Convair PB2Y-3, -3B, -3R, -5, PBY-3
R-1830-90	1200/2700	1100/2550	1495	48.19	63.41	Grumman F4F-3A, -4A, -6, G-36B
R-1830-90B	1200/2700	1100/2550	1490	48.19	62.63	Bristol Beaufort 11; Comm. of Australia Boomerang C-11; Douglas C-47B; Short Sandringham, Sunderland V; Vickers Willington IV
R-1830-90C	1200/2700	1100/2550	1492	48.19	62.63	Convair OA-10; Douglas R4D-6, -7, C-47B, C-117A; Vickers OA-10A
R-1830-92	1200/2700	1050/2550	1465	48.19	60.78	Boeing PB2B-1, -2; Budd RB-1; Convair PB2U-3, -3R, PB2Y-5, -5R, -5H, -5Z, PBY-6, -6A, PBY-5, -5A, -5B, OA-10; Can. Car & Foundry CBY-3; Curtiss YC-76, A, ZC-76; Douglas C-47, C-47A, C, C-48A, C-53, XC-53A, C-53, A, B, C, D, XC-53A, C-68; Higgins C-76A; N.A.F. PBN, PBN-1; Sikorsky JR2S-1; Vickers OA-1A, PBV-1A; Waco YC-62; Savoia Marchetti SM95
R-1830-94	1350/2800	1100/2600	1573	48.40	59.63	Convair Model 39; Budd Rb-2; Convair RY-3, XPB2Y-6, XPB4Y-2, C-87C
R-1830-96				48.19	57.94	Convair PB2Y-3 (Navy Bulletin AERE-431-AWL May 8, 1944)
R-1830-98	1350/2800	1100/2600	1573	48.40	61.02	Convair (Consolidated) PB4Y-2

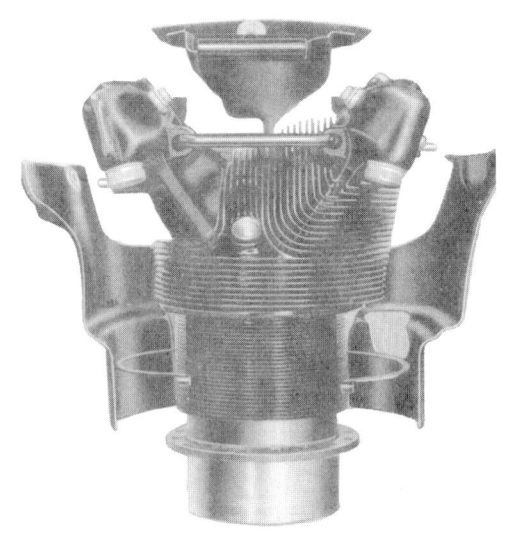

Fig. 14 Baffles direct flow on the piston head (courtesy of Pratt & Whitney Archives).

comparable displacement. These smaller and lighter reciprocating parts permitted higher engine rotational speeds without excessive stresses in the rotating parts. Of great concern was the cooling effectiveness of finned cylinders in the second row. The solution was to use baffles to control the flow between the first-row cylinders and direct it around the second-row cylinders (see Fig. 14). The hardened steel cylinder barrels with machined cooling fins were screwed and shrunk into cast aluminum heads, which were extensively finned for effective cooling of the cylinder and rocker boxes. The baffles were made out of duralumin and were mounted on the aluminum heads. The R-1830 case is a three-piece aluminum alloy forging, jointed in the planes of the front and rear cylinder rows (see Fig. 15).

The tappet guides are inserted in each end of the crankcase. The crankshaft is machined from a solid steel forging and is supported on three bearings in

Fig. 15 R-1830 case (courtesy of Pratt & Whitney Archives).

the crankcase. The link rods are of the I-configuration, and along with the master rods, are machined from steel forgings. The master rods are split because of the one-piece steel crankshaft. The pistons have flat heads with sculptured indentations to make room for the inlet and exhaust valves when the piston is at top dead center. There are five piston rings to maintain high compression in the cylinder as well as to minimize oil consumption. The underside of the piston is ribbed and finned, which not only contributes to strength per weight but also helps to cool the piston.

The propeller reduction gear is a planetary type, with the fixed outer gear fastened to the crankcase. Figure 16 shows the R-1830 propeller planetary reduction gear parts, demonstrating the simplicity of construction with spur-type gears. Six pinion gears rotate in a cage attached to the propeller shaft. The engine crankshaft drives the inner gear. The gears have hardened teeth, ground to fine tolerances. The pinion gears rotate in bronze bushings. This complete reduction gear assembly can be removed from the engine in one section.

Another feature introduced in engines in the early 1930s was automatic valve lubrication (see Fig. 17). The hollow exhaust valve is filled with sodium to speed the conduction of heat from the hotter end to the cooler stem. The moving parts in the valve rocker boxes had to be greased as part of regular engine maintenance until this automatic lubrication system was provided. Engine oil, under pressure, circulates continuously through the valve tappets

Fig. 16 R-1830 propeller planetary reduction gear parts (courtesy of Pratt & Whitney Archives).

The Later Piston Engine Years

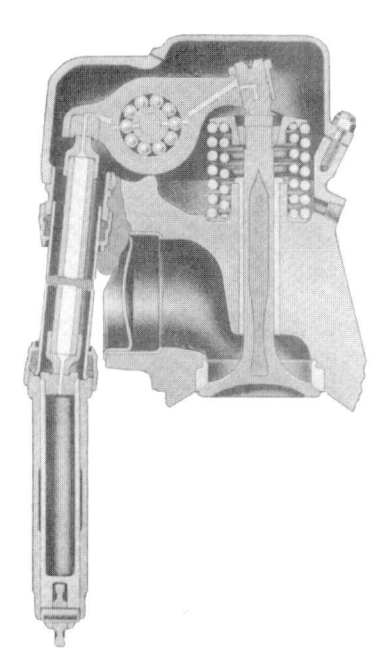

Fig. 17 Hollow exhaust valve (courtesy of Pratt & Whitney Archives).

and hollow push rods to the parts of the rocker assembly needing lubrication. The oil returns to its tank through the push rod cover tubes or through scavenging pipes to a cylinder head sump from which it is returned to the oil tank by means of a suction stage in the oil pump.

The automatic mixture control simplifies the fuel–air mixture and power output of the engine. What the control does is to provide the minimum fuel flow for any power setting over the range of aircraft operational altitudes. The pilot has only three settings for the automatic mixture control: Auto Rich, Auto Lean, and Idle Cut-Off. This device reduces the pilot's workload by eliminating the need to continuously adjust the fuel–air mixture at the selected power setting. This automatic device measures the temperature of the oil returning from the engine. If the engine oil temperature is above the maximum, the control sends the oil to an oil cooler to bring the temperature down to an acceptable level. If the oil from the engine is colder than the minimum acceptable level, then the device permits the oil to bypass the oil cooler. The device helps in cold-weather starting, where the oil heats up quickly and then stays within the acceptable limits.

R-1535 Twin Wasp Jr.

The R-1830 features are also typical of the R-1535. The argument that kept coming up in the early 1930s was that air-cooled engines would have more drag than liquid-cooled engines because the shape of the liquid-cooled engine conformed more closely to the fuselage contours than did the air-cooled

Fig. 18 R-1535's best-known application was in Howard Hughes' H-1 Racer, which set a land speed record in 1935 of 352.3 miles per hour (courtesy of Pratt & Whitney Archives).

engines. Therefore, it was tempting to take a small diameter engine such as the Wasp Jr. and make a double row out of it. Hobbs was thinking in terms of a much higher powered engine. The company's second production twin-row engine was the R-1535, again with 14 cylinders in two rows (Fig. 18, Tables 13 and 14). The cylinders have the same bore and displacement per cylinder as the R-985.

R-2180A Twin Hornet A

There are two Pratt & Whitney engines with 2180 cubic inches displacement. The first is the R-2180A and the second, which was the company's last piston engine, is called the R-2180E. By mid-1935, experience with twin-row engines was sufficient to let the company design another engine. The Wasp

Table 13 R-1535 Specifications for the Twin Wasp Jr.

Horsepower	700–825
rpm	2500–2630
No. of cylinders	14
Weight (lb)	1122–1130
Bore (in.)	5.1875
Stroke (in.)	5.1875
Design start	
First run	October 1931
Production quantity	2,880

TABLE 14 APPLICATIONS OF R-1535 TWIN WASP JR.

Model	hp/rpm take-off or military	Normal hp	Weight (lb)	Diameter (in.)	Length (in.)	Aircraft installations
S1A1-G	660/2400	660/2400	999	43.88	54.50	Boeing 247A
SA7-G	700/2500	650/2500				P&W's Boeing 247A
S2A4-G	700/2500	700/2500	1070	44.13	53.25	Vought SBU-1, Vought V-142
SB4-G	825/2625	750/2250	1124	44.13	53.27	Fokker G-1; Vought V-143; Brequet 695-AB2, 699-B2; Bristol Bolingbroke; Miles Master III; Potex 63-12
S3B4	700/2500	650/2500	1136	44.13	53.27	P&W's Boeing 247A (1933–1947)
R-1535-2	825/2630	750/2250	1142	44.19	56.06	Vought SB2U-3, SBU-3
R-1535-7	725/2500	725/2500	1102	44.12	54.50	Douglas XO-46, O-46A
R-1535-11	750/2500	750/2500	1102	44.12	54.50	Northrop RA-17
R-1535-13	825/2580	750/2500	1130	44.13	53.25	Northrop RA-17A
R-1535-66	700/2500	700/2500				Great Lakes SBG-1
R-1535-72	700/2250	650/2200				Grumman F2F-1, XSF-2, XF3F-1
R-1535-82	750/2500	700/2500				Great Lakes XBG-1; Vought SBU-1, XSB3U-1
R-1535-84	700/2250	650/2200				Grumman F3F-1
R-1535-94	825/2630	750/2550	1122	44.13	53.25	Curtiss SBC-3; Northrop BT-1
R-1535-96	825/2630	750/2550	1122	44.13	53.25	Vought SB2U-1, -2, XSB2U-3.
R-1535-98	750/2500	700/2500				Vought SBU-2, SB2U-2

TABLE 15   ENGINE PROGRAMS

P&W	Wright Aero
R-985	R-975
R-1535	
R-1830	R-1820
R-2180	
R-2800	R-2600
R-4360	R-3350

cylinders were stacked up in two rows of seven to produce the R-2180A Twin Hornet. There is a little confusion here in the naming of engine models. The R-1830 has a cylinder bore of 5.5 in., which is between the R-985 (5.1875 in.) and the R-1340 (5.75 in.). Yet the 1830 is called the Twin Wasp. On the other hand, the Twin Hornet has the same cylinder bore (5.75 in.) as the Wasp. What is in a name? It is not an indication of the cylinder bore.

The R-2180A was a case of being too little too late. If you look at the inventory of engines, compared to what Wright Aeronautical had in development or in the planning stages, the situation becomes a little clearer in the 1930s (see Table 15). Pratt & Whitney had too many programs compared to Wright Aeronautical. Wright's three main engines were covered by a slightly more powerful Pratt & Whitney engine. There was no need to invest in other engines to fill the gaps in between the covered engines. These "gap engines" (R-1535 and R-2180) were a drain on the company's engineering resources. So it is not surprising that Hobbs cancelled the R-2180 (see Table 16) and put minimum effort into the R-1340, R-985, and the R-1535.

The applications of the Twin-Hornet were quite limited (Table 17). About half the production of only 30 engines from 1937 to 1939 was for the DC-4

TABLE 16   SPECIFICATIONS FOR THE R-2180A TWIN HORNET

Horsepower	1200–1500
rpm	2500–2600
No. of cylinders	14
Weight (lb)	1565–1647
Bore (in.)	5.75
Stroke (in.)	6.0
Design start	1933
First run	1936
First flight	
Production quantity	30

TABLE 17   APPLICATIONS FOR THE R-2180 TWIN HORNET

Model	hp/rpm take-off or military	hp/rpm normal	Weight (lb)	Diameter (in.)	Length (in.)	Aircraft installations
S1A-G	1400/2500	1150/2350	1675	51.63	62.63	Douglas DC-4 (Sold to Japan)
R-2180-1	1200/2500	1000/2350				Boeing B-20, P-44; N. American XB-21
R-2180-5	1400/2500	1150/2350				Boeing Y1B-20, XB-20
R-2180-7	1400/2500	1150/2350				Stearman (Boeing) XA-21
R-2180-9	1400/2500	1150/2350				N. American B-21 (Sold as Twin Hornet S1A-G)

and the rest was for the Army Air Corps' Boeing P-44, North American XB-21, and Stearman XA-21, as well as three engines to the U.S. Navy. The Twin-Hornet paved the technical path to the Double Wasp (R-2800). Not only did the Twin-Hornet have the same bore as the Wasp, it also had solid master rods–which became a characteristic of the R-2800. However, it differed from the Wasp in that it had plain journal bearings for the crankshaft instead of roller bearings.

R-2800 Double Wasp

This engine was Hobbs' brainchild. Perry W. Pratt (no relation to Francis Pratt of the Pratt & Whitney Tool Company) was the development engineer in charge of the R-2800C series engines. The Double Wasp had the same cylinder bore as the R-2180 and the R-1340 Wasp. It obtained 2800 cubic inches displacement by clustering two rows of nine cylinders. It makes sense to call the engine a Double Wasp because that is exactly what it is. The 2800 had a slightly higher stroke (6.0 instead of the Wasp 5.75 in.).

Around the Engineering Building at Pratt & Whitney, I used to hear the expression, "the engine that won WWII" when the R-2800 came up in conversation. An exaggeration? Perhaps, but there is a certain element of truth to the statement. Pilot Corky Meyer, founder of the Society of Experimental Test Pilots, has selected what he considers the best fighters in the European and Pacific theaters: the Republic P-47 in Europe and the Grumman Hellcat in the Pacific [7]. Both of these aircraft were powered by the R-2800 (Fig. 19).

The Double Wasp (Fig. 20) returned to the Wasp design philosophy, a solid master rod for each row and therefore a three-piece shaft (because of the double row). However, its crankshaft bearing configuration differed from the Wasp, which used roller bearings, in that the Double Wasp used plain journal bearings on the crankshaft. The propeller ball bearing for thrust and load was

Fig. 19 From left to right are the Republic P-47, the Vought Corsair, and the Grumman Hellcat (courtesy of Pratt & Whitney Archives).

Fig. 20 The R-2800 engine (courtesy of Pratt & Whitney Archives).

retained. The crankcase was similar to the R-1830, a forged aluminum case in three pieces. The cooling area of the cylinder heads was increased because of advancements in head manufacturing techniques, deeper cuts, and more (thinner) fins. This engine was the first in the family of piston engines to achieve a specific weight of one pound per horsepower and produce one horsepower per cubic inch of displacement (Table 18). It powered about 55 types of aircraft in military and commercial applications (Table 19).

R-2000 Twin Wasp

The R-2000 (Fig. 21 and Table 20) was a further development of the R-1830 with a few differences: 1) the front view slightly resembled the R-2800 in that the magnetos were mounted as in the R-2800; 2) the crankshaft bearings were more like the R-2800 (plain journal type instead of the roller bearings as in previous engines); and 3) the bore was just a 0.25 in. larger with the same stroke. The R-2000 was a very conservative extension of the R-1830, so modest that the engine serial numbers blend seamlessly as an additive to the R-1830 Serial Number Book. The applications were significant (Table 21), including the Douglas C-54E flight around the world from Washington D.C., a distance of 23,147 miles, in 149 hours and 44 minutes on September 28, 1945.

Bob Fitzgerald, a company test engineer, tells how the early R-2000 engines used the same crankcase as the R-1830s, with just a bigger opening for a slightly larger cylinder. Cracks in the case developed during the endurance testing program. One day, when Fitzgerald was on the Experimental Assembly floor with the center section crankcase in a vise, Wright Parkins happened to walk by and look at the case. Parkins took a hammer and gave the crankcase a gigantic blow. The result was a badly cracked crankcase and one broken hammer, which he threw under the table. Instantly one could hear the snapping sounds of toolboxes being closed throughout the Assembly Floor as Parkins

TABLE 18   SPECIFICATIONS FOR THE
R-2800 DOUBLE WASP

Horsepower	1800–2800
rpm	2600–2800
No. of cylinders	18
Weight (lb)	2150–2560
Bore (in.)	5.75
Stroke (in.)	6.0
Design start	1937
First run	September 1937
First flight	July 1939
Production quantity	125,443

TABLE 19 APPLICATIONS FOR THE R-2800 DOUBLE WASP

Model	hp/rpm take-off or military	hp/rpm normal	Weight (lb)	Diameter (in.)	Length (in.)	Aircraft installations
S1A4-G	1850/2600	1500/2400	2300	52.50	75.72	Vickers Warwick 1
2SC13-G	2100/2800	1700/2600	2355	52.80	78.39	Convair 110
CA15	2100/2800	1800/2600	2355	52.80	78.39	Convair 110, 240; Douglas DC-6, -6A Martin Mercury 202
CA15A	2400/2800[a] 2100/2800	1800/2600	2350	52.80	78.39	Douglas DC-6
CA17	2400/2800[a] 2300/2800	1800/2600	2355	52.80	78.39	Convair 110
CA18	2400/2800[a] 2100/2800 2400/2800[a]	1800/2600	2350	52.80	78.39	Aero Sud Ouest SO-30-P; Brequet BR-763; Convair XT-29, 240; Douglas DC-6; Martin Mercury 202 DC-6
CA18A	2400/2800 2400/2800[a]	1800/2600	2350	52.80	78.39	Hamilton Standard Test, Aero-Nord 2501
CB2	2300/2800 2500/2800[a]	1900/2600	2357	52.80	81.40	Martin 404 (EAL)
CB3	2050/2700 2400/2800	1800/2600	2357	52.80	81.40	Chase XC-123
CB14	2300/2800 2500/2800[a]	1900/2600	2390	52.80	81.40	Douglas DC-6A, B; Martin 202A, 404, Convair 340
CB16	2050/2700 2400/2800[a]	1800/2600	2390	52.80	81.40	Douglas DC-6A, B, R6D-1
CB17	2200/2800 2500/2800[a]	1800/2600	2390	52.80	81.40	
R-2800-1	1800/2600	1500/2400				Convair XA-19B

The Later Piston Engine Years

R-2800 X-2	1800/2600	1500/2400	2500	52.50	88.81	Vought XF4U-1
R-2800 X-4	1850/2600	1600/2400	2270	52.06	75.72	Vought XF4U-1
R-2800-5	1850/2600	1500/2400				Douglas B-23; Martin B-26, A, B, XB-26B; Curtiss XC-46
R-2800-6	1850/2600	1500/2400	2300	52.06	75.72	Vought XTBU-1
R-2800-8, -8W	2000/2700	1675/2550	2480	52.50	88.47	Brewster F3A-1; Goodyear FG-1; Vought F4U-1, -1C, -1P, -2
R-2800-10, 10W	2000/2700	1675/2550	2480	52.50	88.47	Curtiss P-60A, XP-60E; Grumman F6F-3E, -3F, -3H, -3N, -3P; Northrop P-61, A, XP-61, YP-61 (-10W as follows: F6F-5, -5E, F6F-5N, -5P); Northrop F2T-1
R-2800-11	2000/2700	1625/2250	2415	52.00	80.00	N. American XB-28, XB-28A
R-2800-12	2000/2700	1625/2550				Hamilton Standard Exp.
R-2800-13	2100/2800	1000/2400		52.00	75.72	Wright Field Sea Level Test Engine
R-2800-14W	2100/2800	1700/2600	2315	53.00	78.50	Goodyear FG-3; Northrop XP-61D; Republic YP-47M; Vought F4U-3
R-2800-15	2000/2700	1625/2550	2430	52.00	80.00	N. American XB-28, XB-28A
R-2800-16	2000/2700	1625/2550	2265	52.50	75.72	Grumman XF6F-2; Vought F4U-3
R-2800-18W	2100/2800	1700/2600	2560	52.80	93.77	Curtiss YP-60E; Goodyear FG-4; Grumman XF6F-6; Vought F4U-4, -4E, -4N, -4P, XF4U-4, -4B, F4U-7
R-2800-21	2000/2700	1625/2550	2265	52.50	75.72	Curtiss P-47G; Republic P-47C, D, RP-47B, C, XP-47E, F, K
R-2800-22	2100/2800	1700/2600	2359	52.80	78.13	Convair (TBU) TBY-2; Fairchild
R-2800-22W						C-82, A; Grumman XF7F-1; Martin XPBM-5, PBM-5; Vought XTBU-1 (-22W as follows: Grumman F7F-2, -2N, -2P, F7F-1N, XF8F-1, F7F-3)

(Continued)

TABLE 19 APPLICATIONS FOR THE R-2800 DOUBLE WASP (CONT)

Model	hp/rpm take-off or military	hp/rpm normal	Weight (lb)	Diameter (in.)	Length (in.)	Aircraft installations
R-2800-27	2000/2700	1600/2400	2300	52.50	75.72	Douglas JD-1, B23, A-26, B, C, XA-26A, B, C, XFA-26C; Fleetwing XA-39; Grumman XF6F-1, -4, F7F-1; N. American XB-28A
R-2800-28	2100/2800	1700/2600	2364	52.80	78.13	Curtiss SB2C-6
R-2800-29	2000/2700	1675/2550	2377	52.50	90.00	Northrop XP-56
R-2800-30W	2300/2800	1720/2600	2585	52.80	93.50	Grumman F7F-5, XF8F-2 (semiproduction engine)
R-2800-30W	2250/2800	1720/2600	2560	53.00	93.75	Grumman F8F-2, XF8F-3 (production engine)
R-2800-31	2000/2700	1600/2400	2280	52.50	75.72	Lockheed PV-1, PV-2, A, B, C, D, RB-34, RB-34A, B
R-2800-32W	2300/2800	1900/2600	2705	53.00	98.50	Vought F4U-5
R-2800-34	2100/2800	1700/2600	2359	52.80	78.13	Martin RM-1 (404), PBM-5A; Convair TBY-2, Curtiss XC-46B; Fairchild C-82
R-2800-34W	2100/2800	1700/2600	2359	52.80	78.13	Martin RM-1 (404); Curtiss XG15C-1 (-34W in nose and de Havilland jet in fuselage); Douglas XA-26D; Eastern XTBM-5; Fleetwings BTK-1; Grumman XTB3F-1, F7F-4, F8F-1, -1B, XF8F-1
R-2800-35	2000/2700	1625/2550	2355	52.06	75.72	Republic XP-47B
R-2800-37	2100/2800	1700/2550	2300	52.80	78.13	Brewster XA-32, A
R-2800-39	1850/2600	1500/2400	2300	52.06	75.72	Martin B-26A, B, XB-26D
R-2800-41	2000/2700	1600/2400	2300	52.50	75.72	Martin B-26B-2
R-2800-42W	2300/2800	1800/2600	2543	52.80	94.03	Vought F4U-4B

THE LATER PISTON ENGINE YEARS

R-2800-43	2000/2700	1600/2400	2300	52.50	75.72	Curtiss C-46; Martin AT-23, A, B, Martin B-26B, B1, B3, B4, B-10 through B-75, Martin B-26C, XB-26D, B-26E, F, G, TB-26H
R-2800-44	2300/2800	188/2600	2344	53.00	78.50	N. American AJ-1, -2, -3
R-2800-44W						
R-2800-46W	2300/2800	1800/2600	2327	53.00	78.50	Grumman AF-1S, -2S, XAF-1S, -2S
R-2800-47	2000/2700	1600/2400	2300	52.50	75.72	Vickers Warwick II
R-2800-48	2300/2800	1900/2600	2367	53.00	81.50	Grumman XTB3F-1S, AF-2W, AF-2S
R-2800-49	2000/2700	1600/2400	2300	52.50	75.72	Hughes (D2A) XA-37
R-2800-50	1900/2600	1900/2600	2310	53.00	81.50	Bell HSL; Sikorsky S-56, HR2S-1
R-2800-51	2000/2700	1600/2400	2300	52.50	75.72	Curtiss R5C-1, -2, C-46A, D1, D5, E, F, G
R-2800-52W	2500/2800	1900/2600	2390	53.00	81.50	Douglas R6D-1, C-118A; Convair C-131B
R-2800-53	2000/2700	1675/2550	2580	52.50	104.18	Curtiss P-60A, XP-60A, XP-60C
R-2800-55	2100/2800	1700/2600	2650	52.80	106.16	Curtiss XP-60F
R-2800-57	2100/2800	1700/2600	2315	52.80	78.39	Northrop P-61D, C; Republic XP-47J, L, N, P-47M, P-47M.
R-2800-59	2000/2700	1625/2550	2290	52.50	75.72	Republic P-47D, C, XP-47L
R-2800-61	2100/2800	1700/2600	2400	52.80	90.26	Republic XP-47J
R-2800-63	2000/2700	1675/2550	2265	52.50	75.72	Republic P-47D, C, XP-47L
R-2800-65	2000/2700	1675/2550	2508	52.50	88.47	Northrop F2T-1, XF-15, A, P-61A, B, XP-61E
R-2800-71	2000/2700	1600/2400	2325	52.50	75.72	Douglas JD-1, A-26B, C, XA-26C, XFA-26C.
R-2800-73	2100/2800	1700/2600	2351	52.80	78.39	Northrop F-15A, XF-15A, P-61C, XP-61F; Republic P-47N

(Continued)

TABLE 19 APPLICATIONS FOR THE R-2800 DOUBLE WASP (CONT)

Model	hp/rpm take-off or military	hp/rpm normal	Weight (lb)	Diameter (in.)	Length (in.)	Aircraft installations
R-2800-75	2000/2700	1600/2400	2325	52.50	75.72	Curtiss C-46A, C-46D-10, C-46E (R5C-2), C-46F, G, XC-113.
R-2800-77	2100/2800	1700/2600	2321	52.80	78.39	Northrop P-61C, XP-61D Republic P-47N
R-2800-79	2000/2700	1600/2400	2325	52.50	75.72	Douglas JD-1, A-26B, C, XA-26C, XFA-26C
R-2800-81	2100/2800	1700/2600	2345	52.80	78.39	Republic P-47N
R-2800-83, A	2100/2100	1700/2600	2384	52.80	79.13	Chase XC-123; Curtiss XC-46B; Douglas A-26D, F, XA-26F, DC-6; Lockheed C-69E; Vought AU-1 (had -83W)
R-2800-83W						
R-2800-85	2100/2800	1700/2600	2376	52.80	78.13	Fairchild C-82A, N
R-2800-85XA	2100/2800	1700/2600	2376	52.80	78.13	Douglas XC-112A
R-2800-95						Douglas C-118—DC-6 designated by Truman as "The Presidential White House" (engine sold as CA15)
R-2800-97	2400/2800	1800/2600	2350	53.00	78.50	Chase XC-123; Convair XT-29, A, B

[a] Water-injection rating.

Fig. 21 On the R-2000, the magnetos are mounted on the nose (courtesy of Pratt & Whitney Archives).

walked away. No one ever knew what he had in mind when he conducted his high-impact test. But that was Parkins. Years later when a ceramic blade manufacturer tried to sell Parkins on ceramic turbine blades, he took the salesman's sample and threw it on the floor of his office. The blade broke into a thousand pieces. Parkins told the salesman that when his blade could pass that test he could come back to sell Parkins ceramic blades.

R-4360— THE WARTIME ENGINE

This engine was conceived during wartime. Pilots refer to it as the "corncob engine" because of its resemblance to a twisted ear of corn. Its four rows of seven radial cylinders produced a total displacement of about 4360 cubic inches. The solid, one-piece crankshaft was ordered initially as a forging with an extra throw to confuse the enemy spies. The bore and stroke were the same as in the R-2800.

TABLE 20   SPECIFICATIONS FOR THE R-2000 TWIN WASP

Horsepower	1300–1450
rpm	2700
No. of cylinders	14
Weight (lb)	1570
Bore (in.)	5.75
Stroke (in.)	5.5
Design start	1939
First run	1939
First flight	
Production quantity	12,966

TABLE 21 APPLICATIONS FOR THE R-2000 TWIN WASP

Model	hp/rpm take-off or military	hp/rpm normal	Weight (lb)	Diameter (in.)	Length (in.)	Aircraft installations
2SD1-G	1450/2700	1100/2550	1590	49.10	61.02	Douglas DC-4 Skymaster
D3	1450/2700	1200/2550	1575	49.10	61.02	Douglas DC-4 (N.A.L.)
D5	1450/2700	1200/2550	1585	49.10	59.66	Douglas DC-4 (N.A.L.)
D8	1200/2550	1200/2550	1585	49.10	59.66	Military helicopters
2SD13-G	1450/2700	1200/2550	1605	49.10	59.66	Cancargo CBY-3; Douglas DC-4; Svenska Aeroplane Aktrebolaget; SAAB-90A1
R-2000 X-1	1350/2700	1100/2550	1550	49.10	60.74	Wright Field Exp.
R-2000 X-2	1350/2700	1100/2550	1410	49.50		Vought XF5U-1
R-2000-3	1350/2700	1100/2550	1570	49.10	59.65	Douglas C-54, A, R5D-1
R-2000-4	1450/2700	1200/2550	1605	49.25	59.75	Douglas R5D-1, -2, -3, -4
R-2000-7	1350/2700	1100/2550	1570	49.10	59.65	Douglas R5D-1, -2, C-54A, B, C, F
R-2000-9	1450/2700	1100/2550	1590	49.10	61.02	Douglas R5D-2, -5, -6, C-54G
R-2000-11	1350/2700	1100/2550	1580	49.10	59.62	Douglas R5D-3, -4, C-54G; Chase YC122

Since the Army Air Corps in the 1930s was convinced that the future in aircraft propulsion was with liquid-cooled engines, Pratt & Whitney was asked to develop high-output liquid-cooled engines. Hobbs considered this a drain on the company's engineering resources, which could be invested more productively in air-cooled engines. Rentschler convinced General "Hap" Arnold in 1940 that if Pratt & Whitney could drop the liquid-cooled engine programs, it would develop a higher power air-cooled engine, which became the R-4360.

In mid-1940, the company planned how to support President Roosevelt's demand for 50,000 aircraft per year. In September 1940, United Aircraft President Eugene Wilson, asked Major General George Brett for his views on Pratt & Whitney putting its R-985 engine on outside production while the company concentrated on the R-1340, R-1830, and R-2800 engine programs. Pratt had already expanded its production facilities for a French order, then again for a British order, and made a further expansion to meet President Roosevelt's requirements. The French, and to some extent the British, had let their air forces' strengths deteriorate after WWI, compared to Germany, which had not. The French financed an $8 million expansion of Pratt & Whitney's facilities to produce about 300 R-1830 engines a month for French aircraft, starting in October 1939. Soon after the French commitment, the British were talking about a further expansion of about half that of the French.

Also under discussion at this time was the dropping of the liquid-cooled engine programs in which the Army Air Corps was interested, the X-1800 and X-3730 programs [8]. Wilson called General Brett a few days later to follow up on the general's discussion with his staff about the company dropping the liquid-cooled engine programs.

His reply to Wilson was as follows [9]: "We have made many plans for future aircraft around that engine and to pull that engine out of the picture at the present time would be rather disastrous." The government had paid nothing of the $500,000 contract to the company as of that time, and Pratt & Whitney continued its pursuit of dropping the contract. In November 1940, Mr. Rentschler wrote to General "Hap" Arnold expressing his views on the value of air-cooled engines relative to liquid-cooled engines [10]:

> The plain facts of it, of course, are that 15 years ago this company believed that an air-cooled radial could equal or exceed the high-speed of a liquid-cooled engine of approximately the same weight. The Wasp proved this in the air against the Curtiss D-12 and the facts have not changed one bit right down to this present minute. If we can assume that the air-cooled radial can equal the high-speed of the liquid-cooled, then every other factor is vastly in favor of the air-cooled type, namely, cost, simplicity, durability, reliability, etc.
>
> On the other hand, it is my opinion that the whole argument about the 1000–1200 liquid-cooled–air-cooled engines is purely academic so far as the forward picture is concerned, because the field, in our opinion,

must move now into the 2000-hp type. You may know that our 2800 engine, now rated at 1850 hp, has just passed an outstandingly successful Army-type test at 2000 hp. Our primary interest in developing the Navy Vought fighter, which is now flying at Anacostia, was to establish the drag-coefficient of this large air-cooled type. This not only compares favorably with the lower horsepower air-cooled engines but is actually less. Now that this fact is proved we are more sure than ever that the whole field of pursuit will move in to the 2000-hp class, and, as usual, the air-cooled type has arrived there years before any other type is available.

The final scene in this drama occurred when Robert B. Patterson, Undersecretary of War, wrote a memorandum to the Secretary of the Navy [11]: "This project had many advantages and looked like a distinct advance along liquid cooled lines. However, since the Pratt and Whitney Company are no longer enthusiastic in regard to this development, it is not believed that that organization can be expected to carry it to a successful conclusion."

No one seemed to realize at that time that the United States would go to war with Japan, Germany, and Italy by the end of 1941, and the immense effort Pratt & Whitney was about to expend to provide air-cooled engines for bombers, transports, and fighters. The company had conducted a liquid- vs air-cooling experimental program with a fly-off between a Curtiss P-40 with

Fig. 22   A standard Curtiss P-40 next to the same aircraft model (technically a Curtiss H-81A) with the R-1830 air-cooled engine instead of the Allison liquid-cooled engine (courtesy of Pratt & Whitney Archives).

Fig. 23   The first R-4360, known as the X-Wasp (courtesy of Pratt & Whitney Archives).

the standard Allison engine and the same aircraft with an R-1830 (see Fig. 22). The results were in favor of the air-cooled engine. At 20,000-ft altitude and with 1100 hp, the R-1830 powered aircraft was actually 5 mph faster than with the Allison engine.

The net result of Rentschler's intervention with General Arnold was the cancellation of all liquid-cooled engine programs. That was the beginning of the R-4360 program. While Eugene Wilson and Mr. Rentschler were petitioning the government to drop the liquid-cooled engine programs at Pratt & Whitney, Hobbs was already making plans for a 3000-hp air-cooled engine. His studies began in the summer of 1940, and looked at three configurations: 1) three rows of nine-cylinder power sections (27 cylinders); 2) four rows of seven-cylinder power sections (28 cylinders); and 3) six rows of five-cylinder power sections (30 cylinders). Three rows of nine would be like adding another row to the R-2800. However, there might not be enough room between cylinder heads to get cooling flow back to the last row. From a cooling viewpoint, one would select the six rows of five-cylinder power sections.

In the end, Hobbs selected the four rows of seven-cylinder power sections, not the best cooling design but good enough, with a more compact engine. His selection was like having two R-1830s in series with R-2800 cylinders and with each successive row indexed away from the first rows top cylinder to provide the twist in the cooling airflow. Hobbs authorized the design of a 3000-hp air-cooled R-4360 on November 11, 1940. Hobbs set an ambitious timetable for the first ground test (done April 28, 1941) and the first flight test (run on May 25, 1942). The official 150-hr type test was successfully completed on February 6, 1945. The first production engine (R-4360-4) was shipped in January 1945.

It was an ambitious, optimistic engineering leap. The first R-4360, known as the X-Wasp, demonstrated the feasibility of a four-row, air-cooled engine with 2800 hp (Fig. 23). The engine used cast magnesium front and rear cases. The crankcase was made up of five pieces of forged aluminum alloy. The one-piece forged steel shaft had four throws and was supported by five plain journal bearings. The R-4360 cylinder was built up from a steel cylinder barrel with a shrunk-on forged aluminum cooling muff and head, both with the maximum fin cooling area possible within the limits of the manufacturing process. The R-4360 cylinder looked different from previous engine cylinders because of the orientation of the rocker boxes (see Fig. 24). In the B-36 airplane, the engine was in the pusher orientation (that is, engine backwards to flight direction). In order to avoid a cooling problem, the flow of cooling air over the cylinders was increased significantly by the addition of a fan (Fig. 25).

Pratt & Whitney found that there was no particular advantage in directing fuel injection into each cylinder for one- and two-row engines. The usual company method of spraying fuel into the impeller was very effective in producing a uniform fuel–air mixture to the cylinders. However, the fuel-injection program for the four-row, R-4360 made more sense. Pratt & Whitney engineer Phil Hopper pointed out to me that the cylinders at the end of the intake manifold (first row as viewed from the front of the engine) tended to have their spark plugs "leaded up" more so than the cylinders at the rear of the engine. One of the test engineers on this program, Cliff Horne, kindly gave me these comments on fuel injection, with the caveat, "just remember it was over 50 years ago":

> There were 28 pistons in the pump that provided fuel to each of the 28 cylinders through an injector nozzle set for 350-psi discharge pressure. There was a fuel line from each pump to an interface block that connected to the individual lines from each cylinder. One of the big development

Fig. 24 The R-4360, note the orientation of the rocker boxes (courtesy of Pratt & Whitney Archives).

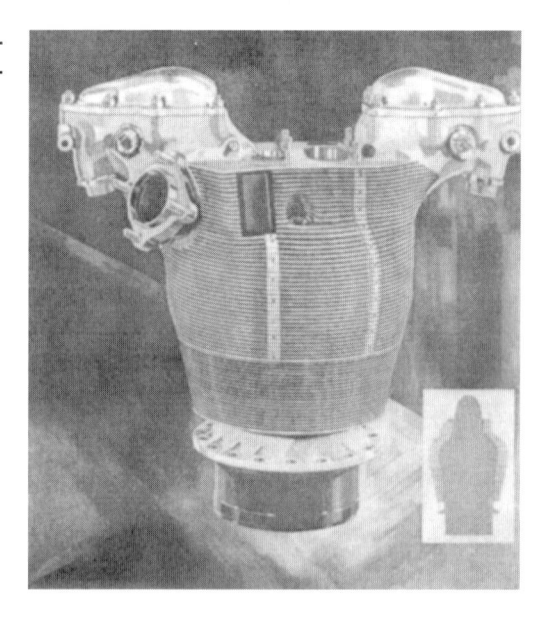

THE LATER PISTON ENGINE YEARS 113

Fig. 25 The R-4360-5 for installation in the B-36 as a pusher with the propeller shaft on the right at the wing trailing edge and with fan on the left (courtesy of Pratt & Whitney Archives).

problems was binding and seizing of the pump pistons as they had a running clearance of five millionths of an inch. Material development, piston surface treatment and incorporation of a 10-micron fuel filter finally produced satisfactory durability. Bendix built the carburetor-injection pump unit and Bendix and P&WA worked very closely on its development. The R-4360 engine appeared to run much smoother with the fuel injection system, engine acceleration was more responsive, and as I recall there was a decided improvement relative to detonation characteristics. The fuel injections system was used on the R-4360 VDT [Variable Discharge Turbine] engine. This engine didn't go anywhere but it was intended to be used on the B-50 bomber. An interesting and efficient engine that, if so desired, could carry take-off power to 40,000-ft altitude.

The first flight took place at the Vultee plant in Downey, California. The Army Air Corps provided the aircraft and Pratt & Whitney shipped the experimental engine X-108 to Vultee for installation in the aircraft. Howard Sargent, Pratt & Whitney test pilot, made the first flight at Downey on April 25, 1942 for about 30 minutes. Then the aircraft was shipped to East Hartford after the engine was removed and shipped separately. The aircraft with the R-4360 Wasp Major engine was ready for testing in East Hartford on July 27,

1942 (Figs. 26 and 27). The engine (Table 22) was a success with many applications (Table 23), including a Boeing B-50A, the Lucky Lady II, making the first non-stop flight around the world on March 2, 1949. Later in July of that year the jet-powered de Havilland Comet made its first flight—an omen of a significant change in commercial aviation, implying the end of piston-powered aircraft.

R-2180E–The Last of the Great Piston Engines

In August of 1944, a year before the hostilities of WWII ended, Pratt & Whitney started the design of a commercial engine that incorporated the best of what had been learned over its 20 years of experience. The result was the R-2180E. The E was added to minimize the confusion with the previous R-2180, Twin Hornet. The following is the background for the decision to launch another new engine program.

The early DC-4 aircraft had a gross weight of 65,000 lb. This established a propulsion requirement for an engine with a displacement close to that of the

Fig. 26 The R-4360 under test; the "corncob" effect is obvious (courtesy of Pratt & Whitney).

The Later Piston Engine Years

Fig. 27 This model V-72 Vengeance was modified to take the Wasp Major (courtesy of Pratt & Whitney Archives).

Twin Hornet (2180 cubic inches displacement). Pratt & Whitney built about 30 engines for this size of DC-4 and other potential applications at that time, the Stearman X-100 and the North American NA-21. Then the airline customers decided that they preferred a smaller gross weight aircraft. The R-2000 was created to satisfy them. However, the military in WWII took over the DC-4 before it reached the airlines, and the gross weight started its inevitable upward climb (a trend from concept to operational use in just about every airplane I ever heard of). The final C-54 ended up with a gross weight of about 73,000 lb but was still powered by the R-2000. Pratt & Whitney believed that was time for another engine, the R-2180, using the latest technology (Fig. 28).

This engine had some unique features (Tables 24 and 25). Dual oil transfer bearings permitted the use of reversible-pitch, full-feathering

TABLE 22   SPECIFICATIONS FOR THE R-4360

Horsepower	3000–4300
rpm	2700–2800
No. of cylinders	28
Weight (lb)	3325–3892
Bore (in.)	5.75
Stroke (in.)	6.0
Design start	November 1940
First run	April 1941
First flight	May 1942
Production quantity	18,679

TABLE 23 APPLICATIONS FOR THE R-4360 WASP MAJOR

Model	hp/rpm take-off or military	hp/rpm normal	Weight (lb)	Diameter (in.)	Length (in.)	Aircraft installations
TSB1-G	3000/2700	2500/2550	3325	52.50	96.75	Vought F4U-1
TSB3-G	3500/2700	2650/2550	3482	54.00	96.50	Boeing 377; Model 10 & 19; Republic RC-2 (engines converted to CB1, B6, CB2)
B5	3250/2700	2650/2550	3482	54.00	96.75	Boeing 377; Model 10 & 19
B6	3500/2700	2650/2550	3584	55.00	96.50	Boeing 377
B7						Boeing 377 (B6 equivalent with 100/130 fuel)
VSB11-G	3000/2700	2500/2550	3498	54.00	96.75	Aero Sud-Est SE-2010
B13	3500/2700	2650/2550	3520	54.00	101.76	Aero Sud-Est SE-2010
CB1						Boeing 377 (two engines—converted TSB3-G with "C" cylinders for service test by Pan Am)
CB2	3500/2700	2650/2550	3670	55.00	96.50	Boeing 377
R-4360-2						Goodyear F2G-1, -2; Martin XBTM-1 (YR-4360-4 with Bendix carb., changed to R-4360-2)
R-4360-2A						Goodyear F2G-1, -2; Martin XBTM-1 (YR-4360-4 with CECO carb., changed to R-4360-2A)
R-4360-4	3000/2700	2500/2550	3352	52.50	96.75	Goodyear F2G-1, Martin XBTM-1, JRM-2 (Semi-production engines)
R-4360-4	3000/2700	2500/2550	3400	52.50	96.75	Goodyear F2G-1; Martin AM-1, XBTM-1, JRM-2, XP4M-1 (production engines)
R-4360-4W	3000/2700	2500/2550	3413	52.50	96.75	Martin AM-2
R-4360-4A	3000/2700	2500/2550	3390	52.50	96.75	Hughes XF-11, HFB-1 (H4)

THE LATER PISTON ENGINE YEARS

Model					Application
R-4360-8	3000/2700	2500/2550	3525	52.50	114.25	Douglas TB2D-1 (semiproduction)
R-4360-8	3000/2700	2500/2550	3570	52.50	114.25	Douglas TB2D-1 (production engines)
R-4360-9	3000/2700	2500/2550	3425	52.50	101.75	Convair A-41
R-4360-10	3000/2700	2500/2550	3785	52.50	130.75	Boeing XF8B-1
R-4360-13	3000/2700	2500/2550	3685	52.50		Republic XP-72
R-4360-14	3000/2700	2500/2550	3584	52.50	111.00	Curtiss XBTC-2
R-4360-17	3000/2700	2500/2550	3308	52.50	87.00	Northrop B-35, XB-35, YB-35 (direct drive engine)
R-4360-18	3000/2700	2500/2550	3355	52.50	96.75	Lockheed XR60-1; Model 89
R-4360-20A	3500/2700	2650/2550	3540	54.00	102.00	(-20A): Fairchild C-119B, XC-120, C-120; Martin P4M-1, XP4M-1.
R-4360-20WA						(-20WA): Douglas C-124A; Fairchild R4Q-1
R-4360-21	3000/2700	2500/2550	3308	52.50	87.00	Northrop B-35, XB-35, YB-35 (direct drive, pusher)
R-4360-22W	3250/2700 3500/2700[a]	2650/2550	3490	54.00	96.75	Lockheed R6V-1
R-4360-24	3000/2700	2500/2550	3411	52.50	96.75	Martin JRM-2
R-4360-25	3000/2700	2500/2550	3483	52.00	109.75	Convair B-36A (Model 37), XC-99 (pusher engine)
R-4360-27	3000/2700	2500/2550	3404	52.50	96.75	Douglas C-74 (DC-7), XC-74
R-4360-31	3000/2700	2500/2550	3506	52.50	114.25	Hughes XF-11; Republic XF-12
R-4360-33	3000/2700	2500/2550	3595	52.50	109.25	Boeing XB-44
R-4360-35	3250/2700	2650/2550	3490	54.00	96.75	Boeing TB-50A, D, H, B-50A, B, D; Fairchild XC-119A
R-4360-35B	3500/2700[a]					
R-4360-35A	3250/2700	2650/2550	3490	54.00	96.75	Boeing C-97A, C, KC-97E, YC-97A, YC-97B; Douglas XC-124A
R-4360-35C	3500/2700[a]					
R-4360-37	3000/2700	2500/2550	3346	52.50	96.75	Hughes XF-11; Republic XF-12, XR-12
R-4360-41	3000/2700	2500/2550	3567	53.50	109.75	Convair B-36B

(Continued)

TABLE 23 APPLICATIONS FOR THE R-4360 WASP MAJOR (CONT)

Model	hp/rpm take-off or military	hp/rpm normal	Weight (lb)	Diameter (in.)	Length (in.)	Aircraft installations
R-4360-41	3250/2700	2650/2550	3567	54.00	109.75	Convair B-36B, D, E, RB-36, XC-99
R-4360-41A	3500/2700[a]					
R-4360-43	4300/2800	3150/2600	3720	55.00	103.50	Boeing YB-50C, B-54A, RB-54A
R-4360-45	3000/2700	2500/2550	3308	52.50	213.85	Northrop B-35, XB-35, YB-35. (Direct drive length = 87.00)
R-4360-47	3000/2700	2500/2550	3308	52.50	326.40	Northrop B-35, XB-35, YB-35. (Direct drive length = 87.00)
R-4360-49	3250/2700	2650/2550	3490	54.00	96.75	Douglas C-74
R-4360-49A						
R-4360-51	4000/2800	3100/2600	4020	55.00		Convair B-36 study (tractor installation)
	4300/2800[a]					
R-4360-53	3800/2800[a]	2800/2600	4040	55.00	117.00	Convair B-36D, E, F, H, RB-36D, E, F, H
R-4360-59	3800/2800[a]	2800/2600	3689	55.00	102.00	Douglas C-124 Study
R-4360-59	3500/2700[a]	2650/2550	3691	55.00	96.50	Boeing C-97D
R-4360-59B	3500/2700[a]	2650/2550	3811	55.00	96.50	Boeing KC-97F
R-4360-63A	3400/2800	2800/2600		55.00	103.75	Douglas C-124C
	3800/2800[a]					
R-4360-65						Boeing C-97A, C, KC-97E

[a] Water-injection rating.

THE LATER PISTON ENGINE YEARS 119

Fig. 28 Right front view of the R-2180E, note that the magnetos are mounted on the front as in the R-2800 and the R-2000 (courtesy of Pratt & Whitney Archives).

hydromatic propellers. An automatic, two-position valve overlap mechanism increased the detonation-limited power by about 100 bhp. The increased power came from better scavenging of the residual exhaust gases and the resulting cooler cylinder charge. This feature was described in an engineering brochure [12]:

> All of the exhaust valves are actuated by a cam at the front of the power section and all of the intake valves are actuated by a cam at the rear of the power section. When a specified engine speed is reached, the valve overlap is automatically changed by moving the fixed gear in the rear cam planetary gear system. A built-in booster pump, driven from the crankshaft supplies oil pressure to the hydraulic pistons which rotate the gear. When the engine speed becomes less than the specified value, the mechanism automatically returns the fixed gear to its original position.

The crankcase was in three pieces, same as the R-2800, and the master rods were solid pieces. The crankshaft bearings were steel sleeves plated with

TABLE 24 SPECIFICATIONS FOR THE R-2180E TWIN WASP

Horsepower	1650
rpm	2800
No. of cylinders	14
Weight (lb)	1870
Bore (in.)	5.75
Stroke (in.)	6.0
Design start	August 1944
Production quantity	75

TABLE 25 APPLICATIONS FOR THE R-2180 TWIN WASP

Model	hp/rpm take-off or military	hp/rpm normal	Weight (lb)	Diameter (in.)	Length (in.)	Aircraft installations
E1	1650/2800	1300/2600	1870	54.00	76.20	Scandia SAAB 90A-2, 90B-2
R-2180-11	1650/2800	1300/2600	1830	54.00	76.00	Piasecki H-16

silver-lead-indium. Its ignition system was low tension, the benefits of which were described in the engineering brochure: "It's relatively immune to weather and altitude conditions; it reduces spark-plug erosion by 65 percent; it has superior ability to fire fouled spark plugs; and it minimized flash-over, corona, and electrical loss problems." The cylinders were from the R-4360 with an important change: the intake was on the side and the exhaust was on the top of the cylinder.

THE CONTRIBUTIONS OF TWO ENGINEERS

T.E. TILLINGHAST—FROM WWI PILOT TO COMPANY EXECUTIVE

One of the most effective voices promoting Pratt & Whitney engines in the 1930s through the 1940s and into the 1950s was Theose E. Tillinghast–universally known as "Tillie" (Fig. 29). Tillinghast, after graduating with a degree in Mechanical Engineering from Rhode Island State College in 1917, joined the U.S. Army Air Service. After his flight training at Selfridge Field in Michigan, he was assigned as a flight instructor. When he was assigned to a British air unit, he flew a Sopwith Camel. Shot down on September 22, 1918 behind the German lines, he managed to escape from the internment camp and make his way back to England by way of Holland. By then the war had ended, so he returned to the United States and became commander of the 57th Aero Squadron at Selfridge Field. Later he became Chief of the Powerplant Branch at McCook Field as a Captain from 1926–1929.

In 1929, he resigned his commission and joined Pratt & Whitney as Executive Engineer and in 1932 was named Assistant to the President of Engineering. In 1934, he became Sales Manager, effectively promoting the company's engines. Toward the end of 1943, the company was in the process of reorganizing to make communications among the top-level executives more effective. When this news reached the level of General in the U.S. Army, Pratt & Whitney heard about the Army's concern right away [13]:

> One of the Army Generals demanded to know how Tillinghast was affected and whether Tillinghast was being shoved aside. Lyman [United

THE LATER PISTON ENGINE YEARS 121

Fig. 29   T.E. Tillinghast (courtesy of Pratt & Whitney Archives).

Aircraft public relations] explained quickly that Tillinghast was in a stronger position than ever and that he was now in a position to give the Army better sales service than ever before. The general was greatly pleased and after the conversation was over Lyman went to the telephone, called Rentschler and told him that he, on his own responsibility, had put Tillinghast in a most favored spot. Rentschler said fine, 'call Wilson and tell him what you have done,' which he did and a few days later, Wilson reported that he was just beginning to appreciate what a great job Tillinghast has done over the years for the corporation.

Tillinghast received the General William E. Mitchell Memorial Award in 1958 from the Mitchell Air Force Association as "the United States citizen making the outstanding individual contribution to aviation progress" [14]:

When water injection was developed at Pratt & Whitney Aircraft for its R-2800, engines used in the Republic P-47 Thunderbolt fighters then fighting the Messerschmitt and Focke-Wulf German fighter aircraft, Mr. Tillinghast went to Europe to consult personally with his old colleague General Jimmy Doolittle and the men of the P-47 fighter squadrons.

Mr. Tillinghast has long been known as a man who shouldered the problems of Pratt & Whitney Aircraft's customers with intense personal interest and in perennial good humor.

PERRY W. PRATT—DEVELOPER OF R-2800C ENGINE

Perry W. Pratt (Fig. 30) was not only the driving force behind the success of the R-2800C but was also the man, along with Luke Hobbs, most responsible for the company's success in getting into the jet engine business with the J57/JT3. I was with the company only a few weeks when I was asked to make a preliminary stress analysis on a blade being considered for the J57 high-pressure turbine. Someone showed me the layout and stood by patiently

Fig. 30 Perry W. Pratt (courtesy of Pratt & Whitney Archives).

while I measured the blade chords at root and tip as well as the blade span. I explained to him that I was only looking for ballpark precision, although scaling measurements off a drawing is normally punishable by "hanging from the yardarm until dead." However, this fellow simply went down the aisle after I handed him the drawing. Much to my surprise, I found out later I was speaking to Perry Pratt.

Perry's uncle, Howard Pratt, had his own machine shop and taught his teenage nephew how to use all of the machine tools. Paul Fisher relates this time of Perry's life [15]:

> Howard Pratt checked Perry out on the shop machine tools with the same severity and care he had experienced in his apprentice days. At 17, Perry was also sandwiching his time between his senior year in high school and serving as the chief mechanic for the Corvallis Airways, whose fleet essentially consisted of an American Eagle and a Waco 10, both powered by OX-5s, which Pratt occasionally overhauled from scratch.
>
> One day a Zenith, only two or three of which were built, rolled up and Perry saw his first Pratt & Whitney engine, a Wasp A whose rocker box required repacking. Perry recalls that he just stood back and stared. For several years he had owned a Wasp overhaul manual engine whose components and parts he had largely memorized. He recalls today that before he went for the grease for the rocker box, he told himself, "Someday I'm going to work for the outfit that built this engine, if it's the last thing I do."

When Perry graduated from Oregon State College in 1936, the market for young engineers was not promising. He settled with his new bride for an instructorship at New Haven Junior College and graduate study at Yale. In the following spring, Pratt got a call from Wright "Parky" Parkins, who also came from the Northwest [15]:

> Parky had a couple of openings for test engineers and Ed Leader, among others, his old associate in his crew days at the University of Washington, was now coaching at Yale. From what Leader could learn, he told Parky, this Oregon kid was theoretically brilliant but more interesting to him, the kid had a basic appreciation of tools and machines. He might couple the brilliance with the potential to become "one hell of a working engineer," which turned out to be true, Parky recalled twenty years later.

Perry came to Pratt & Whitney in 1937, just in time to see the company in a tight financial bind as the market for engines decreased due, in part, to the Army Air Corps' emphasis on liquid-cooled engines rather than on Pratt & Whitney's or Wright Aeronautical's air-cooled engines. Perry and his wife were now expecting their first child and the most attractive financial opportunity appeared to be an instructorship at New York University. So Perry left Pratt & Whitney with regrets, but he was back in 1939 when the company hired professors for the summer. At the end of the summer, Parkins called Perry in for a talk, resulting in Perry's return to the Engineering Department. Perry and Luke Hobbs jumped from the piston era into the jet era, and they are the two men most responsible for the company's success in aircraft gas turbines in the early days.

REFERENCES

[1] Fisher, P., "A Man Called Luke," *The Bee-Hive*, United Aircraft Corporation, Pratt & Whitney Aircraft Division, United Technologies Corporation, 1954; reprint of Feb. 1954 issue of *U. S. Air Services* magazine.
[2] Fisher, P., "Conversation with Hobbs," interview, Pratt & Whitney Archives, Feb. 8, 1950.
[3] Fisher, P., "Parky," *The Bee-Hive*, United Aircraft Corporation, Pratt & Whitney Aircraft Division, United Technologies Corporation, Jan.1955.
[4] *The Bee-Hive*, United Aircraft Corporation, Pratt & Whitney Aircraft Division, United Technologies Corporation, May 1931.
[5] Hobbs, L. S., "Aircraft Engines and National Defense," Leonard Hobbs files, Pratt & Whitney Archives, 1939.
[6] Gunston, B., *The World Encyclopedia of Aero Engines*, 3rd ed., Patrick Stephens Ltd., Haynes Publishing, Somerset, U.K., 1995, p. 27.
[7] Meyer, C., "The Best WWII Fighter," *Flight Journal*, Vol. 8, No. 4, Aug. 2003.
[8] Brett, Major General G. H., "Notes on Wilson Visit," Box 8/98-225-15A, Pratt & Whitney Archives, Sept. 10, 1940.
[9] "Notes on Phone Conversations between Eugene Wilson and General George H. Brett," Pratt & Whitney Archives, Sept. 13, 1940.
[10] "Frederick Rentschler to Major General Henry H. Arnold," letter, Box 8/00-399, Pratt & Whitney Archives, Nov. 15, 1940.
[11] Patterson, R. B., "Memorandum for the Secretary of the Navy," Pratt & Whitney Archives, Jan. 10, 1941.

[12] *The New R-2180 Twin Wasp E Engine*, Pratt & Whitney Aircraft, Pratt & Whitney Archives, 1947.
[13] Harrower, C. R., "Memo to Saga File," Pratt & Whitney Archives, Dec. 7, 1943.
[14] Tillinghast file, Box 8/98-226, Pratt & Whitney Archives.
[15] Fisher, P., "Notes on Conversations with Perry Pratt, Late Fall & Winter," interview, Cabinet F22, Pratt & Whitney Archives, 1965–1966.

Chapter 4

THE PISTON ENGINE EXPERIENCE

INTRODUCTION

There are hardly any piston-engine powered commercial or military aircraft on the flight line these days. The younger traveling public does not know the romance of flying in a piston-engine powered aircraft. First there is the startup ritual. Going back to WWII, this ritual started with one of the crew putting his shoulder to the prop and slowly rotating the engine crankshaft through a couple of revolutions. Next, the pilot engaged the starter. The prop started to rotate slowly and then suddenly slowed down as if afflicted with shortness of breath, sounding somewhat like an inveterate smoker's hacking cough. Unexpectedly, as the engine seemed to gasp for air, there would be a sudden explosive blast of black, sooty effluent from the exhaust stacks. All at once the engine seemed to come to life and began a soft purring like a kitten. Then there was the magneto check, left and right at high power. The flat-pitched propeller moaned a soulful whine as the pilot, moving the power lever toward take-off power, released the brakes. The lumbering hulk of metal gradually accelerated down the runway while every shred of metal in the aircraft seemed to resonate at its natural frequency. Then rotation and the slow climb to altitude at climb power. Finally at the cruise altitude, all was at peace in the world as the constant-speed propeller pitch switched smoothly to its cruise position. You just do not get that kind of excitement and thrills from those new-fangled gas turbines!

In this chapter, we will meet some employees and customers of Pratt & Whitney and let them tell their stories about their experiences with the company's piston engines. The chapter continues detailing the work done on direct fuel-injected and liquid-cooling engines. This chapter concludes our history of Pratt & Whitney's piston engines, and Chapter 5 ushers us into the gas turbine era.

PRATT & WHITNEY FLIGHT TEST ORGANIZATION

When Pratt & Whitney first started out in 1925, there was no such group as a flight test organization [1]. Engines were shipped to the military and the aircraft manufacturers. These customers did their own flight testing to evaluate the effectiveness of the engine–airframe combination. Any shortfall in performance was blamed on the engine–propeller combination. Any remarkable aircraft performance demonstrated in flight was due, of course, entirely to the aircraft and had nothing to do with the engine. It became wise for Pratt & Whitney personnel to get involved in flight testing of engines before they were shipped to customers.

The company's flight test operation started in 1927, when Pratt & Whitney built a large hangar at Brainard Field in Hartford, Connecticut for the company's transport and test aircraft. The first company aircraft was Rentschler's personal aircraft, a Vought O2U-1 that he bought for his own pleasure. After a year, he turned the aircraft over to the company to be used as a test aircraft. Next the company bought a Boeing mail plane, the Model 40B. This was a more practical aircraft because its cabin space (normally for mail and/or passengers) could be used for flight and engine instrumentation and was a more comfortable environment for an engineering observer. The first company test pilot was William B. Wheatley, an Army Reserve pilot from the experimental test department. He conducted flight tests on a part-time basis.

In 1928, A. Lewis MacClain (Fig. 1), a former Army pilot and a graduate of MIT with a Masters Degree in Engineering, was hired to set up an experimental test organization. He built a first-rate organization from the ground up, and ran this group until the middle of WWII. Initially, the flight-testing was under the direction of the Engineering Department. During WWII, it became obvious that the installation of the engine in an aircraft introduced problems apart from engineering, such as the electrical, hydraulic, fuel, and

Fig. 1 A. Lewis MacClain, test pilot, next to the Vought O2U-1 (courtesy of Pratt & Whitney Archives).

Fig. 2 The Vultee YA-19 was a sleek-looking aircraft that began testing the R-2800 on August 8, 1941 and ended testing in 1946 (courtesy of Pratt & Whitney Archives).

oil connections to the aircraft, as well as the cowling to minimize drag and provide the required engine cooling at all flight conditions. MacClain was asked to set up a separate installation engineering department and installation flight test department to work on these aircraft-related problems. The company also occasionally borrowed aircraft from the Army Air Corps for engine testing. One such aircraft was the Vultee YA-19 (Fig. 2), which was used specifically for the flight test of the R-2800.

R-4360 TEST FLIGHT

In early 1942, Pratt & Whitney was looking for a single-engine aircraft in which to test the R-4360. A modification of the Vultee Model 72 (Vengeance with Wright R-2600) looked like it could do the job. The changes to the V-72 included an extended fixed landing gear to accommodate a larger propeller, a fuel tank in the bomb bay, a reinforced dive brake flap, flap well covers, and a new engine mount to take the heavier, more powerful engine. This new aircraft model was called the V-85.

The engine was shipped to Vultee at Downey, California in January 1942. The first flight of the aircraft, a 3-minute flight, was made there on April, 25, 1942, with Howard Sargent as the pilot and Dick Smith as the flight-test engineer. The aircraft and engine were then shipped to East Hartford for the flight test program (Fig. 3).

The first flight in East Hartford took place a little after 6 p.m. on July 24. About 5.5 hr were accumulated in 13 test flights up through September 14. Then on September 15, a 24-minute flight ended in a tobacco field not too far from the Bradley Field Army Air Corps Base (Fig. 4). Andrew Willgoos, Chief Engineer at Pratt & Whitney, reported to Vultee on November 11, 1942 that, "The forced landing which resulted in the loss of the Vengeance V-72 airplane was the result of a fuel supply stoppage in a part associated with the engine and carburetor system and did not involve the failure of any airplane structure or engine installation so far as we were able to determine" [2].

Fig. 3 From left: William Elhart, crew chief; Warner Gaines, test engineer; George Matusik, pilot; Dick Smith, engineering flight observer; pilot Howard Sargent; pilot Fred Borsodi; Ed Granville, supervisor of hangar; and Charles Arnold. The aircraft is a modified Vultee Vengeance (courtesy of Pratt & Whitney Archives).

Dick Smith (shown in Fig. 3) described the event to me in 2005 more vividly:

> The flight progressed nicely for a little over 20 minutes when all of a sudden the engine stopped. Howard [the pilot] tried to start the engine but met with no success while the aircraft was approaching the ground. The altitude was not great enough for a jump so Howard resolved to make the best of a dead-stick landing in a field that looked like it had adequate real estate for such a maneuver. Things were looking up as he

Fig. 4 It is a miracle that the pilot and observer survived this accident with relatively minor injuries (courtesy of U.S. Air Force).

touched down on the brief stretch of paved road and then he was looking forward to a gradual slowing down in the field. However, those invisible wires converted a textbook dead-stick landing into a nightmare.

Dick said he remembers being held by his shoulder harness in the vertical position upside down and being aware that he should carefully unbuckle the harness so that he did not fall straight down, breaking his neck in the process. But the fuel vapors got to him and he lost consciousness. After a brief stay in the Army Air Corps medical facility he was shipped to the Hartford Hospital and finally home that night. Howard escaped with just a broken finger.

R-4360 FLIES FROM HAWAII TO CHICAGO

In August 1938, the Navy ordered from Martin a prototype of the world's largest—and heaviest—flying boat, a patrol bomber. It flew in July 1942, was designated JRM-1, and was powered by four Wright R-3350 engines. An upgrade of this aircraft, JRM-2, with P&W R-4360 engines was built in 1947. A typical situation in which the engine company helps the customer achieve its objective is described by Bob Baer, from Installation Engineering [3]:

> Early in 1948 VR-2 [Naval Air Group] at Alameda received the one and only JRM-2 ... ever built ... with P&W R-4360 engines at 3500 hp each, 500 more than the R-3350s....
>
> The Navy now had no money to pay for performance data or a flight manual so they gave it to VR-2 and told them to do the best they could and just operate it like a JRM-1. That was a little silly because it wasn't a JRM-1... Sid Perry was at the time helping me fly speed-power polars on the aircraft from which we could construct a cruise control.... The cruise control data that resulted enabled VR-2 to operate the aircraft with good efficiency, which they did.
>
> One day in late July, I received a phone call telling me to be in Captain Gurney's office at the Sunnyvale Naval Air Base at Moffett Field on the south end of SF Bay at 11 o'clock.... Gurney [commander of the Fleet Logistics Air Wing Pacific] had received a telex from Washington, which ... said: "We have determined that your JRM-2 aircraft is capable of flying a seaplane non-stop distance record. You will therefore fly from HNL [Honolulu] to Chicago and land in Lake Michigan at 12 noon on August 28th where suitable Naval rank and publicity will be arranged." Now it's one thing to fly a record flight but it's another thing to be told you will do it to arrive at a stated date and time. They were also told to carry a 10,000-lb payload.
>
> Gurney ... said [to me] he understood that I and my people had developed the cruise control for the aircraft. "Yes," says I. "Well," says he, "do you think this aircraft can make such a flight?" I said it could if they followed my cruise control. Gurney ... [said] "Tell Washington we'll do

it." Then he ... pointed a finger at my chest and said, "And you be on board to see that we do it!"

On August 25th we flew to HNL on the JRM-2, checking out the long-range procedures as opposed to the usual higher power operation. We did that in 12 hours of daylight. The data checked out okay.

We spent the 26th resting, and refueling and provisioning the aircraft. Our 10,000-lb payload consisted of some 40 newspapermen with typewriters plus food and liquor for them.... The payload tallied out to 14,445 lb....

The next morning ... at 8 a.m. we took off from Kihi Lagoon next to the airport, but the wind required us to take off landwards. At 165,000-lb gross ... the aircraft rate of climb after takeoff was just about the same as the rise of the land. The pilot, Lt. Commander Bob Hunt finally eased her into a slow left turn over Hickam Field and Pearl Harbor and off we went.

En route, the newspapermen were filing radio reports and the two Naval radiomen almost dropped from exhaustion. Over SFO [San Francisco], there were radio broadcasters to commercial stations who aired them live.... The overnight portion of the flight was both interesting and weird. Here we were in a huge seaplane threading our way through the Sierras and Rockies on a clear moonlit night. We threaded our way because we got our best range at about 10,000 ft and the aircraft was unpressurized ... so I didn't want to sacrifice range by using higher altitudes...

We quietly overflew Chicago by almost 200 miles to Ann Arbor, Michigan, and then returned to Chicago to land at 12:05. Everyone was happy—Chicago, the ... Navy, the press, and all. The crew and I were pretty tired but happy too.

To my surprise ... I received a letter from [aviation pioneer] Glenn L. Martin dated September 28th thanking me for the work I had done to successfully execute the flight. Copy to H. M. Horner didn't hurt me either.... The official flight distance was 4745 miles in 24:13 flight time. That's a hot 195 mph with tail winds most of the way!

R-1830 Brings the Crew in Safely

By the end of WWII, about half the American's air horsepower was supplied by Pratt & Whitney. U.S. Army Air Corps Captain David C. Burton, pilot of a B-24 in a raid on Regensburg, Germany on February 23, 1944, shared the reason for his "love affair" with Pratt & Whitney's craftsmanship [4]:

This was my crew's seventh combat mission and our group of four squadrons put up about 50 planes for this mission. We were on the bomb run when our Number 3 engine was hit by shrapnel from an anti-aircraft shell that exploded near the copilot's side of the plane. I pushed on more power to the remaining three engines in order to stay with the formation on the bomb run. A bomber with one engine feathered and lagging behind the formation always attracted enemy fighters because a bomber that was already crippled is easier to shoot down. Our three engines responded but then I noticed that the Number 1 engine's oil pressure was dropping and the rpm was fluctuating. By the time we released our bombs and were headed for home base, Number 1's prop governor was failing and the prop started to runaway or overspeed. It had to be feathered before the engine froze-up and couldn't be feathered.

It was now decision-making time. If we kept flying, we could not keep up with the formation and would lose the protection of the other plane's gunners. We were also losing altitude and expected to become an easy target for the enemy fighters. On the other hand, if we bailed-out into the enemy-held, snow-covered mountains below, we might not survive anyway.

Lady Luck was with us because by the time we could no longer stay close to the formation, the enemy fighters had left the area. It was now time to concentrate on how to keep our B-24J in the air as long as possible.

The two 1200-hp R-1830 Pratt & Whitney engines were now being "sacrificed" for the few minutes of flying time we thought we could expect from them. Normally, the allowable take-off throttle settings of 49-in. manifold pressure and 2700 rpms were to be used for only a maximum of 5 minutes. To keep from losing too much altitude and flying speed, we turned up the supercharger past these settings and when the engines would get too hot, we would cut back on the power and let them cool off. We wondered how long these engines could take this abuse.

At normal settings we lost altitude and even though we started at 23,000 ft, a descent of 200 to 300 ft/min meant we would be out of sky before we could reach the Adriatic Sea where the Air-Sea Rescue Units waited. Switzerland was another option but we would be interned for the remainder of the war. The entire crew wanted to try to get back to home base.

Over the next couple of hours the engines responded every time we applied full power to them. The plane was becoming lighter because of the large amount of fuel we were burning so we were still at about 6000-ft altitude by the time we reached the Adriatic Coast. Everyone of the crew was up on the flight deck watching the engines and instruments, especially the altimeter and rate of descent. It appeared that we were almost holding our own. If the engines held up to this tremendously high manifold pressure and rpm, for another two hours, we would make it home.

I yelled over the sound of the engines, "We're approaching an Air-Sea Rescue boat so now is the time to decide if you want to bail out. They say the sea is not very rough and they shouldn't have any trouble picking you up." The general consensus was, "Hell no; we stay with the plane and crew and trust that those two beautiful Pratt & Whitneys keep running!"

Those engines were treated with tender loving care, if it can be said that running them at higher than take-off power settings was TLC. They ran too hot all of the time and the high oil temperature and low oil pressure told us we could lose either engine at any time. We knew our safety depended on both engines continuing to run because some lives would be lost if we had to ditch in the sea. Even when we passed another Air-Sea Rescue boat, the crew wanted to stay with the plane and ditch if we had to.

Later, as we passed over the eastern shore of south Italy, we had a number of opportunities to land at other airbases. They looked good to use, but there is a masculine pride that comes over men at times like this, "We've come this far and the engines are still working, let's go for it!"

We flew over the base at about 1500 ft, firing red flares and hoping they would see that we had two engines feathered. The tower did recognize our plight and in turn fired red flares—telling other planes to get out of the landing pattern and let us in. We had to crank down our landing gear and flaps by hand because we had lost our hydraulic pressure when the Number 3 engine was hit.

Many times in practice, we would bet among ourselves that I would not have to touch the throttles after I cut the engines in the landing pattern. It was a game of flying to a certain point on the downwind leg, cutting off all the engines on the B-24 and coming in "Dead Stick." On this mission, it was no game; I was afraid that if I used the normal approach pattern, these tired engines would quit when I pulled back on the throttle and we would hit short of the runway.

All the crew took crash-landing positions, in case we came in short. With half-flaps, the B-24 glided just as other planes had done and as we turned on to the final approach, I called, "Full flaps! We've got it made! Thank God for Pratt & Whitney!" Everyone let out a "yah-hoo" when the landing gear settled on the runway.

Our ground crew chief couldn't believe what we did to those engines and how we still kept flying. It was the last flying those engines did. They were so hot when we shut them down that some of the parts literally melted.

That night, as we sat around bragging about our mission, we paused and toasted, "Here's to Pratt & Whitney and those wonderful men and women in their factories that make those magnificent engines. Hear! Hear!"

R-1830 AND SUPERCHARGERS

Walter Doll joined Pratt & Whitney right out of the University of Rhode Island in June of 1938 and worked through the WWII years. He described to me his early experiences as only one of five experimental test engineers, responsible for 30 engineers:

> The Experimental Test Office, at that time, was virtually the hub of the Engineering Department.... We had enormous authority and responsibility. We could go anywhere in the company and change anything with a red pencil and a note: designs, shop and test instructions, etc.
>
> We were expected to work many nights and weekends, for no extra pay. We were given a small meal allowance for overtime work, which

amounted to $0.75 for three hours and $1.50 for five hours of continuous overtime. I threw my meal allowance money in a coffee can, and Mary [my wife] and I bought our first washing machine out of the can...

We all were totally dedicated to the success of the company.... We all had the idea that if we did what was best for the company ... the company would take care of us and that is ... what happened.... At lunch hour we would play touch football and softball in the parking lot outside our office, and, when payday came on Friday, we would go out and have a party.

My first job ... was in the newly created Supercharger Group (one project engineer, three test engineers). The major engines at that time were the 500-hp Wasp, R-985, the 1200-hp Twin Wasp, R-1830, with development just starting on the 2400-hp Double Wasp, R-2800.

Each of these engines had an internal supercharger consisting of a relatively crude centrifugal impeller (11-in. diameter in the R-1830), rotating at very high speed, and discharging through a vaned diffuser into the cylinders through the intake pipes. The supercharger compression ratio was approximately two to one. The amount of power that an engine can develop is proportional to the amount of fuel–air mixture that you can get into the engine per unit time. Doubling the manifold pressure thus essentially doubles the engine power. Or, the supercharger would let you take a given amount of power to a much higher altitude, which is an advantage for flight efficiency or military combat.

At the time of my arrival, the company was just starting to develop a special version of the R-1830 engine which had an auxiliary supercharge that could be turned on and off (with two speeds, operated with clutches and a hydraulic coupling). This was for use in a Navy fighter plane being built by Grumman, I believe, which was used in the Pacific in the war. When we first ran this engine, it lost power when the auxiliary supercharger was turned on.... Management ... decided that we needed to know more about superchargers and formed the aforementioned Supercharger Group, of which I was the newest and youngest member.

At that time, we bought the superchargers as a complete package unit from the General Electric Company branch in Lynn, Massachusetts. They were responsible for all of the engineering and design. We found that the R-1830 auxiliary supercharger was performing poorly and quite differently from the General Electric specifications. The main problem was that the flow capacity was too large for the rest of the engine and it was stalling out, thus causing the engine to lose power. Mainly, by intuition, we adjusted the flow capacity, and got the engine operating.

Management then decided that we should learn how to design our own superchargers, and not depend ... on one vendor who had just let us down. So I began to work on superchargers for the R-2800 and the newer R-4360 engines. I was also involved in other miscellaneous jobs, including responsibility for several of the R-1830 engines.

Because of the very small passages in the superchargers and the very high flow velocities (at times supersonic) it was very hard to make internal measurements of any significance and usefulness. The most successful technique we developed was a visual flow procedure. We mixed ... carbon lampblack in some very gooey linseed oil. I would pour this into the carburetor and it would go right through the supercharger. At that time, before EPA, we just blew it out into the rig room. Nobody complained if we didn't do it too frequently. Where the flow inside the supercharger was good, and doing what it was supposed to, the lampblack would blow right through without leaving any marks. Where the flow was bad, and separating or recirculating, the lampblack would stick to the wall with a very distinct pattern. By looking at the pattern I could ... deduce what was wrong and ... fix it.

One day I was testing a new supercharger in a new test rig with one 6-in. diameter exhaust pipe discharging horizontally right across the rig room aisle at the height of about 3 ft. I was just about to pour a can of my glop into the carburetor when I saw some movement out of the corner of my eye. There were some people coming down the aisle, about to walk in front of my exhaust pipe. They were Frederick Rentschler, founder and corporate Chairman of United Aircraft, Luke Hobbs, corporate Vice-President for Engineering, Charles Lindbergh, the famous aviator, and General "Hap" Arnold, head of the Army Air Corps (before the Air Force became a separate service). If I hadn't looked up when I did! ...

LIFE AS A TEST ENGINEER

Dana Waring graduated from Cornell with a degree in Mechanical and Electrical Engineering and an M.S. from the California Institute of Technology. He told me of his experience at the company beginning in 1940:

After finishing my training session I was assigned to the R-2800 Group. The test engineers at that time had their desks in the corridors of the rig rooms where it was noisy enough to discourage much work. The rigs were testing component parts and were generally run by Duesenberg engines (X-Test having cornered the market on old Duesenbergs since they put out the most power at that time). However, we were soon moved into the newly constructed and not yet finished Experimental Test Office, which was two stories with the test design group upstairs and the test engineers on the first floor. There were only maybe twenty test engineers then and we rather rattled around in all that space.

Responsibility came very fast. In addition to their programs ... on the first shift, test engineers were assigned to the second and third shifts in rotation and had to be Jacks-of-All-Trades since they took over on whatever tests were being run. My first night I was assigned to run an engine to determine its detonation characteristics. I went out to the test house and told the operator I was ready to start it up. He informed me that either the foreman or the engineer had to start the engine and the foreman was busy elsewhere. So I boldly told him to hold my hand and I would do it.

Wow! It was a much bigger thrill than starting the little engine in the Piper Cub that I had learned to fly! ... Detonation runs were ... the most dangerous runs to make since conditions were on the edge of burning a hole in a piston if you missed the telltale puffs of smoke from detonation and went too lean with the mixture.

It was all very new and exciting ... you built your test engine with what you had plus any parts that you could find in Used Stores plus the special test parts, but that usually left you without such critical parts as a magneto or carburetor ... that were in short supply. So when your test was completed it behooved you to inspect these critical parts immediately since other test engineers were eager to scavenge them to complete their engine builds and they would just vanish, even while being trundled from the test stand to the shop.

Another thing you had to watch out for was the habit of project engineers, especially Perry Pratt, who would wander through the shop at odd hours and change the instructions that you had written out for the experimental assembly crew. It could be ... disconcerting to criticize the foreman for not following your instructions, only to have him point out that they had been changed. The shop foremen ... were very knowledgeable, however, and would ... guide us, once we got to know them.

One thing that made us feel important was Luke Hobbs' habit, as chief engineer, of getting the test engineer up to his office for a first-hand account of a test that he was interested in. We had to be really prepared for such occasions since he would quiz us down to the last detail.

It also helped ... to have Perry Pratt give us lectures on the fine points of gasoline engine technology. However, between three nights of second shift and two nights of Perry's lectures we did not have much time at home. This was hard on those of us who were married. I was lucky to live so close to the plant that I could bicycle to work and slip home during lunch and dinner times.

During the war, the guards at the gates carried shotguns. One night about three o'clock in the morning I was ready to go home ... when the guard ... told me I could not leave at that hour, it wasn't a shift change. I was in no mood to argue and told him that engineers could do as they pleased and drove on. The shotgun came down aimed my way but he did not fire.

I remember [test engineer] Jim Wheeler coming into the test office one night on second shift shaking and white as a sheet. He had been running an early run with water injection. At the time the rating for the R-2800 was 2400 hp but with water injection Jim had just pulled 3400 hp without realizing it. He said it was just as smooth as silk.

Our shop foremen were quite phenomenal. I once asked Nick the grinder for a hemispherical end on a half-inch rod. He ... took it over to an ordinary grinding wheel and freehanded it and told me to check it with a gauge. It was so accurate that I could not see light between the part and the gauge.

I spent several months in the Philadelphia Naval test facility while they ran qualification tests on our engine. They were running water injection runs on our engine and it required ethyl alcohol so it would not freeze at altitude. They had a 55-gallon drum of it for the purpose but the level in it went down rapidly since the crew and many "visitors" would take a shot in their paper cup before going to the coke machine.

Across from our test stand they were testing a Curtiss Wright engine. One day its magnesium gearbox caught fire. The operator did as he should have and stopped the engine and rolled down the curtains to cut off the air supply. Unfortunately the fire department had not been trained for magnesium fires (magnesium burns in water) so when they rolled up the curtains and hosed the case, it was really spectacular.

There were several spectacular engine failures.... In one the crankshaft counterweight broke across the damper holes while running at take-off speed. The crankshaft then made just one turn and jammed against the broken piece resulting in the whole engine shearing its mounts and rotating until the carburetor jammed on the mounting structure. The propeller shaft then sheared and the propeller went forward into the test house wall where it spun on its nose cone until it dropped to the floor and then circled around the inside walls, floor and ceiling until it lost its momentum. Viewing this from the operator's window was quite dramatic. All of a sudden the engine was upside down without a propeller and with fuel and oil squirting all around from the broken lines ... we could just see the propeller spinning against the wall and then drop down and gyrate around. Fortunately there was no fire.

In another accident the reduction gear failed and the nose housing broke, sending the magneto, a sizable chunk, through the outer of the two windows to the operating room, landing on the space between. Fortunately, it came no further since we had no time to duck out of the way.

I [once] calculate[d] the altitude performance of an R-2800 engine with a couple of exhaust gas turbines directly connected to the engine. Using data from exhaust gas turbosuperchargers and a calculator (we did not have computers then), I produced a report about one and a half inches thick. It indicated an attractive increase in performance for an engine with turbines mounted on either side of the rear case much as engine-driven superchargers were. A design project was started for such an engine, the R-2800E, that I was overseeing, but the end of the war terminated this program, and shortly thereafter I was transferred to handle the analytical part of a newly formed Burner Group to do component development work for our budding gas turbines.

R-2800C Program

We met Perry Pratt at end of Chapter 3. Here he is in his own words recalling some of his experiences with his R-2800C Program [5]:

> Orv Mohler, the All-American running back at Southern California ... was not a technical man but twice ... in WWII he bailed Pratt &

Whitney—and me in particular—out of real tough situations. The first was on the Martin B-26s. The fuel discharge nozzle diaphragms on the early R-2800s were giving way. To get some jet thrust from the exhaust, the discharge area of the flexible exhaust pipe had been necked down. The hot gases were backing up and accumulating in the nacelle where they ignited the excess fuel.

Major Mohler came over as an Air Corps trouble-shooter to Martin's in Baltimore. Twelve new B-26s were on the line; there had been some bad accidents with those already out. Mohler and I went out on the line ... and had the engines run up. I was certain that it was a fuel discharge nozzle but Orv had to be in a position to demonstrate it to the airframe people. After a minute or two, he turned to one of the Martin guys and asked, "Any idea when it will be really dark?" The Martin engineer looked puzzled, then said, "Maybe in three hours." "Let's all meet here again in three hours," Mohler said.... So at dark we ran the engines up again, saw the exhaust pipe rapidly heat up, and determined that it was hot gases backup from the restricted discharge area.... The fix was simple—putting in a piston as a redundant part of the system.

The C-46s flying the hump were overheating badly and ... one Sunday General Saville called military and industry people together in Washington to discuss the problem. Craigie opened the meeting with a discussion of the problem [then] asked for discussion.... Mohler stepped up, outranked by nearly everyone and said crisply, "This is not an engine problem. It's an installation problem that falls back on the airframe maker. If Curtiss will cut a hole in the nacelle cowling to let the hot air out, the problem is solved." It was done, it worked, and Mohler had scored again very big in my eyes.

CONTROLLABLE-PITCH PROPELLER

Boeing sold 60 of its 247s to United Air Lines as the first modern airliner, equipped with Pratt & Whitney Wasp engines and Hamilton Standard propellers. Eugene Wilson, former President of United Aircraft Corporation, described a situation with United Air Lines, Boeing, Pratt & Whitney, and Hamilton Standard in an early demonstration of the Boeing 247 airplane with a full load at a high-altitude airfield in Cheyenne, Wyoming [6]. When the propeller pitch was set to be optimum at take-off, it was not as efficient at cruise power at altitude. However, when the propeller pitch was set to be optimum at cruise power, the airplane could not take off with full load. Wilson said then the standard blame game started: The airframer started to blame the engine, then the engine manufacturer blamed the airframe, and later they both blamed the propeller. Frank Caldwell of Hamilton Standard then unraveled a drawing of a variable pitch propeller that could be effective at both take-off and cruise power. Caldwell later received the Collier Trophy for this invention.

Wilson recalled another story concerning Caldwell. Prior to WWII, the propeller branch of the de Havilland Aircraft Company, Ltd. had a license to manufacture Hamilton Standard variable-pitch propellers. John Parkes, the propeller manager at de Havilland, was the driving force to get the Hamilton Standard propellers into the Spitfire. Mr. Churchill's tribute to the "few who did so much" in the Battle of Britain also owed a little to Caldwell and Parkes for their performance improvement of the Spitfire.

R-2800-32W's Rough Idle

The Chance Vought F4U-5 was nearly the last of the piston-powered Vought Navy fighters. The F4U-5 had better performance than its popular Navy fighter, the Hellcat, because of its high-altitude capabilities. Pratt & Whitney's carburetor representative to Chance Vought, Bill McGaw, told me the story of R-2800-32Ws and their rough idle. Since more than half a century has gone by since the event, Bill prefers to use X and Y to distinguish the two types of plugs rather than take the chance of incorrectly naming the manufacturers. Bill's story is an example of effective communication between the men on the flightline and senior management.

> The delivery of the F4U-5 was ... held up because of what Vought felt was an unacceptable idle performance of the R-2800-32W. This was a strange situation because this idle problem did not exist in other installations of the engine. One day a Chance Vought crew chief approached me with a suggestion to change the spark plugs to X from the Y plugs mandated by the Navy. After the shift left, I witnessed a F4U-5 engine running with the X plugs the chief installed. The idle was smooth as silk and engine acceleration was perfect. I was forever grateful to that crew chief.
>
> Then I high-tailed it back to Pratt & Whitney for a hurried meeting with Art Smith in which I told him about the successful performance with the X plugs. Art got on the phone right away to talk to his counterpart in the Navy on the F4U-5 program. A short time after Art discussed the problem ... Pratt & Whitney got permission to switch to the X plugs....
>
> The Navy had a huge supply of Y spark plugs and wanted them used in the F4U-5 program. The problem was the Y plugs fouled at low engine speeds, affecting idle performance. The substitution of X plugs, though the Navy accountants perceived a cost increase in the program budget, solved the problem and Chance Vought was back shipping aircraft to the Navy.

R-1830s and the Luftwaffe

Was there a Luftwaffe bomber with a Pratt & Whitney R-1830s and Hamilton Standard propellers? In August 1939, South African Airways

(SAA) ordered two Junkers J-90 civil transport aircraft. W. B. Scott represented United Aircraft Corporation in South Africa before WWII. He tells an interesting story about SAA trying to take delivery of its J-90 aircraft [7]:

> Orders had already been placed on Junkers for the new four-engined JU-90 aircraft.... The aircraft were to be installed with P&W Twin Wasps (R-1830s) and Hamilton Standard propellers. This demand did not go down too well at Dessau. SAA again appealed to Hartford for help to ensure a no-hitch installation.... Pratt & Whitney sent over a crackerjack installation engineer, Rudy Wallace, whom I joined in Dessau.... We were able to mow down any opposition to the installation of this foreign engine.
>
> South African Airways never got the aircraft. About three weeks before the war started on September 3, 1939, the first JU-90 aircraft was ready and test flown to my satisfaction and acceptance for delivery. We arrived at the airport all set for departure to South Africa, two days later only to find both wings off and on trestles alongside....
>
> While in the Middle East during the war we were entertaining a couple of RAF Intelligence blokes. During the course of conversation, these chaps heard that I had been in Dessau working on the JU-90 aircraft being built for South African Airways. It appears that during the Blitz, a four-engined German bomber was brought down just outside London. As is usual, the RAF intelligence examined the wreckage and to their surprise found the bomber had Pratt &Whitney engines and Hamilton Standard propellers.

R-1830 Bearing Problem

Pratt & Whitney faced a "technically and financially desperate" situation in 1938 when the U.S. military refused R-1830C engines due to a master-rod bearing problem. Bob Baer related how the company remedied the situation [3]:

> For many years the master rod bearings, which rode on the crankpin, were made of silver which was bonded onto the steel shell clamped into the master rod. This had been an extremely successful bearing.... However, as engine speeds increased the bearing reached its limit. The take-off speed of the R-1830C series had gone up to 2700 rpm. At this point the bearing of each of the two banks generate about 800 or more horsepower. The bearing was perhaps 3 in. in diameter and a bit over 3 in. long. Furthermore in military operations the rpm increased to around 3000 rpm in a dive. The bearings started to fail in military aircraft with disastrous regularity. The military finally refused to accept any more engines....
>
> One afternoon Chuck McKinnie ... said that Buz Ryder [Earl Ryder, one of the original engineers and chief idea man for problems like this] thought that coating the bearings with lead would improve their load-carrying

ability. "How much lead?" I asked. "Buz didn't say," answered Chuck. "How about 0.005 in. of lead?" I asked. "Try that," said Chuck.

I was given unlimited overtime in the plating shop to plate lead on two bearings for old [engine number] 193. Johnny Chapelle, the night man in the shop, and I ... must have plated a dozen bearings because plating will adhere heavier on the surface lowest in the bath. Consequently our initial attempts produced out-of-round bearings.

We finally developed a method of plating a few minutes with the bearing in one position and then rotating a few degrees in the bath and plating a few more minutes, then rotating a bit more. We finally got two bearings, which were by micrometer measurement evenly plated on the entire surface. It was 7 a.m. I took the bearings to the Experimental Shop for installation in 193. The engine was then mounted on a dynamometer.

Dive tests required that the engine be driven up in rpm by the dynamometer to something like 3200 rpm with the throttle almost closed. This produced centrifugal forces, which placed a greater load on the bearing than with full power. We had to drive the engine up to "dive speed" 25 times for 30 seconds each dive. This simulated a fighter aircraft diving with the propeller in flat pitch, which would drive up the engine rpm. After 25 dives the engine was still running. No bearing failure.

I phoned Chuck.... He told me to repeat the test but turn off the oil supply when the dive speed was reached. This would simulate a zero or negative g situation in a diving aircraft. After 25 dives, still no failure despite no oil pressure.

"What now?" "Increase your dive rpm to 3300 and repeat 25 dives with no oil pressure. Then increase rpm by 100 and do 25 dives. Keep increasing until failure." He jokingly added, "And let me know when you get to 4000 rpm."

Away went old 193. My dynamometer operator, Phelps Lane, and I got increasingly nervous as the rpm increased. We were happy that the bearings didn't fail [but] worried that the speeds we were reaching might result in catastrophic failure of the engine, the engine mounts, or the dynamometer. To the best of our knowledge no engine or dynamometer had ever been run that fast. A day or so later reached the 4000 rpm speed, did 25 dives but had no failure.

I called Chuck and reported our success.... He ... soon phoned back.... "Luke Hobbs wants to come down and see a run." Down came Luke and we started 25 more dives without bearing failure but on one of the last dives there was a mighty clatter from the test cell. After we shut down and went into the cell we found pieces of intake and exhaust valves all over the floor. Disassembly showed that about half of the 28 valves had failed due to the speed—but the bearings were in perfect shape. With jubilation and sighs of relief Pratt & Whitney went back into production with lead-plated master-rod bearings.

But our troubles weren't over. Soon reports began to come in that the military were failing leaded bearings. Inspection showed that the lead was washing off the bearing surfaces.... In service the lubricating oil became acid and the military in the field had to periodically test the acidity of the oil and change oil if necessary....

The chemical geniuses in the ivory tower determined that to test the oil properly the sample had to be diluted with alcohol. Not any old alcohol. Only grain alcohol would work. The supply of grain alcohol was kept under lock and key by Ken Bird in Production Test[ing]. I would go over to Production and sign out a quart of alcohol for experimental purposes. You guessed it. As we got near payday and we were down to our last dollar we had nothing to drink at home so I'd bring the remains of a bottle home, dilute it to 90 proof, add flavoring which we purchased at the drugstore and presto—gin or bourbon, which do you want? Then I'd go back to Ken Bird and sign out another quart for experimental purposes.

However helpful to our drinking and financial problems this situation was untenable technically.... Buz Ryder came up next with the idea of flashing the lead bearing with indium, an element which would inhibit the effects of oil acidity. This worked and alcohol testing became unnecessary. To my knowledge piston engine aircraft bearings are still lead plated and indium flashed. When the war came, Pratt & Whitney gave the process to all engine manufacturers so it became universal practice.

R-1535 AND HOWARD HUGHES

I got to know Edmund L. "Skip" Eveleth at age 93 while he was advising a group of engineers dedicated to making a duplicate of the Hughes H-1 airplane that set world speed records in the 1930s. Skip called me one day when I was at the Pratt & Whitney Archives. He needed performance curves for the R-1535, which I supplied. Skip was thrilled to be talking to his clients working on the R-1535, which had been refurbished for the airplane. Skip was a real character with a wealth of aviation stories. Jesse Hendershot and I visited him in Ozark, Alabama to conduct a video interview with him. What follows is a story he told me about Howard Hughes visiting Pratt & Whitney's General Manager, Mr. H. Mansfield Horner in the 1930s [8].

Hughes knew that Pratt & Whitney engines had greater power capability than indicated in performance guarantees. He wanted to know what the real power output was for the R-1535. Horner had the actual performance curves from Skip when he was meeting with Hughes but it was not company policy to hand out those curves. So, Horner told Hughes that he was going to step out of the room, but "if a breeze should blow the performance curves to the floor and Hughes picked them up and put them in his briefcase," Horner would publicly deny he gave them to Hughes. This was a Shakespearean ending where all was well. Hughes set a speed record at 352 mph in 1935 and two years later set a transcontinental speed record of 327 mph.

Boeing 247's Wasp Engine Whine

Furlonge H. Flynn, better know in the industry as "Tiny" Flynn, was Pratt & Whitney's man in the 1930s racing scene. Tiny wrote down his reminiscence of a worrisome whine heard in 1933 [9]:

> I was ... based in Seattle. I was making a trip to San Diego, but just before my departure, Wright Parkins arrived at Boeing to be present during the Department of Commerce aircraft certification flight of the Boeing 247.
>
> Upon my return to Seattle ... around 8:00 o'clock in the evening [I] was quite surprised to see Wright Parkins waiting for me ... they had run up the engines in the Boeing 247 installation for the first time and both engines had a peculiar high-pitch whine, particularly at the lower engine speeds.... Parkins was quite ... worried about this condition and asked me to go to the Boeing flightline with him and listen to the engines.... During the run-up of the engines, the high-pitch whine was very similar to a whine that I had heard on a previous occasion where I was trying to determine the cause for low oil pressure in another engine....
>
> By cobbling up an oil system and driving the blower and rear section by means of a belt from a lathe, we could simulate normal cruising speeds of an engine and for the first time, I experienced the very high-pitch sounds caused by gear contact which are normally ... drowned out by the lower and heavier sounds of propeller and engine exhaust noises. I drew the attention of Parkins to this similarity and although he believed that this could be a possible cause ... he made arrangements with an inspector of the Boeing Company to bring in a banjo the next morning. That evening, he sent a telegram to Luke Hobbs, asking for the number of teeth on each gear in the blower gear train.
>
> The following morning, with banjo in hand, the Boeing people started up one engine, running it at an rpm at which the noise was most noticeable. While the engine was running, [Parkins] then tuned one string on the banjo to the same pitch of the whine. After being satisfied that his tuning was correct, we then proceeded to the University of Washington and procured the use of the Physics Department's tuning forks. These were set up on a long table. I then took a small rubber mallet and continued to strike the various tuning forks while Parkins picked the string he had tuned until the sounds of the tuning fork and the banjo string were the same.
>
> From that point forward, Parkins had the frequency causing the same whining tone and, with the known rpm of the engine and the gear tooth information (which Luke Hobbs had wired him in the meantime), he was able to definitely confirm that the sounds came from the impeller gear trains.

Andrew Willgoos, responded to Parkins' call for help and contacted him at Boeing with these comments: "If the noise which you are hearing is a vibratory one, produced by the impeller, it has probably always been present in 10:1

engines and perhaps has not been noticed because of the fact that the present ship and exhaust manifold may be quieter, or perhaps the shape of the carburetor intake is sufficiently different so that this resonance is magnified. I will be glad to have your slant on this after you have had a chance to check into it."

Flynn completed the story, "... within a few hours of running, the tooth form smoothed up and the whine disappeared and was no longer a cause for concern."

PRATT & WHITNEY ENGINES WITH FUEL INJECTION

Pratt & Whitney experimented with direct fuel injection engines. The theory was that direct injection gives a more uniform fuel–air mixture in each cylinder and therefore an engine with direct fuel injection would have better performance. Tests did not bear out that assumption and there was little improvement. However, there is another factor to consider, the cooling of the mixture as the fuel evaporates in the impeller section. At a fuel–air ratio of 0.08, the evaporation of fuel could lower the mixture temperature by about 40°, assuming that it takes 130 BTUs to evaporate a pound of fuel. This cooler mixture is farther away from the detonation temperature limit and could permit in the best of circumstances about a 10 percent further increase in power.* The following engines were tested with their fuel-injection systems: R-1340, R-1535, R-1830, R-2800, and R-4360 (Fig. 5). Wasps were tested in airline operation. Direct cylinder fuel injection offered no benefit over the company's standard system of injecting fuel into the slinger ring in the supercharger where the evaporative cooling provided a little extra margin from detonation.

WATER INJECTION

When a pound of water evaporates, it extracts almost a thousand BTUs from the air, resulting in cooler air. The evaporation of water can be used to cool the fuel–air mixture and then one can use a good portion of that sudden increase in margin from detonation to pull more power out of each cylinder. Pratt & Whitney experimented with water injection as early as 1938 when Art Smith and Don Hersey applied for a patent on a water-injection device [10].

During the summer of 1942, the company's sales manager, T. E. Tillinghast (Tillie), discussed the effect of improved fuels on engine output with Opie Chenoweth of the Army's Air Technical Service Command [11]. Chenoweth suggested that water injection would be a way of pulling more power out of the engines. Tillie responded that he would see if Pratt & Whitney's engineering staff could provide a device for the R-2800 for a power surge capability.

---

* This percentage was obtained by eyeballing the effect of mixture temperature on power from Fig. 28 in *Aviation Fuels and Their Effects on Engine Performance*, NAVAER-06-501, USAF T.O. No. 06-504, Ethyl Corporation, Richmond, VA, 1951.

Fig. 5 R-1340 Wasp with fuel injection in a 1931 United Air Lines aircraft (courtesy of Pratt & Whitney Archives).

General Oliver Echols, head of Materiel Command, asked Republic and Pratt & Whitney to provide emergency power capability for the P-47. This is the same Oliver Nichols who was against Pratt & Whitney dropping the H-2600 liquid-cooled engine program in 1940. There was an urgent need for the P-47 to pull away and climb faster than the Focke-Wulf and the Me-109 in a combat situation.

Wright Parkins put his team together right away. Sam Fox and R. C. Palmer developed a water regulator, which would be mounted on the rear of the R-2800. In the final solution, all the pilot had to do was to press a button on top of the control stick with his thumb to get a sudden surge of power. Once the pilot pressed that button, the automatic water-injection control performed the following actions: It turned on the water; it leaned out the carburetor fuel–air mixture because water was being used now instead of fuel to cool the mixture; and it increased the pressure of air from the supercharger, which is normally held back to prevent detonation.

Remember Dana Waring's account of seeing a pale Jim Wheeler tell him that, while running a water-injection test on an R-2800, he suddenly realized that the 2400-hp engine was putting out 3400 hp? Mechanically the engine could put out such power for a short time, but with normal supercharger exhaust temperatures the throttle was limited to the maximum detonation-free power, about 2400 hp. By cooling the air entering the cylinders by water evaporation, the throttle could be advanced until the engine failed mechanically. The guaranteed ratings were conservative enough so that much higher power was available for a short time.

In November 1942, Bill Closs and W. D. Carlson ran calibrations tests on an R-2800 to determine water system setting requirements. Carl Bristol, with Stu Conley and Chuck Roelke, field engineers stationed at Republic, helped to install the water-injection device on an engine at Republic in early January 1943. Flight testing started there on January 7. Pratt & Whitney was confident of the results because it had already installed a water-injection system on its R-1830 in a company flight test with a Grumman F4F Wildcat flown by company test pilot, G. A. McKusick. Water-injection kits were moving into mass production. Installations overseas started around July 1943.

The benefit of water injection was noted by Luftwaffe pilot Heinz Knoke in his diary on April 28, 1944 [12]:

> Brand new aircraft arrive straight from the factory. They are equipped with supercharged engines and the new methane device. The latter is something which I myself tested. It makes it possible for us to obtain from the engine a power boost of as much as 40 percent for several minutes in case of emergency. This power boost is obtained by the injection into the cylinders of a mixture of methyl alcohol and water.

George Slye described the P-47s with water-injection engines [10]:

> [A] Thunderbolt pilot who, finding himself the target of an overwhelming number of diving Focke-Wulfs, flicked on his water-injection control, pulled his fighter into an incredibly steep climb and maneuvered his P-47 into a position above and behind the enemy, from which he knocked down three of his erstwhile attackers before returning to his British base.
>
> Soon similar jubilant reports came in from the Pacific theater … where Navy Corsairs and Hellcats, as well as Army Thunderbolts, were powered with Pratt & Whitney "Twenty-Eight Hundreds" equipped with the magic of water injection. Whether or not the enemy figured you had 2000 hp under the cowling when you could actually pull 2800, 3000, or even 3200 hp for periods up to fifteen minutes, you had him either out-guessed or out-performed—or both.

LIQUID-COOLED ENGINES—SIDE TRIP TO NOWHERE

There was no question in the minds of Rentschler and Mead in 1925 that the future of commercial and military aviation was in dependable air-cooled radial engines. At that moment very few engineers had the wide spectrum of experience in both air-cooled and water-cooled engines as had George Mead. In the first six years of Pratt & Whitney, seven models of air-cooled engines

Fig. 6 The R-2060 Yellow Jacket liquid-cooled in-line engine (courtesy of Pratt & Whitney).

had been tested: single-row R-1430, R-1690, R-1860, R-985, and double-row R-2270, R-1830, and R-1535.

R-2060 YELLOW JACKET

The question came up about a light-weight liquid-cooled engine. The Yellow Jacket R-2060 was the result, and the engine was tested in 1932. Instead of a cast block, Mead selected a common crankcase with five banks of four cylinders. From the front, a quick look at the engine reminded one of a radial engine. Its displacement was 2060 cubic inches (Fig. 6) and the target power output was about 1000 hp. After about 35 hours of testing the project was cancelled. The final run produced 1116 hp. It was clear that the air-cooled, twin-row R-1830 was the better approach to an engine in the 1000- to 1500-hp range.

H-3130 SLEEVE VALVE ENGINE

In the late 1930s, the Navy sponsored the development of a 2300-hp air-cooled engine. It was surprising that George Mead convinced the Navy that a liquid-cooled engine with sleeve valves (instead of the more conventional poppet valves) could meet the Navy performance objectives. A sleeve is a concentric cylinder with holes between the piston and the cylinder. The sleeve rotates to expose the holes for exhaust and intake functions.

Mead became fascinated with sleeve-valve engines in 1937. The feeling was that poppet valves would always limit the time between overhauls for the conventional air-cooled engine. The British engineers, principally Sir Roy Fedden of Bristol, were enthusiastic about the potential benefits of sleeve valves. Mead, who had been in contact with these engineers, directed the design and testing of a new liquid-cooled engine, the H-3130. This engine was first tested in 1938 as experimental engine X-86. It was of the

H configuration—two banks of six cylinders on top and two banks of six cylinders on the bottom. The engine was like two opposed 12-cylinder engines in the same crankcase. The engine was quite thin with the frontal projection like a small door, tall but narrow. The thinness was expected to be an advantage for a buried installation in a wing. The engine was first built with 3130 cubic inches displacement, but then modified to accept larger cylinders to boost the displacement to 3730 cubic inches. This larger H-engine was tested in 1940 as experimental engine X-97.

H-2600 Sleeve Valve Engine

In the late 1930s, the Army Air Corps was also convinced that the future was liquid-cooled engines. A huge program was sponsored by the Air Corps for the development of large liquid-cooled engines, the Hyper Engine program. The company's entry was the H-2600 (Fig. 7).

H-3730 Sleeve Valve Engine

This engine (Fig. 8) was built from the H-3130 experimental engine with larger cylinders to obtain a displacement of 3730 cubic inches. The engine had a two-stage supercharger and an aftercooler to lower the supercharger discharge air temperature before going into the cylinders.

Comparison of Liquid-Cooled Engines to Air-Cooled Engines

The Wasp (R-1340), the Hornet (R-1690), and the Wasp Jr. (R-985) were adequate to meet the propulsion needs of the 1920s. In the early 1930s more engine power was required for the 1000–1500-hp market. Pratt & Whitney

Fig. 7 The H-2600, experimental X-1800 on the left and H-3130/H-3730 on the right (courtesy of Pratt & Whitney).

Fig. 8 The experimental H-3130/H-3730 (courtesy of Pratt & Whitney).

went to twin rows, first with the experimental R-2270, which paved the way for the production Twin Row Wasp (R-1830). The liquid-cooled Yellow Jacket, R-2060, was another approach but was abandoned.

The next market segment to enter was the 2000–3000-hp range. The Navy was interested in the lower part of this range. This time period was before the Navy or the Army Air Corps could imagine the potential of the R-2800 that, of course, filled the lower part of this range effectively.

The largest power requirements in the 3000+ hp could be satisfied by either the liquid-cooled H-3730 or the air-cooled R-4360. Hobbs was so confident of the air-cooled approach that he put all his eggs, so to speak, in the 4360 basket and managed to get Rentschler to convince the Army Air Corps and the Navy to cancel the liquid-cooled engine programs in order to use the Pratt & Whitney engineering resources more effectively.

COMPARISON OF WWII LIQUID-COOLED ENGINES

Figure 9 compares three Pratt & Whitney engines with the Rolls-Royce Merlin, Daimler-Benz, and Allison liquid-cooled engines. The Pratt & Whitney engines (H-2600 and R-2060) have smaller bores and smaller cylinder displacement than the Rolls-Royce Merlin or the Allison V-1710. Pratt & Whitney's largest liquid-cooled engine was in the ballpark of the excellent Daimler-Benz 601 engine, which was used in the Me-109.

By the end of WWII, the statistics were in on air-cooled vs liquid-cooled engines. In fighter aircraft, the two types of engines were comparable. There were outstanding fighter aircraft that were liquid-cooled, such as the P-51, P-38, Spitfires, Hurricanes, and Me-109s. On the other hand, there were also

THE PISTON ENGINE EXPERIENCE

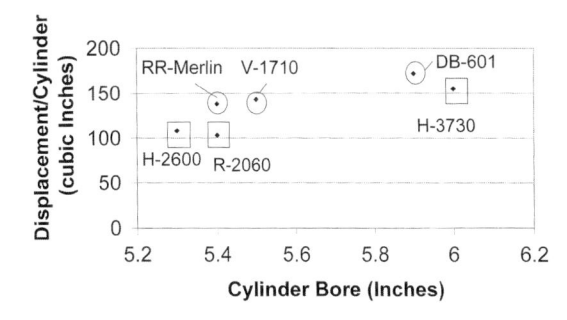

Fig. 9 Cylinder sizes of liquid-cooled engines made by principal WWII manufacturers.

outstanding fighter aircraft whose engines were air-cooled, such as P-47, Corsairs, Hellcats, and Focke-Wulfs. So one might say it was a draw, air-cooled vs liquid-cooled engines in fighter aircraft.

However, when one discusses transport and bomber aircraft, the overwhelming choice is air-cooled engines. So you might say that Rentschler, Mead, and Hobbs were correct in developing air-cooled engines in spite of the critics who held onto the belief in liquid-cooled engines up through the end of the war.

LOOKING BACK ON PISTON ENGINE DEVELOPMENT

The full potential of aircraft piston engines was probably never realized because of the sudden appearance of gas turbines. It was as if the Piston and Propeller entertainment duo was pushed off the stage when a surprise appearance of a glamorous Gas Turbine celebrity jumped into the limelight. The Piston and Propeller duo had more to their act but the audience would never see it because the curtain came down on them a few years after WWII.

W. D. Gove, a Pratt & Whitney development engineer, wrote an interesting summary of piston engine development in 1947 [13]. He looked from 1927, the era of the Wasp and Hornet, to the R-4360 Wasp Major in the 1940s (Fig. 10), discussing the progress in cylinder power capability over about two decades of experience. An engine's power output is proportional to the product of displacement, rotational speed (rpm), and brake mean effective pressure (BMEP). Just increasing the total displacement accounted for over a quarter of the gain. Another increment in power came from running the

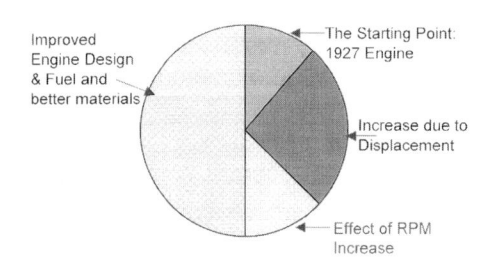

Fig. 10 Engine horsepower grew from 400 hp in the 1927 Wasp to 3500 hp in the early Wasp Major (data from [13], courtesy of Pratt & Whitney Archives).

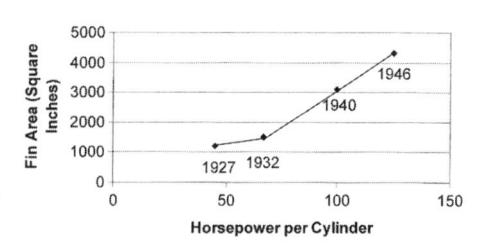

Fig. 11 Progress in cylinder development (data from [13], courtesy of Pratt & Whitney Archives).

engine at a higher rotational speed. The major increase was due to clever design, improvement in materials' strength, and production of better detonation-resistant fuels.

From 1927 on, there was improvement in the ability of a cylinder to transfer heat from the cylinder fins to the atmosphere (Fig. 11). The latest cylinders had more than six fins per inch, cut as slices into the forged head. This more effective cooling method allowed the cylinder output to triple compared to the early Wasp engine. Figure 12 is a schematic of the four-row R-4360, which shows a straight passage for the cooling air from the entrance all the way back to the rear row. When the cooling air is channeled to flow around the cylinders, then each cylinder is more likely to be equally cooled. Note how the baffles direct the cooling air around the opposite sides of the cylinders.

Because the horsepower of an engine depends upon the cylinder inlet pressure, the engine output drops off with altitude. Supercharging was a way to compensate to some degree for this drop-off. Superchargers were centrifugal compressors run at higher rotational speeds than the engine crankshaft. Figure 13 shows the progress made in supercharger design. *Higher* efficiency means not only less power to pump the air up to a certain pressure level but also a *lower* discharge temperature. This lower temperature helps prevent detonation. Pratt & Whitney put the effort into improving the efficiency levels of their own superchargers by paying more attention to the design of flow passages and using special test techniques to minimize flow separation in the passages. Walt Doll described some of this effort earlier in this chapter (Sec. "R-1830 and Superchargers").

During this time period, improvement in fuels also contributed to higher efficiency. In fact, from 1937 to 1947, the improvement in fuels permitted a 50 percent increase in take-off power (rich rating) and about an 18 percent increase in cruise power (lean rating), as shown in Fig. 14.

Fig. 12 Making cooling air effective with baffles ([13], courtesy of Pratt & Whitney Archives).

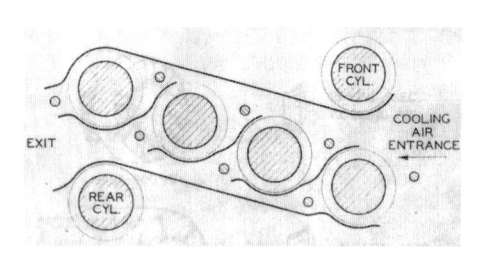

THE PISTON ENGINE EXPERIENCE

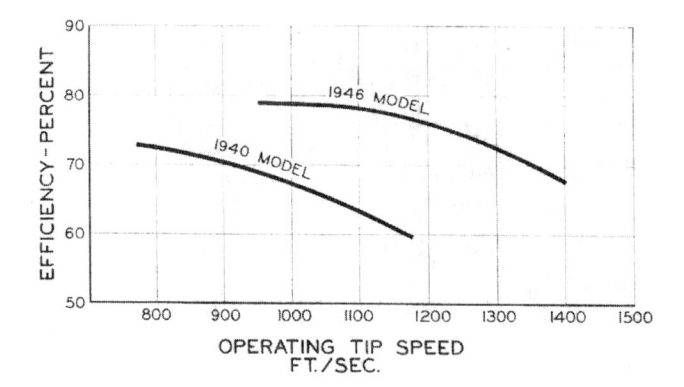

Fig. 13 Improvement in supercharger efficiency ([13], courtesy of Pratt & Whitney Archives).

It is interesting to notice how both liquid-cooled and air-cooled high-powered engines have evolved over time. Originally the argument was that the radial engine had too much frontal area, and did not install as well as the sleek-looking in-line, liquid-cooled engines. Gove pointed out how they both evolved into similar cylindrical forms [13]. Figure 15 shows the distinctly different external shapes of lower-power liquid-cooled vs air-cooled piston engines. High-powered liquid-cooled engines, however, took on the same cylindrical shape as air-cooled, multi-row engines.

I spoke to Gove* in the early 1950s about the potential reliability of turbojet engines. He showed me a graph of data points on time between overhauls

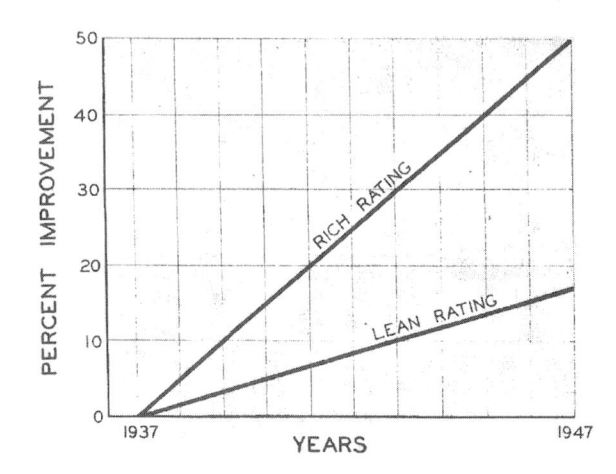

Fig. 14 Improvement in fuels ([13], courtesy of Pratt & Whitney Archives).

---

*Other than senior management, Gove was one of the few men we never called by first name. Watching Mr. Gove turndown a meritless propulsion proposal pitched by an inventor was one of the smoothest business interactions I've ever witnessed.

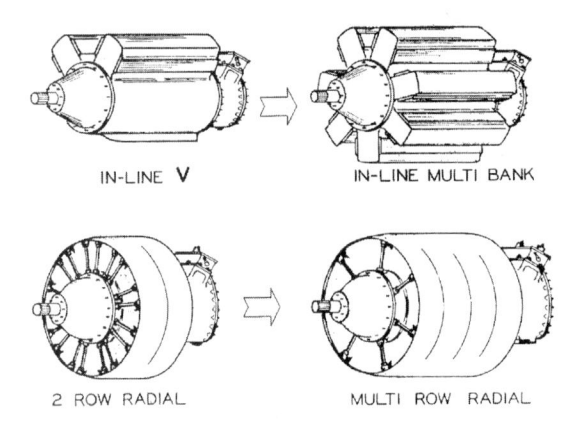

Fig. 15 Piston engines evolved into similar shapes ([13], courtesy of Pratt & Whitney Archives).

(TBOs) for piston engines. From 1936 to 1946, TBOs had doubled, but the best point was up at 1800 hr by the early 1950s (Fig. 16). Jet engine TBOs, at that time period, were considerably less than 1800 hr. The commercial R-2800 later would achieve about 2500 TBOs. We both wondered if gas turbines would ever have the reliability of piston engines. My British friends tell me that the Bristol sleeve-valve engines are in the 3000-hr ballpark for TBOs. However, we know now there is no wondering about the durability of gas turbines. They have greatly exceeded the old piston engine benchmarks.

Figure 17 illustrates why only few engine manufacturers exist worldwide. It takes a huge array of experimental test equipment, a highly trained group of development engineers, and a bulging bankroll. In the 1940s it took four

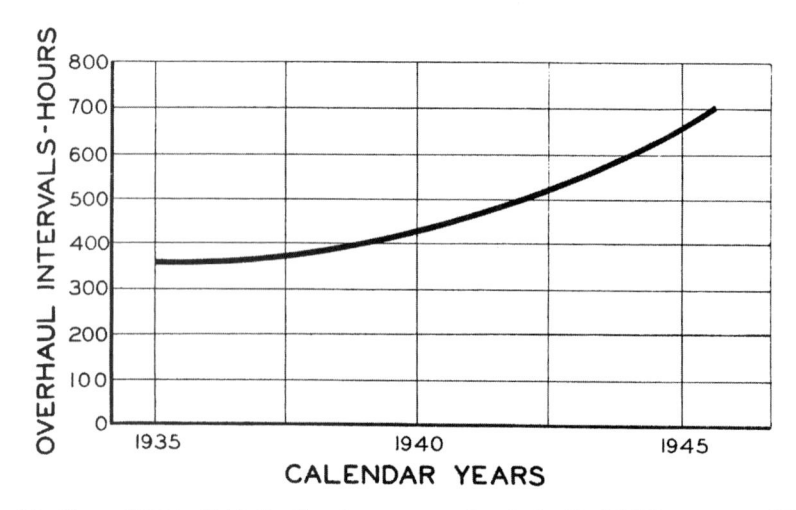

Fig. 16 From 1936 to 1946, the time between overhauls doubled ([13], courtesy of Pratt & Whitney Archives).

The Piston Engine Experience 153

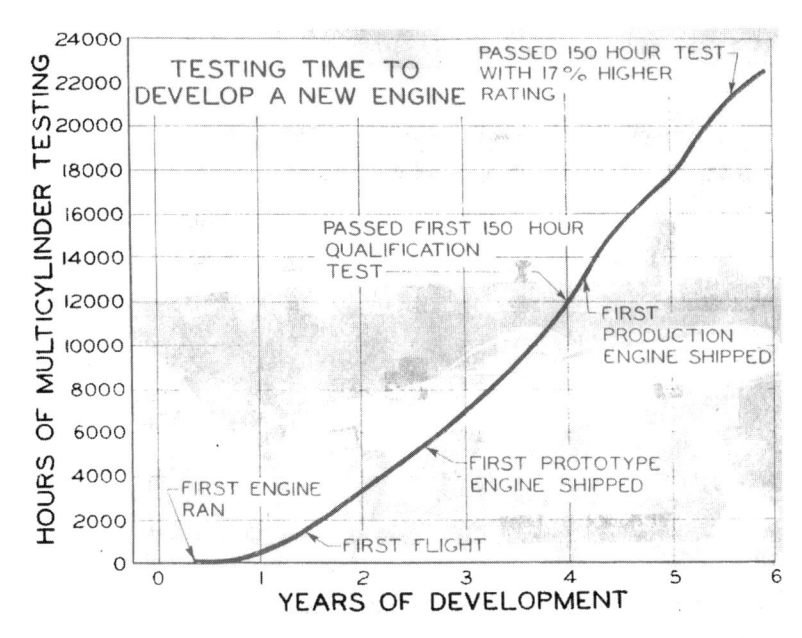

Fig. 17 Test hours required to develop an engine ([13], courtesy of Pratt & Whitney Archives).

years to pass a 150-hr qualification test [13]. It is not the type of business for a quick return on investment at a relatively low risk.

Thermodynamicists must have wondered why engine designers did not use the energy in the exhaust of piston engines to put more power into the crankshaft. Sanford Moss made a significant step in using this energy in the exhaust with his turbosupercharger, which bolstered the density of the engine intake air at high altitudes. Figure 18 shows where the energy goes when a piston

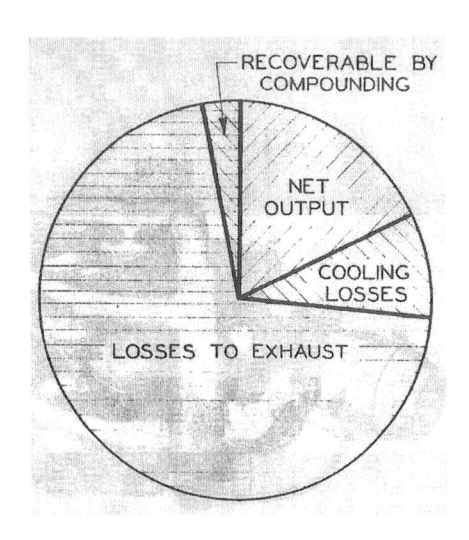

Fig. 18 Energy distribution in a piston engine operating at military power ([13], courtesy of Pratt & Whitney Archives).

engine is operating at military power and in a fuel-rich condition. Most of the energy is thrown out in the exhaust and in the cooling air. The net power output, as a fraction of the energy input, is about a third of the energy input when the engine is operating with a fuel–air ratio for optimum cruising.

Also shown in Fig. 18 is how much energy could be recovered by compounding. *Compounding* is the process of directing the piston engine exhaust to a turbine, whose output is fed back to the crankshaft through a suitable transmission system. Wright Aeronautical converted its R-3350 engine to a compound engine with three power turbines. This engine powered the Douglas DC-7C, which acquired the name "Seven Seas" because of its capability of crossing oceans. Pratt & Whitney conducted studies of compounding its R-2800 and R-4360 after WWII.

R-4360 EXHAUST GAS ENERGY RECOVERY STUDIES

A first step in compounding is to add a power turbine (or turbines) to an existing engine with an internal supercharger. This results in about a 20 percent improvement in power because of the increasing pressure ratio across the power turbine, which feeds power back into the main crankshaft and could provide some useful thrust with a suitable exhaust tailpipe. The altitude power of such engines is not as high as those with turbosuperchargers. Figure 19 is a comparison of power output for a conventional internally supercharged engine with a simple compound engine, also internally supercharged.

Figure 20 is a compound example of an R-4360 with an internal supercharger and an external turbine powered by the engine exhaust. The power

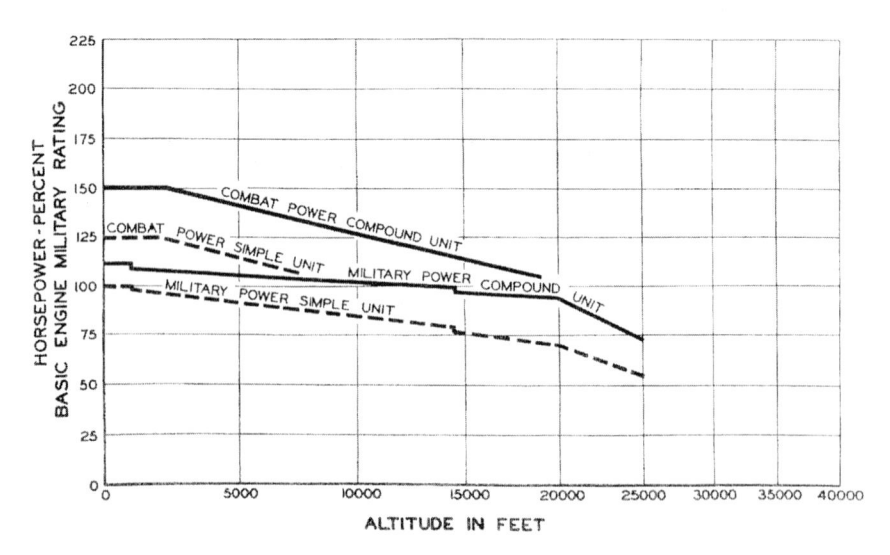

Fig. 19 Conventional vs supercharged performance ([13], courtesy of Pratt & Whitney Archives).

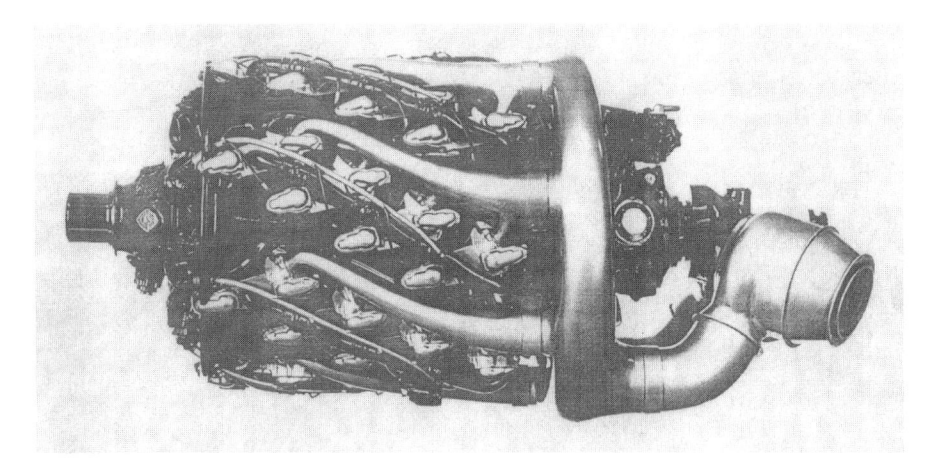

Fig. 20 A compound R-4360 ([13], courtesy of Pratt & Whitney Archives).

from the turbine is directed to the engine crankshaft through suitable transmission gearing. In a turbosupercharged compound engine, the exhaust goes into two feedback turbines, which put power into the engine crankshaft (Fig. 21). Then the flow enters two parallel turbines, which drive compressors that supercharge the inlet air and pump this air through intercoolers before entering the engine.

The ultimate effort in capturing the energy exhaust of the R-4360 is shown in Fig. 22. This compound R-4360 directs the exhaust into the burner of a

Fig. 21 A turbosupercharged compound R-4360 (courtesy of Pratt & Whitney Archives).

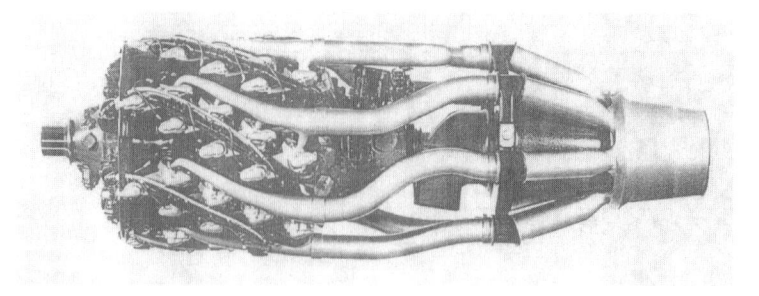

Fig. 22 A compound R-4360 where the exhaust from the turbine produces jet thrust (courtesy of Pratt & Whitney Archives).

turbojet and additional air is compressed separately to join the piston engine exhaust stream in the burner. Then the output of the turbine is directed back into the R-4360 crankshaft, and the exhaust from the turbine produces jet thrust. The supercharging is handled by a three-speed, engine-driven supercharger. The combustion process in the piston engine is not complete at high-power settings. Therefore, one could imagine a compressor, burner, and turbine downstream of the piston engine in the form of a simple turbojet. The fuel rich exhaust, now at about 1400° F, flows into the burner where it mixes with new air from the compressor. These combined flows then power the turbine, whose power provides the compounding to the crankshaft, power to the compressor and thrust power from the tailpipe. This is a complex mechanical arrangement that must have made the gas turbine much more appealing, especially in the 1946 time period when a gas turbine was still described as "a simple rotating element." This promising technical achievement was never put into production because of the diminishing market for reciprocating engines in the gas turbine era.

Figure 23 indicates that compounding is good for the turbosupercharged engine by comparing (at military power) a turbosupercharged engine, a turbosupercharged compound engine, and a compound engine with a turbojet unit. At high altitudes, the reduction in specific fuel consumption is about 20 percent. On the other hand, the combination of the R-4360 with compounding and the turbojet unit was better than the simple compound engine. This is because the pressure ratio across the turbine increases with altitude, providing more power back into the R-4360 crankshaft and providing more jet thrust. However, capturing the maximum energy in the exhaust was a complicated mechanical challenge.

These studies, made during 1945–1947, of the future possibilities of the R-4360 with various forms of compounding offered attractive performance improvements. However, the engineering management wisely shifted priorities towards getting established in the jet engine business with its own design and developments rather than obtaining the last ounce of piston engine

THE PISTON ENGINE EXPERIENCE   157

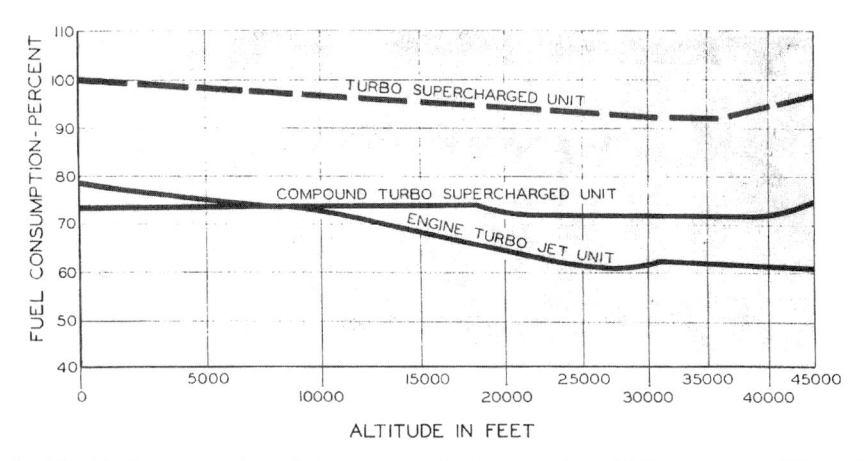

Fig. 23 Fuel consumption of three types of piston engines ([13], courtesy of Pratt & Whitney Archives).

exhaust energy. In contrast to this viewpoint, Wright Aeronautical took exactly the opposite approach. It developed its compound R-3350 engine and preferred to build British jet engines under license.

FINAL OBSERVATIONS ON PISTON ENGINES

In 1942, Sir Roy Fedden visited the United States with a contingent of British engineers to assess the activities of the American aircraft and engine companies in support of the WWII effort. He was a friend of George Mead and had convinced Mead of the merits of sleeve valves over poppet valves. When Fedden returned to England in early 1943, his group put out a report under Fedden's authority with a very tight circulation list. The following are some quotations from the engine part of this extensive report [14].

> During the last 15 years there has been considerable competition between Pratt & Whitney and Wright [Aeronautical], and one firm has been ahead of the other in cycles throughout this period.
>
> It would seem that Pratt & Whitney are now in the ascendant. They have an engineering and design organisation of 1750 people and Mr. Hobbs, their engineering manager, has now got the layout of experimental and development section in the form for which he has been planning for some years. Mr. Willgoos has undoubtedly been one of the master forces on engine design.
>
> It is characteristic of the versatility of outlook of this organisation that the two sleeve valve engines referred to previously, which Sir Roy Fedden saw in 1938, were dropped some two years ago after considerable promising development running, in favor of the "Wasp Major"....

When Sir Roy Fedden was in America in 1938 and again in 1940 he felt that there was a strong tendency towards liquid cooling.... A day's talk with the powerplant sections at Wright Field and the Navy Bureau ... indicated that air-cooling is now re-established, owing to the considerable reduction in cooling drag ... as the result of the good work that has been done finning, baffling, and cowling. There would appear to be the feeling that there are problems in greatly increasing the size of the liquid-cooled engine, some of which difficulties are not so pronounced with air cooling; in other words a reversion to the American viewpoint of some years ago.

It is characteristic of the Americans, however, that they go up two roads rather than one. This line of thought is typified by the development work done by Pratt & Whitney on the two liquid-cooled sleeve valve engines.... These engines were of the vertical "H" type and were about 2400 and 3000 cu. in. capacity respectively.... The firm decided to discontinue their development in favour of the air-cooled Wasp Major. A further example is that the Navy claim to have developed the finest liquid-cooled engine in the world in 2470 cubic inches. Lycoming which has been run at powers in excess of one hp per cubic inch. Although it is understood that the firm is being given an educational order for 100 engines, the Navy regard this type purely as an insurance policy and propose to continue with air-cooled engines. The Mission found that Lycoming was also developing a larger engine on similar lines but of about 3300 cu. in. capacity.

The Mission saw two or three interesting new engines and there is no question that two soundly developed large engines, the 24-cylinder Allison of 2600 hp and the 28-cylinder Pratt & Whitney of over 3000 hp will be available in production within the next 12 months.

No ... developments are going forward on this engine and the only comment is to record the impression ... that this type has built up a reputation second to none in America for reliable service.

This mission to America was mentioned in Bill Gunston's book about Sir Roy Fedden. Before WWII, Fedden had traveled throughout Europe and the United States and visited most of the engine manufacturers. In Munich he was a friend of Bruno Bruckmann, Chief Engineer of BMW. Soon after the war he tried to find Bruckmann and when he did, Gunston relates the outcome as follows. The story starts with Fedden's thoughts about how the war would have gone if the Me-262 had more time [15].

He wondered how the air war would have gone had there been a further few months of fighting. He discussed the matter with Bruckmann as they sat at the roadside eating a snack lunch of K rations. Suddenly Bruckmann turned to Fedden and said, "Did you know that on Christmas Day 1943 I was reading a copy of your Mission to America Report?" Fedden was thunderstruck. Every copy of the highly secret report had been numbered and signed for, yet one had reached Germany within weeks of publication.

Fedden showed his horror, but Bruckmann cut him short: "The horror was on our side. On that Christmas Day Goering himself had the report and an interpreter; the rest of us—most of the industry and Luftwaffe leaders—were told what it said. We knew that, with your name on it, it was exact and beyond dispute. We realised at that meeting we could not win the war, and that within a year or 18 months your air power would overwhelm us. From that time on, our whole outlook was different—though Hitler would have us shot if he had known. So your report played no small part in your winning the war."

What stunned Bruckmann was the overwhelming avalanche of production airplanes and engines from the United States into the Allies' military offensive power. Who made it happen? George Mead's personal efforts with the major aircraft engine companies in organizing the automobile companies to produce aircraft engines had to be a very significant factor in bringing WWII to a successful conclusion.

REFERENCES

[1] "Flight Testing at United Aircraft," Pratt & Whitney Archives; much of the information in this section of this chapter is from this document, probably written by Harvey Lippincott.
[2] Willgoos, A., "Report to Consolidated Vultee Aircraft Company," Willgoos Files, Box 7/98-172, Pratt & Whitney Archives, Nov. 11, 1943.
[3] Baer, R., "Memoirs," Pratt & Whitney Archives.
[4] Burton, Captain D. B., "Letter to Bob Schofield," Fisher Saga file, Box 8/98-278, Pratt & Whitney Archives, 1988.
[5] Fisher, P., "Conversations with Perry Pratt," interview, Pratt & Whitney Archives, 1965.
[6] Wilson, E., *Slipstream*, Whittlesey House, McGraw-Hill, New York, 1950.
[7] Scott, W. B., "Letter to Robert Daniell from W. B. Scott (former member of South African Airways)" Pratt & Whitney Archives, July 23, 1990.
[8] Eveleth, E. L., *Achievers: Memorable Moments and Anecdotes of American Pioneers*, Aviation Publishing, Inc., 2000.
[9] Flynn, F. H., "Memo to Paul Fisher," Pratt & Whitney Archives, Dec. 30, 1954.
[10] Slye, G., "Water Magic," *The Bee-Hive*, United Aircraft Corporation, Pratt & Whitney Aircraft Division, United Technologies Corporation, Jan. 1947.
[11] "Notes on Improved Fuels," Tillinghast file, Box 8/98-226, Pratt & Whitney Archives, 1942.
[12] Knoke, H., *I Flew for the Fuehrer*, Greenhill Books, London, U.K., 1997, p. 152.
[13] Gove, W. D., "Reciprocating Aircraft Engine Development," lecture to Air University, Maxwell Field, AL, Pratt & Whitney Archives, Jan. 1947.
[14] Fedden, R., *The Fedden Mission to America, December 1942–March 1943: Final Report*, Sec. 4, Great Britain Ministry of Aircraft Production, London, U.K., June 1943.
[15] Gunston, B., *By Jupiter, The Life of Sir Roy Fedden*, Royal Aeronautical Society, London, U.K., 1978.

Chapter 5

TRANSITION TO GAS TURBINES

LOOKING BACK

At the Centennial Exhibition held in Philadelphia in 1876, there was a 1400-hp Corliss steam engine on display. It had two high-pressure cylinders of 40-in. in diameter and a stroke of 10 ft. The volume of such a monster was about the size of a standard Cape Cod house with six rooms (four down with two up on the second floor). The horsepower per cubic inch displacement was about one-half percent of that of the R-2800!

The word *turbine* has a Latin root, *turbare*, which means *to stir up*. Professor Claude Burdin (1790–1873) is credited with using the word *turbine* circa 1837 to refer to water wheels that converted hydraulic energy into useful horsepower [1]. One of his students, Benoît Fourneyron, successfully developed a water turbine that produced 60 hp with a wheel about a foot in diameter. The point is that piston engines have practical economic size limits and turbomachinery offers a more economically attractive approach to higher power levels.

Piston engines produced thrust power for aircraft by driving a propeller, which was the mechanism to accelerate the air from the front towards the rear of the aircraft. There were several practical limits for the piston–propeller combination in the production of thrust, including the limit on propeller disk diameter (affecting aircraft landing gear height) and the amount of horsepower a propeller could absorb (blade area and revolutions per minute), and the limit on the power output of an aircraft internal combustion engine because of the complexity of stacking cylinders and transferring power to a propeller shaft.

Here is an interesting exercise in back-of-the-envelope analysis. The Wright Brothers flew a 700-lb aircraft at about 30 mph. I believe the lift/drag ratio was about 7, which indicates a drag or thrust of about 100 lb. The thrust horsepower was 100 lb multiplied by 44 fps (feet per second) (which is 30 mph) and divided by 550 foot pounds per second per horsepower. The answer is 8 hp, which agrees with the Wright engine output.

The transition to gas turbines started in early 1941. The war in Europe had been going on for more than 16 months. Germany controlled most of Europe.

Hitler and Stalin were carving up Europe and amassing the most powerful military machines in history. Mussolini was fantasizing about his new Roman Empire and the Japanese were piling up victories in the Far East. The United States had not yet entered the war but it was clear that the question was not *if* but *when*.

The country was gearing up to be "the arsenal of democracy." In 1939, President Roosevelt asked for 50,000 aircraft to be manufactured each year and the aircraft industry was beginning to respond. George Mead was organizing automotive engine manufacturers to handle engines for aircraft production. The Ford Motor Company was preparing to produce the R-2800. Buick and Chevrolet were getting ready to build the R-1830. Jacobs, Nash-Kelvinator, and Continental were also being drawn into the Pratt & Whitney engine production team. Pratt & Whitney's most powerful production engine at that time was its 2000-hp R-2800. The R-4360 program was in the early stages of development and had not yet run its first experimental engine. Its first flight test would not be until May 1942, about a year away.

When I started work at Pratt & Whitney in 1948, my first assignment was in the Turbine Group working on the PT2 turbine. One of my questions was, "What was the PT1?" The answer then as well as in the later years was, "It was a free-piston compressor with a turbine. It was an abomination!" That negative response is understandable if you have ever seen the test hardware of the PT1 at the Pratt & Whitney Museum. The free-piston compressor is made of cast iron. It looks like something out of *Industrial Power Magazine* and as far from aviation as you could imagine. However, when one focuses on what was state of the art in early 1941 for aircraft engines and what it would take to produce a 4000-hp piston engine, then the idea of a free-piston compressor with a turbine makes more sense.

In 1941, Pratt & Whitney was focused on having its spectrum of piston engines built by automotive and smaller aircraft engine companies for the war effort. United Aircraft Research Department and some U.S. government agencies had evaluated jet propulsion before WWII and concluded (correctly) that turbojet propulsion was of no interest at then current aircraft flight speeds. Not only did Pratt & Whitney seem to have no knowledge of the British and German efforts on jet propulsion, it was also unaware that General Electric and Westinghouse were developing aircraft gas turbines (Fig. 1). Neither did Pratt & Whitney know that Bell was preparing to fly a jet-powered aircraft in late 1942.

BEGINNINGS OF *PT1*

In the late 1930s and early 1940s Pratt & Whitney sponsored small research programs at the Massachusetts Institute of Technology (MIT). Pratt & Whitney was conducting its own supercharger development with Andrew

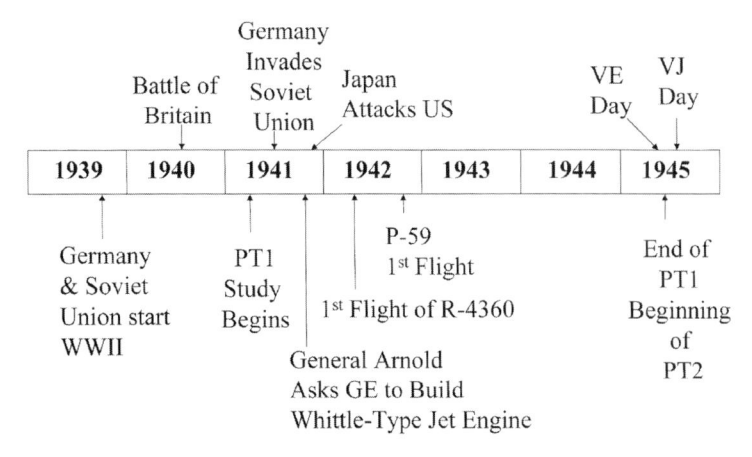

Fig. 1 Time line of engine development during WWII.

```
                    Germany
                    Invades
         Battle of  Soviet    Japan              VE    VJ
         Britain    Union     Attacks US         Day   Day
            ↓         ↓         ↓                 ↓     ↓
    ┌─────┬──────┬──────┬──────┬──────┬──────┬──────┐
    │ 1939│ 1940 │ 1941 │ 1942 │ 1943 │ 1944 │ 1945 │
    └─────┴──────┴──────┴──────┴──────┴──────┴──────┘
       ↑            ↑         ↑                   ↑
    Germany       PT1       P-59              End of
    & Soviet     Study    1st Flight            PT1
    Union start  Begins  1st Flight of R-4360 Beginning
    WWII                                         of
                         General Arnold          PT2
                         Asks GE to Build
                         Whittle-Type Jet Engine
```

Kalitinsky of MIT instead of buying the units from General Electric. Kalitinsky talked to his Pratt & Whitney engineering contact, John Marchant, about a new concept of a propulsion system. Air was compressed in a free-piston compressor driven by a diesel cylinder integral with a double-acting, positive displacement compressor. The exhaust from the compressor entered the diesel cylinder for a power stroke and subsequently into a duct leading to a turbine, which drove a propeller through a reduction gear and provided some jet thrust from the turbine exhaust.

The principle of the free-piston gas turbine sounds complicated. However, if one reviews the Hyper Engine Program (liquid-cooled engines) that the U.S. Army Air Corps was funding, or looks at a cutaway of the air-cooled R-4360, this free-piston gas turbine concept appears more interesting as an approach to 4000+ hp. The following are some of the large Army Air Corps liquid-cooled engines:

1) Lycoming XR-7755: 5000 hp with 36 cylinders

2) Wright Tornado, R-2160: 2365 hp with 42 cylinders with a design study to achieve 3945 hp with 70 cylinders

3) P&W H-3730: about 3100 hp with 24 cylinders

John Marchant discussed Kalitinsky's idea with Val Cronstedt, Executive Engineer at Pratt & Whitney (Cronstedt's report [2] provides much information for this section). Shortly thereafter, Professor Jerome Hunsaker of MIT, contacted Pratt & Whitney and C. H. Chatfield, United Aircraft Executive and MIT alumnus, and arranged a meeting at MIT with Professor Hunsaker and Professor C. Richard Soderberg, who had formerly been Chief Engineer at Westinghouse. Luke Hobbs and Val Cronstedt attended for Pratt & Whitney. The outcome was a contract with MIT to study the concept of the free-piston gas generator plus turbine and complete a report on the project by October

1941. The assumptions for the study were a minimum of 4000 hp at sea-level take-off, a maximum altitude of 40,000 ft, a flight speed at Mach 0.6 to 0.75, and a maximum turbine inlet temperature of 1600° F.

Soderberg and Kalitinsky submitted a report to Pratt & Whitney titled *Free Piston Gas Turbine Power Plant for Aircraft* regarding the engine concept on September 6, 1941 [3]. The engine consisted of eight free-piston gas generators, a burner for emergency power, and a turbine that powered a propeller through a reduction gear and whose turbine exhaust discharged through a variable jet nozzle. The report centered on two specific aircraft applications, a single-engine pursuit aircraft capable of Mach 0.75 at 40,000-ft altitude, and a four-engine bomber capable of Mach 0.6 at 40,000-ft altitude.

SINGLE-ENGINE PURSUIT PLANE

The aircraft used in the evaluation had a gross take-off weight of 20,000 lb, about the operational weight of the P-47D. The speed was 100 mph higher than the P-47. The engine was capable of providing 4500 hp from sea level up to the maximum altitude, 40,000 ft. The thrust was divided between the propeller and the turbine exhaust. At sea level and at low flight speeds, the propeller provided practically all of the thrust, but at the maximum altitude, the propeller provided only 67 percent of the thrust and the turbine exhaust jet the remaining 33 percent of the total thrust. The propulsion system weight, including the propeller, was 1.77 lb/hp. At the Mach 0.75 speed, the propeller efficiency was only 66 percent. Where this engine really looked attractive was in its fuel consumption, estimated to be 0.41 pounds per hour per pound of thrust!

FOUR-ENGINE BOMBER

Four engines in this application developed 18,000 hp at sea level for a 226,000-lb bomber. The flight speed of this bomber was Mach 0.6 at 40,000 ft with a propeller efficiency of 85 percent. At the 40,000-ft altitude, the propeller was providing 75 percent, while the turbine exhaust jet provided the remaining 25 percent. The outstanding advantage of this type of powerplant was its very low fuel consumption, about 0.36 pounds of fuel per hour per pound of thrust (about a 30 percent reduction over conventional piston engines).

FREE-PISTON GAS GENERATOR

Figure 2 is a simplified sketch of a free-piston compressor, which consists of a double-acting reciprocating compressor piston powered by an integrally mounted diesel piston. Note the two diesel pistons in the center of the figure—identical cruciform-shaped objects. The vertical part of the shape

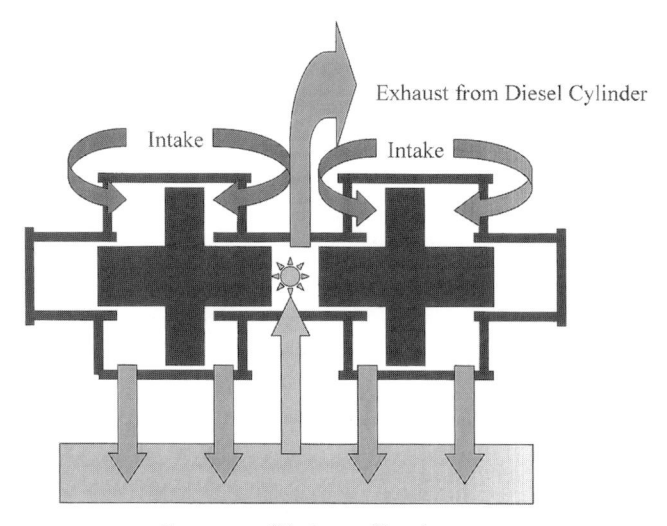

Fig. 2 Free-piston compressor.

(the thinner part of the object) represents the double-acting compressor piston, which moves left and right. The small volume to the right of the left piston is the position when the annular compressor piston is in its extreme right position. As it moved to this position, it compressed the air in front of it. The cruciform structure on the left moved to the right, and at the same time, the structure on the right moved to the left. In both cases, the moving annular pistons compressed the air in front of them. In the center of the figure is what looks like a star with projections around its periphery. This shows the diesel cylinder where the fuel is injected.

We have here one diesel cylinder with one set of intake ports and one set of scavenging ports for the two diesel pistons. The two diesel pistons in the center of Fig. 2 are part of the same reciprocating structure. The opposite ends of the diesel pistons have pistons that push against trapped air to cushion the power stroke and act as a pneumatic spring to push the diesel pistons back toward the center of the figure. We are looking at the moment where the diesel pistons have just been driven to the center by the energy stored in the air cushion chambers on each end of the compressor.

Imagine that the fuel is injected and burning takes place. Then the pressure and temperature build up as a result of the heat release, and the two cruciform structures start to move away from the center. The air that had just been compressed before fuel injection exhausts into a chamber below. The fuel has been injected and the diesel cylinder causes the two diesel pistons to separate. The air in front of the separating compressor pistons is now being compressed. Also, the little chamber at the extreme left and extreme right of Fig. 2 is loaded with a permanent charge of air to act as a cushion when the cruciform structures are violently forced outward from the center. In other words, these two air chambers act as powerful springs to store energy to push the structures back toward the center to start another cycle.

By now it should be obvious that we are talking about a two-stroke diesel cycle. The air cushion chambers push the floating structures toward the middle in the intake and compression part of the cycle. The diesel free-floating pistons (left and right) push the floating structures away from the center in the fuel injection and exhaust part of the cycle. These cycles occur at a high frequency of about 2800 cycles per minute. The high-pressure air goes into a storage chamber that supplies high-pressure air to the diesel cylinder. In other words, the diesel is supercharged. The work output of the diesel piston–cylinder combination is just enough to compress the air. The exhaust from the diesel cylinder then goes to a gas turbine, which powers a propeller through a reduction gear. The normal diesel exhaust temperature in this engine is about 1200° F. It is possible to get a sudden burst of power from the turbine by increasing this gas temperature in a burner to about 1600° F for a brief time.

WHY A FREE-PISTON GAS TURBINE?

Here are some of the attractive features of the free-piston gas turbine compared to 1940-era piston engines:

1) Improved fuel economy compared to other aircraft engines.
2) Reduced weight for high-altitude applications.
3) Reduced cooling requirements and absence of intercoolers.
4) Flexibility in installation.
5) Reduced propeller size for high-altitude applications (some of the thrust is provided in the turbine exhaust jet).
6) Absence of unbalanced inertial forces in the compressors and the absence of torque fluctuations on the propeller (less fatigue stress on the aircraft structure).
7) Use of a safer fuel than aviation gasoline.

Pratt & Whitney's first engine program in the gas turbine era was the PT1. A new numbering and engine designation system had to be introduced. In the gas turbine era, the letter P meant propeller. The letter T stood for turbine. So, PT1 meant Prop Turbine number 1.

EXPERIENCE WITH THIS CONCEPT

Figure 3 illustrates the concept outlined in Sodenberg's report [3]. This is the engine arrangement used in railroad and naval applications. The internal combustion engine is supercharged by the compressed air, and the exhaust

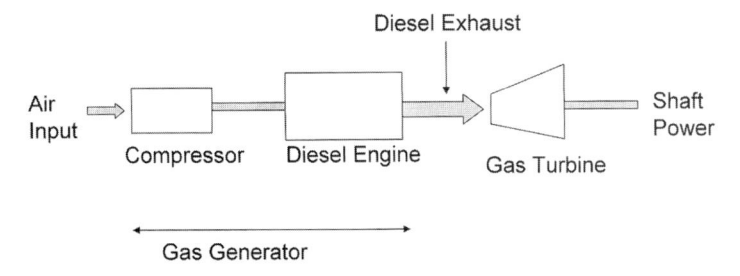

Fig. 3 Soderberg's concept for a supercharged diesel and turbine.

gas from the engine powers the turbine, which provides power for the application. Soderberg showed that the trend in locomotive and marine power fields was towards gas turbines. He mentioned a Navy study that said gas turbines were preferred over steam because of safety reasons. The Navy claimed that battle damage to the steam systems caused more casualties from escaping hot steam than from direct hits on the crew.

Engineers grouped "generators" to feed gas turbines, which provided the shaft power. This could build a variety of powerplant power sizes. The "generator" consisted of a two-cycle diesel engine whose output was completely consumed in compressing air, which was then used to scavenge the diesel cylinders. The exhaust was the input to the gas turbine. The power turbine would be sized to the required horsepower, and then as many generators as were needed to provide the gas flow would be assembled from standard size generators. This was a way of minimizing development costs.

PT1 Program

The PT1 program was officially launched with an engineering order dated October 27, 1941, as a company-funded program. The Navy was aware of the program, which was small enough not to interfere with Pratt & Whitney's primary contributions to the war effort. The company rented space in the Williams Brothers Silver Company in Glastonbury, Connecticut. Kalitinsky was part of the staff and Soderberg participated on a part-time basis. Howard Sprenkle was the project engineer, and Ralph Hooker, from the Supercharger Group, participated in the design of the 4500-hp engine. Val Cronstedt was listed as Executive Engineer in the project. In his history of the PT1, Cronstedt gave credit to Sam Heron, who helped with high-temperature materials advice [2].

The free-piston compressor concept had a certain appeal because pistons and cylinders were Pratt & Whitney's expertise, especially as the PT1 diesel cylinder bore (5.5 in.) was the same as the R-1830. Soon reality set in when the design got underway. The team decided to build the unit out of heavy cast iron to avoid any mechanical failure problems while wrestling with the aerothermodynamic challenges in getting two mirror image diesel-driven, double-acting compressor pistons to oscillate at high frequency. The first testing began about ten months later, in August 1942. At this time the United States was at war and the PT1 program was on the bottom of the priority list. The PT1 program suffered from huge technical challenges and a slow turnaround time for test hardware.

It was not until March 1943 that the PT1 team got both sides of the free-piston compressor working. It was encouraging to the engineers when they learned that Junkers had been building these compressors for operation in submarines. They finally got the free piston compressor to work satisfactorily

Fig. 4 a) The free-piston compressor, which was meant to solve compressor operating problems before worrying about flight weight hardware, and b) exploded view (courtesy of Pratt & Whitney Archives).

Fig. 5 PT1 turbine under test: the hot gas inlet (against wall) comes into the turbine case perpendicular to the shaft (courtesy of Pratt & Whitney Archives).

and then ran into piston scuffing problems due to friction between the piston and the cylinder wall. In 1942, the turbine was designed. Soderberg pitched in during the summer of 1942 to design the gas path hardware for the turbine. He left his legacy at Pratt & Whitney in the form of a design system that would later be used in the PT2 (T34) and JT3 (J57) that distinguished Pratt & Whintey's turbines from the competition. He believed in shrouded turbines because of their higher efficiency.

Pratt & Whitney started to get into engine performance analytical studies [2] because the PT1 engine was more amenable to analysis than were radial piston engines. The learning process on high-strength, high-temperature materials included contacts with the National Advisory Committee for Aeronautics High-Temperature Materials Subcommittee, the Haynes Stellite Division of Union Carbide, and the Battelle Memorial Institute. Pratt & Whitney started a program on investment casting with the Metallurgical Research Laboratories.

Experimental work began on two types of burners, one between the gas generator and the turbine as well as another as an afterburner. The burner just upstream of the turbine was first run in June 1943. Again, Pratt & Whitney was getting direct experimental experience on the elements of a turbojet engine. The organization was growing, with 74 personnel in 1943. More than 400 test hr were accumulated on the free-piston compressor (Fig. 4), and over 280 test hr were accumulated on the turbine (Fig. 5). This was a great accomplishment for a program at the bottom of the priority list.

In addition to the experimental program, Pratt & Whitney was studying various applications for the PT1. The engine had the greatest advantage over

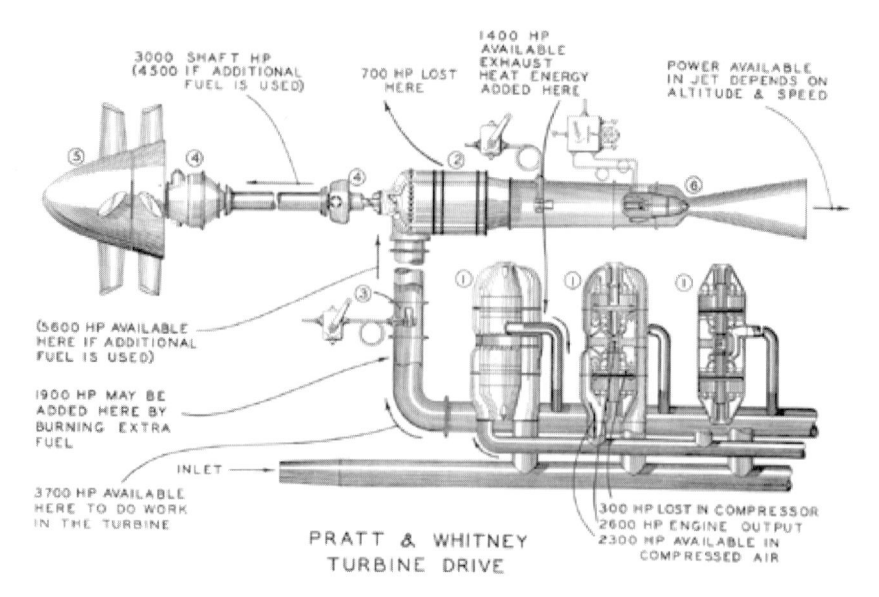

Fig. 6 Schematic of the complete powerplant for the PT1 (courtesy of Pratt & Whitney Archives).

competing engines in the bomber application. Figure 6 is a schematic of the complete PT1 powerplant. Air enters the inlet and is directed to the eight free piston air compressors [Fig. 6 (1)]. The exhaust from the eight units (essentially a supercharged diesel) is piped to the power turbine [Fig. 6 (2)]. The engine thrust is made up of the propeller thrust and the jet exhaust. It is possible to obtain a sudden boost in thrust by using the auxiliary burner [Fig. 6 (3)] just upstream of the turbine. The turbine shaft power moves through two reduction gears [Fig. 6 (4)], and then to contra-rotating propellers [Fig. 6 (5)]. The turbine pressure ratio and the jet velocity are controlled by the variable nozzle area [Fig. 6 (6)].

COMPARISON WITH PISTON ENGINES

The United Aircraft Research Department compared the use of a scaled PT1 type of engine with conventional piston engines in the B-29. Table 1 shows the improvement in aircraft performance expected with the PT1 configuration.

TABLE 1   RANGE WITH BOMB LOAD

	Bomb load	2000-mile mission	2900-mile mission
Piston engine	5250 miles	20,000 lb	7600 lb
Scaled PT1	8980 miles	32,500 lb	25,600 lb

THE END OF THE PT1 PROGRAM

Performance of the PT1 has specific fuel consumption that can be compared to conventional piston engines (Fig. 7). The Wasp and Hornet engines had sea-level specific fuel consumption values of 0.55 at full power, and thus the PT1 appeared to be a great improvement over existing engines. Willgoos sent reports about the PT1 experimental program to the Navy in March and May of 1945, wanting to proceed with a demonstrator 4500-hp free-piston gas turbine powerplant. Instead, the PT1 program (T32) was cancelled and the PT2 (T34) program was started.

There was no question that the PT1 program was a useful warm-up for Pratt & Whitney to step into the gas turbine business. The engineers working on PT1 got instant training in turbine design and to a lesser extent, burner design. Knowing what the challenges were in gas turbines was a great help in jumping into the next engine design, a turboprop (PT2). Instead of being an "abomination," the PT1 made sense for its time period.

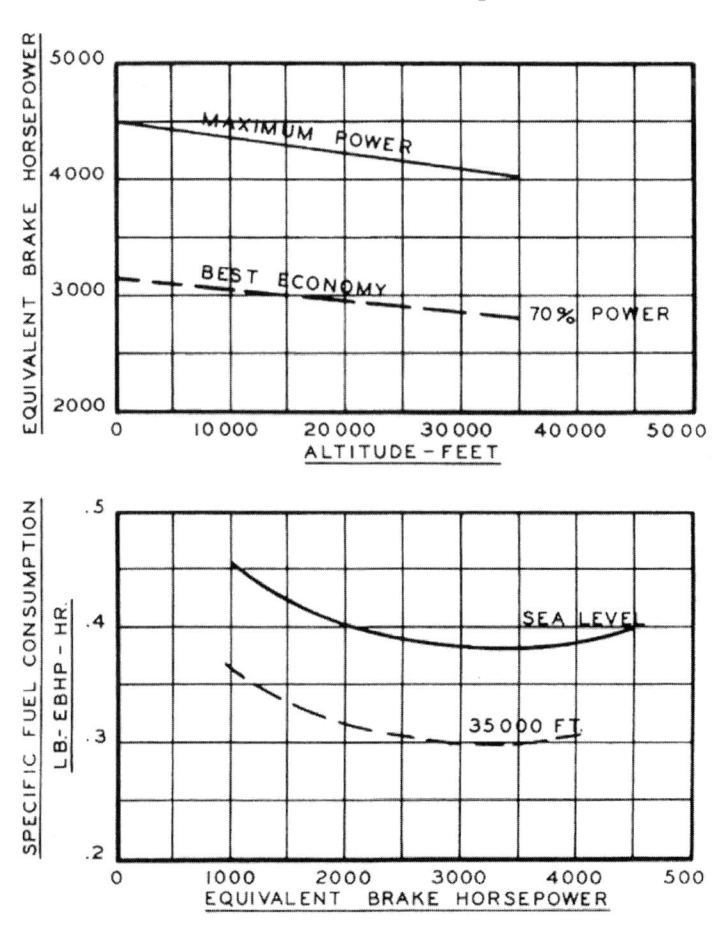

Fig. 7   Estimated performance of the PT1 (courtesy of Pratt & Whitney Archives).

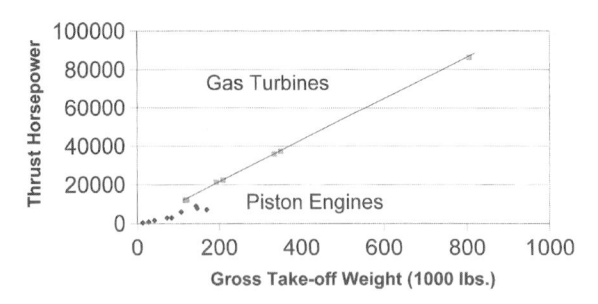

Fig. 8 Aircraft cruise horsepower vs GTOW.

However, once the jet age dawned, the relative simplicity of the axial-flow turbojet made more sense than a piston engine because of the jet's higher flight speed potential and lighter weight. Piston-powered aircraft were usually limited to less than 200,000 lb for gross take-off weight (GTOW) (Fig. 8). The B-36 had six R-4360 piston engines and four J47 turbojets, making it the largest operational piston-powered aircraft. The end of the line for production piston-powered aircraft is about 170,000-lb gross take-off weight: the Martin Navy flying boat, JRM-2. The huge flying boat built by Howard Hughes had a maximum take-off weight of about 360,000 lb. It flew briefly in November 1947 without payload. Gas turbines can supply more thrust power for a given GTOW and furthermore can provide power for higher cruise speeds and greater GTOW up to the ballpark of a million pounds at Mach 0.9.

HOBBS MOVES TO GAS TURBINES

In late 1944, Luke Hobbs felt he must make a commitment to aircraft gas turbines for Pratt & Whitney. General Electric and Westinghouse already enjoyed a tremendous technical lead in their new business of aircraft engines, and were being funded by the U.S. government. They and Allis-Chalmers were the logical choices to get into gas turbines because of their extensive experience with turbomachinery. Also, the piston-engine manufacturers had their hands full in producing powerplants to guarantee air superiority in Europe and the Pacific operations. Three new aircraft engine competitors were enjoying the party for gas turbines while Pratt & Whitney, Wright Aeronautical, and Allison were wondering apprehensively about their financial futures. Pratt & Whitney was a one-product company: aviation was United Aircraft's (Pratt & Whitney's parent company) only business while at General Electric, Westinghouse, and Allis-Chalmers, aviation was just a tiny entry in their financial reports. Those companies had huge worldwide operations in the power generation field.

Hobbs was a serious student of history, especially engine history. In a 1939 memo, he looked ahead at the probability of the United States entering the war and wrote the following comments that also apply to Pratt & Whitney's post-1945 situation [4]:

> Aviation's achievements have been spectacular and have captured the public's imagination. The actual accomplishment, however, has been by means of the slide rule and the drafting board, the wind tunnel and the metallurgical laboratory. It is a story of painstaking research and methodical engineering progress—yet paradoxically carried on at a dizzy pace. It is a story, also, of precision manufacture and skilled craftsmen, for the manufacturing standards are almost unbelievably high.

A few weeks after the end of the hostilities in Europe on May 8, 1945, Hobbs wrote a situation paper for H. M. Horner, General Manager of the Pratt & Whitney Division [5]:

> The time has arrived ... when we must make a concrete move on the turbine and jet program.... The deeper we get into the matter ... the more it is obvious that, as concerns the practical utilization of these units, experience is going to be essential to make even the roughest final decisions....
>
> It is ... as if we had been required in 1905 or '06 to make at one time the major powerplant decisions ... which we have gradually made through the years ... we cannot move with any real confidence, such as we have in the past....
>
> We must acquire experience in the design and development of these units as well as their application as rapidly as possible.... It should, however, be realized that the choice is inherently somewhat of a tentative one....
>
> There will be some place in the picture for [gas turbines] and even if ... the present general interest only temporary ... we should consider them if a powerplant larger than the Wasp Major were required.... Designing and developing something in an orthodox engine, an appreciable step in advance of the Major, would be an engineering undertaking of great magnitude. On what seem to be reasonably conservative assumptions as concerns the turbine, it would not appear that developing a unit of 10,000 hp would be exceedingly difficult....
>
> Overall ... it does seem that the turbine–propeller combination is more flexible and ... more efficient [than a turbojet]. In addition, this is the field in which we are the least behind and could possibly ... catch up the more quickly, so that I am inclined ... to stick to this although I am convinced that it is probably just a matter of getting started in this first and then moving into the jets so that both fields will be covered.
>
> On the purely technical side ... we can estimate, I think, fairly closely. In fact, I have been agreeably surprised at our ability to quickly assemble a competent technical group. It is, of course, not yet seasoned, but our Glastonbury work [PT1] is, I believe, already paying dividends if it never gets any place in itself.

At this point, Hobbs started a design of an axial-flow turboprop engine, just to get some experience. He reasoned that the next engine should be a turbojet after his design teams learned how to design an axial-flow compressor. Here is how he described the post-WWII environment to Paul Fisher in 1953 [6]:

> What I did was split [the engineering department] into two parts. One part was put on the so-called "production engines" ... which was under Parkins, who was given the job not only of carrying through the engines on the real heavy production that we could see ahead on immediate piston engine production, but for the further development of both the 2800 and the 4360 engines.... We were certain that the end of the piston engine ... was not here. There was good business ahead for those and so both engines were to be developed....
>
> Then a separate engineering department was set up under Andy Willgoos ... to get the jet, the turbine engines going. As a part of this setup, I transferred (literally stole) Pratt from Parkins, and put him in charge of the technical group.... But the big difference [between piston and jet engines] is the knowledge that has to go into them, and we didn't have that knowledge. But we did have Professor Soderberg, whom we had hired as the consultant on the original diesel engine and turbine job, (PT1) and we had Pratt who had done an excellent job on not only supercharging his own R-2800 when he was project engineer on it, but had also shown an aptitude for the technical side of this thing, and so we built up under him a technical group. Well, now to show you ... they worked hard.... We made a fast conversion of some of our piston engine facilities in the back end of the regular test area and then we started out on the layout of a complete turbine laboratory which has become the Willgoos Laboratory....
>
> Now in one of the ... experimental facilities ... one of the first things that was started was some research work on this compressor business, to try and find out how we should build an axial-flow compressor ... the first PT2 that ever ran would not only not turn itself over, it wouldn't even run; it wouldn't turn itself over, but when we tried to motor it over, we couldn't motor it over. It took us actually a year to get the first compressor really working. Now, as in a lot of things in this game, that probably was fortunate; it was awfully discouraging at the time, but we learned the hard way so that when the J57, the big jet job, came along, we had something to work on.

Hobbs' first move was to transfer Perry Pratt from the R-2800 operations to a new position vaguely defined as "to plan a path to the future in aircraft engines." The transfer, according to Hobbs, was made amidst the unmistakably forceful and clearly articulated objections of Wright Parkins [6]. Pratt's new research group was known officially as the Technical & Research Group (T&R Group). It consisted of compressor, burner, turbine, performance, and control groups. They developed the component design systems and ran their own test facilities. There was a heavy analytical effort in support of the experimental effort in contrast to earlier piston engine design, which had a relatively small amount of analytical effort accompanying experimental testing.

The conclusions that Hobbs reached after consideration of the commercial and military markets boiled down to this: There were two turboprop sizes that would bracket the market requirements, 3200 hp and 6000 hp; and there was a need for a military turbojet in a larger size than what existed in 1945. He knew Pratt & Whitney would have to start climbing the gas turbine learning curve. His extensive piston engine experience taught him that "speed was king," and the way to get speed was through more horsepower, and to stay on top, you have to deliver a bigger engine to the airframer than the competition and get there before the competition for the first flight!

Hobbs, in his 1945 memo to H. M. Horner, continued [5]:

> What we are doing now is to lay out the simple 3200-hp turbine ... to get it under development as quickly as possible.... As soon as the manpower is available, we will then start the 6000-hp design.
>
> Actually as far as the market is concerned, I ... believe that we should proceed with the larger job first, but there are two reasons why I do not think we can do this. The first is we are looking to the Navy for our initial contract and the smaller size is the one they desire. Secondly, we can probably proceed a little bit more rapidly with the development of the small unit and, as it is, we are probably going to be handicapped by not having the requisite development facilities, which situation would only be aggravated if the larger were started first.
>
> On the matter of facilities ... I have recently requested approval for the Pratt & Whitney Division to secure the services of an engineering firm so that this can now be gotten into some sort of definite shape...this is going to be terribly expensive due to the immense amount of power required to operate these units (the Navy's laboratory is estimated at approximately twelve million dollars, and I doubt that they will need as complete equipment as we will in the end). Our situation is somewhat aggravated by the fact that it seems definitely indicated that we should concentrate on component testing as opposed to the orthodox engine condition where the bulk of the development is necessarily accomplished on multi-cylinder units.
>
> There are, however, compelling reasons why this development must take this form. The first is that because of their simplicity, only a relatively minor amount of trouble should be encountered due to connecting the units together so that the development can undoubtedly move much more rapidly (and in the end more cheaply) by means of the separate testing. Secondly, all of the experience which has so far been obtained with these units, both here and in England, has shown that the combustion

chamber requires by far the longest time in development. On the 19XB, after three or four years, the combustion chamber is the only part completely unreleased to us and the one on which Westinghouse state that alterations will probably be required right up until the time production units come out. The assembly cannot be tested unless the combustion chamber is operating properly so that if the development were to depend upon the assembly testing, it might be delayed beyond all reason because Westinghouse had not completed its production facilities at that time.

Because Westinghouse had not gotten its production facilities in operation in time, the Navy asked Pratt & Whitney to build 130 19XB engines.

PT2 Program Begins

The T34 program (PT2 is the Pratt & Whitney designation) started in the summer of 1945. The design goal was a power level, selected by the Navy, at 3550 hp with 1000 lb of static thrust. This was at a somewhat higher power level than Hobbs had wanted, but he had no choice at this point.

Engine Description

The engine was an axial-flow gas turbine whose shaft output drove a front-mounted propeller through a reduction gear. The compressor pressure ratio was a respectable 6.7:1, which was high for the state of the art in 1945 (in the range of 4:1–5:1). The PT2's 13 stages had an average pressure ratio of 1.16 per stage. At stages 6 and 7, there were bleed valves which opened at low powers to prevent surge.

Engine surge is a complex phenomenon. One might look at it as starting with a separation of flow on the suction side of a compressor blade. This could be due to a sudden drop in airflow, caused for example by a rapid maneuver by the aircraft. This low flow area approaches the compressor rotor with a higher than design angle of attack that initiates blade stall in much the same way that a wing experiences stall at high angle of attack. The net effect is the compressor system not providing the proper flow to the burner and turbine. At this point several scenarios are possible, none of which are encouraging to the pilot. There could be just a loud bang with a quick recovery to normal operation. At the other end of possible scenarios is a complete stoppage of the engine, which subsequently winds down to become a drag on the aircraft. The engine designer makes sure there is sufficient margin from compressor stall to handle the normally expected inlet flow distortion. Bleed valves and variable stators provide additional margin when needed at the lower power settings.

The diffuser section, between the compressor and the burner, had provision for a single-plane mounting structure. The burner consisted of eight

TRANSITION TO GAS TURBINES   177

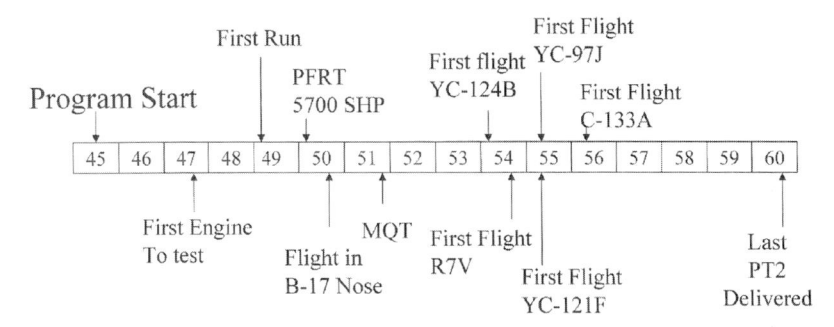

Fig. 9 PT2 time line.

combustion cones that led into an annular burner section. The fuel nozzles in each combustion cone were of the dual-nozzle type. The turbine was a shrouded three-stage turbine with labyrinth air seals on the case meshing with the shrouded tips. At take-off power the turbine power output to the shaft was about 9500 hp. The compressor absorbed about 5500 hp and the remaining 4000 hp drove the propeller through the reduction gear. The discharge from the turbine through a fixed nozzle provided about a 1000 lb of thrust at sea-level static.

It took almost four years before the engine could run properly. At one time it seemed the engine required more power input for starting than the engine could put out while running! The PT2 program (Fig. 9) had its problems in the beginning but the net outcome was respectable, especially as this was the company's first attempt.

In his unpublished memoirs [7], Walt Doll, Chief Engineer of the Technical and Research Group, tells us why Pratt & Whitney had those technical surprises:

> In the latter part of 1944, Management told Perry Pratt, the Assistant Chief Engineer, to see what the gas turbine business was all about, and to determine what Pratt & Whitney ought to do about it.... We conducted a lot of studies of different kinds of designs, and parameter choices, including turboprops, turbofans, and turbojets; various possible engine applications; and size requirements. We also looked at what the Europeans had done and were doing. Then we tried to decipher from all of this what was going to be really important in the future, what we were capable of doing at this time, and what Pratt's most effective role might be. Then it was time to say: "Damn the competition and let's get in there and do it."
>
> Perry picked a team of six young engineers, of whom I happened to be one, and we rented an office in Boston where we worked with Professor Richard Soderberg, who was then the head of the Mechanical Engineering Department at MIT. He was a former Chief Engineer of the Westinghouse

Steam Turbine Division and knew a lot about fluid mechanics and turbine design, as well as the structural design of high-speed rotating machinery.

The other five members of our team were Reeves Morrison, Ken Bodger, Frank McClintock, Frank Lockhart, and Dick Foley... . In Boston we continued our engineering studies and worked on developing suitable design systems and criteria. We finally decided that our first gas turbine would be a turboprop of about 3200 equivalent shaft horsepower, for driving a military transport airplane. This engine had the internal designation of PT2, military designation T34. We had ... no good information for designing the engine, particularly the axial-flow compressor, which was the hardest part of design. We ... laid down the aerodynamic and thermodynamic design of the engine, using slide rules, as there were no computers. Then we brought the design back to East Hartford and the Design Department took over.

While we were in Boston ... I shuttled back and forth to Pratt & Whitney to set up various kinds of test rigs to get the data we wished we had when we were designing the engine. We expected to be in deep trouble when we first ran the engine, and we were.

Our Management in all of this was superb. They put their trust in us young engineers and gave us everything we asked for in the way of tools, test rigs, and equipment. They even surprised us by throwing in the Willgoos Lab ... which was a big help ... in the long run. I believe that Management felt that they were in a do-or-die position, and, God bless them, they threw in all the marbles that they had. That left it strictly up to us engineers. We had no room for excuses. Fortunately, we were able to make the whole dream come true.

The PT2 itself was initially a dog.... When we put it on the test stand we couldn't start it. It is very hard to tell what is wrong with an engine if it won't run and provide you with some useful information. We finally put it on a dynamometer stand and got it started using about 800 hp to crank it over ... the compressor flow capacity was too large for that of the turbine and the compressor was stalling out, creating bad aerodynamic and thermodynamic conditions. So we made a larger flowcapacity turbine and ended up with a 6000- to 6500-hp engine instead of the 3200-hp engine we thought we had designed. It is usually true in the aircraft engine business that more power is better, because airplanes always get heavier than intended. So, if you are going to make a mistake, it is better to err high, not low. This engine was eventually developed and flew successfully in military transports. As far as I know, it did a serviceable job. However, it didn't set any great records.

Once the aerodynamic and thermodynamic problems were under control, the PT2 went on to a successful flight testing program in the nose of a B-17, a military qualification test in the latter part of 1951, and about five flight applications in the Douglas YC-124 Globemaster, the Lockheed R7V-2, the Lockheed YC-121F, the Boeing YC-97J, and the Douglas C-133A. In 1965, 20 years after the initial design, the engine was installed in the Super Guppy.

TRANSITION TO GAS TURBINES 179

Table 2 PT2 Take-off Power

Aircraft	Take-off shaft hp (lb)
C-133	6950 (wet)[a]
	5950 (dry)
Super Guppy	5500 (dry)

[a]Wet = water injection to increase air density by lowering the temperature of air entering engine.

The primary applications for the PT2 were in the Douglas C-133 and the Boeing Super Guppy. The Douglas C-124 and the Lockheed C-121F and R7V2 were essentially one-shot demonstrations. Table 2 gives the take-off powers of the PT2 in its principal applications.

Pratt & Whitney's first attempt at getting into the gas turbine business was not a howling success. The production run was only 485 engines. However, what was learned was enormous not only in the design process but also in developing the Technical and Research Group as a competent preliminary design organization.

How Aggressive Was the Company with T34?

Ivan Driggs, Chief Scientist of the U.S. Naval Development Center, and Otis Lancaster, Assistant Director of the Research Division in the Bureau of Aeronautics, described state-of-the-art gas turbines during the time period when the PT2 was conceived—engines were already running at power levels from 1000- to 4000-eshp (Fig. 10) [8]. When the Technical and Research (T&R) Group jumped into the T34/PT2 preliminary design, they bet that they could do better than the current state of the art. In 1945, Pratt & Whitney had a design objective of about 4000 eshp (equivalent shaft horsepower). In 1945,

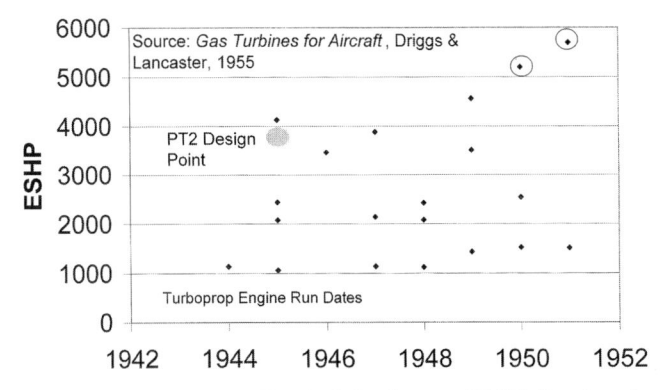

Fig. 10 Equivalent shaft hp vs year. The circled points are T34/PT2 ratings five to six years after 1945, achieving a power output of 5000–6000 equivalent horsepower [data from 8].

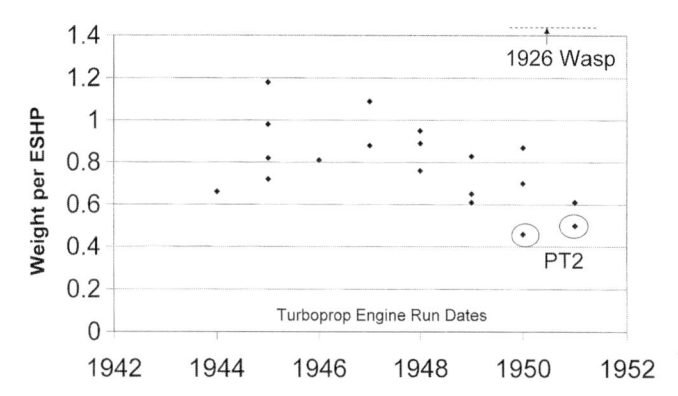

Fig. 11 The specific weight of the PT2 was quite competitive with the rest of the industry and far below that of the 1926 Wasp [data from 8].

General Electric was running its TG-100 at 2440 eshp. Aiming at around 4000 eshp was an optimistic target for a company that had essentially no technology for axial-flow compressors and combustion chambers (Fig. 11).

The T&R Group and the Design Department under Willgoos did surprisingly well in achieving an excellent specific weight compared to the rest of the industry. Just for reference, I put the 1926 Wasp engine specific weight in Fig. 11 to illustrate the weight advantage of turbomachinery compared to piston engines. Piston engines headed for the exit when the gas turbine era came on stage. The gas turbine offered a substantial installed weight advantage and a fuel consumption about 10 percent higher than that of the piston engine at full power (Fig 12).

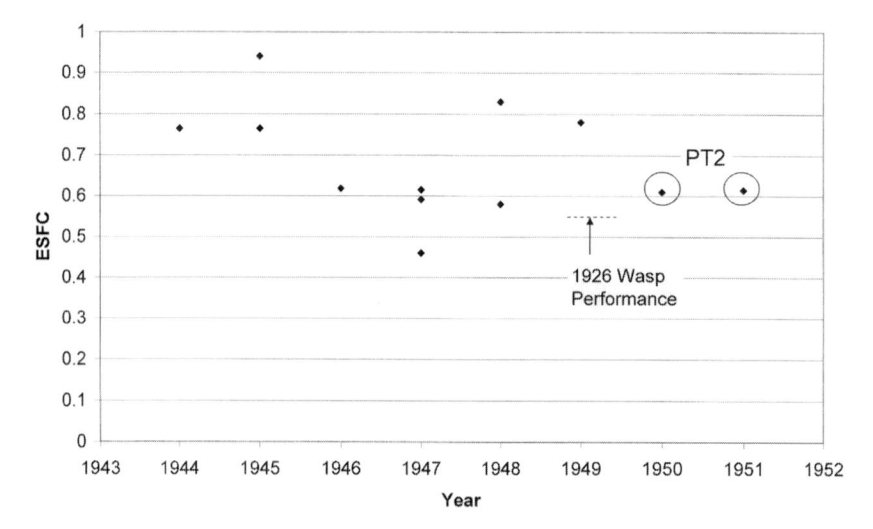

Fig. 12 The equivalent shaft specific fuel consumption compares quite well with the competition: PT2 fell close to the ballpark of the 1926 Wasp shaft specific fuel consumption [data from 8].

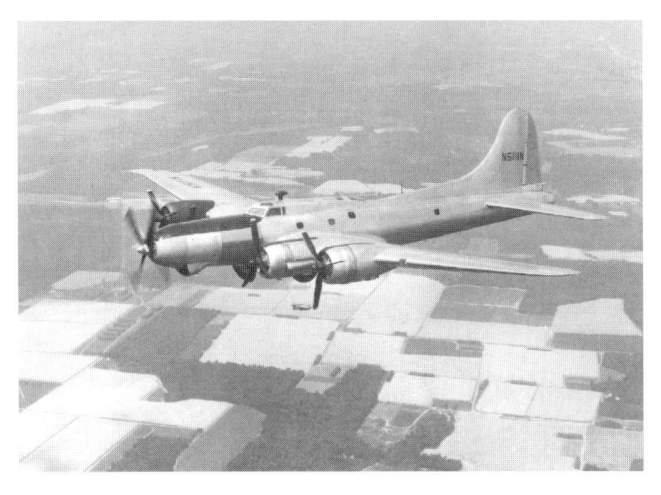

Fig. 13　The B-17 powered by the PT2 while the Wright Cyclones are on "break" (courtesy of Pratt & Whitney Archives).

B-17 FLIGHT TEST DEMONSTRATION

The famous Wright Cyclone (R-1820) produced about 1000 hp when four of them pulled the B-17 into the air in WWII. This rugged airframe was an attractive choice for Pratt & Whitney as a flying test bed for the PT2. The Boeing modification of the aircraft (Model 299Z) was made ready for the PT2 in 1950 (Figs. 13 and 14). The flight test operations with the B-17 were under the direction of Harold Archer, Pratt & Whitney's chief engineering test pilot. One of his spectacular demonstrations was to fly the aircraft on the PT2 and have the other four engines feathered. On a certain occasion a tobacco farmer and his migratory workers were startled one day to see the aircraft with four engines "dead." They expected a terrible crash and reported the perceived impending disaster to the authorities [9].

In August 1950, Archer flew the B-17 across the field at the Naval Air Station at Patuxent River. His demonstration was spectacular because he

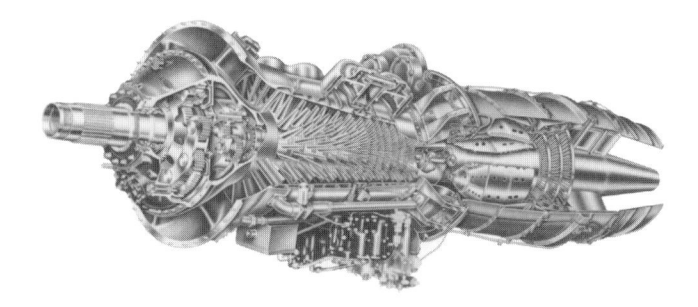

Fig. 14　The PT2 (T34) Turboprop with a design take-off power of 3550 hp and 1000 lb of static thrust (courtesy of Pratt & Whitney Archives).

buzzed the field at around a 100-ft altitude and a speed more than 100 mph greater than what you would have expected out of a B-17, especially with all four piston engines feathered! On another occasion in Connecticut, Archer flew the aircraft over the rolling hills on PT2-power alone. By this time PT2 solos were routine. I am not sure what happened, but it was reported that the PT2 suddenly stopped providing thrust [9]. At that instant, the sage words of Boeing's famous test pilot, Tex Johnston, might have come to Harold's mind, "Altitude above you and runway behind you do you no good!" Archer coolly fired up the four Wright Cyclones, and no doubt set an unofficial world record in starting up piston engines as an aircraft glided toward impending disaster. The flight log made no mention of the need for replacement underwear for the flight crew.

PT2 in the Lockheed Constellation

One of the early potential applications of the PT2 was in the Lockheed Constellation. Its 5000-hp output appealed to Lockheed. However, when a failure-mode analysis was undertaken, a startling situation came to light. What would happen should an engine suddenly stop? This stoppage would result in a windmilling drag so great as to cause an instant asymmetric force on the failed-engine side of the aircraft because the windmilling drag exceeded the positive thrust from the adjacent engine on the wing. The PT2 was a single-spool turboprop, which required a lot of power to start and absorbed a huge amount of power in windmilling. There was a worst-case scenario speculation at Lockheed that such a sudden force could cause the fuselage to snap in two!

Win Gove was the Project Engineer in charge of developing a 100-percent reliable fix to this potential problem for PT2 installations. The approach Pratt & Whitney developed in conjunction with the Hamilton Standard hydromechanical propeller and the Curtiss electric propeller was the use of a negative torque sensor, whose signal would feather the propeller. This approach worked well in the PT2 applications.

Two Engineers Rise to Prominence at Pratt & Whitney

Andrew Van Dean Willgoos

On the eastern side of the Connecticut River and just a little south of Hartford today stands an impressive facility built by Pratt & Whitney (Fig. 15). It is a gas turbine laboratory capable of testing full-size aircraft engines throughout their altitude-speed operating envelopes. When the facility was dedicated in the fall of 1949, it was the largest privately owned test facility in the world. It represented an investment of about twice the value of the facilities in aircraft piston engines, a gigantic financial gamble.

TRANSITION TO GAS TURBINES 183

Fig. 15 Willgoos Engine Laboratory (courtesy of Pratt & Whitney).

This huge facility was named after Andrew Van Dean Willgoos, who was a key member of Rentschler's team at Wright Aeronautical Corporation back in 1919 (Fig. 16) and by 1949 he was in charge of all gas turbine engine programs. Andy's parents came from England with the surname "Wildgoose." This name was changed later to "Willgoos around 1911 after the parents came to the United States. Andy's daughter, Marjorie, tells us about the family background:

> Dad was the twelfth child in a family of thirteen.... We have been told that Andy, as he became known worldwide, was drawing as soon as he

Fig. 16 Young Andy Willgoos at the start of his career (courtesy of Majorie Betts).

could hold a pencil. And, when he lingered over his meals and his sisters were pushing him to hurry so they could complete their assigned chores of clearing and washing up, his mother would say, "Leave the boy alone. He's thinking."

Willgoos' formal education stopped when he graduated from high school in 1905. His first full-time job was at age 17 as a messenger in the blueprint department for Western Electric Company. At night he studied mechanical engineering and mechanical drawing, and before long he was a draftsman. After a couple of years he joined Viele, Blackwell and Buck Consulting Engineers in New York City, and later the Standard Plunger Elevator Company.

In 1911, the Simplex Automobile Company selected Willgoos from a large number of applicants for a draftsman position. This put him into the engine business. The company later became the Crane Simplex Automobile Company manufacturing Hispano-Suiza engines during WWI. Willgoos became its chief draftsman. When Wright-Martin Aircraft Corporation took over the Crane Simplex Automobile Company, Andy was chief design engineer. When Rentschler brought Andy to Wright Aeronautical, Henry Crane wrote a letter to Andy in 1920 in response to Andy's letter that expressed concern about Crane's health. Crane wrote back to Andy in 1920, when Andy was chief design engineer (unpublished letter dated April 13, 1920):

> I certainly appreciate having such an expression from you. It means a whole lot more to me than a perfunctory letter of commendation that I received from the Board of Directors. I have always hoped that you and the other boys felt that way because I have always had absolute confidence in your support and ability and have known that mutual feeling of this kind is the only way to really get results.... I surely hope that the time will come when I can have such assistance again. In the meantime you are capable of being of great service in your line.

In 1949 Andy was in charge of all gas turbine engine programs and Wright Parkins was in charge of the production engines (piston engines). I remember how shocked we young engineers were on a March morning when we learned that Andy Willgoos had died of a heart attack while shoveling snow in his driveway. The Board of Directors of United Aircraft Corporation in March 28 1949, under the signatures of Frederick B. Rentschler (Chairman) and Charles H. Chatfield (Secretary), unanimously adopted the following resolution (excerpted from an unpublished letter by Chatfield to Willgoos' widow dated March 6, 1949):

> Every Pratt & Whitney engine developed during his 23 years as Chief Engineer bore the unmistakable stamp of his genius as a designer. He

had a large personal share in the creation of the aircraft powerplants which contributed so importantly to victory in World War II.

His achievements received world-wide recognition, and his utter devotion to his work, joined with his unique mastery of the practical aspects of his profession, won him the respect and admiration of all his fellow workers. He was universally beloved for his absolute integrity and for his simple friendliness.

In his death, the Corporation has lost one of its ablest engineers, and his associates, an inspiring leader.

Mrs. Grace Willgoos received a great number of letters expressing sorrow at Andy Willgoos' untimely death, which their daughter, Marjorie Betts shared with me in 2002 along with the other letters quoted in this section. The following sentiments are typical of the admiration people had for Willgoos:

"I want to repeat what you already know that everyone who knew him through the years loved and respected Mr. Willgoos."

"Andy to me was a truly great man—one whose fine character shone through his every thought and action. I know of no one whose good opinion I have treasured more highly. A leader in his profession, his work was respected by all for its complete integrity and soundness. The world is the richer for the standards of conduct as a man and an engineer which he set before the young engineers who were so fortunate as to serve with him in their impressionable years. The most sincere tribute to Andy's character is the manner in which these young engineers who have worked under him refer to him in their most casual conversations."

"This news filled me with the greatest pain. I had the privilege to be near your husband at the time we were licensees of Pratt & Whitney engines and I fully appreciated his skill and deep knowledge and moreover the goodness of his character, his humanity and noble heart. I shall always keep the long letter he wrote to me at the end of the war, which touched me deeply."

ARTHUR E. SMITH

Art Smith (Fig. 17) graduated from Worcester Polytechnic Institute in 1933 during the hard times of the Depression. He wanted to work for Pratt & Whitney, but had to wait about two years before he could be hired. In the meantime, he worked as a test engineer at International Motors in Allentown, Pennsylvania and later as a sales engineer for Manning, Maxwell & Moore in New York.

His first job at Pratt & Whitney was as a test engineer in 1935. Then he moved up the organization as an assistant project engineer and then project engineer. In 1942, he became the chief engineer at the Kansas City plant in Missouri, where he supervised the production of R-2800C engines. He

Fig. 17   Art Smith (courtesy of Pratt & Whitney Archives).

returned to East Hartford in 1944 and was promoted to chief engineer in 1949. He started 1952 as assistant engineering manager, and became engineering manager in 1956, and in 1972 became the chairman of United Aircraft Corporation. When he retired in 1973, he could look back on 38 years of achievements.

Art, a quiet, unassuming gentleman, was the only one from Pratt & Whitney Engineering to achieve the top position in the corporation. When asked by a reporter about the secret of his professional success, he said he was merely at the right place at the right time. Harry Gray, who succeeded Art at the chairman of United Aircraft, said, "He was an outstanding human being. He was a wonderful individual, very understanding and compassionate and clearly one of the brilliant engineers and engineering managers that Pratt & Whitney had" [9].

I found an amusing anecdote while preparing this chapter [10]. Art was an avid golfer when he managed to find some spare time from his job. Art was hosting Arnold Palmer around one day and assured Palmer that the company helicopter would take him to the airport so that he could continue his tight golfing schedule. All was going well until the helicopter's engine would not start. But Palmer put his frustrated host at ease when the golfer assigned the blame for the misfortune on Pratt's competitor.

References

[1] Kirby, R., Withington, S., Darling, A., and Kilgour, F., *Engineering in History*, McGraw-Hill, New York, 1956.

[2] Cronstedt, V., "An Appraisal of the Pratt & Whitney Free Turbine Drive," report, Cabinet 32, Drawer 4, Pratt & Whitney Archives, June 1944.

[3] Soderberg, C. R., and Kalitinsky, A., "Free Piston Gas Turbine Power Plant for Aircraft," report, Pratt & Whitney Archives, Sept. 6, 1941.

[4] Hobbs, L., memo, Pratt & Whitney Archives, 1939.
[5] Hobbs, L., "Memo to H. M. Horner," Pratt & Whitney Archives, May 30, 1945.
[6] Fisher, P., "Interview with Hobbs," Pratt & Whitney Archives, 1953.
[7] Doll, W., "Pratt & Whitney Aircraft Memories," Pratt & Whitney Archives, 1999.
[8] Driggs, I., and Lancaster, O., *Gas Turbines for Aircraft*, Ronald Press Company, New York, 1955.
[9] *The Best of the Bee Hive,* United Aircraft, Pratt & Whitney Division, Pratt & Whitney Archives, 1950.
[10] Art Smith files, Cabinet 25, Drawer 5, Pratt & Whitney Archives.

Chapter 6

WWII Ends and Turbojet Development Begins

Pratt & Whitney Service Organization

As Pratt & Whitney grew to meet the demands of WWII, the company began training the customer to take care of his investment in his aircraft engine. The following is a summary of how Pratt & Whitney taught the customer (military and commercial) on how to take care of the engine and how to operate the engine in service.

In 1928, Lionel B. Clark, aged 26, came to work at Pratt & Whitney Aircraft in Hartford. He had experience as a machinist and as a mechanic. His first assignment was in the engine assembly department, working on Wasp and Hornet engines so that he would become familiar with these engines. The Service Department at that time consisted of two men (service manager and assistant service manager) and a part-time secretary. Here's how Clark described his job [1, pp. 18–20]:

> Now the way it was done—engines built green were put through engine tests. They were disassembled and laid out for inspection ... then rebuilt. The crankshaft and blower section were assembled by one man [who] specialized on the work. Cylinders were assembled by another group. There was no assembly line like the automobile people had. Two men built that engine from start right up to finish. It would take about eight hours for the two men to assemble a Wasp engine. All the nuts and bolts were tied together with bronze safety wire.
>
> They had signs up, "There's no substitute for quality." There was none of this throwing-things-together like you see in automotive assembly lines. They were good mechanics. They'd take time to polish a part or stone it to take off a burr before they put it together. It wasn't just a case to see how fast they could build it but to see how good they could build it.

That experience on the engine assembly line was enough to qualify Clark as a service representative. When the customer had a problem, Clark said he would get a call to see Hobbs [1, p. 33]:

> They would get you to go from the assembly floor up to the office. Luke Hobbs handled some of the first service problems. He was the best one

to brief you on a trip. He had a little pad with a very coarse pencil and he would talk to you for a little while and make two or three notes on this piece of paper and then pass the piece of paper to you, and then continue to talk and do the same thing. You had three or four pieces of paper in your hand and at least you know where you are going and what you were supposed to do and what it was all about. He was the best when it came to briefing.

When the service representatives were not making their calls, they went back onto the assembly floor. While on the road, representatives would carry engine parts as baggage. Clark continues: "An airplane would be forced down somewhere and we'd get into [it] with tools and parts and try to fix it up and get back to home base." [1]

The Service Department expanded to where there were men strategically placed around the country. They were Bill Chamberlin, Navy at Norfolk, Virginia; Gus Soderling, New York; Charlie Runyan, Chicago; Charlie Bunce, Anacostia, Washington, D.C.; and Wilbur Thomas and Paxton, West Coast.

In 1930, Clark was at the National Air Races and the Thompson Trophy Race in Chicago [1]:

> All the planes were Lockheed Vegas, one with a Hornet; all the rest had Wasps. Wiley Post won that race competing against Roscoe Turner and some of the best fliers in the country. That's the first thing that attracted my attention when I got there. I was there when the airplanes were landing.
>
> That's the first time I met Wiley Post. I contacted him there. He never had a mechanic. He didn't have any money. We told him we were at the field. In case he had any problems, we would be available.

That day Emil ("Matty") Laird had an airplane in the race, the *Laird Solution*. The airplane was a snappy looking biplane powered by a Wasp Jr. (R-985). Laird had his airplane company in Chicago. For $5000 he built a fast little biplane for the Goodrich Rubber Company in 28 days. Goodrich supplied the R-985. Charles "Speed" Holman got into the cockpit right after the aircraft came out of the shop, flew on a short hop to Curtiss Reynolds Field for the Thompson race and within 30 minutes after his landing he was an entry into the famous race [2]. Lionel Clark picked up the story regarding "Speed" Holman, who was racing the only aircraft with a Pratt & Whitney engine [1, p. 69]:

> The minute he landed in the field, I went up and ... said, "If there is any way we can be of any assistance, we'll be glad to help you." He said, "This cockpit is mighty hot. Can you get me a drink of water?" So I did, some way or another, get a hold of a thermos bottle and brought him some cold water. He drank quite a bit of it. He thanked me more than you can imagine. Nothing was done to that airplane before the race....

Our entry seemed to be running rougher than any other engine on the line. Speed called me over and asked about the carburetor setting. The carburetor idle needed to be adjusted. There wasn't time enough to adjust it. I assured him that the jets were correct for top speed. A moment later the gun was fired and they were off.

The race went on for twenty laps [after the Navy pilot crashed]. Holman held the lead and finished 20 seconds ahead of Haizlip [in a Travel Air Mystery Model 2R]. A small Howard plane with lots of pep was third.

Anyway, the Laird beat the two Travel Air Mystery ships and that's the thing that we got a kick out of. They had the Wright engine about the same size as the Wasp Jr. The Laird Solution beat those two airplanes without any trouble whatsoever. Speed Holman wrapped that Solution around those pylons. Brother, he could fly an airplane!

Harvey Lippincott, archivist for Pratt & Whitney, knew how the airplane got its name, "Solution." The airplane was competing with two Travel Air "Mystery" airplanes powered by Wright engines. He had asked Matty Laird why the plane was called the Solution. Laird told him, "As I was watching the race with this chap, he said, 'Well, Matty, that's the solution to the Mysteries.' So it stuck." [1, p. 70]

Clark would later be the Pratt & Whitney representative who would make sure Wiley Post's Wasp engine (S/N 3088) would be in the best of condition for Post's round-the-world flight. Post used this engine in 1930 when he won the National Air Races. Subsequently he flew around the world in his Winnie Mae with Harold Gatty in 1931, and then two years later he made the trip around the world as a solo. In both cases Clark inspected the engine and gave the okay for the flights. "The Wasp performed perfectly," Post said about the R-1340 Wasp [3].

BEGINNINGS OF THE CUSTOMER TRAINING FUNCTION

Clark was Pratt & Whitney's first instructor in an organization initially known as the United Aircraft School. Pratt & Whitney of Canada wanted to have three of their men trained on engines. Clark took them off to one side in the engine assembly area and delivered the first training program. The second training session was with George Pranaitas, whose job with the State of Connecticut was to take care of the engine in the state's Corsair airplane. Clark pulled an engine that needed an overhaul off the normal line and pushed it into a corner of the room where he, Pranaitas, and another gentleman from Connecticut, Clarence Knox, pulled it apart and did the overhaul. Training could never be any better than that!

The formal founding of the Pratt & Whitney Aircraft Service School took place in 1935 in a small section of a hangar at Rentschler Field. The school evolved into a state-of-the-art building with up-to-the-minute learning facilities at Rentschler Field under the name of Customer Training Center after several moves to increasingly larger facilities in East Hartford and in Wethersfield. The successful development of the training function was principally under the able guidance of Lionel Clark, William Doolittle, and Dick Wellman.

The spectrum of training covered the early piston engines of about 400 hp up to the latest gas turbine engines of about 100,000-lb of thrust as well as rocket engines, fuel cell powerplants, and stationary gas turbines. Today, the Customer Training Center's mission is to train all Pratt & Whitney engine customers on the operation and maintenance of their commercial and military engines. The customer training comprises lectures, interactive classroom work, computer-interactive learning, hands-on training, and engine testing.

ARMY AND NAVY ENGINE TRAINING SCHOOL

H. Mansfield Horner, General Manager of Pratt & Whitney Aircraft, established the Army and Navy Engine Training School in January 1942. He looked at this as a further Pratt & Whitney contribution to the war effort. This activity was separate from the customer training function just described. A Pratt & Whitney publication described the school [4]:

> Engines, propellers, and airplanes—produced at top speed—are United Aircraft's major contribution to the all-out United States war effort. Beyond this, however, it was realized early that the effective use of this equipment in combat would depend entirely upon the knowledge and ability of those responsible for its operation and maintenance. To carry on the tremendous job of maintaining thousands of fighting airplanes in condition for combat on warfronts all over the world, personnel of the armed forces must be familiar with the time-tested methods of installation, overhaul and maintenance which manufacturers of aircraft equipment find necessary for its ultimate performance. Since a large portion of the firstline fighting airplanes are equipped with either Pratt & Whitney engines, Hamilton Standard propellers, or both, United Aircraft has a real part to play in the tremendous technical training program facing our armed forces in their emergency expansion.

Pratt & Whitney developed an eight-week training program for Squadron Engineering and Base Overhaul officers. The company leased the largest hanger at Brainard Field (in Hartford across the Connecticut River from the company). The school used 19 military piston engines and six functioning airplanes to make the instruction directly relate to the real world. The students learned how to assemble and disassemble engines as well as how to maintain them (Fig. 1). The instructors were all highly qualified, experienced, and motivated individuals chosen from various departments in Pratt & Whitney and Hamilton Standard (Fig. 2). The company also provided living quarters for the

Fig. 1　It is hands-on learning for students at Pratt & Whitney's Army and Navy Engine Training School (courtesy of Pratt & Whitney Archives).

Fig. 2　Bill Doolitle (ninth from left) assisted Lionel Clark (sixth from right) in customer training. They stand with the school's faculty in front of a B-26 (courtesy of Pratt & Whitney Archives).

officers by leasing five dormitories from the Federal Public Housing Authority at Brainard Field. Fritz H. Gometz was the school supervisor who reported to F. H. "Tiny" Flynn, Assistant Service Manager. The school closed in November 1944 after 3275 officers from Army, Navy, and Marines had been trained.

PRATT & WHITNEY "TECH REPS"

Not only did Pratt & Whitney provide training for the Army and Navy officers, it sent more than 200 service representatives overseas to help Pratt & Whitney-powered aircraft stay in top condition. United Aircraft had 236 field service representatives overseas in combat zones during WWII. Unfortunately six were killed and one died of natural causes. In early 1946, T. E. Tillinghast, President of United Aircraft Service Corporation, talked to a group of senior service representatives about the wonderful dedication these men demonstrated [5]:

> Discomfort and dangers are factors one must necessarily associate with the kind of work the field men did. But, there were two qualities the great majority assigned to combat areas displayed that stood above raw courage and the good-humored ability to withstand discomforts.
>
> The first of those qualities concerned your inherent American ability to improvise at your work. The second was the frank and critical evaluation you placed upon equipment in battle action. You transmitted to us the urgent demands of military combat as you learned of those demands from the men in action. You were the link who saw to it that our stuff was improving every day while the Germans went into mass production of stereotyped models. You have established a fine tradition for the cooperation between the armed services, on one hand, and yourselves as volunteer civilian technicians, on the other, in the defense of our country.

Harvey Lippincott (Fig. 3), archivist for the company later in his career, was one of the men who served overseas in the Pacific and European theaters. Harvey described how he assisted the pilots in Italy [6]:

> To increase pilot confidence in the engine it was good product public relations to gather together pilots and show them specimens of badly damaged engines which still brought the pilot home such as one R-2800 on a P-47 that had a major flak strike which wiped out one of the two magnetos, cutting out one of the two sets of spark plus, punctured several cylinder walls, severed two ignition leads and blew off completely one cylinder head leaving the barrel standing. Only fifteen of the eighteen cylinders were working. The pilot said though the engine was smoking and leaking oil badly, "It only ran a little rough!" (In two days we had the engine back in service.) Confidence in the engine rose dramatically with the pilots who examined this battered hulk.

Fig. 3 Harvey Lippincott, tech rep and company archivist (courtesy of Pratt & Whitney Archives).

Once a tech rep had successfully established himself, it was amazing how he could assist a military unit in ways other than technical. Often he found himself as mediator or go-between the officers and enlisted men between whom certain lines of communication were often difficult. Both groups could "talk" to the civilian tech rep who could help both sides reason out their problems. Many a morale problem was solved or ameliorated through the unique offices of the tech rep. Quite often the tech rep could help smooth over ruffled feelings, correct annoying working conditions, help fill deficiencies and generally help effect better working relations between diverse military personnel.

One of the most famous tech reps was Charles Lindbergh, who showed the Marines in the Pacific Campaign how to get the most out of their Corsairs [7]:

> His mission was to study the performance of fighter planes under combat conditions with a view to improvement in design and the design of new types ... and he did just that.... Lindbergh went on many missions in Corsairs, took part in strafing raids, flew cover for bombers, and did some special bombing to show just what a Corsair could do.

FLIGHT OPERATIONS ENGINEERING

Any history of Pratt & Whitney could not possibly be complete without mentioning Flight Operations Engineering. The original version began in the early 1940s under the name of Airline Engineering. It was essential to operate a commercial aircraft engine within strict guidelines if it was going to be a successful venture. The most effective approach was to use trained powerplant engineers to work directly with the flight crew personnel to be sure that they understood how the engine should be operated for safety, dependability, performance, and economy.

Flight Operations Engineers accompanied many of the initial training flights as members of the flight crew. This placed the engineers in a position to answer engine operational questions and help assure compliance with approved procedures during in-service engine operation. As a result, the airline crews developed personal relationships with the Pratt & Whitney engineers, which benefited both the airlines and Pratt & Whitney. The competition followed Pratt & Whitney's lead and established similar groups.

Early in the transition to turbine-powered aircraft, Flight Operations Engineering was assigned the task of writing the specific operating instructions for each engine model. This required coordination of procedures between Pratt & Whitney and the various aircraft manufacturers to avoid any conflicts.

In the 1970s, Flight Operations Engineering also developed an engine-monitoring program, the first in the industry. A computer processed the engine data taken during scheduled airline operation. The printouts would show any deviation from the "norm" that might indicate performance deterioration. Most major airlines have since adopted this beneficial program.

Turbojet Development Begins

How would you like to have been in Luke Hobbs' shoes right after WWII? Your company made a gigantic contribution toward establishing air superiority in the military operations of the war. This undoubtedly shortened the period of hostilities and thereby saved lives on both sides of the conflict. Your military customers now comprise a new crew because the guard has changed in both the Army Air Corps and the Navy. New relationships with decision-making top leaders have to be developed from scratch, because the American system of military procurement differs sharply from that of the British in "relationship stability" between aircraft industry and government. Professor Schlaifer described this concept about a half century ago [8]:

> There was a great stability in Britain both in the post of Director of Scientific Research and in that of Assistant Director of Technical Development for Engines. The former was held by two men between 1925 and the Second World War, and the second of the two had been Deputy Director during the entire tenure of the Directorate by the first.... This stability is again in striking contrast to the American practice, where the officer at the head of each section at Wright Field or in the Bureau of Aeronautics has usually held his post for a tour of duty of only a few years at the most. Stability of personnel was of the greatest importance, since it accomplished the larger part of all that could have been gained from authority to make legal contracts running several years in advance. If a British official promised that a project would be supported until completion, the firm could rely on that promise; the corresponding American official could not even be sure that he would be in charge of the project a single year later.

AMERICAN COMPANIES ENTER THE JET AGE

This changing of the guard did not apply to General Electric and Westinghouse, who had been working with the new government teams throughout WWII. Westinghouse, General Electric, and Allis-Chalmers were no strangers to gas turbines. The need for special high-temperature, high-strength materials was also great in the field of industrial power. It was such a lack of materials that slowed the growth of industrial gas turbines as well as aircraft gas turbines. These companies, after established aircraft engine companies, were the logical choices to develop aircraft gas turbines, but they had to get used to lightweight structural design and conduct a huge amount of development testing.

GENERAL ELECTRIC LEADS THE WAY

Figure 4 describes how quickly General Electric responded to General Arnold's request to get into the jet engine business, starting with Frank Whittle's centrifugal-flow engine. General Electric was brought into the aircraft engine business in the latter part of 1941. There was concern that the Germans would invade England and the British gas turbine technology would be captured.

General Electric started off with the Whittle technology, including a full-size engine and all the drawings for building the engine. British Royal Air Force (RAF) Wing Commander Whittle made a visit to ensure all was going well. General Electric came through in time for the Bell XP-59A flight to become the first American turbojet-powered aircraft, three years behind the Germans. In early 1944, Lockheed flew its first turbojet-powered aircraft, the XP-80, with special thanks to the British for the Halford-designed engine, H-1. Later the aircraft was fitted with the GE I-40, which was the most

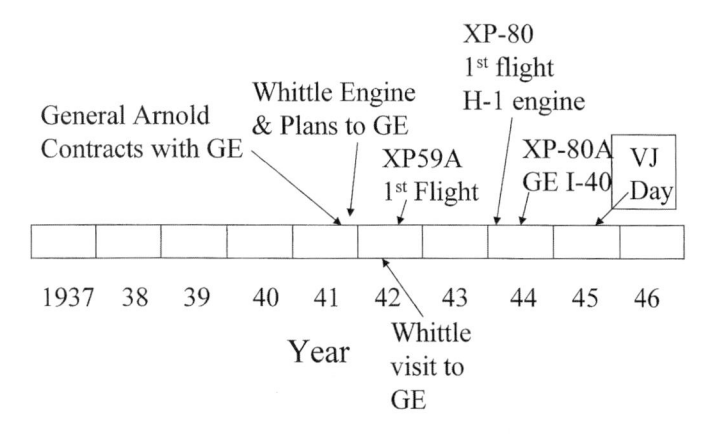

Fig. 4 A time line of General Electric's great progress in gas turbines.

powerful American turbojet at that time. General Electric was running its TG-100 (T31 at about 2400 eshp), the first American turboprop, in early 1943. Shortly thereafter, General Electric started the design of the J35 (about 3750-lb thrust), its first axial-flow turbojet. In a little more than a year after the end of the war, General Electric had its turboprop in flight test and its J35 axial-flow turbojet set a U.S. speed record in an XP-84.

WESTINGHOUSE JUMPS IN

Luke Hobbs, in addition to observing General Electric's progress, also saw some startling achievements from Westinghouse (Fig. 5). Westinghouse developed two Navy turbojets, the J30 and the J34. The J30 (19A) first flight test in an F4U was in late 1943. The Navy contracted with Westinghouse to develop axial-flow gas turbines for aircraft in late 1941.

Westinghouse started with its 19A model (J30 at about 1500-lb thrust), which was an axial-flow engine. It was first used on a Corsair as a booster engine and then powered the XFD-1 Phantom, the Navy's first jet-powered aircraft to operate off a carrier. Their next engine was the J34 (WE model 24C at about 3000-lb thrust) for the Navy and powered the XF6U, the Vought Pirate. The Navy appeared to have grabbed on to Westinghouse as its engine supplier for the future. Things were moving along so well that Westinghouse formed a separate Aviation Gas Turbine Division in early 1945 under Rein Kroon, father of the Westinghouse aircraft gas turbines.

PRATT & WHITNEY RESPONSE TO ITS COMPETITORS

Both General Electric and Westinghouse lacked substantial engine manufacturing facilities for aircraft engines at that time, and were looking for help in production capability. Hobbs turned down offers to license the Westinghouse

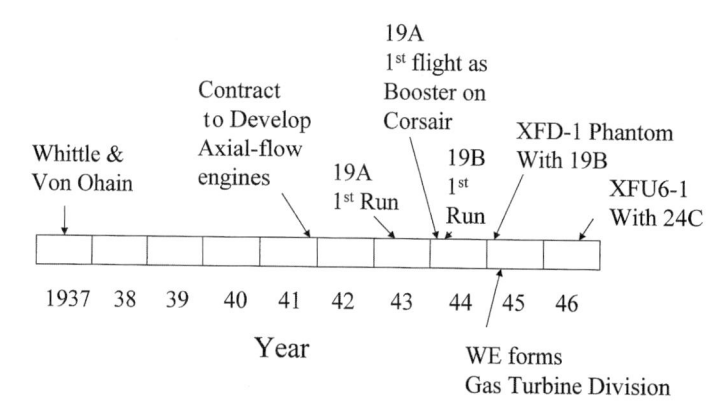

Fig. 5 A time line of Westinghouse's development of axial-flow turbojets.

19XB, the Rolls-Royce Nene, and the General Electric J35. He described these events as follows [9]:

> We fought off the manufacture of the 19XB for months and months to the point that we were alienating the Navy which we could not afford to do.
>
> Shortly after the war was over the government notified General Electric that they needed a considerable quantity of TG-180s (J35s) produced. General Electric came to us and asked us if we would take on this production. Although it was not remarkable in any way, this was an axial flow design and possibly we could have learned a little something from it but we turned it down without any hesitation.
>
> Five or six months after the end of the war Rolls sent Pearson over to offer us a license on the Nene and Derwent engines. This proposal was turned down. The Nene was of the centrifugal type and we were already well on our way as far as turbines were concerned so that there was nothing to be gained. As I have pointed out we took on the Nene under different circumstances when it offered some guaranteed production business.

Allison took on General Electric's production role for a while. It looked as though General Electric would design and develop an engine, and then it would go to Allison for production. For example, General Electric's I-40 became the J33-GE in 1944, and the following year at Allison it became the J33-A-21. On the other hand, the Navy approached Pratt & Whitney to ask if it would produce the initial production run of Westinghouse J30s, about 500 although that was subsequently reduced to 130. This was against Rentschler's and Hobbs' philosophy of being a licensee, but they wanted to be a major supplier to the U.S. Navy, so they went along with it. In retrospect, it was a good idea because it gave the Pratt & Whitney Manufacturing Department some immediate experience with axial-flow gas turbines.

EUROPEAN DEVELOPMENT

Both Frank Whittle and Hans von Ohain had been running their demonstrator engines back in 1937. The Whittle engine was closer to flight because it ran on hydrocarbon fuel while the von Ohain engine ran on hydrogen to demonstrate the principle of a turbojet engine. However, the German effort proceeded at a greater pace because of the lack of bureaucracy. Heinkel was demonstrating a real prototype of a jet fighter (He-280) in mid-1941 while the British were flying the equivalent of Heinkel's 1939 flight demonstration. At that stage of the game, the British were about two years behind the German effort. However, things changed on the German side, and Messerschmitt became the winner with its Me-262, which entered operational service in 1944. In late 1943, Metro-Vickers put its F2 axial-flow turbojet in the Gloster

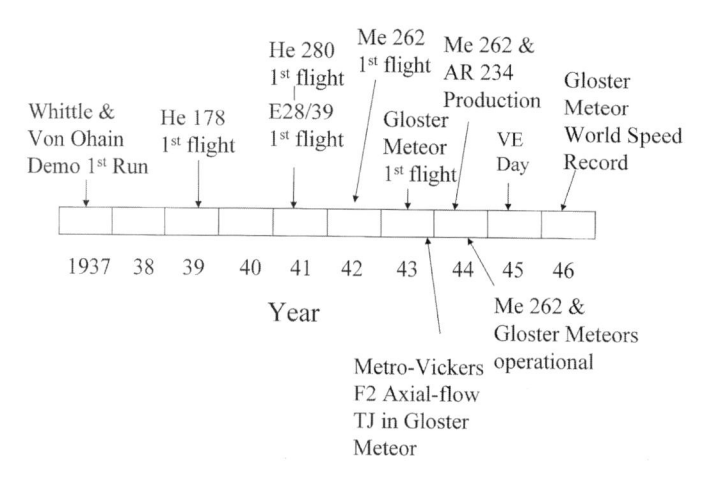

Fig. 6 A time line of the highlights of the British and German developments of jet propulsion. Dates have been rounded to the nearest half-year; in 1941, the He 280 and the E28/39 flew within six months of each other.

Meteor for a flight demonstration. By September 1946, the Gloster Meteor (powered by R-R Derwent engines) had set a new world's speed record of 616 mph. Figure 6 highlights this progress.

In early 1941, Rolls-Royce started to build jet engine test facilities. Then in rapid succession (Fig. 7), after working on Whittle engines, it broke away from the pack with its own engines, the Derwent (2000-lb thrust) and the Nene (5000-lb thrust). The Nene in 1945 and 1946 was the world's most powerful jet engine. The Trent (at a little over 1000 eshp) was the first turboprop to fly. By the end of WWII, Rolls-Royce was one of the principal leaders in aircraft gas turbines.

What gave Hobbs the confidence that Pratt & Whitney could compete successfully with these established companies when it had practically no jet engine experience? Hobbs, with more than a quarter century of aircraft engine experience, understood what it took to produce a dependable engine.

GRUMMAN OPENS THE DOOR TO ROLLS-ROYCE

Grumman and the U.S. Navy had been working on a new single-engine fighter aircraft in late 1946—the XF9F-2. Hobbs explains how the Nene got into the Navy planning for this aircraft [9]:

> It was Grumman that basically brought about the manufacture of the Nene in this country. Grumman had been badly hurt several times by having powerplants fail to come through, or fail to perform after they had come through, leaving them with new airplanes but no engines and therefore no business.... However, the I-40 was the only engine actually available in this country at the time in the power they wanted so they were compelled

WWII ENDS AND TURBOJET DEVELOPMENT BEGINS   201

to use it. They were the Navy's favorite producers and when Phil Taylor [owner of Taylor Turbine] came along with his Nene promises it was exactly what Grumman wanted. They went to the Navy and sold the Navy on the idea, and it was a good one, of having direct engine competition for this airplane.

Contrary to the case of the 19XB, we did not try to fight this off. It presented some immediate certain production which we could use and we were not required to develop or sell the engine.... However, the actual fact is that the Navy selected us and officially advised the Taylor interests that there would be no deal unless they could make arrangements to have Pratt & Whitney build it. The point is that although we certainly were willing to take on the job we did not ask for it. We obviously did not want the Nene to interfere with our own developments and since it was essentially a production engine, it was turned over to Parkins' group and this was their first jet experience.

Hal Andrews, in his 1976 article for *Aeroplane Monthly*, related the background for the Panther [10]. In 1946, Grumman could see its long line of Navy fighter aircraft coming to a premature end in the jet age. The most attractive aircraft coming out of the Grumman studies, Grumman study aircraft 79D, was the XF9F-2 powered by a single Rolls-Royce Nene engine, which became the Panther. Grumman obtained its engines for the prototype aircraft program in September 1947, with the Nene designated as XJ42-TT-2. The Panther flight test program began in early 1948. Grumman and Pratt & Whitney had a long history of working together to meet the Navy's needs for high-performance piston engine powered fighter aircraft.

In late 1946, Phil Taylor persuaded the Navy to test the Nene in its facilities [11]. Then Rolls-Royce sent two engines to Philadelphia for the Navy to test in its Naval Air Experimental Station. The Taylor Turbine Corporation

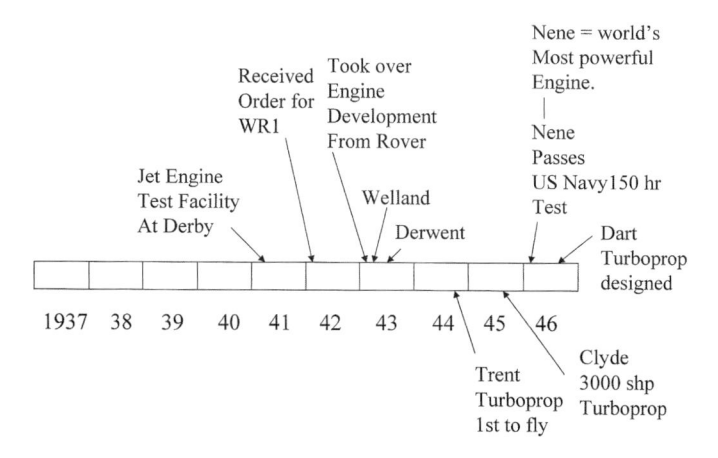

Fig. 7 A time line highlighting Rolls-Royce's rapid progress at the end of WWII.

wrote the final report on the test in February 1947 [12]. The 150-hr test was run on engine number 1004 at a maximum thrust of 4500 lb. This model engine had already passed the British 100-hr test. The Navy waived the normal military requirements in the following areas: 1) no bleed from the compressor, and 2) accessories were not loaded. The Navy was impressed and wanted the engine to be manufactured in the United States.

The Navy asked Pratt & Whitney to "Americanize" the engine for the F9F-2 at 5000-lb thrust. Pratt & Whitney negotiated with Phil Taylor and Rolls-Royce for the Nene and signed an agreement in April 1947 with Rolls-Royce. The Navy issued the designation of J42 and Pratt & Whitney internally referred to the engine as the JT6. Hobbs was concerned about the future potential of the Nene because he knew from his experience in aviation that the customer is never satisfied with the level of thrust that he swears is all he needs! Hobbs commented [9], "What actually happened before we took on the Nene, I told Hives [Ernest W. Hives, later Lord Hives and Rolls-Royce Manager] that we could not buy "a dead horse" and that unless he had a follow-on development of a larger and better engine (which he did not have at the time) I would not approve the licensing. After a delay of only a week he agreed to produce the larger engine for us."

During a visit to Rolls-Royce in early October 1947, Bill Gwinn (Pratt & Whitney General Manager) and Wright Parkins (Pratt & Whitney Chief Engineer) discovered that Rolls-Royce was not continuing with the Nene. This was a disappointment for Hobbs, who promptly called Hives and followed up with a cable on October 7, 1947. The opening sentence summarized the Pratt & Whitney & Rolls-Royce agreement: Pratt & Whitney would manufacture the Nene under license and Rolls-Royce would develop a more powerful Nene, which Pratt & Whitney would also manufacture under license [13].

Hobbs went on to say that the J33 engine competition was closing in on the Grumman airplane with predictions of much higher thrust than the Nene. Without a more powerful version of the Nene, the engine application for the Nene in a Grumman aircraft would be much less likely. In other words, the whole Nene program with Grumman would end in a financial disaster. Hobbs was right. The J48 (a 30-percent scaled-up of the Nene called the Tay) accounted for 78 percent of the total of the J42 and J48 engine production run. In response, Hives spent 19 days trying to divert the U.K. government's support for the advanced Derwent VII back to the Nene upgrade. In an October 26, 1947 letter to Hobbs, Hives assured him that Rolls-Royce would have an upgraded Nene running in about 10 months and would provide Pratt & Whitney with a new design, which became the Tay [14].

This larger engine's development fell upon Pratt & Whitney. There was no application for the engine in the United Kingdom, and Rolls-Royce was not interested in developing an engine with no application. However, one of the benefits to Rolls-Royce was their freedom to sell the developed engine to other customers outside the United States. One such customer was Hispano-Suiza, who produced the Nene, the Tay, and an uprated Tay (renamed the Verdon in France for a French swept-wing aircraft).

In February 1947, the U.S. Navy and Grumman brought the Allison J33 into the program as a backup for the XF9F-3 model. The Nene and the J33 looked similar (both centrifugal-flow compressor engines with a technical ancestry going back to Whittle) and were able to fit into the fuselage of the aircraft. The Allison J33 (previously the GE J33) was rated at 4000-lb thrust.

PRATT & WHITNEY'S FIRST TURBOJET SUCCESS

J42 ENGINE DESCRIPTION

The Nene was a centrifugal engine (Fig. 8). It had a double-sided centrifugal compressor with about a 4:1 pressure ratio. A single-stage turbine drove the compressor, a cooling air impeller, and the accessories. Compressed air at the discharge of the compressor flowed directly into nine combustion chambers and then into the single-stage turbine. Air entered the double-sided centrifugal compressor, flowed into the burner cans, and then on through a single-stage unshrouded turbine and exhausted to a fixed tailpipe. Figure 9 shows air entering the J42 at the double-entry centrifugal compressor through screens at the left and then into nine combustion chambers exhausting to a turbine and through a fixed area tailpipe. This photograph can be compared with the cutaway in Fig. 8. The engines Whittle built had reverse-flow

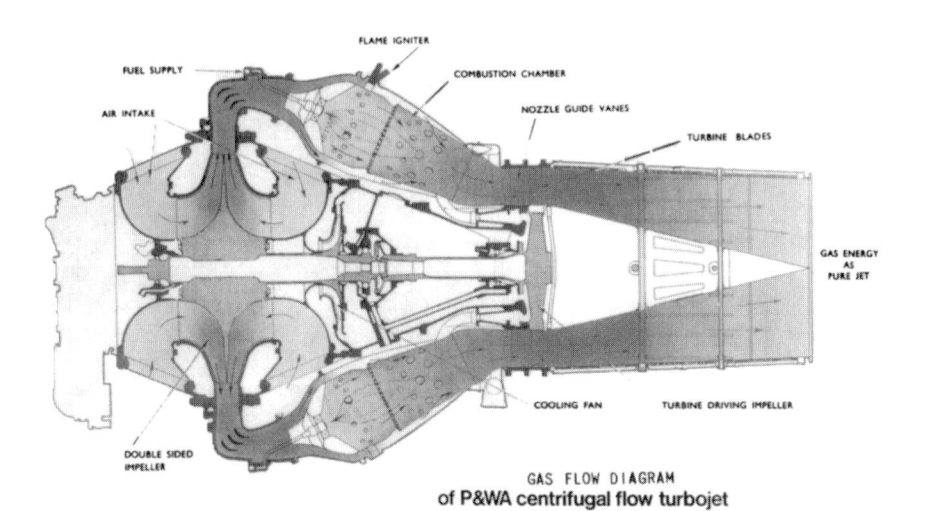

Fig. 8   A cutaway of the Nene flow path (courtesy Pratt & Whitney Archives [12]).

Fig. 9 The Americanized version of the Nene engine (courtesy of Pratt & Whitney Archives).

combustion chambers, which resulted in a larger diameter engine than this "straight-through" combustion system. The engine rotated counter-clockwise when viewed from the rear, the opposite of Pratt & Whitney engines.

AMERICANIZING THE NENE

During AIAA's 1989 celebration of 50 years of jet propulsion, Pratt & Whitney engineer Bill Brown (Fig. 10) spoke about the Nene. What Pratt & Whitney thought "Americanizing" meant was changing the threads from Whitworth to American standards and redesigning accessory drives so that American accessories could be used. However, the Navy said Pratt & Whitney must meet all American specifications including fuel and control systems, materials, and direction of rotation. The Navy eventually waived the direction of rotation requirement and accepted the British preference for counterclockwise rotation as seen from the rear of the engine.

Fig. 10 Bill Brown, Senior Project Engineer for the J42 Program.

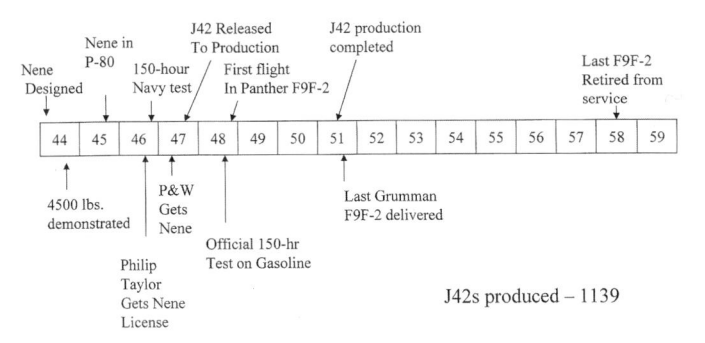

Fig. 11 A time line for the J42 (JT6).

Other surprises were that the burner had to work with a standard sparkplug and ignition system (Rolls-Royce used a torch igniter), and the engine had to operate with leaded gasoline or with JP-4 without any control adjustments. The Navy wanted to have an aircraft that could fly to airports where either aviation gas or jet fuel were available. "Americanizing" included redrawing about 1000 British drawings to standards and nomenclature familiar to the Pratt & Whitney Production Department.

Brown and his team ran the first American version in March 1948, just eight months after receiving the drawings (Fig. 11). They completed the military 150-hr official model test at 5000-lb dry and 5750-lb wet in October 1948, only 17 months after receiving the drawings! Water injection increased the mass flow into the engine to increase the thrust. Brown also did some development work on an afterburner for the J42 but it was not used in the Panther. The afterburner application would come later with the J48 in the F-94.

OVERHAULS AND FOREIGN OBJECT DAMAGE

The J42 had been authorized to go 1000 hr between major overhauls in September 1951. This was the first turbojet to achieve this milestone. It meant that the J42 was quickly put into the class of operational piston engines, which in this time period went from 800 to 1400 hr, depending upon the engine model. Before this, old timers in the business had wondered if turbine engines would ever hold up as well as piston engines in combat. As a result of the J42's impressive durability, the Navy was able to cut back on spare engines and spare parts.

Yet some of the Navy and Marine pilots reported that the engine was ingesting rocks, three-inch tree trunks, and other debris from their rocket and cannon fire without serious damage or failures. This initial engine operating experience demonstrated that the simplicity of the centrifugal flow compressor engine was a strong plus in operational durability at this stage of gas turbine evolution. There was no question that this type of engine was coming to a halt as the axial-flow compressor engines came on line. They were definitely more vulnerable than centrifugal engines because the air stream flowed directly into those engines, compared to the screen-protected centrifugal engines.

Early axial-flow jet engines were very susceptible to foreign object damage (FOD). This meant that a line of men would have to walk down the runway and pick up any objects that might be sucked into the engine inlet. Some inlets had retractable screens that would protect the engine from FOD during take-off. The screens were retracted during climb-out, when there was no risk of ground object being ingested. This meant, of course, that take-off thrust was decreased because of the pressure loss across the screens.

GETTING THE BUGS OUT

The significance of the J42 program was that Pratt & Whitney demonstrated competence in aircraft gas turbine development and made an impressive entrance onto the aircraft gas turbine stage in the Grumman Panther. At this early phase of aircraft gas turbines, companies were not fully aware of the aircraft engine development effort that would be required. Their engineering staffs did exceptional work at designing and building a steam or gas turbine installation where limited development work was done at the site on an engine fixed firmly in place. This kind of work was usually referred to as "getting the bugs out."

On the other hand, aircraft engine development consists of a considerably greater effort than getting the bugs out of a stationary engine installation. Unforeseen design problems crop up because of movement of the engine about three rotational axes and motion in three directions, and the engine gulping inlet air whose quality (in total pressure and velocity profiles) is dependent upon the attitude and speed of the aircraft. Westinghouse and Allis-Chalmers headed toward the exit after they made projections of where aircraft engines and attendant business risks fit in with their overall business plans. General Electric stayed the course.

J42 SEES ACTION

Lieutenant Commander W. T. Amen, on an operational mission in Korea on November 9, 1950, shot down the first MiG-15 in his F9F-2. The engines in the F9F-2 and the MiG-15 were of remarkably similar configuration. Another pilot shared his story via Pratt & Whitney's Public Relations Manager, Bob Holtz [15]:

> Captain George Keys, a Marine Panther pilot, was hit by enemy machine gun fire in Korea while bombing and strafing at tree top levels. One bullet

entered a J42 combustion chamber, tearing a 2.5-in. hole in the cover and then passed through the nozzle guide vanes and entered the turbine, which was whirling at about 12,000 rpm. The bullet swirled around the turbine several times, curling out through the tailpipe. Captain Keys reported he felt only a slight jar, and made several more bombing and strafing runs at full throttle without any engine malfunction or loss of speed. He flew 20 minutes more at 100 percent power while in the target area and then flew for an hour and 20 minutes at cruising speed to reach his base safely with only a slightly higher fuel consumption and a two percent increase in engine revolutions to indicate the major damage suffered by the engine.

The Marine pilots asked P&WA service representative J. D. Young on duty with the squadron in Korea, to pass on their admiration for an engine that "swallows assorted missiles as though it were part of its regular job of dependable operation."

PARKINS FLY-OFF

Table 1 lists specifications of the J42, but something additional was requested by the customer. The Navy wanted to bring the J33 along as an alternate engine for the Grumman Panther. It was similar to the J42 and could fit into the same installation space. General Electric developed the J33 and Allison produced it since General Electric at that time was also into the development of its J35 axial-flow turbojet.

Pratt & Whitney was used to competition, and the threat of a competing engine did not intimidate Luke Hobbs or Wright Parkins. One day Parkins was visiting the Navy, and heard what appeared to be uncharitable comments that the competition was alleged to have made to the Navy and in the media about the Pratt & Whitney engine in the Panther and the positive advantages to the Navy if all the aircraft were powered by the J33. Parkins, short-fused, had no patience with "Excrementum Tauri." Parkins suggested that the Navy have a

TABLE 1 J42/JT6 DATA AT SEA LEVEL

Model	Take-off thrust (lb)	Military thrust (lb)	Cruise thrust (lb)	Weight (lb)	Diameter (in.)	Length (in.)	Aircraft
J42-TT-2	5750 wet 5000	5000	4000	1761	49.5	96.8	XF9F-2
J42-P-4	5750 wet 5000	5000	4000	1721	49.5	101.5	F9F-2
J42-P-6	5750 wet 5000	5000	4000	1729	49.5	101.5	F9F-2
J42-P-8	5750 wet 5000	5000	4000	1729	49.5	101.5	F9F-2 F9F-2P

Fig. 12 F9F-2 Panthers in flight (courtesy of U.S. Navy).

fly-off to settle the matter! According to Larry Carlson (Pratt & Whitney executive), Parkins exclaimed, "Whoever gets to 40,000 feet first wins!" Those of us who worked for him in the Engineering Department could imagine that the transcript of Parkins' remarks most assuredly had to be stamped higher than "Parental Guidance Recommended." Customers appreciated Parkins' straightforwardness; the Navy was elated with the prospect of a real air show.

Remember earlier, in the piston engine era, when the conventional wisdom was that air-cooled engines could not compete with liquid-cooled engines, which were believed to install with less drag than the air-cooled engines? Pratt & Whitney settled that argument by a fly-off between the Allison-powered P-40 and the R-1830 powered P-40. The air-cooled engine actually was a few miles per hour faster than the standard P-40!

When Parkins returned to East Hartford, you can imagine the ensuing turmoil. Pratt & Whitney got a J-42 powered Panther and the engineers went to work refining the installation. They discovered that the engine was sucking in more air through the inlet than the engine used or what was needed for adequate engine compartment cooling. They installed an annular blocking device to minimize the excess airflow. Harold Archer, Pratt & Whitney test pilot, ran test flights to 40,000 ft. He took off from Rentschler Field and minutes later would be someplace in New Hampshire at 40,000 ft.

In November 1948, the J-42 powered Panther (Fig. 12) reached 40,000 ft before the competition. However, the final test was to see what each engine could do in an altitude test facility, the Willgoos Laboratory. The J42 was the winner again and the Navy had to be pleased. Also, in Derby, U.K., Rolls-Royce must have had a joyful vicarious experience that cried out for a celebratory pint!

THE J48 (JT7)

In December 1950, Ray Dorey, of Rolls-Royce, explained the design philosophy of the Tay to Art Smith (who we met at the end of Chapter 5) [16]:

> Regarding your query on the rating of the above engine, when the design of this was first conceived, the engine was based on a 30 percent scale up

of the Nene. If you check with your Nene drawings you will find that the diffuser area and nozzle guide vane area are all 30 percent up on your standard production Nene engine. Taking the Nene at a basic rating of 5000 lb, this gives a normal rating for the Tay of 6500 lb.

Since that date, in order to improve the surge characteristics at high altitude ... we chose to increase the capacity of the engine still further by 5 percent to put the working line further away from the surge line. The engine as it stands today therefore should, on a comparable basis to the Nene, give a rating of 6800-lb plus, as indeed, most of the engines that we have both produced have done.

As you know, the Tay engine in the U.K. has no direct application to a military aircraft, but we are committed to complete a Type Test. In order to do this satisfactorily, we decided to adhere to the original folder rating of 6250 lb. This we consider necessary to avoid further expense and development to establish it at what has now become its natural rating of 6800 lb. The fact that we are doing this test at 6250 lb is only a matter of expediency and does not represent in any way our views on the engine's potential.

ENGINE DESCRIPTION

The J48 started out at a thrust level of 6250 lb. It was developed up to 7250 lb of thrust through a combination of increases in airflow and turbine inlet temperature. It was rated at 8750 lb with an afterburner. Table 2 lists the

Fig. 13   J48 with afterburner (courtesy of Pratt & Whitney Archives).

Table 2 J48/JT7 Data at Sea Level

Model	Take-off thrust (lb)	Military thrust (lb)	Cruise thrust (lb)	Weight (lb)	Diameter (in.)	Length (in.)	Aircraft
J48-P-1	8000 A/B 6000	8000 A/B 6000	5000	2070	50	202	XF-93A YF-94C
J48-P-2	7000 Wet 6250	6250	5000	2075	50	106.75	F9F-5
J48-P-3	8000 A/B	6000	5000	2600	50.25	202	F-94C
J48-P-5	8750 A/B	6350	5250	2760	50.25	202	F-94C F-94D Dassault MD 452-08
J48-P-6	7000 Wet 6250	6250	5000	2040	50.25	106.75	F9F-5 F9F-6
J48-P-7	8750 A/B	6350	5250	2725	50.0	202	F-94C
J48-P-8	7250	7250	5600	2080	50.5	109.75	F9F-6 F9F-6P
J48-P-8A	7250	7250	5600	2101	50.50	109.75	F9F-8, -8T, -8P

WWII ENDS AND TURBOJET DEVELOPMENT BEGINS 211

Fig. 14 J48 without afterburner—note the screens protecting the inlet to the centrifugal compressor (courtesy of Pratt & Whitney Archives).

specifications for this engine. The J48 came in two configurations, with and without an afterburner (Figs. 13 and 14). The non-afterburning version was used in the Grumman Cougar F9F-6 model with swept wings (Fig. 15). The afterburning J48 was used principally in the Lockheed F-94C (Fig. 16) and the North American XF-93A (Fig. 17).

Fig. 15 The Cougar looked a lot like the Panther, but with swept-back wings (courtesy of U.S. Navy).

Fig. 16 Air Force F-94C with the J48-P-7 (courtesy of U.S. Air Force).

J48 APPLICATIONS

The Air Force was interested in an all-weather interceptor (F-94) and a penetration fighter (YF-93A), both of which could use the J48 thrust level with an afterburner. Bill Brown had the challenge of developing an afterburner for those applications. Sir Frank Whittle and Dr. Hans von Ohain said that the most challenging problem in a turbojet engine was to make the combustion chamber produce its design temperature rise in the compressor discharge stream before the flow entered the turbine. The afterburner was a similar technical challenge plus variable geometry (the jet nozzle had to open properly to avoid a compressor surge). Brown had to scour the country for information on the combustion systems that resembled what goes on in an afterburner. The first experimental afterburner tests were on the J42. Brown claimed that the J48 afterburner was the first in the free world that could ignite and operate to the ceiling of the aircraft.

Fig. 17 Air Force F-93A (courtesy of U.S. Air Force Air & Space Museum).

Playing Leapfrog in Late 1947

In the first two years following WWII, Pratt & Whitney was engaged in a "total immersion" educational course in aircraft gas turbines that comprised design, development, and production efforts:

1) *Design*: Perry Pratt's Technical & Research Group mapped out the design of a turboprop (T34/PT2). The greatest challenge was in the axial-flow compressor. The first engine to test was in late 1947. It was not until early 1949 that the engine would run under its own power.

2) *Development*: Also in 1947, the company stepped up to the commitments of "Americanizing" the world's most powerful centrifugal turbojets. The development effort provided two excellent products—the J42/JT6 and the J48/JT7.

3) *Production*: The manufacturing department was getting experience with axial-flow gas turbines of Westinghouse design, about 130 engines. In addition, the centrifugal J42 had a production run of 1139 engines and the J48 amounted to 4108 production engines.

The company was accumulating experience with turbines, as evidenced in Fig. 18 showing the growth in turbine blade size since the first Whittle engine in 1937.

By the end of 1947, Pratt & Whitney was still struggling with what seemed to be an almost vertical learning curve as far as design of a new turbojet was concerned. The Westinghouse J30 production engines were underway. The commitments to the J42 and J48 were not very far along. The "all Pratt &

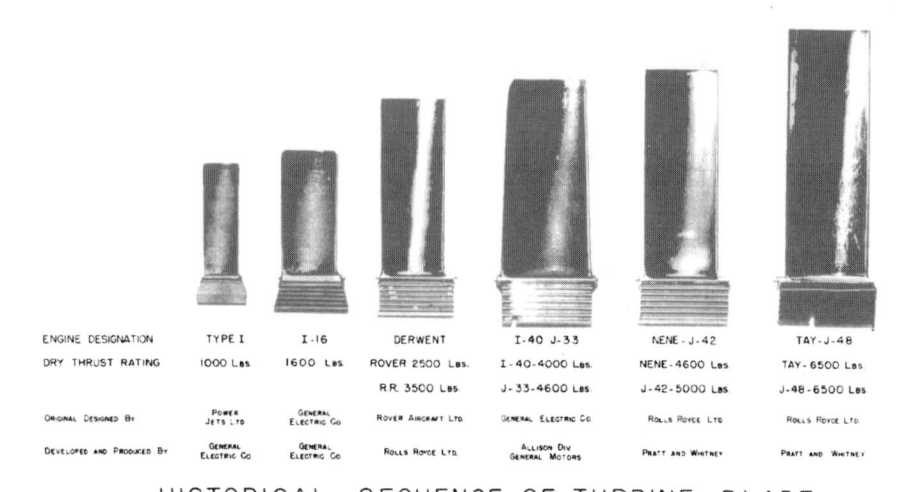

Fig. 18 Progression of turbine blades in centrifugal engine evolution (courtesy of Pratt & Whitney Archives).

Whitney design" for the T34 was not inspiring. There was no question of fulfilling these commitments. The question was: Can we survive until then? The situation was analogous to that of the French army in WWI. General Foch sent this telegram to his commander regarding the defense of Paris: "My center gives way. My right recedes. The situation is excellent. I shall attack."

Pratt & Whitney "attacked" in late 1947, taking the first step in the evolution of an engine that would seize the world's attention!

REFERENCES

[1] Lippincott, H., "Transcription of Interviews with Lionel Clark," Pratt & Whitney Archives, April 1961–Feb. 1985.
[2] Gywnn-Jones, T., *Farther and Faster*, Smithsonian Institution Press, Washington, D.C., 1991, p. 158.
[3] Clark, L. B., "The History of Post's Wasp 3088," *The Bee-Hive*, United Aircraft Corporation, Pratt & Whitney Aircraft Division, United Technologies Corporation, Pratt & Whitney Archives, Sept. 1933.
[4] "United States Army & Navy Engine Training School," Pratt & Whitney Archives, 1943.
[5] "Home from the Wars," *The Bee-Hive*, United Aircraft Corporation, Pratt & Whitney Aircraft Division, United Technologies Corporation, Pratt & Whitney Archives, Jan. 1946.
[6] Lippincott, H., "Recollections of a Service Representative," Pratt & Whitney Archives.
[7] Lyman, L. D., "Lindbergh, 'Tech Rep,'" *The Bee-Hive*, United Aircraft Corporation, Pratt & Whitney Aircraft Division, United Technologies Corporation, Pratt & Whitney Archives, 1949.
[8] Schlaifer, R., "Development of Aircraft Engines," Division of Research, Graduate School of Business Administration, Harvard Univ., Cambridge, MA, 1950.
[9] Hobbs, L., "Memo to Paul Fisher Regarding History of the Pratt & Whitney Turbine Development," Pratt & Whitney Archives, March 13, 1962.
[10] Andrews, H., "Grumman Panther: Fighters of the Fifties," *Aeroplane Monthly*, London, U.K., Aug. 1976.
[11] Hooker, S., *Not Much of an Engineer*, Airlife Publishing, Crowood Press, Wiltshire, U.K., 1984, p. 99.
[12] Deacon, W. K., and McCaul, G. J., "150-Hour Qualification Test on Rolls-Royce Nene 1 Turbojet at U.S. Navy Aero Engine Laboratory," report, Taylor Turbine Corporation, New York, NY, Pratt & Whitney Archives, Feb. 1, 1947.
[13] Hobbs, L., "Telegram to Ernest W. Hives," Hobbs Files, Pratt & Whitney Archives, Oct. 7, 1947.
[14] Hives, E. W., "Letter to Luke Hobbs," Hobbs Files, Pratt & Whitney Archives, Oct. 26, 1947.
[15] Hotz, R., "News from Pratt & Whitney Aircraft," Pratt & Whitney Archives, Sept. 16, 1951.
[16] Dorey, R. N., "Letter to Art Smith from Rolls-Royce Limited," Derby, U.K., Pratt & Whitney Archives, Dec. 21, 1950.

Chapter 7

BIRTH OF THE TWO-SPOOL TURBOJET

INTRODUCTION

In the spring of 1949, Luke Hobbs recommended to Pratt & Whitney's Engineering Department that they scrap the twin-spool turbojet[*] they had under development for the B-52 and start over with a different compressor design. Of course the target date of delivery of prototype engines to Boeing remained the same—now less than 36 months to the XB-52 flight test! If the J57 turned out to be a miserable failure, Hobbs would be the one to blame. However, on the happy occasion of the Sperry Award to him and Perry Pratt, on January 8, 1973, Hobbs gave credit to those responsible for the victory: "It is not only pleasing, but is further made a most satisfying thing to me to see Perry Pratt receive the credit long due him for pulling out of almost thin air the advanced technology required at that time, and the recognition ... of the contributions of Andy Willgoos and Dick Soderberg" [1].

As WWII came to a close, Luke Hobbs was aware of Pratt & Whitney's serious deficiency in the aircraft gas turbine marketplace. He presented the management with three engine projects that the company should take on, in the following sequence: a small turboprop (the PT2/T34, see Chapter 5), a small turbojet, and a large turbojet.

As Pratt & Whitney moved into aircraft gas turbines, it had to create a new vocabulary for company designations of its engines (Table 1). The letters PT referred to prop turbine (turboprop) and the letters JT represented a jet turbine (turbojet). At the same time, Pratt & Whitney carried over from the piston engines its tradition of designating the "maturity" of a specific engine model as A, B, or C. The A was the initial model, the B was an improvement, and the C was the most recent. This designation system had been created during the period of the early Pratt & Whitney aircraft gas turbines.

A variation of the system was the JT3-6, the engine proposed to the Navy for a 7500-lb thrust turbojet. The designation meant that this engine was turbojet number 3 with a 6:1 compressor pressure ratio. The next engine was JT3-8, which had a dual compressor on separate shafts. The number 8 refers

---

[*]*Twin spool* and *two spool* are used interchangeably throughout, but "twin" should not imply that the two spools look alike. Each spool has a different number of stages.

TABLE 1 ENGINE DESIGNATIONS

Company designation	Military designation	Comments
PT1 (Turboprop)	T32	Supercharging piston compressor driven by a free-piston diesel whose exhaust drove a turbine turning a propeller. No application.
PT2	T34	Turboprop in C-133.
PT3	T52	Twin-spool turboprop. No application. Study engines only—in the 5700- to 8300-hp range.
PT4	T45	10,000-hp turboprop for B-52. Study engine only with component testing.
JT1 (Turbojet)		Designation not used.
JT2		Designation not used.
JT3-6		7500-lb thrust turbojet proposed to the Navy and rejected. Overall pressure ratio was 6:1 in a single-spool configuration.
JT3-8		First U.S. twin-spool turbojet. Overall pressure ratio of 8:1.
JT3-10		Twin-spool turbojet with 10:1 overall pressure ratio.
JT3-10A		Same as above with repackaged accessories to reduce overall diameter of nacelle.
JT3A (JT3-10B)	J57	Turbojet engine for the B-52 with wasp waist design.

to an overall pressure ratio of 8:1. The JT3-10, also with a dual compressor, had an overall pressure ratio of 10:1. Once the final design of the J57 had been established, the designation became JT3A.

GENESIS OF THE J57

Figure 1 maps out the various steps to the final JT3A (J57) for the B-52 flight test program. Keep in mind that the turbojet program was one continuous effort toward a high-performance engine. The PT4 (T45) was a design study and experimental effort focused on high-pressure-ratio compressors as well as the burners.

The series that led up to the JT3A (J57) started with the turbojet studies that in turn led to the JT3-6 (7500-lb turbojet) proposal to the Navy. The Air Force sponsored the PT4, which was a 10,000-hp turboprop targeted to an aircraft such as the B-52. The turbojet evolution continued with the JT3-8 twin-spool engine. Subsequently this engine was upgraded to a 10:1 overall pressure ratio turbojet (JT3-10), which was expected to be the B-52 engine. The final and the successful step was the "wasp waist" JT3A twin-spool turbojet that went on to more than a 20,000+ engine production run.

Hobbs had his next step in mind after the PT2/T34. The analytical design studies pointed to a 7500-lb thrust engine started in March 1946. By November, Pratt & Whitney had completed design studies of engines with pressure ratios in the range from 4 to 6 with a thrust rating up to 8200 lb. At this time the state-of-the-art in compressor pressure ratios was in the 4 : 1 class and the most powerful engine was about 5000-lb thrust (the Rolls-Royce Nene).

The 6:1 pressure ratio design was selected. Walt Ledwith, in the Design Department, made the first layout in early 1947. The detail design began in May and the engine was designated as the JT3-6. The 6:1 compressor had an outer diameter of 36 in. and a hub/tip ratio of 0.66. The hub/tip ratio is the ratio of the diameter of the blade root to the blade tip diameter. Hobbs proposed the development of this turbojet to the Navy, but the initial reaction was disappointing [2]:

> I carried two engines down to the Bureau of Aeronautics recommending that they start experimental projects on both the 7500 and 10,000-lb units. The chief technical man in the Bureau was called in and his advice to the Bureau was to buy neither engine. He was certain that the Navy would never need an engine of 7500-lb of thrust.
>
> I then put on about a two-month selling campaign on the topside of the Bureau to convince them by simply citing past history that they should have, and eventually would have to have, these larger engines. So finally it was decided by those in authority that the 7500-lb thrust engine was a fairly good bet—they should go for it.

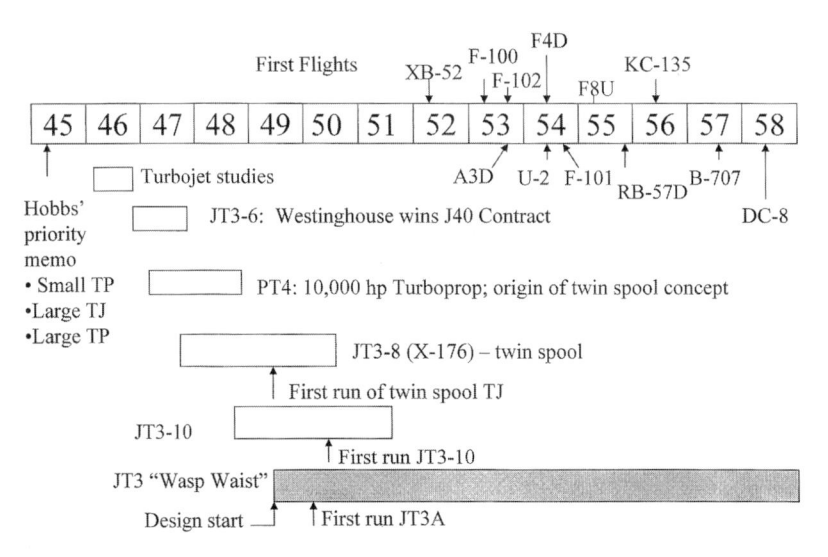

Fig. 1 Evolution of the JT3A (J57).

The Navy asked Allison, Westinghouse and others to bid on a 7500-lb thrust turbojet. Westinghouse won the contract for what became the J40 turbojet. Hobbs continued his story [2]:

> Pratt & Whitney lost out and, of course, having met difficulty in selling them 7500 lb, you couldn't sell them 10,000 lb.... The Research and Development Board, their watchword was economy, economy ... so it meant that even if ... we could sell the Air Force on a big jet, we probably could not get it by the Research and Development Board. They would count it as being a duplicate of this thing the Navy was going into [the J40].
>
> We knew that the Air Force was ... strong for long-range bombardment. It was one of their basic elements in all of their strategy.... They had a B-52 coming through at that time around a turbine effort that the Wright [Aeronautical] Company had started—propeller-turbine (T35 at 5500 hp). We knew that would be a high-consumption engine just by its very design and so we thought our best bet would probably be to sell the Air Force on a high-compression, low fuel consumption turboprop engine for this bomber. At the same time, we would design it so that we could convert it to a jet engine at any time and, in fact, our hope was that after we once got it established, then we could get it converted to a jet.

As we look back now, the Navy did Pratt & Whitney a gigantic favor by rejecting the JT3-6.

HOBBS AND PRATT APPROACH THE AIR FORCE

Perry Pratt met with Air Force engine expert, Opie Chenoweth, at Wright Field. That is where he learned that Wright Aeronautical was developing a turboprop (T35) for the B-52 in the early days of the program. Chenoweth told Pratt, "If you want to get in the jet field with any strength, you've got to aim at the B-52" [3]. Perry began working with George Schairer of Boeing, who was leaning toward a jet-powered aircraft in place of a turboprop-powered bomber. Also, at this time, the feeling of some in the Air Force was that if the B-52 became turbojet-powered, the J40 would be the engine because it was already under development under Navy sponsorship.

Pratt & Whitney's policy was not to produce other manufacturer's engines under license. Rentschler wanted the Navy and Air Force to understand that Pratt & Whitney was in business to build and sell its own engines. Hobbs went to Wright Field to meet with General Craigie and deliver the Rentschler message, which was well received. Subsequently, Hobbs met with General Powers, who was the head of the Air Force Engineering and Development in Washington. Powers mentioned that General "Tooey" Spaatz, who replaced General "Hap" Arnold as head of the Air Force, had asked Powers if he had anything going with Pratt & Whitney. Spaatz was, of course, aware of

Pratt & Whitney's immense contribution to the WWII aircraft engine effort in such aircraft as the P-47, C-47, C-54, B-24, and B-26. In addition, perhaps he recalled a much earlier meeting with Chance Vought and Frederick Rentschler who invited a young Major Spaatz to fly the first Wasp-powered Navy fighter at the Naval Air Station at Anacostia.

General Powers quoted Spaatz to Luke Hobbs, saying, "This is the best engine outfit in the business and you haven't a single project with them. This is wrong, I think." Then Powers asked Hobbs, "Haven't you a turboprop on the boards or anything worthwhile that we might back?" Hobbs came back recommending a 10,000-hp turboprop engine, which later became the T45 (PT4) [4]. The go-ahead for the large turboprop program was received August 1947.

Meeting the Challenge of Creating Enough Propulsion

Cliff Simpson, at the Air Force Aero Propulsion Laboratory at Wright Field, had some interesting observations about Pratt & Whitney's bold jump into the unknown [5]:

> Pratt & Whitney had one opinion, everyone else had another. The opposing opinion to Pratt & Whitney was most clearly presented by General Electric, so their version is used. The opinions can be summarized in that Pratt & Whitney considered 12.6:1 pressure ratio to be optimum and everyone else said 6:1 which would get a 10 percent improvement in fuel consumption. The opinion for 6:1 was based on a study of the effect of leakage on Cycle Selection.... If leakage characteristics were selected compatible with then current state-of-the-art leakage, the optimum pressure ratio was about 6:1.
>
> This, therefore, was the expressed ... opinion of 4/5 of the industry. Pratt & Whitney's conclusion was that better air sealing was essential and 12.6:1 was the best pressure ratio.... The engines resulting from this initial attack were the J57, J52, and J75 and the two fan versions used by most airlines today [1950s and 1960s]—the JT3D and JT8D.
>
> The old Power Plant Laboratory decided to pursue the T45, which was altered to the J57 about one year later. So, for the next three years we took gaff from the remaining big four in industry.

Hobbs' philosophy in the T45 program was to design the 10,000-hp turboprop and test those components that would be compatible with a turbojet, using the same compressor, combustion system, and high-pressure turbine (Fig. 2). In the end, no PT4 engine was ever built. The T45 program emphasized component testing.

It is easy to see why the Air Force was interested in turboprop power for the B-52. First of all, the British were already demonstrating long-range aircraft with turboprops. The Bristol Britannia was underway in 1948, aiming

Fig. 2 Mock up of the PT4/T45 (courtesy of Pratt & Whitney Archives).

for a first flight in 1952, using Bristol Proteus turboprops, which eventually developed 3900–4100 hp and had ranges up to 5000–6000 miles. The Bristol Brabazon was flying in 1949 with eight Bristol Centaurus piston engines, installed as four pairs. A planned Brabazon development would have been powered by eight Proteus turboprops (also in pairs) but that was not to be. This aircraft was targeted for long-range applications. Also underway at that time was the huge Saunders-Roe Princess flying boat that was powered by 10 Bristol Proteus turboprops to carry 220 passengers. All of these aircraft, of course, were not in service at the time the Air Force was thinking about the B-52 and the B-60. However, the conventional wisdom of the day preached that long-range aircraft had to be powered by turboprops rather than turbojets, because turbojets had poorer specific fuel consumption compared to turboprops. The conventional wisdom seemed to make sense until a new element was brought into the discussion—a more efficient turbojet that enabled a much faster aircraft.

Here is an easy way to look at this propulsion challenge. An aircraft gas turbine converts the chemical energy in the fuel to propulsive energy in two steps. The first step is to convert the chemical energy to a high-temperature, high-pressure gas, which can be used in the second step in either of two ways. The efficiency of this energy conversion process is called the *thermal efficiency*. This high-temperature gas could be used to power a turbine–propeller combination to provide thrust. Or it could be used to provide a high-speed jet, which results in thrust. The efficiency of this second step is called the *propulsive efficiency*. At propeller-powered flight speeds, it makes sense to go with the turboprop, because the propulsive efficiency of the propeller is much higher than that of the jet. However, if one designs the aircraft for high-subsonic flight speeds and high altitude, then the propulsive efficiency of the jet becomes of more interest to those who like high-speed, long-range aircraft. Pratt & Whitney increased the thermal efficiency by use of a high pressure-ratio compressor system and provided a high cruise speed, beyond the reach of efficient propeller operation.

After WWII, George Schairer of Boeing, aware of the German pre-war work on aerodynamics, recommended that the B-47 have swept-back wings. The B-47 was a great step in bomber design but was too small to meet the Air Force requirements for a heavy bomber. When the engineers at Boeing working on the B-52 saw that the J-57 had a much higher thrust level and a much higher thermal efficiency than the then state-of-the-art turbojet, they could see a higher speed turbojet-powered B-52 as a real possibility.

Concerns Regarding High-Pressure-Ratio Compressors

Carl Kopplin, in an engine performance group in early 1947, was making a study of the starting horsepower of a turboprop with an 8:1 compressor pressure ratio. The power requirement was unreasonably high. Kopplin suggested a two-spool compressor. Dick Smith (the project engineer in charge of the Turbine Design and Research Group) and Bill Sens (the project engineer in charge of the Advanced Engine Performance Group) had conducted general studies of twin-spool compressors in the Spring of 1946. Their work was of great interest for this PT4 study exercise. In November 1947, the JT3-6 compressor was modified to become part of a two-spool turboprop with an 8:1 overall pressure ratio [6].

I asked Walt Doll (the project engineer in charge of the Compressor Design and Research Group) why Pratt & Whitney decided on two spools for the compressor instead of variable geometry. Walt put it succinctly: When a high-pressure-ratio compressor is operating way off its design point, such as during starting, the geometry of each compressor stage is far from optimum—too much angle of attack on the blades and vanes in the early stages compared to the design point. Varying the stators in the early stages helps a great deal. On the other hand, one could divide the compressor into two lower pressure ratio compressors that run at different speeds. This configuration, of course, required more bearings because of the two independent spools. Either approach was expected to work. Pratt & Whitney knew a lot more about bearings than it did about variable stators in axial-flow compressors. Therefore, Pratt & Whitney selected the two-spool approach to an 8:1 compressor.

Technical Concerns of a Twin-Spool Turbojet

Dr. Stanley Hooker of Bristol, previously with Rolls-Royce, was familiar with two-spool turboprop engines. Rolls-Royce's first all new turboprop design was the Clyde, which had two spools in 1946. The low spool consisted of the propeller, reduction gear, a nine-stage axial compressor, and a low-pressure power turbine. The high spool had a centrifugal compressor and a high-pressure turbine. The power output was about 3000 shaft hp with 600 lb of jet thrust and was demonstrated in 1945 on a test stand. Flight tests were made in 1947.

The Clyde operated satisfactorily with its two separate spools. In the starting days of the JT3, there was a great deal of communication between Rolls-Royce and Pratt & Whitney because of the J42 (Nene) relationship. In one of the communications, Stanley Hooker expressed the opinion that a two-spool turboprop made sense but not a two-spool turbojet! Naturally such an opinion from a highly respected "engine man" drew a lot of attention at Pratt & Whitney, especially as the JT3-8 was moving irrevocably down the two-spool path. Hobbs asked Perry Pratt to follow up on this, without revealing that Pratt & Whitney was betting the farm on a twin-spool turbojet. Hooker, with his typical openness, sent Perry an elaborate mathematical analysis in an April 1949 letter [7]:

> If a jet engine incorporates split compressors, then at the expansion end there will be three controlling areas, namely the stator area before the first turbine, the stator area before the second turbine, and the final jet nozzle. If the turbine wheel is situated such that there is sonic velocity in the area ahead of it and also in the area behind it, then the temperature ratio across that turbine wheel is fixed. That is to say, the temperature drop across the wheel is proportional to the total temperature ahead of it. Since the heat drop across the turbine must be equal to the heat rise across the compressor which it drives, it follows that the temperature rise across the compressor must also be proportional to the flame temperature. Normally, the heat rise across the compressor is proportional to its rpm squared.
>
> Thus in a two shaft engine, when sonic velocity occurs at the final jet nozzle both turbines will be situated between the two areas in each of which sonic conditions exist. It follows, therefore, that the speed of both will be proportional to the flame temperature, and that they will run at a fixed speed ratio. Once this state of affairs arises the compressors might just as well be on one shaft.
>
> Actually, the compressors run such that the temperature rise across each is in a fixed ratio, once sonic velocity occurs at all of the three areas mentioned earlier.... This difficulty can be overcome by a variable final nozzle area, or by bleeding air from between the two compressors.

Stanley Hooker was not the only one concerned about twin spools. Some at Pratt & Whitney were talking about "aerodynamic lockup" of the two spools. Andy Willgoos, Chief Engineer in charge of gas turbines, was taking no chances in the design and insisted on a design that could bolt the two spools together if necessary.

Perry asked Don Jordan (Fig. 3), development engineer for the project, to make sense out of Hooker's analysis. Don pointed out that Hooker was using a relationship between the temperature rise across a compressor as proportional to the square of the rotational speed. This relationship was all right for centrifugal compressors but did not hold for axial-flow compressors where

Fig. 3 From left to right, these are the principal engineers of the J57: Gil Cole (chief, Engine Design), Gus Hasbrouck (chief design engineer), Don Jordan (development engineer), Ross Begg (development engineer), Barney Schmickrath (assistant chief engineer), Dick Baseler (project engineer), and Karl Fransson (chief, Design Analysis) (courtesy of Pratt & Whitney Archives).

the temperature rise was also a function of the pressure ratio. At any rate, just a couple of months later in June of 1949, the JT3-8 ran and demonstrated that all was well with a twin-spool turbojet. Hooker also must have been relieved in May 1950 when his twin-spool Olympus turbojet spun over without the dreaded "twin-spool lockup phenomenon."

AIR FORCE APPROVAL FOR PT4 TO JT3 SWITCH

In 1948, when the XB-52 was coming up for review at Wright Field, the configuration included the Wright Aeronautical XT35 Typhoon turboprop. This engine had already been flight tested in the nose of a B-17 in 1947. Pratt & Whitney, of course, was somewhat behind Wright Aeronautical with the PT4/T45, which was just getting underway at that time. Turboprops, as mentioned earlier, were the engines of choice for long-range aircraft. In 1948, the Rand organization issued a report promoting the idea that the B-47 should be powered by turboprop engines to achieve intercontinental range.

However, Colonel Warden (discussed in more detail later in this chapter), from the Air Force Project Office, was not happy with propeller engines. In a program review in Dayton in October 1948, he asked about jet power. That was no problem because the Westinghouse J40 was coming along in development and was expected to be available when needed in the aircraft development program. The Boeing team of Ed Wells (vice president of Engineering), George Schairer (aerodynamics specialist), Bob Withington (performance engineer), and Vaughn Blumenthal (performance engineer) were at Wright

Field for the briefing. This spectacular team spent the subsequent week in Dayton to prepare a new proposal to the Air Force for the B-52 [8].

A month earlier, Hobbs had convinced the Air Force that the T45 program should be restructured as a turbojet to produce about 10,000-lb thrust. The Air Force agreed and the engine designation became the J57, which became a backup for the J40.

The Boeing team in Dayton sketched out a huge bomber to be powered by eight J57 engines, which had a greater thrust than the J40. Boeing's William Cook in his book, *The Road to The 707*, said the decision to go ahead with the J57 was based upon the compressor test data that Perry Pratt showed George Schairer earlier [8]. This recommendation certainly must have been considered somewhat of a risk because the JT3A was not even designed yet, and it would not be until about 15 months later (January 1950) when the first J57 (JT3A) experimental engine would run.

DESCRIPTION OF JT3-8

The first twin-spool turbojet in the JT3 development program (JT3-8) first ran in June 1949. This engine (Fig. 4), with an 8:1 overall pressure ratio, was experimental engine X-176, which eventually went to the Smithsonian for display. The JT3A would have a 12:1 overall pressure ratio. The engine assembly began in April and was completed on June 11, 1949. The engine was installed in an X-31 stand and made a preliminary run on June 28 for a half hour at only 2600 rpm when a fuel leak occurred. The engine was run on the next day for a little over 20 minutes. Then, after teardown and inspection, it was reassembled with a new fuel pump for the third run. The engine got

Fig. 4   JT3-8 (right front 3/4 view) (courtesy of Pratt & Whitney Archives).

Fig. 5 High rotor from JT3-8 (six compressor stages and one turbine stage) (courtesy of Pratt & Whitney Archives).

up to about 98 percent of the high rotor design speed without the predicted aerodynamic rotor lockup occurring. The test program continued into the subsequent year with different sized tailpipe nozzle areas. In addition, the challenges of off-design compressor stall were studied and bleed valves were installed. The experience on this 8:1 dual-compressor engine helped to make the J57 configuration proceed more smoothly. Most of this testing was done before the J57 (JT3A) was tested in January 1950. The engine accumulated about 145 hr before being put into storage. The turbine was shrouded, which minimized aerodynamic end losses at the blade tips. This was a distinguishing mark of Pratt & Whitney since the early days of the PT1 and the PT2. It was also an indication of Soderberg's influence (see Chap. 5). The engine never encountered the feared "aerodynamic lockup" where the two rotors synchronized their speeds (Figs. 5 and 6).

Fig. 6 Low rotor from X-176 with a seven-stage low compressor and a single-stage low-pressure turbine (courtesy of Pratt & Whitney Archives).

Hobbs knew that to be successful he would have to offer the Air Force a more powerful engine than the competition, which was now Westinghouse's J40, and provide engines for the initial flight test program before the competition could deliver its engines. He also realized that he would have to offer an engine better than then current state of the art, and moved in the direction of higher overall pressure ratio for lower specific fuel consumption. The procession to a higher overall pressure ratio was already under way in March 1948 when a supercharged version of the JT3-8 was designed by adding two compressor stages to the low compressor. These stages increased the engine airflow to achieve about 10,000-lb thrust. This new engine was the JT3-10, to be built and tested as the engine for the B-52.

DESCRIPTION OF THE JT3-10

The JT3-10 began in March 1948 in a series of studies stemming from the JT3-8, with the addition of two supercharging stages added to the front compressor. The JT3-10 engine had nine stages on the low compressor and six stages on the high compressor (Fig. 7). Each compressor was driven by a single-stage turbine. Six months after the JT3-10 studies began, in September 1948, the Air Force modified the T45 contract to a JT3-10 type of high-overall pressure-ratio turbojet. This engine configuration was first run on February 1950. This particular configuration, however, did not end up as the J57 and ran a month after the J57 (JT3A).

Hobbs says that his original specification thrust number for the J57 was about 8700 lb, but the B-52 kept growing. Then Pratt & Whitney pushed it up to 9700 or 9800 lb. When Mr. Rentschler heard about it, he exclaimed, "Hell,

Fig. 7 Experimental engine X-184 (JT3-10) (courtesy of Pratt & Whitney Archives).

you people aren't salesmen! What the hell is 9700? He started calling it a 10,000-pound engine." [9].

Hobbs now had concrete proof that Perry Pratt and his team had achieved a remarkable technical renaissance in aircraft gas turbines. At first, he was satisfied to have engines on the second flight test airplane (YB-52) because the timing was too tight to get into the XB-52. He knew now that he had a more powerful engine than the J40 coming along and it was "an Air Force engine." The J40 was to be the engine for the B-52 after the turboprop was ruled out. This would put more pressure on Westinghouse, which was beginning to have trouble in J40 development. Then Hobbs was determined to get into the first airplane, the XB-52, to knock out the competition. It turned out the YB-52 actually became the first test aircraft because of a mishap with the XB-52.

Hobbs thought the risk was worth it. His great depth of knowledge and experience about engines and the men who made them gave Hobbs the resources to make the right decision to get the J-57 into the B-52. His piston engine experience told him the secret to success was to get a more powerful engine into flight testing before the competition did. This same scene would be played later in the Republic F-105 when the P&W J75 got there "bigger and sooner" than the designated engine, the Wright Aeronautical J67. Little could he have ever imagined that Pratt & Whitney engines would still be in the B-52 more than a half century later!

SODERBERG RECOMMENDS A NEW START

In September 1948, the Air Force gave Pratt & Whitney its approval to get on with a 10,000-lb thrust turbojet for the B-52. The low and high compressors of JT3-8 were indicating poorer performance than expected. The low compressor of the JT3-10 engine would not run until November 1949. After the turbojet program had been going along for about seven months, Soderberg came to Perry Pratt and told him to stop. Soderberg strongly recommended a completely new approach to the compressor design that would result in a weight-saving of more than 500 lb and achieving the design performance objectives.

This news arrived at a time of upheaval in the Engineering Department. Andy Willgoos, who was Perry Pratt's boss, had died suddenly. He had run half the Engineering Department on the advanced engines while Wright Parkins was in charge of the engineering of current piston engines. Now Parkins had to take over the entire engineering function. So, with less than a week on the job, he was faced with a decision that could make or break the company. Soderberg convinced Perry Pratt and then Perry convinced Parkins that they must start over again if Pratt & Whitney was to get the J57 into the B-52. Parkins then made the case with Hobbs. A memo from Perry Pratt and Dick Soderberg, which later became the basis for a memo from Parkins to

Hobbs, recommended a new course of action for design of a twin-spool turbojet for the B-52.

The essence of the Soderberg and Pratt memo is as follows, regarding the JT3-10A that had been committed to the B-52. The JT3-10A differed from the earlier JT3-10 in the arrangement of the accessories around the engine waist to improve the installation in the Boeing nacelle. The engine was the same in the JT3-10 and JT3-10A. The JT3-10B is what Soderberg and Pratt were proposing be the engine configuration for the B-52 in place of the then current JT3-10A. This 10B engine would later be named the JT3A where the letter A implies "the early model" [10].

> These to whom copies of this memo are addressed met in conference yesterday with sectional layouts of two versions of the JT3 production type engine. One of these is called the 10A and the other the 10B. As you know the design of the 10A is well along. Detailing of the engine started May 1 and is scheduled to be finished August 1, with the first engine parts scheduled to be ready by December 1, 1949. The first 50-hour development test on the production type engine is scheduled to be finished by December 1, 1950.
>
> A shift at this late date from the 10A to the 10B design will delay the program four months if we assume that the same amount of time will be required for developing either engine into a satisfactory production article. This ... is difficult to analyze. All of us are sure that the 10B will present much less mechanical problems than the 10A and many of the thermal problems should be easier to solve. From these standpoints there is little if anything to be said in favor of the 10A.

Parkins mentioned other points brought up by Soderberg and Pratt [10]. The 10B had turbine disc diameters of about 21 in. compared to the 33 in. for the 10A. This smaller diameter moved the natural frequency of the low-pressure shaft out of the operating range. The 10B low-pressure compressor would be a constant inner diameter (ID) type scaled from the PT2, a compressor that was performing satisfactorily. The 10B high compressor would still be a constant outer diameter (OD) type, but brought inward closer to the engine centerline to match the new low-compressor flow path. The average blade heights for the high compressor would be greater than the JT3-10A, another factor to improve efficiency. The last stage compressor blades for the JT3-10A had an aspect ratio less than one. The JT3-10B increased the aspect ratio. The 10B engine, because of being of the wasp waist configuration was expected to be at least 500 lb lighter.

Another factor, not mentioned by Soderberg and Pratt, was the installation advantages of the wasp waist design, because the nacelle diameter could be reduced from that of the 10A design by packaging the accessories closer to the engine centerline. Parkins ended his memo to Hobbs with a brief paragraph that rests the basis for the engine design change solely on the shoulders of Perry Pratt and Professor Soderberg [11]:

> I said in the beginning of this memo that this involves a delay of four months in the schedule. I am sure, however, that ... time will be saved provided ... that Pratt and Soderberg are correct in their contention that the time and effort we have already invested in the component testing of the 10A compressors does not work to the advantage of the one design over the other. The compressors of the 10B design are patterned after the PT2. Actually the low-pressure compressor is a scaled-up version of the PT2 on which we have the most experience and knowledge ... we know from test data that the 10A high-pressure compressor is not good enough to do the job and must be redesigned anyway. The only thing that Pratt and Soderberg can recommend for improving this compressor sufficiently to meet the ratings is to employ the construction laid out for the 10B.

JT3A WITH A WASP WAIST DESIGN

PARKINS AND THE DESIGN PROCESS

Hobbs gave Wright Parkins the job of developing the prototype wasp waist J57s in time for the B-52 and B-60 flight programs. The Convair B-60 was theoretically in competition with the B-52, but it certainly looked as if the Air Force preferred the Boeing approach. Nevertheless, Pratt & Whitney had to get prototype engines to Convair for its flight test program. Luke Hobbs knew what he was doing when he gave the task to Parkins.

Hobbs related that Parkins initially agreed to do the "impossible job." Then after a few months he came back to Hobbs accompanied by Bill Gwinn, who was the Factory Manager with control of the production and the semi-production facilities. Parkins was in Engineering and had only the Experimental Production Shop. The message to Hobbs from Gwinn and Parkins was the job was impossible to get prototype engines out on such a short schedule. Hobbs' message was this [10]:

> Those dates are going to be met because that is going to make or break Pratt & Whitney, that one airplane and that one engine installation. Either we are going to get by or we are going to pass out. That date will be met. Every engineering change that is made in those engines, from now on, the experimental shop must supply the parts to semi-production. Semi-production will go right ahead, no matter what they have to throw out of the place, and will build those prototype engines to the specifications!

Sometime after the go-ahead on the wasp waist JT3, the analytical designers discovered that the natural frequency of the low rotor was located in the operating speed range. One fix was to shorten the shaft, but that made the burner section shorter than the burner specialists liked. Here was a typical engineering trade-off, where a shorter burner meant a shorter shaft and a higher natural frequency, comfortably out of the operating speed range. However, the shorter the burner, the more likely is the greater development effort to smooth out the temperature profile entering the high-pressure turbine.

One day during the early summer of 1949, I noticed a small group of engineers around a desk in the Burner Group close to the Turbine Group where I was located. Wright Parkins was talking to a couple of mechanical layout designers he had brought with him and a group of burner engineers. Parkins listened to the arguments pro and con regarding the length of a burner vs the critical frequency of shafts. Then, with his typical decisiveness, he held up his hands, as if describing how small the fish was that his buddy fisherman caught, and said, "Make the burner this long!"

Another problem that came up was the rapid wearing of the number 5 roller bearing on the JT3. In January 2003, test engineer Niles Brook gave me some insight into how this problem was solved:

> We were having a roller bearing skidding problem with the JT3 in the number five position.... Parkins held a weekly meeting on the JT3A (X-185) engine. There was Perry Pratt, Gil Cole, Dick Baseler, Gerry Beardsley, Gill Barnes, Dave Phinney and I.... to report on the experimental engine program. I was not a bearing expert but I did have all the bearing manufacturers thinking about this skidding problem. They didn't know much more than we did.... We tried ... various things. The secret was to get a load on the bearing. I thought if we make the bearing out of round as a means of preloading it, the bearing mounting structure, clamped down on the bearing, would put the desired load on the bearing.
>
> When the bearing wear problem came up in Parkins' meeting, I told the group about preloading the bearing and I had it in a test rig. Parkins yelled, "Oh, s—t! That will never work!" That was the end of the meeting. So, we're going out the door and I say to Baseler, "What the hell do we do now?" Baseler says, "He didn't say to stop." So two weeks later we had the results and we won!

The rest of the number 5 roller bearing story is told by Paul Fisher with Perry Pratt [3]:

> As testing on the J57 intensified, the number five bearing wore out after about 150 hours of time. Fix after fix failed. Finally Parkins summoned Pratt and he remembers that Parkins' grim opening line went: "You've had your chance, Perry. Now I'll take over. I'm going to get Gus Hasbrouck and tell him to get hopping on a whole new re-design of the main bearing layout."
>
> Pratt's plea for more time and his point that a re-design was tampering with the heart of the engine failed. A dozen or so engineering changes

were under way before Parky left on a business trip that afternoon. Pratt hung around until he was sure Parkins had pulled out of the garage and then, "I went to the guys releasing the drawings, urging that they be sure to leave me enough metal on the forgings so that we could go back to the old designs. They did; they were released so that the parts could be made Parky's way or mine.... Niles Brook was new with us, a test engineer. He ... suggested that we make the outer race of the bearing one-thousandth of an inch out-of-round. Unknown to Parky, we got an engine built with Niles' fix. It worked. Meantime, Parky's re-design couldn't get up to speed. So here we had the case of a young engineer solving a problem that already was grave and could have been very, very tough."

Parkins, despite his gruff manner, was well respected because of his fairness to people. Some (who insist on anonymity) described him as a bull in a china shop. Perhaps a fairer way to describe him would be to borrow an expression from Sir Winston Churchill. Parkins' manner in achieving an objective was "direct, resolute, and effective."

JT3A Evolves During the Design Process

Bruno Fiori was one of the principal designers of the JT3A (the final configuration after the new go-ahead), and he summarizes the sequence of events as follows [12]:

> In mid-1948 a jet version of the PT4 was introduced. The configuration of the jet engine, designated the JT3-8 was similar to the PT4 except for omitting the reduction gear and a turbine stage. The external configuration of the compressor section was straight which gave the engine for the most part a cylindrical appearance. The designs were worked in parallel up to the Spring of 1949. The cylindrical configuration for the jet engine did not afford the best installation characteristics. Even though the accessories were placed in separate packages around the circumference of the engine the frontal area was too large for installed efficiency. On or about January 1949, a new engine model was introduced designated the JT3-10A. This model had virtually the external appearance of the JT3-8 except for the accessory arrangements. Its engine accessories and starter were arranged in one cluster and mounted at the bottom of the engine off the intermediate case. The airframe accessories were mounted in the inlet cone and mounted off the inlet case. The configuration offered reduced frontal area, consequently improving installed efficiency.
>
> In March of that year, Andrew Willgoos had passed away and consequently the group had to be restaffed [Fig. 3]. Wright Parkins was assigned Willgoos' position. Richard T. Baseler was assigned as project engineer, and Gus Hasbrouck was assigned the design chief position when Moncrieff was made head of the patent group. Parkins had

requested a revised configuration designated the JT3. The compressor flow path took on a conical waist shape in appearance. This permitted the accessories to be tucked in radially, in essence behind the wake of the inlet case. This configuration afforded a considerable reduction in frontal area and became very important years later at the introduction of the fan engine. The location of the accessories were essentially the same as the JT3-10A except the gear box was driven from the diffuser case through an elbow gear box and lay shaft instead of being driven through the intermediate case with a tower shaft. There was much emphasis put on the design of this configuration as one would say we "burned the midnight oil."

I want to emphasize that the design experience that we acquired from the PT4, JT3-8, and the JT3-10A was of great significance in the expediency of this design. Within three months the details were released for fabrication. [At] the end of the year the engine was mounted on the stand and tested. The engine was a success. I have never seen such a state of exuberance at P&WA before and perhaps never will again in my time. Personnel were flying on a cloud.... The success of the JT3 or the J57 was the spirit of P&WA as the "Spirit of St. Louis" was to Lindbergh, and it was one of the important contributions to the security of our nation.

In the meantime, the B-52 needed prototype jet engines in the spring of 1952. Gus Hasbrouck, Chief Design Engineer, started to refine the test engine design for manufacturing the prototype engines (XJ57-P-3) for the B-52 flight test program. The race was on to get the JT3A into the Boeing flight test program before the Westinghouse J40 was ready.

DESCRIPTION OF THE JT3A

The JT3A design configuration was chosen for the lowest engine plus fuel weight for a cruise speed of 500 mph at 45,000 ft for a 10-hr duration. The overall pressure ratio was 12:1, with a turbine inlet temperature of 1500° F. The compressor split was selected to be a 3.75 pressure ratio on the low compressor at 6000 rpm, and a 3.2 pressure ratio on the high compressor at 9500 rpm. This split was determined after careful consideration of minimizing surge problems on both compressors, achieving a rapid "spool-up" from 60 percent speed to full speed, and obtaining the optimum balance in the high and low turbines among the variables of operating stress levels, efficiency, and weight.

The nine-stage low compressor was a constant inside-diameter design, a scale-up of the PT2 compressor aerodynamics. The seven-stage high compressor was of a constant outer-diameter configuration brought in closer to the engine centerline compared to the JT3-10A design. This gave the engine the wasp waist look (Figs. 8 and 9). The turbines were a constant mean-line design

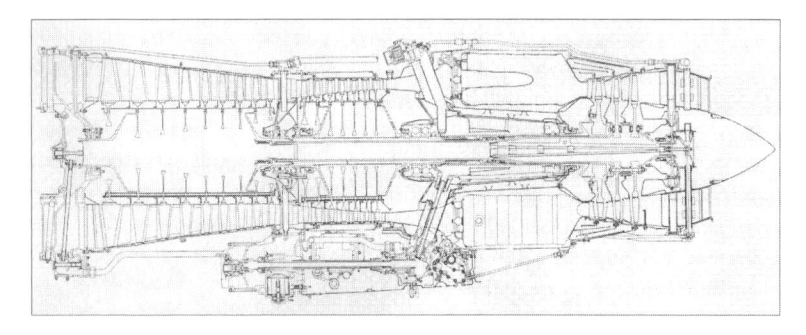

Fig. 8   Line drawing of the JT3 (courtesy of Pratt & Whitney Archives).

configuration. The high turbine was a single stage and the low turbine had two stages. All three stages were shrouded with labyrinth tip seals, which set Pratt & Whitney apart from other engine manufacturers at this point in time.

The combustor was the can-annular type with eight cans, each with six fuel nozzles. What made the burner different from other burners was the presence of a cylindrical tube down the middle of each can. Additional air was introduced through this tube to help bring the combustion temperature down to a level compatible with the turbine maximum temperature limits. The engine accessories were driven from the high spool. The low spool had a single power take-off drive from the low compressor.

RUNNING THE FIRST JT3A

Niles Brook was in charge of testing the first JT3A (J57) wasp waist engine. Its first run was in January 1950. This was the crucial test to demonstrate that the redesign recommended by Perry Pratt and Dick Soderberg was the better way to go. Niles Brook and his boss, Dick Baseler, were there as was Frank Morrison, the test engineer responsible for directing the assembly and test of

Fig. 9   Cutaway of the JT3 (courtesy of Pratt & Whitney Archives).

the engine. It was run in a production test house. Niles Brook described to me in January 2003 how things went on that historic day:

> I remember going in the ... big sound-proof doors. And I'm thinking we've got about seven million dollars worth of engine.... We buttoned the engine, set it up, and finished the installation. We had a compressor bleed valve system because we were having trouble matching the compressors. So these big pneumatic compressed air cylinders operated the bleed valves. You hit a tiny little switch and, "Pow!" This compressed air shot would go through there to open the compressor bleed. You knew what was going to happen. It would scare the daylights out of you if you didn't expect the loud noise it made when it opened the bleed valves. Dick Baseler was looking over the engine in the test cell on that day for the first run. One operator, I don't know who he was, kind of indicated, "Should I hit the bleed valve?" I said to him, "If you got the guts to throw the switch, I got the guts to stand in the door and laugh." Well, he hit the switch and it went bang! Baseler straightened up where he was and took it very well. I didn't realize until later on that he had a heart condition....
>
> Now we start cranking the engine over with the starter, and we got up to our program checklist where it calls for start. Hit the ignition. We did. It went off. Everything just went like clockwork.... The tachometer is going. Everything is working. Nothing is going wrong! So I said, "Bring it up some more." So they brought it up. Normal rated rpm was about 7000 or 8000 on the high compressor. So we ended up close to 7500 and what do you do now?
>
> Perry Pratt was there. Guy Beardsley was there. Everybody was there and we didn't have a program beyond running to the 7500 rpm. So we went up to about 8000 rpm. Everything was working. Then I said it's time we shut down and take a look at the engine. So we did and everything came out all right.
>
> There was one close area—the high-compressor spool. You know the twelve tie bolts? As we were disassembling the engine and they're turning over the shafts, we heard this kind of crinkle and clank, clank. We asked, "What the heck was that noise?" So we finished taking off the windage covers from the rotors and that was when everybody discovered the corrosive effects of cadmium at molten temperatures on heat-treated stainless steel. Heat-treated stainless steel tie rods and the nuts were cadmium plated. We got up high enough and hot enough so we melted the cadmium. That caused inter-granular corrosion and snapped half of the tie rods! A little more testing and you would have a bunch of parts—all scattered throughout the test cell!

Needless to say, cadmium was not used on tie bolts after that.

A RUN IN WITH RICHARD T. BASELER

Dick Baseler joined Pratt & Whitney in 1937 as a test engineer after his graduation as an aeronautical engineer from the University of Minnesota. He was an active member of the R-2800 development team and in 1943 worked for Art Smith in the Kansas City plant producing the R-2800C engines for the war effort. After the war he returned to East Hartford where he took over the development of the T34 turboprop gas turbine. After the development of the J57, in 1958, he became Chief Engineer of all production engines. The JT8D, Pratt & Whitney's most successful turbofan engine, was generally referred to as "Baseler's engine."

He was customarily referred to as "Steely Blue Eyes" because of his stern demeanor (Fig. 10). However, in a less business-like environment, he was a much more affable person. Shortly after Baseler retired I was playing trumpet along with coworkers piano-man Bill Helfrich and drummer Gene Montany at Glastonbury Pool and Tennis Club in Connecticut, when Baseler came up to me and suggested I take off my tie and relax. He said, "You look too much like a Pratt & Whitney engineer!" This was a side of Dick Baseler none of us had seen at the plant. He was one fine engineer dedicated to developing and producing the finest engines. Although there were some in the aviation industry who on occasion would disagree with Baseler, everyone had tremendous respect for him.

Once, when Pratt & Whitney was expecting important engineers from Wright Field, I asked Baseler if he would welcome these visitors, starting off with "expediting the badge process" at the reception area. Apparently my naive suggestion was a first at the company and certainly those who might have overheard the emotional outburst from Baseler would never make such

Fig. 10 Dick Baseler, who rose from test engineer to chief engineer during his career (courtesy of Pratt & Whitney Archives).

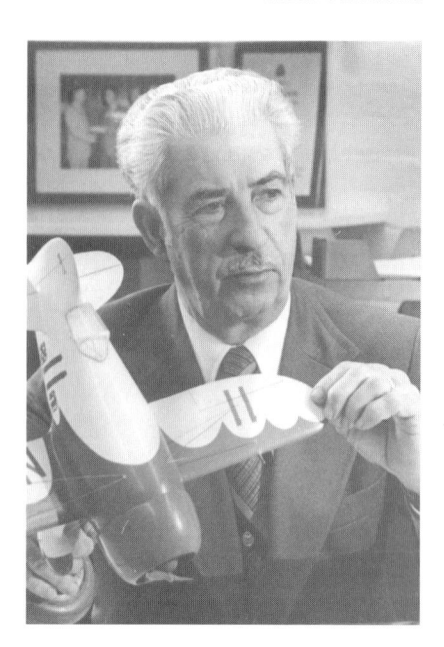

Fig. 11   Ed Granville with a model Gee Bee.

a request again! My next act was to pretend to myself that I could find another job someplace quickly while I proceeded to the reception area in the Engineering Building to get badges for our visitors. The security procedures moved quickly, and as I stepped back to lead the visitors to the conference room, I stepped on someone's shoe—namely, Dick Baseler's! He quickly took over and in the conference room he had all of us completely spellbound with his charm. The meeting went cordially from that moment on and reached a mutually satisfactory conclusion at the end of the day.

Next day, I went back to Dick's office to inform him that he obviously had the wrong guy (me) in this particular assignment, and that I should be replaced by someone who would be more responsive to his way of doing things. He slowly reached for his pack of cigarettes, then the lighter, next ignition, then intake, and finally exhaust. In that agonizingly slow process, a smile developed. As he exhaled the smoke, he was on the verge of laughing as he said to me, "I never did things always the way my bosses wanted and I don't expect you to either." After that incident, my working relationship with Steely Blue Eyes greatly improved.

AN UNSUNG HERO IN THE EXPERIMENTAL SHOP

It is worth mentioning at this point the contribution of Ed Granville (Fig. 11), one of the five Granville brothers. Ed, who was in charge of the Experimental Shop, pointed out to me one day, during our occasional information exchange sessions, that he and his brothers came off the farm and within four or five years built the world's fastest airplane in 1932! Don Jordan, development engineer for the J57 project, in 2005 put it this way:

> Ed was a genius at manufacturing experimental parts and also had the rare ability to motivate the entire shop to accomplish the impossible whenever necessary. Without Ed I really doubt whether the J57 would have made it. Ed could perform any operation as well or better than the operator of the machine or torch. Nobody could tell him it could not be done because he would just do it. He is a major unsung hero in our story.

JT3A IS PRONOUNCED A SUCCESS

When the JT3A ran in January 1950, only nine months after the new go-ahead, the test results justified the change recommended by Soderberg and Pratt. An announcement over the office public address system was made in January 1950 telling of the success of the new JT3A configuration (pictured in Fig. 12), which was an event comparable to the first run of the Wasp R-1340 in December 1925. In both the Wasp and JT3A tests, the viability of the company was on the line. The Wasp, at least, had signs that the Navy would buy a reliable, air-cooled engine. In the JT3A test, there was no such sure-fire application, due to the Westinghouse J40 also under development.

Luke Hobbs had agreed to start over with an all new engine design, the wasp waist JT3A with a 12:1 compressor pressure ratio. This was not an easy decision to make after having spent about four to six million dollars on

Fig. 12 Luke Hobbs with his JT3A (courtesy of Pratt & Whitney Archives).

the present engine configuration. But, as we will learn, it turned out to be the right decision.

THE J57 IN THE B-52

Dana Waring, an Assistant Project Engineer at Pratt & Whitney, was sent to Boeing to represent the company during the flight test program in the spring of 1952. These were the days when electronic controls had vacuum tubes, a source of engine problems. However, these fuel control difficulties were not a surprise to Boeing because the General Electric J47 engines had similar problems. Here is Dana's view of the action at Boeing at that time, as told to me in January 2003:

> The engineers at Boeing could hardly believe the Rotometer measurements of the fuel flow to the J57 because the same instrumentation on the J47 gave somewhat higher readings for an engine with much less thrust. Boeing was greatly impressed with the lower specific fuel consumption of the J57 compared to the J47. [The older J47 had a compressor pressure ratio of 5.4:1 while the J57 had a 12:1 pressure ratio].
>
> There were three major problems in the initial installation of the J57 in the B-52: fuel control; compressor stall as the pilot throttled back for descent; and the number 4 ½ intershaft seal.
>
> One of the first problems we encountered was with the electronic fuel control. At [the] Pratt & Whitney test back in Connecticut, John Rogers [Hamilton Standard Controls] would get out his soldering iron and make changes to the breadboard control right on the spot. But at Boeing, with the prototype controls and their "potted" units, we could only exchange the "potted" units. So we kept all the controls that came our way and Bill Buckingham [Pratt & Whitney Control Group] spent long nights swapping the "pots" around until he got a control to work for the next engine run.
>
> Then, during the flight test when the pilot throttled back the engines, there would be loud bangs associated with compressor stall, indicating the need to revise the bleed control schedule. We fixed that back at East Hartford by testing an engine in our new altitude facility where they could simulate flight conditions. The problem was solved in a few weeks, which amazed the Boeing people.
>
> The intershaft seal was more complicated to fix. It required taking the engine apart in segments to get at the high- and low-speed shafts and the seal. The tight schedule of the flight test program and the scarcity of flight test prototype J57 engines made sending our engines back to East Hartford out of the question. We had to make the hardware changes out at Boeing. So Paul Hagen [Pratt & Whitney Service Representative] had scaffolding built for us to work on and he coached the entire Pratt & Whitney staff in the area, including office people, in how to disassemble and reassemble the engines.

INSIDE THE B-52 TIME LINE

While Pratt & Whitney was directing its efforts to get into the B-52 aircraft program with its J57 engine, it was not at all clear at the time that the company understood the fragility of the B-52 program inside the Air Force. In the world of enterprises in the private sector and in government programs, nothing seems to progress without much pain and anguish. The heroic story of a once little-known RAF officer, Frank Whittle, was not an isolated incident in aviation history. It is repeated every time a visionary tries to accomplish great feats in a bureaucratic organization.

In the B-52 program, Lt. Colonel Henry E. "Pete" Warden was the hero. This story is told in detail in Lori S. Tagg's excellent book on the B-52 [13]. I have included some of the events from Tagg's book to illustrate the "slings and arrows of outrageous fortune" that Colonel Warden endured in his efforts to launch the B-52 program (Fig. 13).

Henry E. Warden (Fig. 14) joined the Army Air Corps in 1939 and served in the Far East during the war. He came to Wright Field in 1944 and was assigned to the Bombardment Branch under Colonel Donald L. Putt. Later, he moved to the position of Chief of the Bombardment Branch working on the heavy bomber to replace the B-36. The jet age was blossoming and the real fear was the B-36 would become obsolete sooner than originally envisioned.

Colonel Warden, with his civilian deputy, J. Arthur Boykin, outlined the requirements for the B-36 replacement aircraft. In 1946 there was a Request for Proposal (RFP) in which Boeing, Martin, and Convair participated. The desired range was 5000 miles and the maximum speed was 450 mph. Boeing won this study contract. Northrop had some success with its piston-powered Flying Wing (YB-49). A J-35 powered YB-49 flew in October 1947 and

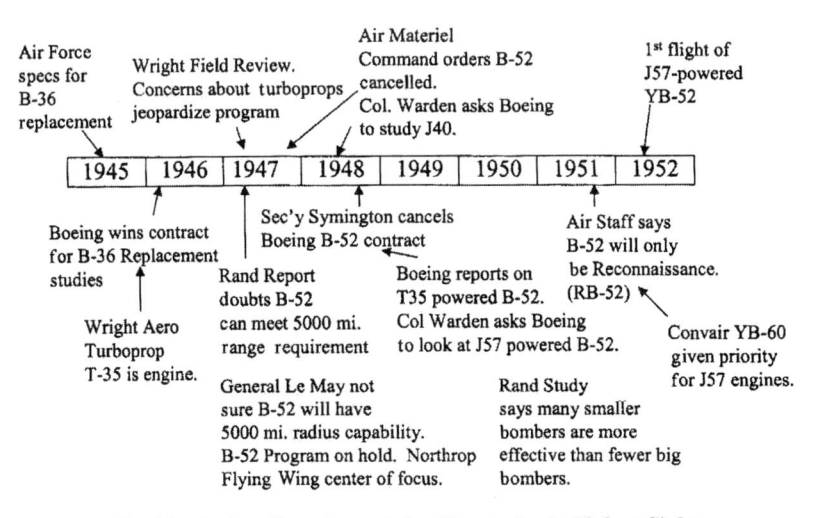

Fig. 13 A time line of events leading to the B-52 first flight.

Fig. 14 Colonel Henry E. Warden, at the Air Force Project Office, championed the B-52 project (courtesy of U.S. Air Force, Wright Patterson AFB).

stimulated great interest because of its low drag. However, in June 1948, a YB-49 crashed; interest declined.

A review at Wright Field in 1947 brought to light some potential problems of a turboprop engine in this application. The B-52 was in the 480,000-lb gross take-off weight size and required a propeller diameter of 23 ft. The B-36 had a propeller diameter of 19 ft. The propeller manufacturers expressed some concern about the greater blade lengths in the B-52. Another concern was that gas turbines rotate at much higher revolutions per minute (rpm) than piston engines. The combination of higher rpm and longer propeller blade lengths meant the development of high-horsepower gearboxes with rpm reductions greater than the state of the art.

The Rand Corporation issued a report that cast doubt on the feasibility of a B-52 meeting the 5000-mile range objective. Subsequently, the Air Materiel Command ordered the B-52 program cancelled. Colonel Warden, in response to the T35 propeller concerns, asked Boeing to study the installation of the Westinghouse J40 turbojet, which appeared to be the logical choice at that time. The B-52 program was still in the study phase, which did not take the huge amounts of money that a full-scale development program would require. This meant it was easier to keep the program going "under the radar screen." However, one of the critics got to the Secretary of the Air Force who subsequently ordered the B-52 program to be cancelled. The program was cancelled several times, but the B-52 has been in operation for over 50 years.

However, the Colonel kept the program going. Boeing continued its efforts, and in October of 1948 reported on the T35-powered aircraft at one of the periodic progress briefings at Wright Field. In view of the potential problems with the turboprop engines, the Colonel asked Boeing to look at the P&W J57, which had been recently transformed from the T45 (PT4) turboprop program. The conventional wisdom in the aircraft gas turbine industry was the Westinghouse J40 level of specific fuel consumptions was just about the level one could expect from a turbojet. It would not be until eight months later when Pratt & Whitney would run its first dual-spool turbojet and not until about 14 months later when Pratt & Whitney would test its engine configuration for the B-52—the JT3A (Fig. 12).

Next, the Rand Corporation, in another of their reports, claimed that it was more effective to have a group of small bombers instead of fewer heavy bombers. General LeMay injected a dose of practical experience into that discussion when he pointed out that if he mentioned the two approaches to Congressmen in the form of 100 smaller bombers vs 50 big bombers, they would in the end cut down the appropriations so that he would end up with only fifty smaller bombers. He said he preferred the big bombers.

In 1949, the Fairchild Corporation management approached the head of the Air Force, General Hoyt Vandenberg, and convinced him of a radical approach to a long-range bomber, one that used canards filled with fuel and later jettisoned. The wings in theory provided the extra lift for take-off. Once again the B-52 program was cancelled.

In 1951, as more oversight began to spread in the B-52 program, the Air Force Staff believed that the primary role of the B-52 was reconnaissance and designated the aircraft as the RB-52. In all of these "on-and-off" cycles, the Boeing team kept soldiering along in response to Colonel Warden, even when it looked as if the whole game was up. Procurement staff wanted to throw the program out for new bids as if the purchase of a huge program such as the B-52 was like buying coffee pots for the Officers Club. The final competition for Boeing was Convair's YB-60 (a J57-powered modification of the B-36). In the end, the Air Force followed the Colonel's recommendations and went ahead with the configuration that had the least challenging problems.

Colonel Warden had to be a man of vision in the same way as Rentschler. He also had to be skilled in the art of persuasion to keep the "warring factions" and the hierarchy from forcing the cancellation of his program. He did this by convincing his superiors of the soundness of his ideas. In turn, his superiors wrestled with the "powers that be" so that the Colonel could keep moving his team toward the goal line. This program is a successful example of the government, an airframer, and an engine company working together as one team. As a result, the U.S. Air Force ended up with an effective weapon system that has performed well for more than a half-century. This achievement was recognized at the time, as the design team for the JT3A won the Collier Trophy in 1952

Fig. 15 President Eisenhower congratulates Hobbs while he receives the Collier Trophy (courtesy of Pratt & Whitney Archives).

(Fig. 15). President Eisenhower attended the ceremony, reminding us of his previous claim that the C-47 engine, developed by Hobbs, was one of the four pieces of equipment that made victory in Europe possible.

APPLICATIONS OF THE J57

BOMBERS. The initial application for the J57 was in subsonic aircraft; first, the B-52 (Fig. 16), and then the KC-135 (Fig. 17) when in-flight refueling became a challenge because the jet-powered aircraft had to slow down to a piston-powered tanker aircraft. These applications, of course, were much faster than piston-powered aircraft but had minor installation developments compared to subsequent high-speed fighter aircraft such as the F-100 and F-101.

FIGHTERS. Bruno Fiori summed up the attitude towards using the J57 in a fighter plane [12]: "Once conferring with Moncrief [Engineering Executive], I had mentioned the possibility of employing the engine for fighter aircraft; he answered me with an emphatic no, and informed me the engine was to be used for bombers only. The tone of his voice told me not to mention it again and I didn't, at least not to him."

The J57 was always considered a "bomber engine" in the early days of development because the bomber was the specific application. A fighter engine has more stringent requirements. In a bomber engine, for example, the maximum compressor exit pressure occurs during the take-off run. On the other hand, a fighter engine could have a requirement for very high speed "on the deck." This

Fig. 16 B52 with J57s and two Hound Dog supersonic air-to-ground missiles (courtesy of Pratt & Whitney Archives).

means higher pressure levels throughout the engine, and consequently higher stresses. As a result, a fighter version of the J57 should be heavier than a bomber version because of the need for a stronger structure.

Perry Pratt said about the J57 [3], "I had not foreseen the J-57 ... as a fighter engine. Bill Sens [Project Engineer in charge of Advanced Engine Performance Group] was the man [suggesting] that by marrying the J57 with an afterburner, its attributes of power and low fuel consumption would find a home."

Paul Fisher, in his interview with Perry Pratt, made the following notes [3]: "Pratt doubts if there ever was an engine designed with the almost instantaneous speed of the powerplant that ultimately went into the U-2. It is his recollection that even before Kelly Johnson of Lockheed had been approached on the

Fig. 17 KC-135 in flight (courtesy of U.S. Air Force).

Fig. 18 Lockheed U-2, initially powered by the J57 (courtesy of the U.S. Air Force Museum).

idea of a very high-flying jet reconnaissance craft, Pratt & Whitney's people at Wright Field were asked what they might propose as a jet for a very high-altitude aircraft." The Lockheed U-2 became the first reconnaissance aircraft powered by the J57 (Fig. 18).

Pratt recalled the response to a very high-altitude engine [3]: "Bill [Sens] said [to] marry the low compressor from the fighter version of the J57 to the high-compressor stages of the bomber version of the J57." The low compressor for the fighter engine had more surge margin to handle inlet distortion more easily than the bomber engine low compressor, which had much less inlet distortion. As the altitude increases, the compressor surge line drops due to Reynolds number effects. The fighter engine's low compressor, with its greater surge margin, would obviously have a greater altitude capability before encountering surge.

The J57-P-23 afterburning engine was developed for fighter applications (Fig. 19). Success was evident from the beginning: the Air Force's F-100 (Fig. 20), first in the Century Series, reached supersonic speeds on its first flight. The Air Force's Convair F-102 (Fig. 21) and the Navy's Douglas A3D Skywarrior (Fig. 22) were also powered by the J57. Even with this early success, no one at Pratt & Whitney could have predicted that its name would be on over 21,000 JT3/J57 engines. Table 2 lists just a few of these applications, along with engine specifications.

Fig. 19 J57-P-23 afterburning engine (courtesy of Pratt & Whitney Archives).

Fig. 20   F-100, powered by the J57 (courtesy of Pratt & Whitney Archives).

Fig. 21   Convair F-102, powered by the J57 (courtesy of Pratt & Whitney Archives).

Fig. 22   Douglas A3D Skywarrior on a Navy carrier (courtesy of Pratt & Whitney Archives).

TABLE 2 JT3 Engine Specifications and Applications

Model	Take-off thrust (lb)	Military or max. cont. thrust (lb)	Normal thrust (lb)	Weight (lb)	Diameter (in.)	Length (in.)	Aircraft
JT3A	8700	8700	7500	4250	41.0	144.7	None
JT3B	9000	9000	7800	4348	41.0	144.7	None
JT3C	9250	9250	8750	4390	41.0	144.7	None
JT3P	11,000 wet[a]	9500	8250	4220	40.5	157.7	B-707 (367-80)
JT3C-4	12,500 wet	11,200	9000	4234	38.9	138.6	Prototype for B-707 (367-80) and DC-8
	11,200						
JT3C-6	13,000 wet	11,200	9500	4350	38.9	138.6	B-707-120 and DC-8
JT3C-7	12,000	10,000	10,000	3495	38.8	136.8	B-720
JT3C-12	13,000	11,500	11,500	3550	38.9	136.8	B-720
J57-P-2	14,800 A/B[b]	9700	8000	5115	40.1	245.6	F4D-1
YJ57-P-3	8700	8700	7500	4250	41.0	145.0	Prototype for XB-52, YB-52, XB-60, YB-60, Northrop XSM-62
J57-P-4, A	16,000 A/B	10,200	8700	4860	39.8	267.7	F8U-1, F8U-1P
J57-P-6	10,000	9500	8250	4125	41.0	159.3	A3D-1
XJ57-P-7	13,200 A/B	8450	7250	5126	41.0	241.0	YF-100A
J57-P-7	14,800 A/B	9700	8000	5075	40.1	245.6	F-100A, F and F-102, TF-102
J57-P-8, 8A, 8B	16,000 A/B	10,200	8700	5060	39.2	245.7	F4D-1
J57-P-9	10,000	9500	8250	3910	41.0	159.3	B-52
J57-P-9W	11,400 wet	9500	8250	3960	41.0	159.3	B-52A, RB-52B
	10,000						
J57-P-10	10,500	10,500	9000	4200	40.5	157.5	A3D-2, 2P, 2Q, 2T
J57-P-11	14,800 A/B	9700	8000	5075	40.5	246.0	TF-102, F-102A, XF8U-1
J57-P-12	16,000 A/B	10,200	8700	5105	39.8	267.7	XF8U-1, F8U-1, F8U-1P
XJ57-P-13	14,000	9220	8000	5025	40.1	216.1	F-101A
J57-P-13	15,000	10,200	8700	4920	40.1	221.6	F-101A, B, C RF-101A, C

J57-P-16	16,900	10,700	4750	40.0	266.9	F8U-2 (with CD nozzle)
J57-P-17	10,500	10,500	4175	40.5	161.7	XSM-62, SM-62A
J57-P-19W	12,000 wet 10,500	10,500	3970	40.5	157.5	RB-52B, C, D, E
J57-P-20, A	18,000 A/B	10,700 (20) 11,400 (20A)	4750	40.0	266.9	F8U-2 N (20), F8U-2 (20A) (with CD nozzle)
J57-P-21, A	16,000 A/B	10,200	5160	40.1	246.6	F-100, A, B, C, D, F
J57-P-23, A	16,000 A/B	10,200	5045	39.8	245.7	TF-102, F-102A, F-101A + F-102A (23A)
J57-P-25	16,000 A/B	10,200	5130	39.8	267.7	YF-105, F-105A
J57-P-29W	11,000 wet			40.5	157.5	B-52
J57-P-29WA	12,100 wet	10,500	4150	40.5	157.5	B-52, B-52A, RB-52B, C, D, E
J-57-P-37, -37A	10,500	10,500	4170	39.8	163.4	B-57D (-37A is the same except for extra generator pad)
J57-P-39	14,800 A/B	9700	5075	40.1	245.6	F-100A, F-100C
J57-P-41	14,800 A/B	9700	5065	40.5	246.0	TF-102, F-102A
J57-P-43W	13,250 wet 11,200	11,200	3870	38.9	138.6	B-52D, E, F, G, KC-135A
YJ57-P-45	16,900 A/B	10,700		39.1	254.7	F-101B
J57-P-53	16,000 A/B	10,200	5055	39.8	248.6	F-101B
J57-P-55	16,900	10,700	5215	40.0	253.0	F-101B
J57-P-59W	13,750 wet 11,200	11,200	4320	38.9	167.33	KC-135A
J57-P-420	19600 A/B	12,400	4840	40.0	266.9	F-8J

[a] Wet indicates the rating with water injection.
[b] A/B means afterburning operation.

FLIGHT TEST OPERATIONS

In the piston engine era, flight testing brought to light certain engine problems and they were solved. Flight testing in the jet era introduced another layer of complexity because of the great increase in aircraft speed. Pratt & Whitney recognized this challenge and obtained several J57-powered aircraft for flight-testing with company pilots. The objective was to discover and solve problems before the Air Force and airframer detected them.

When an engine is put on a sea-level test stand for the initial evaluation, it is a long way from its final configuration. First of all, the inlet air to the engine is uniform and the engine is securely fastened to the test stand structure. There must be a thorough evaluation of the design point performance and its steady-state and transient performance. The next step is to see what the engine performance is at the altitude flight conditions.

This test is either done in a facility such as the Willgoos Lab or at the Air Force facility in Tullahoma. Even this evaluation is not the final answer, because the inlet flow, as in the sea-level testing, is uniform. The real test of the engine is in its aircraft where the engine is no longer in a fixed orientation but can move linearly in three directions and can rotate about each of the axes, six degrees of freedom. This also implies the inlet flow is no longer ideally uniform.

It is a long way from the designer's drawing board conception to an engine operating under full power during a maneuver. Of course, at the same time the engine is about to choke on the digestion of severely distorted inlet flow. This kind of demand on the engine requires the airplane and the engine designers to conduct extensive experiments on the inlet flow through the ducting to the face of the engine.

Different sectors of the inlet stage of the compressor experience a variety of angles of attack as the compressor disk rotates. How the compressor digests the uneven profiles has a significant effect on what the compressor, burner, and turbine do.

These complicated problems with inlet-engine compatibility were attacked head-on. Pratt & Whitney set up its own flight test organization with sufficient facilities and resources to arrive at an engine configuration that would permit aircraft such as the F-100, the F-101, and the F8U (Fig. 23) to perform satisfactorily in their respective flight envelopes.

Harold Archer, Chief Test Pilot for Pratt & Whitney in East Hartford, had the overall responsibility for the flight test operation at Edwards Air Force Base in California located on the Mojave Desert. Jim Peed was the Test Facility Manager with whom I worked in 1948 in the Technical & Research Turbine Group. Cal Christian, a former Turbine Group test engineer, Joel Hitt, and S. T. Swallow were the flight test engineers (Fig. 24). Harry Schmidt, former Air Force fighter pilot, and Walt Allen had the "fun of flying" to the edges of the flight envelopes of the F-100, F-101, and the F8U (Fig. 23).

BIRTH OF THE TWO-SPOOL TURBOJET 249

Fig. 23 Pratt & Whitney flight test aircraft included one North American F-100, two LTV F8U Crusaders, and a McDonnell F-101 (courtesy of Harry Schmidt).

These Pratt & Whitney test pilots were uniquely qualified because of their engineering backgrounds and also probably learned more about the airframe–engine interaction than any of the Air Force or airframer test pilots. One could consider the Pratt & Whitney pilots as "heart specialists" because they focused on what it took to make sure the aircraft's "heart" kept pumping while the aircraft was doing its job.

Fig. 24 Pilot Harry Schmidt is flanked by engineers Cal Christian (left) and Joel Hitt (right) in front of a McDonnell F-101 at Edwards Air Force Base (courtesy of Harry Schmidt).

I learned about the company's flight test operations at Edwards Air Force Base from Harry Schmidt, who did most of the test flying for Pratt & Whitney during the era of the hot Century Series aircraft powered by the J57 (Figs. 20 and 23). Harry Schmidt documented in his book, *Test Pilot*, not only the Pratt & Whitney flight test experience but also that of North American, McDonnell, Convair, Republic, as well as some Air Force activities [14]. I asked Harry Schmidt about the Pratt & Whitney flight testing at Edwards and he told me about it, including setting a new speed record (Fig. 25):

> Although Pratt & Whitney had historically performed engine testing in test-bed aircraft at Hartford [such as the two B-45s in which a J57 and a J75 were mounted in the bomb bay], things changed with the introduction of the J57. During early flight tests in mid-1953 in the first F-100A at Edwards, North American test pilots found that the engine was experiencing severe compressor stall problems, primarily due to the elongated air duct.
>
> A short time later McDonnell, when testing their new F-101, found the same severe compressor stall problems. Everybody involved in the ... programs ... concluded that Pratt & Whitney had to perform their own flight testing in these new Century Series of fighters ... to solve the severe compressor stall problems which were affecting the utility of these new supersonic fighters. Harold Archer, Chief of Pratt & Whitney Flight Test, concluded that Pratt & Whitney should open a flight test facility at Edwards Air Force Base dedicated to the Century Series. Harold hired me, a former U.S. Air Force [USAF] fighter pilot and an engineer, to be the Pratt & Whitney test pilot at Edwards. Jim Peed was selected to be the flight test manager at Edwards and was assisted by several very capable engineers including Cal Christian, S. T. "Brud" Swallow, and Joel Hitt.
>
> In late 1954, Pratt & Whitney received their first F-100A, an early model [a later model F-100 was received about a year later] and company

Fig. 25 Pratt & Whitney's F-101 flown by a U.S. Air Force pilot sets new world's speed record (courtesy of U.S. Air Force).

flight testing started soon thereafter. Compressor stalls were found during all phases of engine operation. Often the power could not be advanced off idle due to compressor stalls on the ground and at the other extreme, compressor stalls were encountered during high-speed flight. Pratt & Whitney engineers worked diligently on the problem, primarily involving changes to the bleed air system. After six months they had solved 90 percent of the compressor stall problems. The flight testing continued involving other issues, such as how to obtain afterburner relights at higher altitudes [a program run by Pratt & Whitney engineer, Walt Hemlock] and how to obtain relights after engine shutdown at higher altitudes. And after every change to the engine to solve one of the above problems we conducted new performance evaluations to see what impact had been made on climb and cruise capability.

[A little] later Pratt & Whitney obtained an early F-101 for similar research.... When Pratt & Whitney mastered the F-101 compressor stall problems, a new engine was installed in the F-101, a J57 engine with a convergent–divergent nozzle [brainchild of Pratt & Whitney engineer Stu Hamilton]. This so-called "supersonic afterburner" increased the speed of the F-101 from Mach 1.6 with the standard afterburner to almost Mach 2.0 with the new afterburner—a tremendous increase in speed due only to the newly designed afterburner.

We were testing the fastest aircraft in the world! When a short time later the English took the world's speed record from the United States, the USAF was obviously intent on reclaiming the world's speed record. The P&W F-101, with the convergent–divergent jet nozzle, was the only aircraft fast enough to regain the record ... the USAF ... sent their own USAF pilot to set a new speed record and put his name in the record books. Pratt & Whitney management was obviously willing to have their F-101 used for such purposes, since Pratt & Whitney only had interest in selling engines—although the flight test personnel were chagrined not to be a part of history-making flights....

Later flight testing of the F-101 involved a special request from the USAF to investigate the new 'pitch-up' phenomena of the F-101 (due primarily to the 'T-tail' design). This was an important flight test event involving airframe problems and having nothing to do with engine problems, more important to McDonnell and the USAF than to Pratt & Whitney. Pitch-up was a new and troubling aerodynamic flight characteristic which ... had serious implications since most aircraft crashed after experiencing pitch-up. We performed this pitch-up evaluation for the USAF without incident and shortly thereafter McDonnell made substantial modifications to the flight control system to eliminate the possibility of unsuspecting USAF pilots encountering pitch-up. [Pitch-up occurs when, at high angle of attack, the wing flow blanks out the flow over the T-tail surfaces, reducing the effectiveness of the horizontal stabilizer with a subsequent loss of control of the aircraft.]

While Pratt & Whitney was conducting flight research on the F-101, General Electric obtained a similar F-101, removed the J57 engines, and installed their own J79 engines. Presumably they hoped that the General Electric-powered F-101 would perform better than with J57 power. The test was probably unsatisfactory for General Electric since we never heard more about that project.

A short time later, Pratt & Whitney received other Century Series fighters for continued testing and Harry Schmidt was later joined by Walt Allen and Bob Denn, who assisted in testing the expanding fleet of Pratt & Whitney fighters.

Another milestone was achieved for this turbojet on July 16, 1957 when Marine Corps Major John Glenn climbed into his F8U Crusader in the Los Angeles area and set a new speed record. He arrived at the Floyd Bennett Field in Brooklyn, New York, a distance of 2460 miles in three hours and 23 minutes, equivalent to about 726-mph average speed. There were three refueling rendezvous with airborne tankers. He was elated with the aircraft performance. After he climbed down from the aircraft he said, "The airplane worked real fine. There was no strain at all."

JT3 COMMERCIAL DERIVATIVE

As the B-52 program moved along, Boeing engineers were becoming more enthusiastic about commercial jet-powered aircraft. Of course, de Havilland was first with a jet-powered aircraft, the 36-passenger Comet in the spring of 1949. At this time, Pratt & Whitney had not even run a twin-spool turbojet and was moving along on a program to provide prototype J57 engines to Boeing and Convair for the first flights of the B-52 and B-60 aircraft programs. This was also when Pratt & Whitney started all over again with the new wasp waist JT3A.

Shortly after the first flight of the YB-52 in April 1952, the Boeing Board of Directors gave its approval to build a prototype aircraft that could serve as an effective sales tool for a military tanker to replace the KC-97 piston-powered tankers and to sell to the airlines as a new high-speed commercial transport. Ten minutes after this board meeting, Wellwood Beall, Executive Vice President, called Stan Taylor (Pratt & Whitney's installation engineer at Boeing at the time) to his office and asked Stan if he could get four free engines for Boeing's new prototype, the 367-80 (Fig. 26). Stan must have given a magnificently persuasive speech when he called East Hartford to get Pratt & Whitney's management to make the commitment, because Stan reported back to Beall about 30 minutes later with the answer, "Yes!"

In 1954, the early days of flying the Dash-80, Boeing was conducting a sales campaign with Pan American and domestic airlines while the Air Force ordered 29 KC-135 aircraft. The standard of airline travel in those days was the piston engine-propeller powered Douglas DC-7, the Boeing Stratocruiser, and the Lockheed Starliner. The domestic airlines were thinking in terms of

Fig. 26 Boeing 367-80 (Dash-80)—the jet-powered demonstrator for the KC-135 and the B-707 (courtesy of Pratt & Whitney Archives).

an intermediate step to turbojets, the turboprop-powered Lockheed Electra. It is hard to imagine how Boeing could have jumped into commercial aviation with the B-707 without the Dash-80 to give the airlines a feeling for handling a jet-powered aircraft. It was an essential tool in convincing the Air Force to buy the KC-135, because Boeing gave a live demonstration of how much easier it was to fill up the fuel tanks of the B-52 from another jet-powered aircraft. It was a ticklish flying situation when the KC-97 struggled to refuel a B-52 on the "ragged edge" of a speed low enough for the piston engine-powered aircraft.

Pan American's Juan Trippe focused on long-range aircraft. His reaction to the Boeing and Douglas jet-powered aircraft was that both were too underpowered to do what he wanted. He wanted a greater payload at a greater non-stop range. He could see that the J57 was too small for his uses but the P&W J75 was just about right. American's C. R. Smith insisted that Boeing offer a fuselage diameter wide enough for 6-abreast seating as in the DC-8. After all the smoke cleared, Boeing went ahead with the B-707-120 (JT3-powered) for domestic airlines and the B-707-320 (JT4-powered) for Pan Am and the foreign airlines interested in long-range flights (Fig. 27).

Boeing recognized that Douglas might have a better chance with the airlines because of its long history of successful piston engine-powered commercial aircraft. However, Boeing was demonstrating a real airplane that would provide a new level in luxury travel with a significant reduction in flight time.

In October 1958, Mamie Eisenhower christened a Pan Am B-707 "Clipper America" powered by P&W JT3C-6 before its scheduled trip to Europe, thus making it the "official launch" of the jet age in the United States. The DC-8, equipped with the P&W JT3-6, was launched domestically by Delta Airlines and United Airlines in September 1959.

Hobbs' Reflections on the Dawn of the Jet Age

Hobbs looked back in 1966 at how he managed to get Pratt & Whitney into the jet engine business right after WWII [4]:

> First, we were in a bad hole. We were not only lacking in experience with the new form of powerplant but ... it was a going thing with tough competitors already firmly established in the field.
>
> Secondly, although we were predominant in the piston engine, it would require the major portion of Pratt & Whitney's available engineering to maintain this position and produce ... the profits necessary ... for us to shift to the new base. Therefore, we would in effect have to build up a new engineering department.
>
> Third, although we probably would have the time and funds required to launch us on our new course, it was very obvious that we would have but one chance. Our first program must be successful or we were through.
>
> There was never any question about going to the wasp waist design once Pratt's group had decided that it was desirable. By this time I had ... complete confidence in the ability of Perry's technical group. ...
>
> Soderberg should be given complete credit for his most major contribution on the technical side.

Fig. 27   JT3C in a Boeing 707 nacelle (courtesy of Pratt & Whitney Archives).

Fig. 28 MIT professor C. R. Soderberg (courtesy of MIT).

Hobbs probably never dreamed in the spring of 1949, when he gave his approval to start over again with the J57 design, that Pratt & Whitney would one day have its name on over 21,000 JT3/J57 engines. Hobbs made it possible for Boeing and Douglas to create a new era in commercial aviation. Soderberg's influence cannot be overlooked as he was the person (Fig. 28) who convinced Pratt & Whitney to scrap what they had for the J57 and start all over again. The new aircraft cut flight times significantly and there was a huge expansion in the number of people flying for the first time. Rentschler said later, "I knew the boys could do it!" [15].

REFERENCES

[1] Hobbs, L., "Remarks of Luke Hobbs at Sperry Award Ceremonies," Pratt & Whitney Archives, Jan. 8, 1973.
[2] Hobbs, L., "Dictaphone Notes," Pratt & Whitney Archives, summer 1953.
[3] Fisher, P., "Interview with P. W. Pratt," Cabinet F-22, Drawer 2, Pratt & Whitney Archives.
[4] Fisher, P., "Interview with Luke Hobbs during Late Fall and Winter 1965–1966," Pratt & Whitney Archives.
[5] Simpson, E. C., *The Memoirs of Ernest C. Simpson*, J. St. Peter (ed.), Air Force Aeronautical Laboratories, June 1987.
[6] Foley, J. R., "The Early History of the JT3 Engine," TDM-1169, Pratt & Whitney Archives, 1954.
[7] Hooker, S., "Letter to Perry Pratt," Pratt & Whitney Archives, April 1949.
[8] Cook, W. H., *The Road to The 707*, TYC Publishing, Bellevue, WA, 1991.
[9] Martin, J., "Interview with Leonard Hobbs," Pratt & Whitney Archives, March 19, 1970.

[10] Soderberg, C. R., and Pratt, P., "Memo to Wright Parkins: Comparison of Proposed JT3-10A and JT3-10B Engines," Pratt & Whitney Archives, May 6, 1949.
[11] Parkins, W., "Memo to L. Hobbs," Pratt & Whitney Archives.
[12] Fiori, B., "Memo to John Martin: Material for the Pratt & Whitney Story," Pratt & Whitney Archives, Oct. 27, 1969.
[13] Tagg, L. S., "Development of the B-52: The Wright Field Story," History Office, Aeronautical Systems Center, Air Force Material Command, Wright-Patterson Air Force Base, Dayton, OH, 2004.
[14] Schmidt, H. P., *Test Pilot*, Mach 2 Books, Shelton, CT, 1997.
[15] Fisher (Paul) files, Pratt & Whitney Archives.

Chapter 8

FOUR MORE TURBOJETS

INTRODUCTION

As the J57 development proceeded, it became clear that Pratt & Whitney's concept of a 12:1 overall pressure ratio with a two-spool compressor was correct. Hobbs and Perry Pratt knew that there was a growing need for more thrust and speed in the aviation business, and they sold the idea of a larger engine, the J75, to the Air Force (Fig. 1). Engine studies for the Navy of twin-spool turboprops and turbojets showed a need for a high-performance turbojet smaller than the J57. This engine became the J52 (JT8 in company model designation). On November 20, 1953, a little over a year after the start of the J75 development program, Scott Crossfield flew the rocket-powered Douglas 558 Skyrocket II to Mach 2, opening up the path for higher speed military aircraft. The Air Force need for a supersonic bomber created the requirement for the J91 (JT9) and General Electric's J93. In Canada, Canadair was responding to a need from the Royal Canadian Air Force for a small turbojet to power a new trainer aircraft. Pratt & Whitney Canada wanted to provide the turbojet, and with Pratt & Whitney Canada's help, Canadair conducted its flight test program with the JT12.

J75 (JT4) TWIN-SPOOL TURBOJET

In the commercial area, Pan American wanted more thrust than its JT3C engines provided so that it could fly fully loaded aircraft non-stop across the Atlantic to Europe. No engine was available. Eddie Rickenbacker, head of Eastern Airlines, was also interested in an aircraft with a greater range than his DC-8 with its JT3 engines. He went to visit Pratt & Whitney to see what it could provide. Perry Pratt recalled that memorable visit [1]:

> When Eddie Rickenbacker came up to discuss engines for Eastern's jet fleet, we planned to show him the J57C. Hobbs was still very dubious about mentioning the J75 since it still was under a military cloak. When Rickenbacker looked at the "C," he shocked both the effete and the security-minded by saying, "B–l s–t, boys! Show me the J75. I know about it and I want it!"

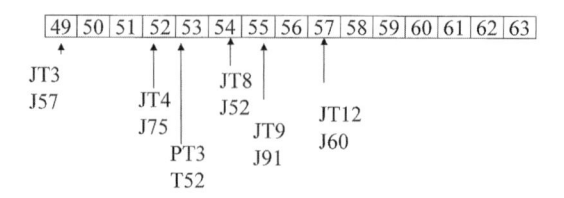

```
  49 50 51 52 53 54 55 56 57 58 59 60 61 62 63
   ↑         ↑     ↑     ↑
  JT3        |     JT8   |
  J57       JT4    J52
            J75         JT12
                   JT9  J60
             PT3   J91
             T52
```

Fig. 1   Time line of engine program start dates.

It was not very long afterwards that Juan Trippe of Pan American was badgering Rentschler about the J75 (or more accurately, its potential commercial version, the JT4). It was Hobb's recommendation that an engine have thousands of operating hours in military operations before being used in commercial applications. At this point, the J75 obviously did not meet the Hobbs criterion. However, its design technology was that of the J57, which was beginning to accumulate thousands of operating hours. Pratt & Whitney's reputation in piston engines was founded on design excellence. Pushing the JT4 commercial version of the J75 into airline service a little early was all right. The PT2 did not meet the Hobbs criterion for two reasons: 1) it was an earlier technology, and 2) it had no extensive military experience to qualify it as a commercial engine. Howard Hughes hounded H. Mansfield Horner, President of United Aircraft, for the PT2 in the Lockheed Constellation to no avail.

Larry Carlson was a Project Engineer on the J75, and sent me a letter about it dated January 23, 2004:

> I believe that the initiation of the JT4/J75 was a contract with the Air Force to develop the J75 for possible use in the F-105 in the event that the Wright J67 didn't work out. We were coming from behind until we ran the first engine and made our performance on the first run. Parkins was en route to the test cell one day with an Air Force general to show the engine. Unfortunately we had an unexpected disaster a short time earlier.
>
> We tried to head him off but this was what he needed to show the general that much development funding would be needed to provide a super engine. He and the general waited at the test cell until the fire was extinguished, went in and examined the smoking hulk. We told them that a turbine disk had burst and pointed to one of its fragments on the floor. Parkins' sales pitch must have worked because we had a generous budget to continue development. The root cause of the problem was inadequate mechanical clearances. After opening up clearances it was a long road to get the performance back.

ENGINE DESCRIPTION

The JT4 (J75) with afterburner, which has a close resemblance to the J57 with afterburner, is pictured in Figs. 2 and 3. The low compressor has eight

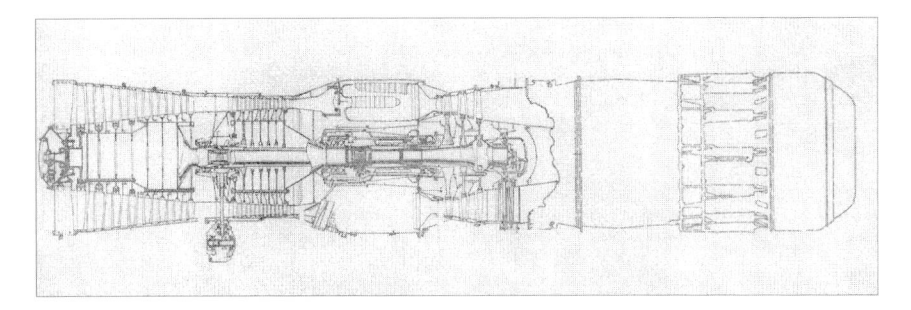

Fig. 2 Line drawing of JT4 (J75) (courtesy of Pratt & Whitney Archives).

stages, one less than the JT3, and the high compressor has seven stages. The can-annular combustor has eight individual combustion chambers, similar to the JT3. The high-pressure turbine has a single stage, while the low-pressure turbine has two stages. People have remarked that if they see either one JT3 or one JT4 separately, they have trouble identifying them. This is understandable, as there is only about two inches in diameter difference between the engines. The J75 low compressor has a lower hub/tip ratio, which tends to keep the compressor diameter smaller than a scaled J57 low compressor. However, when they are side by side (Fig. 4), there is no problem with identification.

Don Jordan was the development engineer on the JT4. Dick Smith became a project engineer on the JT4. I remember when he announced his change of assignment to us. The aviation horizon was expanding so rapidly toward the dream of hypersonic flight speeds and flight beyond the atmosphere that we thought the company would be moving into ramjets as the next propulsion challenge. Some of us thought the JT4 would be Pratt & Whitney's last turbojet.

Walt Doll's compressor group rushed through the aerodynamic design of an eight-stage compressor to reduce engine weight. The following process

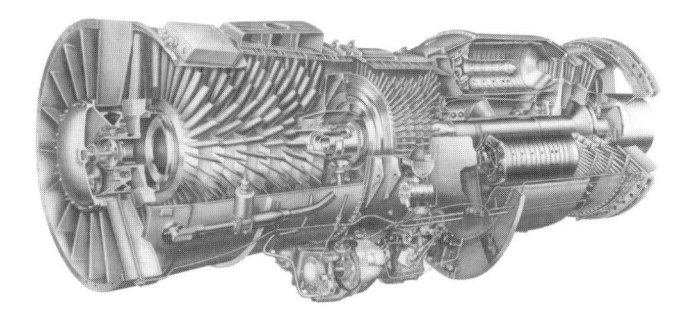

Fig. 3 Cutaway of J75 (courtesy of Pratt & Whitney Archives).

Fig. 4 Compare the J75 afterburning engine (left) to the J57 (right) (courtesy of Pratt & Whitney Archives).

through the mechanical design group so impressed Wright Parkins that he let his team (Pratt and Doll's Technical & Research Group and Gil Cole's Design Group) know that they did a great job [2]:

> I seldom write a letter of appreciation in the department simply because I know from past experiences that they can get to be interpreted as Form Letter No. Blank.... However, in this ... case I will ... pass on to you my sincere appreciation for the way your people knuckled down to the job of getting the J75 8-stage compressor recalculated and redesigned, at least aerodynamically. Needless to say, the J75 is the engine that will make or break Pratt & Whitney and I dare say that if what has been accomplished in Fransson's [Mechanical Design] and Brown's [Technical & Research Group] departments means anything ... we won't have to worry about making a success of this important engine.

In little more than seven years (Fig. 5) after the start of design, the JT4 commercial engine made its first scheduled commercial flight for Pan American World Airways in August 1959. It had made an unscheduled flight the previous month when it powered the Pan Am flight to Moscow, which had the press group aboard to cover Vice President Nixon's visit to the Kremlin. The initial time-between-overhaul (TBO) for the engine was 800 hr. About four years later the TBO was increased to 6000 hr, roughly equivalent to being on the wing for two years of operation.

The total production was 2579 engines (Table 1). The introduction of the turbofan version of the JT3 (JT3D) that offered more thrust at a lower specific fuel consumption limited the production potential of the JT4 engine.

APPLICATIONS

Moving into the jet era, airframers had military and commercial customers with more sophisticated needs than those of the piston engine era. As a result, the airframers made greater demands on the engine manufacturer, who found it difficult to make design trade-offs to meet the new requirements. On the customer side, an airplane manufacturer had to understand the burgeoning field of propulsion choices in addition to understanding military and commercial aircraft propulsion opportunities. The airframer's response to this new business environment was to set up a greatly expanded propulsion group, as well as specialized mission performance groups for military and commercial opportunities.

The JT4 commercial engine was a reliable engine and would have had a great future except for a turn of events brought on by a competitor. The JT4's big daddy, the JT3 (J57), enjoyed a production run of more than 20,000 military and commercial engines. The JT4's production run was only about 12 percent of the JT3 volume. The total production of the JT4 family was 2579 engines, of which 1005 were commercial engines. The bulk of the military engines were in the F-105, CF-105, and F-106 aircraft. Small numbers were in the Martin P6M SeaMaster, as well as the F-102, F-107, and F8U-3 aircraft.

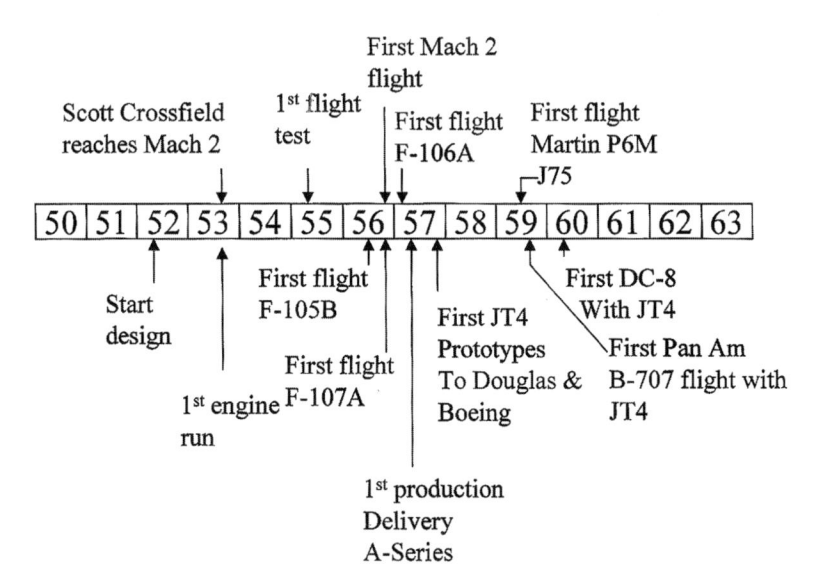

Fig. 5 The J75 program started around the time of the B-52 flight-testing in 1952.

TABLE 1 JT4 SPECIFICATIONS AND APPLICATIONS

Model	Take-off thrust (lbs)	Military or max. cont. thrust (lb)	Normal thrust (lb)	Weight (lb)	Diameter (in.)	Length (in.)	Aircraft
JT4A-1	15,800	15,800	14,000	4900	43.0	146.4	None
JT4A-3	15,800	15,800	12,500	5020	43.0	144.1	DC-8, B-707-220, B-707-320
JT4A-5	15,800	12,500	12,500	4815	43.0	144.1	DC-8, B-707-320
JT4A-9	16,800	13,500	13,000	5050	43.0	144.1	DC-8, B-707
JT4A-10	16,800	13,500	13,500	4845	43.0	144.1	DC-8
JT4A-11	17,500 wet[a] to 90F 17,500 dry	14,900	14,900	5100	43.0	144.1	DC-8, B-707
JT4A-12	17,500 wet to 90F 17,500 dry	14,900	14,900	4895	43.0	144.1	DC-8, B-707
JT4A-20	23,500 A/B[b]	15,500	13,700	6100	43.0	238.1	F-102
JT4A-21	23,500 A/B	15,500	13,700	5800	43.0	231.8	F-102
JT4A-23	23,500 A/B	15,500	13,700	6175	43.0	259.8	F-105
JT4A-24	23,500 A/B	15,500	13,700	5875	43.0	237.6	F-102B
JT4A-25	23,500 A/B	15,500	13,750	5950	43.0	259.3	F-105

Four More Turbojets

YJ75-P-1	23,500 A/B	15,500	13,700	6100	43.0	238.1	YF-102B, YF-106, F-106
YJ75-P-2	15,800	15,800	14,000	4960	43.0		Martin P6M-2, Douglas A3D-1
YJ75-P-3	23,500 A/B	15,500	13,700	6175	43.0	259.8	AVRO YCF-105B Republic YF-105B
J75-P-5	23,500 A/B	15,500	13,700	5950	43.0	259.3	AVRO CF-105 Republic F-105B LTV F8U-3
J75-P-9	23,500 A/B	15,500	13,700	5875	43.0	237.6	F-102B, F-106A,B
YJ75-P-11	23,500 A/B	15,500	13,700	6130	43.0	240.7	North American F-107A
J-75-P-13	15,800	15,800					Lockheed U-2B
J75-P-13B	17,000	17,000					Lockheed U-2C
J75-P-17	26,000 A/B	16,100	14,300	5875	43.0	237.6	F-106A, B, C
J75-P-19	24,500 A/B	16,100	14,300	5950	43.0	259.3	F-105B
J75-P-19W	26,500 wet 24,500 A/B	16,100	14,300	5960	43.0	259.3	F-105D, F-105E

[a] Wet indicates the rating with water injection.
[b] A/B means afterburning operation.

The first obvious application was to give the U-2 aircraft a 50 percent boost in thrust over the J57. This greater thrust capacity provided about another 5000 ft in altitude capability. Security regarding the engines for the U-2 was so strict that hardly anyone at Pratt & Whitney was aware of the U-2 application. Larry Carlson was one of the engineers working on the U-2 application of the J75 (Fig. 6), as he explained in his 2004 letter:

> I was the Project Engineer on the JT4/J75 when we put the J75-P-13 engine in the airplane. Dick Baseler and I were the only ones at Pratt & Whitney who had any knowledge of the program. There were questions about the application of the P-13 but the word didn't get out.... The airplane took off and climbed to peak cruising altitude and cruised at the "coffin corner" where the stall speed and the Mach buffet speed were the same—all at maximum turbine inlet temperature. I was amazed to examine engines after this kind of service. The burners and turbines looked like new. Few thermal cycles and low gas load did amazing things for the hot-section durability.
>
> When the Russians developed a better surface-to-air missile, they downed a U-2 flown by Gary Powers. The wreckage was displayed in Moscow with photos in the press, one of which showed the engine data plate—Pratt & Whitney J75-P-13.

When Pratt & Whitney was working with Republic for the installation of the J75 into the F-105, Perry Pratt was convinced that the Technical & Research Group should not only do advanced engine studies, which it was

Fig. 6 Engine J75-P-13 was used in the U-2B and U-2C aircraft (courtesy of Pratt & Whitney Archives).

doing under Bill Sens, but now should also be concerned about aircraft performance. He wanted to understand how aircraft requirements influence the design of engines and be able to evaluate different engine configuration choices by determining their effects on the aircraft. The United Aircraft Research Department had been making such studies, but Perry wanted such studies done in an environment close to real engine experience [3]:

> The decision as to the next engine designed to meet the Navy's future needs should be based primarily on the thrust requirements in the time period during which the number of engines (per aircraft), reduces to one of deciding on the general level of performance and payload requirements of the combat airplanes that will utilize the new engine. Obviously the problem of airplane performance requirements is beyond the province of the engine manufacturer, but, because it is so vital in establishing advanced engine development programs, we must concern ourselves with it.

A group of engineers who had formerly worked on missile technology at the United Aircraft Research Laboratory came to Pratt & Whitney's Engineering Test Department with airplane and aerodynamics skills. In a short time, the Technical & Research Group formed an Airplane Performance Group under Gene Montany that reported to Bill Sens, who was in charge of all advanced engine performance work. It was not long before Republic experienced aerodynamic problems in the design of its engine inlet for supersonic aircraft speeds. One of Gene's men, Skip Bailey, was able to offer significant help to Republic's engineers.

It soon became apparent once Pratt & Whitney got into aircraft gas turbines for supersonic flight that the effort of engine installation in an aircraft was considerably greater than what had been involved in the piston engine days. An axial-flow compressor prefers the relatively uniform inlet velocity profile, for example, as that provided by a bellmouth inlet. As the inlet air density decreases with altitude, the surge-line pressure ratio decreases toward the engine operating line. This makes compressor surge more likely with sudden power lever changes. The effect is greater for fighter aircraft, whose maneuvers provided a variety of air velocity profiles approaching the engine compressor face. Inlet-engine compatibility would become a critical area for development as fighter-type aircraft came into the military inventory.

The Republic YF-105A prototype flew on October 22, 1955, powered by a J57-P-25. The production aircraft was the F-105B, powered by the J75, which first flew in mid-1956. Subsequent models of the F-105 (Figs. 7 and 8) performed a variety of missions not only as fighter-bombers but also going after anti-aircraft installations, the famous "Wild Weasel" missions in North Vietnam.

Extra weight had to be added to the engine rear mount and downstream structure to stiffen the cantilevered afterburner so that its flexing would be

Fig. 7 Aircraft F-105 with the J75 engine—note that the afterburner is cantilevered from the rear engine mount plane to minimize stress on the rear fuselage section (courtesy of Pratt & Whitney Archives).

within the permissible limits specified by the airframe manufacturer. I suspect minimum weight aircraft would result from using the airframe structure to secure the afterburner and nozzle during maneuvering loads. American aircraft designers preferred to have the engine company stiffen the afterburner–nozzle structure so it could be cantilevered from the rear engine mounts.

Fig. 8 Slipping the F-105's aft section in place over the cantilevered J75 (courtesy of Pratt & Whitney Archives).

Fig. 9 F-106 in flight, powered by the J75 engine (courtesy of Pratt & Whitney).

The F-106 (Fig. 9) came into being in June 1956 as a re-designation of the F-102B. It was envisioned as the "ultimate interceptor." The first flight of an F-106 took place on December 1956, with the J75-P-9 replacing the Wright J-67 (Bristol Olympus under license), the powerplant originally selected for the aircraft. The aircraft specifications were for a supersonic, all-weather interceptor that could reach 70,000-ft altitude within a 430-mile radius. The aircraft began service in October of 1959. Eventually 277 F-106As were produced in addition to 63 F-106Bs (two-seater trainers).

In his 2004 letter, Larry Carlson recalls another experience on the J75 program in connection with the F-106:

> We were working on the J75 in the F-105, F-106, and F-107 at Edwards Air Force base. These were all Mach 2 airplanes with variable geometry inlet ducts ... if the inlet flow was not properly matched to the engine flow, the inlet system became unstable—blowing the shock out the front. This was a real thrill in the transonic range. When the inlet and engine flows were properly matched the inlet was stable and the total pressure recovery was good to give great acceleration to Mach 2. They were all good airplanes but the North American F-107 lost out to the F-105 and the Convair F-106 was chosen as the follow-on to the F-102.
>
> During this time ... the Russians held the world speed record over a straight-line course. The Air Force decided to break the record using a souped-up F-106 and planned to announce the new record at the Wright Brothers' Dinner in Washington. This feat would require a speed of Mach 2.4. Convair reviewed the airplane and we provided two engines with better disk materials. Practice flights were ... made over the course at Edwards. We had a time getting all of this to come together. We had to calculate the required inlet setting on the ground. We picked up foreign object damage (FOD) on the first engine, which required a change, and cracked a couple of windshields on the airplane. We were getting close to doing the job when we picked up more FOD—this time on the second special engine!

We thought we struck out until Major Joseph W. Rogers, the pilot, decided to take the stock F-106 he had flown in, had the manual inlet control system installed in it, fuel[ed] it with [enough] fuel to do the job and ... set a new record at 1525.95 mph. He did this with just enough time to fly to Washington and stroll into the dinner meeting to make the announcement!

The JT4 in commercial service and the J75 in the military were compiling an excellent record but the JT3D cut the commercial future and the loss of the F8U-3 competition to the F4H ended the engine's future prospects. We worked closely with Vought at Edwards during the Navy fly-off of the two aircraft and felt that the F8U-3 performed a bit better. However, the customer decided that it wanted the two-man, two-engine aircraft.

Another application was the Martin flying boat, the P6M SeaMaster. The first flight of the SeaMaster prototype took place on July 14, 1955. Four Allison J71 turbojets provided the thrust. The first six production aircraft were Allison-powered. However, the rest of the production was to be with the non-afterburning J75-P-2. Even though this flying boat was the world's fastest, the program was canceled in August 1959, after only three aircraft were built.

REDUCING NOISE

When turbojet and turboprop powered airline operations got started in the 1950s, it became clear that aircraft engine noise needed some form of regulatory oversight by federal, state, and airport operators. The Port of New York Authority took the lead in establishing permissible noise levels for the jet-powered transports. The consulting firm of Bolt, Beranek, and Newman (BBN) was called in to measure the noise levels of piston-powered aircraft to define a baseline of aircraft engine noise generation. John Tyler of Pratt & Whitney summarized the initial regulations [4]:

1) The take-off noise would be measured at a point three miles from the beginning of the take-off roll. The three-mile distance was an estimate of how far away the aircraft on the ground was from the nearby residences.

2) The maximum noise level for the jets was set at a level exceeded by 25 percent of the noisiest piston engine powered aircraft. In other words, the new jets had to be less noisy than about 25 percent of piston-engine powered aircraft.

3) The unit of noise measurement was called perceived noise level in decibels, PNdb). The subjective scale of perceived noise was suggested by Karl D. Kryter and Karl S. Pearsons. The PNdb scale attempted to take into account the fact that some frequencies are more annoying to human ears than other frequencies.

Acoustic engineers measure noise in terms of decibels, measured with microphones. The noise level in decibels is defined as 20 times the log to the base 10 of the ratio of the actual (root mean square) sound pressure (in dynes/ square centimeter) divided by a reference (rms) sound pressure level of (0.0002 dynes/ square centimeter). If the logarithmic scale was not enough to confuse engine designers, the psychological tweaking of the scale into the subjective "perceived noise level" drove them crazy. When an engineer makes a change to an engine design for noise reduction, it most likely affects the weight and/or performance levels. A designer could feel good about making a design modification that achieved a 10 decibel reduction in noise, only to discover that on the PNdb scale the reduction was only 5 decibels.

It is easy to lose sight of the magnitude of sound suppression when the progress is in terms of a logarithmic scale. What is discouraging about noise reduction is that one can cut the sound energy in the exhaust by a factor of two and yet achieve only three decibels reduction in sound as measured by inanimate microphones. The human ear can experience less than the three decibels reduction in terms of PNdb at some frequencies.

The noise from the early turbojets came from the mixing of the high-velocity jet exhaust with the ambient air. The earliest noise suppressors were based on the theory that if the exhaust jet were broken up into many smaller jets, then the mixing to a low-noise jet velocity would occur closer to the engine exhaust nozzle plane and the volume of air making the noise would be smaller.

Using Computers

The dawn of the jet age was aided by new machinery—the computer. Walt Doll describes using computers during the development of the JT4's compressor, as he remembered it from his days in the Technical & Research Group [5]:

> In those early days ... the only computers around were the punched-card calculating machines used by the Payroll Department, which ... weren't designed for series-type calculations as needed in engineering. In the middle 1940s, Stu Crossman, from the Design Department, had used one of the payroll machines to solve a turbine disk stress analysis problem. As far as I know that was the first use in Pratt & Whitney of any kind of computer other than a slide rule ... There were no hand calculators then. I can still remember when the first one showed up and cost about $500....
>
> Because of the complexities of the design and analytic processes ... it seemed to us that you ought to be able to use computers to improve your ability to first decide what engine you wanted to design, then to execute the design, and finally to calculate the performance. We gradually worked into all of these things.

In preparing for the JT4 engine, my team, under the direction of Bill Podolny and Frank Roberts, put together a pseudo-design system that allowed us to design ... many different compressors and then calculate their performance ... We put this system on the payroll machines, between payroll periods, and designed approximately 600 compressors ... using different combinations of design options. Then we calculated efficiency contours and surge lines. The machines were so slow that you could stand there and watch them make a mistake. The simplistic design procedure couldn't simulate some of the bad things that happen in a compressor when stages stall, but the overall results were nevertheless very helpful. We used these results to block out the basic compressor design for the JT4.

The first time we ran the JT4 engine it broke the record for engine efficiency, and had the best specific fuel consumption ever recorded in a jet engine.... Unfortunately, the peak efficiency was still too close to the surge line for practical operation and we spent a large amount of time and effort achieving a satisfactory resolution.... Peak airfoil efficiency always occurs just before the airfoil stalls. Our viewpoint was that we had many rows of airfoils and we ought to be able to balance some against the others in such a way that we could get an average combination of airfoil rows to perform better than any one single airfoil row with regard to what we wanted. This was how we designed the computer study. Anyway, the JT4 was a step forward. As better computers became available, we continually upgraded our design and prediction processes.

J52 (JT8)

The PT3 became the JT8, beginning with the PT2/T34 program, Pratt & Whitney's first attempt at a turboprop design. The T34, a single-spool gas turbine, was sponsored by the U.S. Navy and continued its development while the JT3/J57 program was underway. When the JT3 engineers were looking at a twin-spool compressor, the T34 program engineers started to look at a twin spool for the turboprop. This initial analytical effort was under an engine designation of XT34-P-10 (PT2H). This effort soon evolved into a Navy program called the PT3, which had a military designation of XT52-P-2.

This XT52-P-2 was a turboprop with a sea-level shaft horsepower of 7600 (equivalent shaft horsepower of 8300). The estimated weight was in the range of 3575–3725 lb depending upon whether the engine had a single reduction gear or a double reduction gear. Another variation was the XT52-P-4, which had an equivalent shaft horsepower of 5700 at a weight of 2850 lb. One of the benefits of the T52 program was the design and test of a 7:1 pressure ratio transonic compressor—a technology which would find a home in not only the J52 low compressor and the J91 (JT9) but also in Pratt & Whitney's smallest turbojet, the JT12.

In the early part of 1953, one of the major concerns of the Navy was about Panther and Cougar aircraft taking off from a carrier and returning. The Navy used the term *cycle time* to refer to the time interval between take-off and landing on the carrier. *Loiter time* was important because of the need for returning aircraft to loiter while other aircraft were taking off. Twin-engine aircraft such as the McDonnell F2H had higher cycle times because the aircraft could loiter on one engine. The Chief of Naval Operations asked the Bureau of Aeronautics what could be done to increase the cycle time. As a result of poor partial power performance of low-compressor-pressure ratio engines such as the J42, J48, J46, and J34, the Navy was convinced that it was important to have two engines in an aircraft so that one engine could provide the loiter power at a better specific fuel consumption than what could be obtained with two engines running at lower power. The Bureau of Aeronautics suggested aircraft powered by twin J46s, or converting the T52 to a turbojet; however, the bureau was thinking of a thrust size of about 5000 lb.

In June 1954, Perry Pratt convinced the Navy that a high overall pressure-ratio turbojet for a twin-engine installation should be about 7500-lb thrust. North American and Douglas were among the bidders for a small Navy attack airplane. Pratt & Whitney did not have an engine in the proposed program but worked on engine performance estimates and airplane studies. North American provided Pratt & Whitney with some drag polars for aircraft performance studies. In September, the Navy authorized Pratt & Whitney to proceed with the development of a twin-spool, 12:1 overall pressure ratio engine. Perry correctly anticipated that the customer needed a more powerful engine than originally thought. The high spool was the starting point for the many JT8D commercial turbofan applications.

The T52 concept changed into the J52 reality. This engine ended up as a very successful product for the company for decades (Fig. 10). The total J52 production amounted to 4,567 engines. Table 2 lists the specifications as well as the applications for this engine.

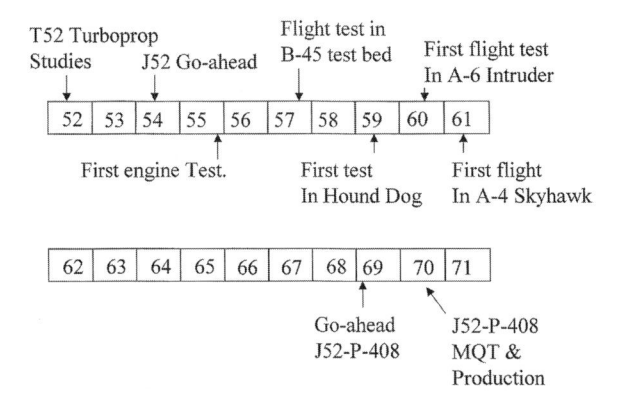

Fig. 10 Time line for the J52.

TABLE 2 JT8/J52 SPECIFICATIONS AND APPLICATIONS

Model	Take-off thrust (lb)	Military or max. cont. thrust (lb)	Normal thrust (lb)	Weight (lb)	Diameter (in.)	Length (in.)	Aircraft
JT8A-20	11,000 A/B[a]	7250	6400	2750	30.0	196.3	None
JT8B-1	8500	8500	7500	2056	30.2	116.9	A2F Proposal
JT8B-20	12,500	8000	7000	2600			None
J52-P-3	7500	3450@55,000 ft altitude		2145	29.8	149.7	NAA AGM-28, GAM-77, -77A
J52-P-4	7850	7850	6950	2000	30.0	127.5	A4D-3 Proposal
XJ52-P-6	8500	8500	7500	2200	29.8	116.4	Grumman A6A
J52-P-8	9300	9300	8200	2118	31.4	116.9	Proposed for Grumman A-6A & Douglas A-4E
J52-P-8A	9300	9300	8200	2118	30.2	116.9	Grumman A-6A, EA-6A, EA-6B Douglas A-4E, F,G, H, TA-4G, H
J52-P-408	12,000	11,200	9900	2300	30.0	116.9	A-4, A-6 Intruder & EA-6B Prowler

[a] A/B means afterburning operation.

ENGINE DESCRIPTION

The engine (Fig. 11) had an axial-flow dual compressor with a total of 12 stages (four fewer stages than the earlier J57). The low compressor had only five stages (compared to nine in the J57) by taking advantage of the new transonic aerodynamics recently developed in the compressor group, and by lowering the pressure ratio enough to get by with a single-stage turbine (Figs. 12 and 13). The high compressor had seven stages (like the J57), driven by a single-stage turbine. The combustion section consisted of a can-annular type burner with nine burner cans (Fig. 14). The turbine was shrouded (Fig. 15).

FOUR MORE TURBOJETS 273

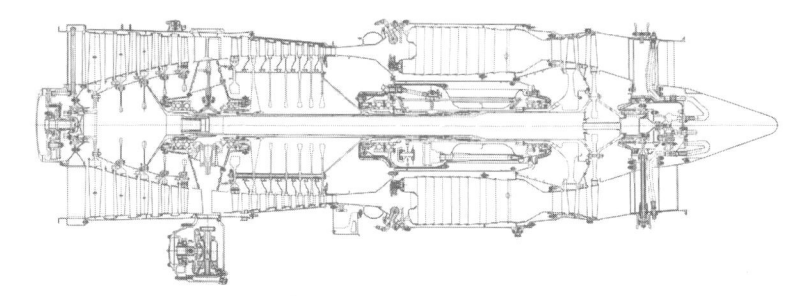

Fig. 11 General configuration drawing of the J52 (courtesy of Pratt & Whitney Archives).

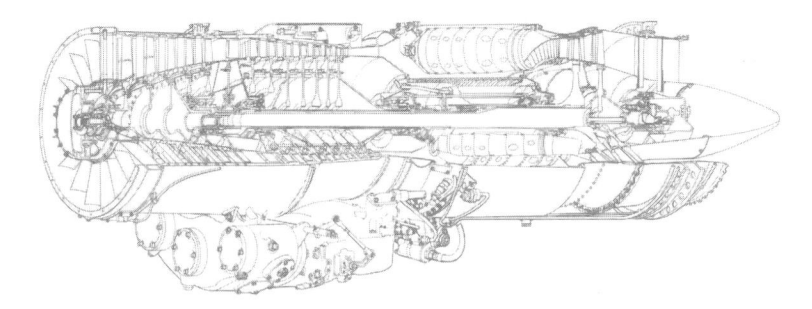

Fig. 12 12:1 two-spool engine design; the first three stages of the low compressor have mid-span shrouds to minimize vibratory stresses (courtesy of Pratt & Whitney Archives).

Fig. 13 J52 non-afterburning engine—note the area around the fourth stage of the low compressor where there are poppet valves for air bleed in the Hound Dog engine (courtesy of Pratt & Whitney Archives).

In the early 1950s, there was an intensive effort to understand compressible airflow in turbomachinery. Pratt & Whitney had sponsored work at United Aircraft Research Laboratories as well as at Pratt & Whitney's own compressor research group. At the same time, NACA Lewis and Langley facilities were deep into aerodynamic research. Lieblein at Lewis and von Doenhoff and Tetervin at Langley were concentrating on boundary layer behavior under pressure gradients. Pratt & Whitney was refining its design methods by moving towards pressure distributions over blade and vane spans as well as boundary layer flow at the end walls.

The analytical engineers in the compressor and turbine groups would frequently discuss their aerodynamic problems. It was quite common, for example, to have someone in the Compressor Group come over to his Turbine Group lunch buddy late in the afternoon with, "Say, I've been thinking about what you said about your blade efficiency problem. Have you tried such and such?"

Two areas led to great improvements in airfoil performance were 1) paying attention to the pressure rise a boundary layer would encounter on compressor and turbine airfoils, and 2) the flow of the boundary layer at the end walls of the airfoils. Ideally what was best for airfoil efficiency was to have much of the airfoil with a laminar boundary layer and then a quick transition to a turbulent boundary layer with a subsequent pressure rise of no more than about half of the dynamic pressure at the start of the pressure rise. It was this

Fig. 14 The burner assembly of the J52 consisted of 9 cans in an annular chamber (courtesy of Pratt & Whitney Archives).

Fig. 15 The low turbine of the J52—note the shrouded turbine design, a standard feature in the company's turbine design (courtesy of Pratt & Whitney Archives).

detailed effort to understand the actual pressure distribution on an airfoil that led to improved efficiency and greater pressure-rise per compressor stage.

APPLICATIONS

There were four major applications of the J52, which were military: the Air Force Hound Dog Missile, the Navy Douglas A-4 Skyhawk, and the Navy Grumman A-2F and A-6 Intruder. In addition, the engine was also in service with the governments of Israel, Kuwait, Australia, New Zealand, and Honduras. By 1981, the J52 had been service for 20 years and accumulated more than 6 million flight hours.

HOUND DOG. North American had been working on a proposal to the Air Force for an air-to-ground missile, which was to be carried by the B-52. North American's approach was to use the J85 with afterburner as the propulsion for the missile. Wright Parkins, who at this time was in the United Aircraft headquarters, called a meeting with engineering and marketing to discuss the situation. Gene Montany, in preparation for the meeting, asked his group for ideas about the propulsion. Dick Staubach had been studying a supersonic missile with a non-afterburning J52 and had some preliminary performance numbers.

Parkins expressed his disappointment (to put it rather mildly) that Pratt & Whitney was not on top of this situation. When he asked for ideas, Gene showed him what Dick Staubach had estimated. Parkins, seeing what low fuel consumption the non-afterburning J52 had compared to an afterburning J85, called the prime bidders on the missile and asked them to consider this approach. North American was enthusiastic about this, and used the J52 instead of two afterburning J85s. Their new missile design exceeded the Air Force range requirement and became the winner in the competition.

After the Hound Dog program got underway, the Air Force wanted Pratt & Whitney to redesign the J52 as a "throw-away" engine. This sounded like a good suggestion until the B-52 pilots found out that they could start these engines up (there were two such missiles on the B-52) on take off and have another 15,000 lb of thrust. Having more take-off thrust and consequently a greater margin of safety trumped the "cheapened" J52, which never came into being.

The first Hound Dog missiles, officially the GAM-77 air-to-surface missiles, began its service on the wings of B-52s in mid-1958 (Fig. 16). The missile was advertised as being capable of carrying a nuclear warhead and flew at "high supersonic speeds" with a range of "hundreds of miles."

The plug nozzle was selected for the Hound Dog application because of the wide range of flight speeds, from subsonic to supersonic. Visible in Fig. 17, the plug nozzle offers a higher performance over the missile's flight spectrum than either a convergent nozzle (which is good for subsonic speeds)

Fig. 16   The installation of the J52 in the Hound Dog missile (courtesy of Pratt & Whitney Archives).

Fig. 17 J52-P-3 pushed the Hound Dog to supersonic speeds; the plug nozzle is at the right (courtesy of Pratt & Whitney Archives).

or a convergent-divergent nozzle (which is good for its supersonic design speed) up through Mach 1.5.

Douglas A-4 Skyhawk. In 1954, both North American and Douglas were in competition for the Navy's A-4. Ed Heinemann's "Hot Rod" from Douglas was the winner, with Wright Aeronautical's British engine, the J65 (Armstrong Siddeley's Sapphire). The single-engine aircraft became known as the Skyhawk. At first glance, the J65 and the J52 seem somewhat similar. The J65 has 13 axial stages of compressor with a pressure ratio of 7.25:1 driven by a two-stage turbine. The J52 has one less compressor stage but has an overall pressure ratio of 12:1 in a two-spool configuration. Both engines were in the 7500-lb thrust class. The first Skyhawk joined the fleet in October 1959.

The Skyhawk was the first application for the J52 when it replaced the J65 because of its better specific fuel consumption (Fig. 18). The Skyhawk (A-4E) is a remarkable aircraft in that it weighs about 9300 lb empty and yet can take

Fig. 18 Douglas Skyhawk returning from a mission (courtesy of U.S. Navy).

off fully loaded with a gross weight of 24,000 lb. In June 1963, two Navy pilots flew two aircraft from Lemoore Naval Air Station in California to Oceana Naval Air Station close to Norfolk, Virginia, non-stop, on the A-4E's internal fuel for about 2100 miles. The Skyhawk II with the J52-P-408 was the aircraft of choice by the Navy's Blue Angels because of its maneuverability and reliability.

Grumman Intruder. The J52 logged more flight hours per month than any other engine in Navy operations, and the Grumman A-6 family of Navy aircraft included a wide variety of models. The Grumman A-6A Intruder (Fig. 19) was introduced into Navy and Marines units in the 1960s. The original designation was A2F-1 for an all-weather attack aircraft. It had to be able to fly long range at low altitudes, find targets, and use both conventional and nuclear ordnance. Powered by two J52s, it had its first flight in April 1960. Its final model, the Prowler, first flew in 1971. The A-6 family had all-weather capability that was not exceeded until the newer F-111 came along with its terrain-following capability.

J52-P-408 — The Last of the J52s

The dash 408 version of the J52 provided 12,000 lb of thrust—about a 60 percent increase from the initial model at 7500 lb of thrust (Fig. 20). The compressor performance was improved by using the latest aerodynamic design techniques of controlling the boundary layer growth on the airfoils, polishing airfoil surfaces, increasing the airflow by five percent, and using abradable seals to control tip leakage. The output was increased by using turbine cooling in the first stage turbine to permit increasing the temperature by about 240° F. The low turbine used higher-strength materials to operate in

Fig. 19 Grumman A-6A Intruder with its two J52s are buried in the lower fuselage (courtesy of the U.S. Navy).

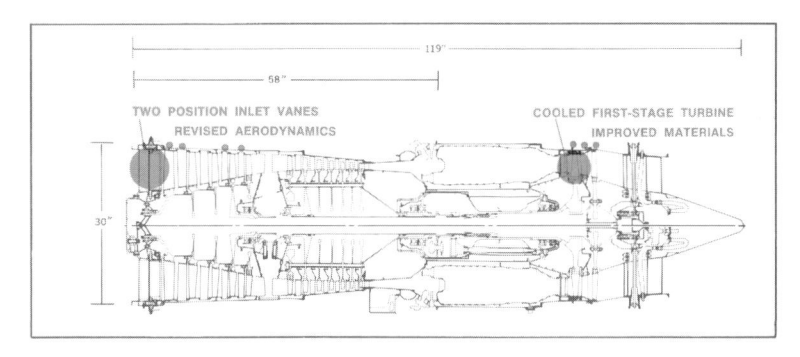

Fig. 20 The most advanced J52 at 12,000-lb thrust (courtesy of Pratt & Whitney Archives).

a hotter environment. The compressor had a variable (2-position) inlet guide vane to facilitate starting and rapid acceleration without compressor stall. On top of these improvements was the added benefit of a five percent reduction in fuel consumption. The J52 was also designed with an afterburner (Fig. 21), but remained an experimental configuration.

J91 (JT9)

If the 1920s was described as the "Roaring Twenties," then certainly the 1950s should be painted in the aviation business as the "Fabulous Fifties" because of the tremendous number of government-sponsored study and hardware programs. Perry Pratt (introduced at the end of Chapter 3) was in close communication with the Department of Defense and the leaders of the airframe industry. Back at the plant, Bill Sens and his group supplied Perry with his necessary ordnance.

In those days, we were looking at turboprops, turbojets, turbofans, turboramjets, and ramjets. We studied the flight spectrum from low-speed vehicles

Fig. 21 J52 with afterburner (courtesy of Pratt & Whitney Archives).

all the way up to the single-stage to orbit Aerospace Plane. The fuels under study were the standard hydrocarbon types, hydrogen, and the exotic types such as Boron hydrides and the overactive Tri-ethyl-aluminum (TEA). The more unconventional powerplants were the hydrogen expanders, hypersonic ramjets, and the Liquid-Air-Condensing-Engine (LACE).

Perry Pratt's influence was important throughout the aviation industry and the Department of Defense. He generated the respect in the industry in the gas turbine era that Hobbs had generated in the piston engine era. Pratt inspired his people to do their best because anything less would be unacceptable. Perry was forever going directly to the engineer grinding out the answer. You might have a great knowledge of a narrow branch of engineering, but after a few questions, he would walk away knowing more about the subject than you did and Perry's questions clarified a few things in your mind.

In the early 1950s, I attended my first American Society of Mechanical Engineers (ASME) meeting in New York City. At the close of the first day's session I ran into Perry. He came right over to me with a couple of other gentlemen and I introduced myself. Perry then introduced me to Willis Hawkins, Chief of Design at Lockheed, and to another Lockheed executive. Pratt then asked, "Jack, we're going out to dinner tonight. Why don't you join us?" I gracefully declined, but I should have gone along because that is the way Perry was, not at all rank conscious.

Pratt & Whitney was making engine studies for high-speed flight as early as 1952. Scott Crossfield's Mach 2 flight in 1953 gave inspiration for even higher speeds. Around the middle of 1954, Pratt & Whitney submitted a proposal to the Air Force for a Mach 3 afterburning turbojet. Shortly thereafter, however, the Air Force Weapon System 110A program got underway and the two engine manufacturers were General Electric and Allison working with North American and Boeing on what eventually became the B-70. The final winners were North American and General Electric. The total procurement was for two aircraft prototypes, one of which made its first flight in late 1965 to a very brief dash to Mach 3. The second prototype crashed in 1966 during a publicity shoot. Pratt & Whitney was not part of that program even though it supported the airframers' studies with engine performance data.

Fig. 22 A time line of the J91.

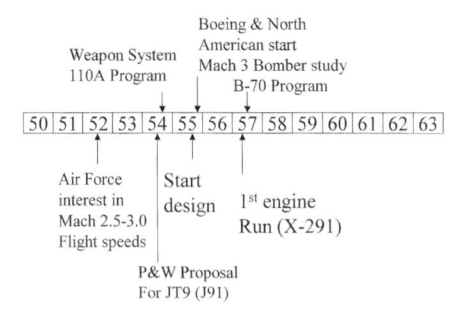

The J91 was designed in late 1955, and the first engine was run under the Air Force contract in mid-1957. There were only two engines in the program. The first engine, X-287, was eventually scrapped and the second test engine, X-291, ended up at the New England Air Museum. The sequence of these events is outlined in Fig. 22.

One outstanding characteristic of the J91 was its compressor. Walt Doll described it, as he remembered it from his days in the Technical & Research Group [5]:

> We had lots of troubles with compressors, mainly because the peak efficiency ... had the unfortunate tendency to occur right on the surge line. If you operated on the surge line, the engine stalled out, lost thrust, [and] burned up the turbine.... You needed a certain "surge margin" to operate the engine stably. Much of our design and development effort was involved in trying to get the peak efficiency and the surge line a reasonable distance apart....
>
> In about 1952 ... we were working [for] the Air Force at Wright Field, thinking in terms of Mach 3 military engines. Mach 3 is a very tough environment for an engine. The air temperature entering the engine is over 650° F, due to the precompression caused by the airspeed. This means that everything in the engine is going to get hotter from there—fuel, lubricants, and controls, as well as some structural parts.
>
> Because of the high aerodynamic drag of the airplane and engines at such high speed, it is essential to get as much power as possible from a given diameter engine. This means getting as high an airflow as possible through the engine. We had a compressor program to learn how to do that. We went to what we call very low hub/tip ratio blading, that is, the hub of the blading was about a third of the tip diameter, compared to about half the diameter in the then current engines. We also wanted a high-speed stage. We designed the stages for transonic flow, which was new. This resulted in very thin wide airfoils at the blade tip and narrow blades at the root. We had to make these blades out of titanium, because neither aluminum nor steel could handle the stresses. This was the first instance of the high throughflow type of blading now used in fan blades, and the first use by Pratt & Whitney of titanium. We built one of these blades to prove that we could, and eventually we built a small scale, 8:1 pressure ratio compressor, with a high throughflow front end, to test out the design principles on a working model. This compressor was about the right pressure ratio and configuration needed for a Mach 3 engine. Bill Podolny ran this program for me.

ENGINE DESCRIPTION

The engine was a single-spool with a compressor pressure ratio of 7:1 in nine stages. The compressor (Fig. 23) had the lowest hub/tip ratio of Pratt &

Fig. 23 JT9 Nine-stage transonic compressor (courtesy of Pratt & Whitney Archives).

Whitney compressors at that time to minimize the engine diameter. The JT9 had a can-annular combustion chamber with eight burner cans and a two-stage turbine. Its size was selected by Perry as the biggest engine that could run at Mach 3 at altitude in the Willgoos Laboratory. The exhaust nozzle was a variable convergent-divergent ejector type.

The airflow capacity at sea level was about 400 lb/sec. Military thrust was 24,500 lb, and with an afterburner the maximum thrust was 35,000 lb (Table 3). The main value of the JT9 was that it was the starting point for the J58 early configuration and as an engine in the Aircraft Nuclear Propulsion program.

The experimental J91 engine X-291 is shown in Fig. 24. The other experimental engine (X-287) was scrapped. Only the J91-1 model got tested and even at that the total test hours on the two experimental engines was less than 500 hr. The dash 5 and dash 7 ratings were projections. The main result

TABLE 3 J91 ENGINE PERFORMANCE (AT SEA-LEVEL STATIC)

Performance	J91-1	J91-5
Max thrust A/B (lb)	35,000	41,500
Military thrust (lb)	24,500	28,700
Weight (lb)	8000	8000
Diameter (in.)	72.5 at ejector	72.5 at ejector
Length (in.)	315.1	309.6

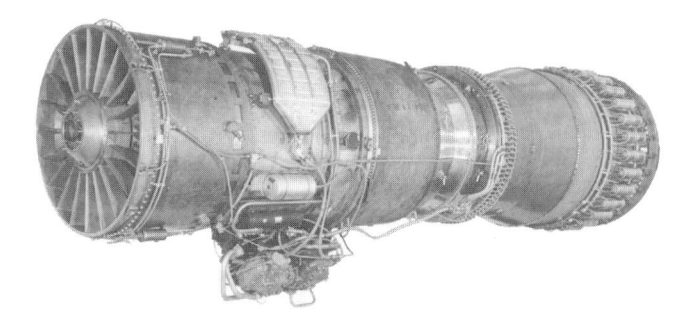

Fig. 24 Experimental J91 engine X-291 (courtesy of Pratt & Whitney Archives).

of the testing was that the transonic compressor in the engine performed well and the overall engine performance was consistent with what was predicted.

Sputnik shook up the populace in 1957, and this little satellite had a profound effect on the Department of Defense's planning. The J91 experimental test program wound down. To Pratt & Whitney's good fortune, the Navy had a requirement for an engine that would provide a high-speed dash at a critical point in its mission. Pratt & Whitney scaled down the J91 engine to 300-lb/sec airflow from the J91's 400 lb/sec. Thus the JT11 was born and was adopted by the Department of Defense as the J58.

AIRCRAFT NUCLEAR PROPULSION PROGRAM

The Aircraft Nuclear Propulsion program began in 1946. The Air Force got together the major engine manufacturers in a program called Nuclear Energy for Propulsion of Aircraft (NEPA). In the end, two engine companies carried on: General Electric worked on the Direct Air Cycle, and Pratt & Whitney took on the Liquid Coolant Cycle. Pratt & Whitney started the Connecticut Aircraft Nuclear Engine Laboratory (CANEL) under Air Force sponsorship in 1955.

Barney Schmickrath was the General Manager of the CANEL facility (Fig. 25). Gene Holtsinger was his executive assistant and Dr. Bob Strough was the chief physicist for CANEL. Barney was a character—he only read memos addressed to him. He made himself available for decisions, which he could make instantly, and had a strong loyalty to his team. One fateful decision he made was to issue a memo threatening dire consequences to any engineer visiting from East Hartford who stopped for coffee at a small diner en route to the CANEL facility in Middletown. He believed those minutes were spent more productively conducting business at the CANEL facility. But one day Barney happened to drop in at the little diner and learned that business had suddenly dropped off. Barney rescinded the memo.

Fig. 25   From left to right, Barney Schmickrath, Clifford E. McColley (Area Manager, Atomic Energy Commission), Congressman Emilio Daddario, Major Paul Baker (U.S. Air Force), Gene Holtsinger, and Bob Strough (courtesy of Pratt & Whitney Archives).

When Barney was Engineering Manager in East Hartford (before his move to CANEL), I had returned from a trip to the Power Plant Laboratory at Wright Field. My boss happened to be in Barney's office when a Dayton representative called Barney to express his dissatisfaction with the results of a meeting. All Barney had to hear was someone was making negative remarks about one of his boys and Barney hung up on him. Nobody trashed Barney's boys! I never heard from Barney about the incident and would never have known about it if my boss did not happen to be in Barney's office at the time of the phone call.

General Electric's Aircraft Nuclear Propulsion engine was the J87 (GE X211). In its direct cycle, air from the compressor exit was ducted into the reactor to pick up heat and then was piped back into the "burner section" of the engine to flow into the turbine. In Pratt & Whitney's Aircraft Nuclear Propulsion engine, J91, the compressor exit air passed through a heat exchanger just before the conventional burner. The fluid on the other side of the heat exchanger was liquid metal that was heated in the reactor core, entered the J91 heat exchanger and then was piped back to the reactor core. Pratt & Whitney used the J91 in the powerplant studies and component tests.

The J91, as a nuclear powerplant, would be used in its application as a "conventional hydrocarbon-fueled engine" for take-off and then at cruise altitude would use the heat from the liquid metal in the heat exchanger to raise the compressor discharge air to an appropriate turbine temperature for the turbojet. However, the powers that be hesitated about having nuclear reactors flying around the world and the program died a natural death. Pratt & Whitney purchased the Air Force facility, which is now used for more conventional aircraft propulsion.

J60 (JT12) Turbojet

In 1957, the Royal Canadian Air Force announced its need for a small jet-powered trainer aircraft. Canadair prepared its proposed aircraft, the CL-41. At that time there were two engine candidates, the Fairchild J83 and the General Electric J85. This potential opportunity stimulated Pratt & Whitney of Canada to get into the competition with its first turbojet design in the 3000-lb thrust class. Pratt & Whitney of Canada had already established itself as a competent builder of aircraft piston engines—the R-985, the R-1340, the R-1830, the R-2000, the R-2800, and the R-4360. Hugh Langshur, head of Pratt & Whitney of Canada's Engineering Department, brought a group of young engineers to East Hartford in June 1957 to work on the design of a small turbojet engine: Jack Beauregard, Elvie Smith, Pete Peterson, John Vrana, Alan Newland, and Fernand Deroschers. The group worked until March 1958 at which time Pratt & Whitney (East Hartford) took over the development. The first engine was delivered to Canadair in September 1959 and powered its CL-41 flight test program (Fig. 26).

Engine Description

The JT12 is a single-spool engine with a 6.5 overall pressure ratio. The compressor has 9 stages. The engineers referred to this JT12 as a junior JT9 because of the similarities in the layout (Fig. 27). The combustion chamber consists of eight cans in a can-annular configuration. A two-stage, shrouded turbine (Fig. 28) drives the compressor. Table 4 lists the specifications and applications for the various models produced. All told, there was a total of 2621 production JT12s and another 352 JFTD12 turboshaft engines.

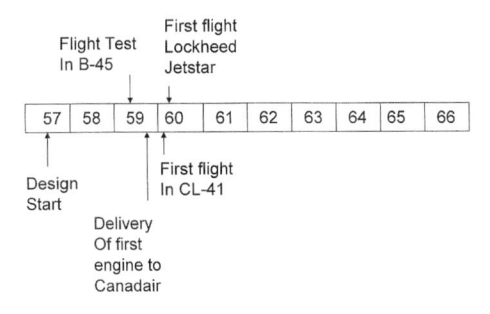

Fig. 26   A time line of the J60 (JT12).

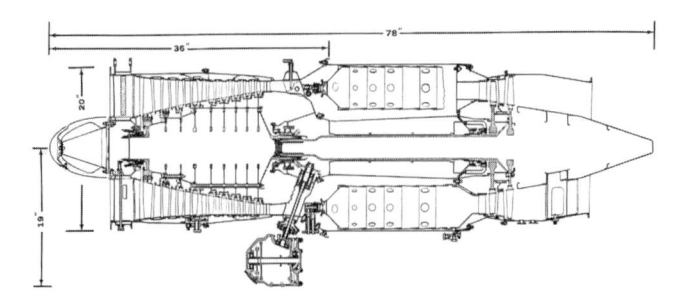

Fig. 27 A line drawing of the JT12 (courtesy of Pratt & Whitney Archives).

APPLICATIONS

The engine was designed for ease of maintenance and for commercial use rather than as a lightweight military engine. This design philosophy was demonstrated in the flight testing of the Canadair CL-41 development program, which proceeded well with the JT12. Unfortunately, the production contract went to General Electric, but the engine found other customers. The JT12 was used in the Lockheed Jetstar (Fig. 29) and the North American Sabreliner (Fig. 30). The engine was used in commercial and military Sabreliners.

PROMOTING FLIGHT SAFETY

Don Brendal was Project Engineer on the JT12 in the 1950s. He related this story to me in 2005 about one of his experiences shortly after a Lockheed Electra crashed in the Boston area while flying into flock of birds:

> Bill Gorton, a Senior Development Engineer, was my boss.... He asked me if the JT12 could ingest a flock of starlings without any problem.... I didn't know. We never ran any such tests because the FAA did not

Fig. 28 The left 3/4 view of the JT12 shows how much of the frontal area is taken up with accessories (courtesy of Pratt & Whitney Archives).

TABLE 4   JT12/J60 SPECIFICATIONS AND APPLICATIONS

Model	Take-off thrust (lb)	Military or max cont. thrust (lb)	Normal thrust (lb)	Weight (lb)	Diameter (in.)	Length (in.)	Aircraft
JT12A-1	2900	2900	2400	436	21.9		CL-41
JT12A-3LH	3300	3000		436	22.0	28.0	
JT12A-3DM	3300	3000	2800	436	22.0	78.0	Dassault Mystere
JT12A-6	3000	2400	2250	453	22.0	70.5	NAR NA-265
JT12A-6A	3000	2570	2570	453	22.0	70.5	NA-265
JT12A-8 Prototype	3300	3000		468	22.0	70.5	NA-265
JT12A-8	3300	3000		468	22.0	70.5 NAR 76.9 LAC	Sabreliner Jetstar
JT12A-20	3900	2830	2350	651	21.6	126.5	NORAIR N-156F
YJ60-P-1	2900	2900	2400	460	22.0	62.7	T-39A
XJ60-P-2	2900	2900	2400	441	21.8	62.7	SD-4 USD-5
J60-P-2	2900	2900	2570	441	21.8	70.5	USD-5 LAC 186-XH-51A
J60-P-3	3000	3000	2570	472	22.0	70.5	T39A,B Martin SV-5J
J60-P-3A	3000	3000	2570		22.0	70.5	T-39A,B,D T-2B
J60-P-5	3000	3000	2570	460	22.0	78.0	C-140A
J60-P-5A	3000	3000	2570	460	22.0	78.0	C-140B
J60-P-6	3000	3000	2570	495.3	22.0	70.5	T-2B

require them and furthermore the JT12 was certified by the FAA. Bill's reaction was, "Well, you better find out."

So, we captured about a dozen starlings and ran a test where we shot the birds into the JT12 inlet using a specially designed gun. After the dust settled, the engine was still running! Subsequent teardown and inspection of the engine parts revealed no damage, which of course was good news. However, this did not satisfy Bill Gorton. He wondered if the engine could take a hit by a sea gull....

So back to the test stand with a JT12 and a shot of a good size sea gull into the engine inlet. Bang! This time the engine broke in two and Gorton exclaimed, "You better fix it!" The wing span of an adult sea gull is

Fig. 29 The Lockheed Jetstar was the JT12's first commercial application (courtesy of Pratt & Whitney Archives).

substantially greater than the inlet diameter of the JT12. This triggered an extensive development test program in which many ... bird ingestion tests were conducted.... The birds were killed in the exhaust of an automobile, immediately loaded into the gun, and then shot into the running JT12.

The changes tested evolved into a configuration that was bird-proof. The two most significant changes made to the engine were to change the inlet guide vanes from aluminum to steel and to increase the spacing

Fig. 30 The JT12 also powered the North American Sabreliner (courtesy of Pratt & Whitney Archives).

between the trailing edge of the inlet guide vanes and first stage of rotating compressor blades. The changes [included] increasing the engine's specification weight and specification length both of which required recertification of the engine as well as tihe Lockheed Jetstar.

We recertified the engine and thought that Lockheed, the builder of the 4-engine Jetstar with JT12s would be happy to recertify the Jetstar with this new bird-proof engine. Lockheed's position was the Jetstar had already been certified in accordance with the requirements of the FAA and Lockheed was not going to spend any more of its resources in another certification that was not necessary.

Negotiations continued for some time. Finally in a meeting in New York with Dick Baseler and me from Pratt & Whitney and representatives from Lockheed and the cognizant FAAs, Baseler agreed to recertify the Jetstar. United Aircraft had a Jetstar on order so we got advanced delivery of the airplane without its interior. Harold Archer and his Experimental Flight Test crew used it to recertify the Jetstar with the new bird proof engines at company expense. Later Dick gave the aircraft certificate to Lockheed at no charge.

This story illustrates some of the "invisible activities" that Pratt & Whitney has been actively involved in for the promotion of flight safety ever since the founding of the company.

MODIFYING THE JT12

When a two-stage power turbine is installed behind the JT12, it is possible to obtain about 4500 shaft hp. In the late 1950s, I was asked to go to Sikorsky to discuss the possibility of modifying the JT12 to become a turboshaft engine for a helicopter. I first had to meet Bill Weaver, the field engineer who covered Sikorsky at that time. Neither of us could imagine that we would be working together eight years later in a Pratt & Whitney assignment with the Department of Defense. Bill and I met in the Sikorsky lobby and then went to the appointment with Ralph Lightfoot, the chief engineer. This was a preliminary discussion to proceed with the turboshaft version of the JT12. Lightfoot seemed pleased

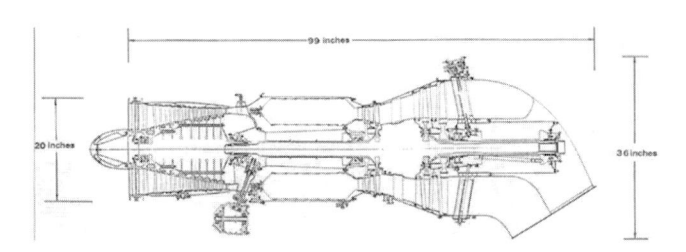

Fig. 31   Line drawing of the JFTD-12 (T73, military designation) (courtesy of Pratt & Whitney Archives).

Fig. 32 Right-side view of the JFTD12 turboshaft engine (courtesy of Pratt & Whitney Archives).

with the JT12's potential, and took action through corporate channels to get Pratt & Whitney to make the commitment for the turboshaft engine. The JFTD12 (Figs. 31 and 32) did get into the Sikorsky S-64 Skycrane and its military version, CH-54A.

REFERENCES

[1] Fisher, P., "Notes on Conversation with Perry Pratt," Pratt & Whitney Archives, late fall and winter 1965–1966.
[2] Parkins, W. A., "Memo to P.W. Pratt and G. N. Cole," Pratt & Whitney Archives, Feb. 20, 1956.
[3] Pratt, P., "Presentation to Bureau of Aeronautics on Navy Jet Engine Program," Pratt & Whitney Archives, June 1, 1954.
[4] Tyler, J. M., "A New Look at the Aircraft Noise Problem," SAE Paper 911B, Los Angeles, CA, Oct. 1964.
[5] Doll, W., "Pratt & Whitney Aircraft Memories," Pratt & Whitney Archives, 1999.

Chapter 9

TRANSITION TO TURBOFANS

INTRODUCTION

At the start of the 1950s, Pratt & Whitney was behind in the aircraft gas turbine business. But by the middle of the 1950s, the company dominated not only the military aircraft engine business with the J57, the J75, and the J52, but also the commercial transport market with high-performance twin-spool turbojets, the JT3 and the JT4. At the end of the decade, a new challenge had developed, one that might be called "bypass mania."

The term *bypass engine* is used to describe an engine that bypasses some of the air from the compressor around the burner and the turbine, and exhausts in either its own separate nozzle or a common nozzle with the turbine exhaust stream. The term *turbofan* is used to describe a bypass engine where the fan exhaust stream divides into two streams—one bypassing the engine and the other entering a compressor, a burner, and a turbine before the exhaust nozzle. In addition, the turbofan could have either separate nozzles or a common nozzle for the two streams.

The aviation community was fascinated by the promise of new aircraft engines that bypassed some of the inlet air. What this enthusiastic group seemed to have forgotten were the theoretical benefits of bypass or turbofan engines were well-known in the engine community, and an engine company does not invest millions of dollars and five years of development time unless there is a relatively certain return on investment in the future. In the early 1950s Pratt & Whitney developed its last turboprop and used that technology in its earliest turbofans.

BIRTH OF THE TURBOFAN

Back in 1930, Sir Frank Whittle had filed a patent in England for a centrifugal-flow turbojet engine. He proposed a way of using the output of the engine to drive a fan-turbine combination, which would bring in a greater mass of air to be accelerated rearward, resulting in greater thrust and decreased fuel consumption. He referred to this added device as a thrust augmenter, also

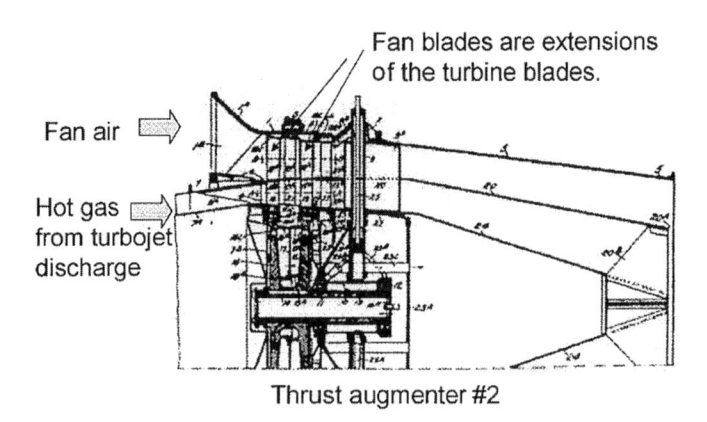

Fig. 1 Whittle's aft fan concept [adapted from 1, Fig. 34].

known as the "aft fan" (Fig. 1). Whittle gave this concept to General Electric in 1942, along with the drawings for the construction of his centrifugal-flow turbojet.

The use of the aft fan was the simplest way to augment the thrust from the centrifugal-flow turbojet in Whittle's time. It involved minimum change to the engine and yet resulted in an increase in thrust. The high temperature and pressure from the turbojet could be used either for a high-velocity jet stream to produce thrust or could be expanded through a turbine whose power would pump additional air from outside and compress it to above atmospheric pressure for subsequent conversion to a jet stream and added thrust. This concept is roughly similar to a gas turbine driving a propeller, except that the propeller (the fan in this case) is accelerating the intake air to greater velocity than what is in the slipstream from a conventional propeller. The aft fan improved specific fuel consumption—an example being about 12 percent improvement in the CJ805 turbojet. Whittle had a creative imagination that went beyond the aft fan configuration. He also developed the concept of a modern turbofan (Fig. 2), in which a multistage turbine drives the fan and the core compressor.

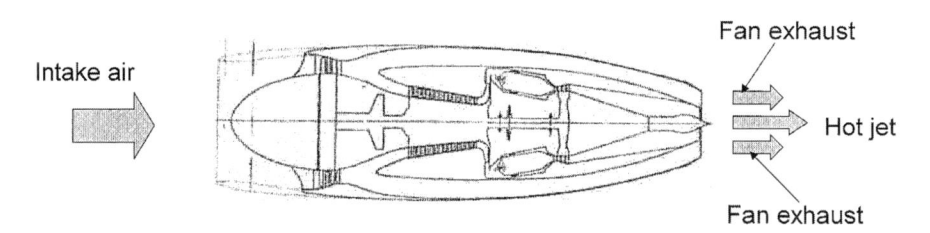

Fig. 2 Whittle's concept of a front-fan turbofan [adapted from 1, Fig. 37].

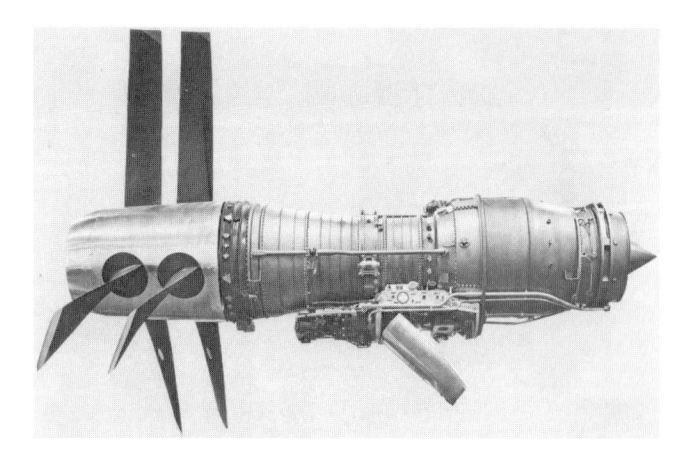

Fig. 3　PT5 turboprop with its propeller (courtesy of Pratt & Whitney Archives).

T57/PT5—PRATT & WHITNEY'S MOST POWERFUL TURBOPROP

In mid-1952, the B-52 flight test program was far enough along to validate the J57 twin-spool configuration that provided a reliable 12 : 1 overall pressure ratio engine for a high-speed bomber. In 1953, while the idea of a turboprop-powered bomber had been ruled out, the concept of a turboprop-powered long-range military transport was still alive (Figs. 3 and 4). The J57 configuration first grew, with a larger thrust size, into the J75.

Fig. 4　Compare the turboprop PT5 (top) to the turbojet JT3 (bottom) (courtesy of Pratt & Whitney Archives).

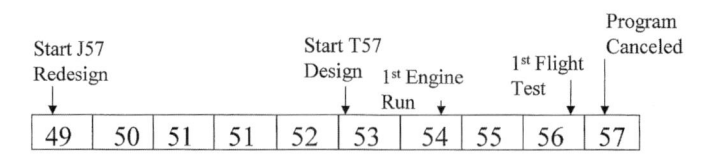

```
        Start J57                Start T57                    Program
        Redesign                 Design   1st Engine  1st Flight  Canceled
            ↓                       ↓       Run ↓     Test ↓        ↓
        | 49 | 50 | 51 | 51 | 52 | 53 | 54 | 55 | 56 | 57 |
```

Fig. 5 Time line for the T57 program.

Then, a more powerful turboprop engine was needed for the huge C-132 long-range four-engine transport. Dick Coar was Project Engineer on the T57, a 15,000-hp turboprop engine. The T57 needed a stronger front end to support the reduction gear and the propeller, which you can see in the figure. Hamilton Standard propellers were to load up the engine. The final propeller would be a single 20-ft diameter propeller with a chord of about 2 ft.

The T57 development program had five active test engines, and over 3100 test hr were completed (Fig. 5). The development program proceeded without any technical surprises with the five development engines. The first engine run of the T57 was in the second half of 1954. Two years later, the T57 was flight tested in the Douglas C-124 flying test bed, powered by four R-4360 piston engines (Fig. 6). The 15,000-hp reduction gear performed well too, and handled about 17,000 hp (Fig. 7).

The Air Force requested that the four piston engines remain at idle power while the T57 provided the required thrust. The T-57 could match the power of the four R-4360 piston engines without difficulty. But the T57 program was canceled in early 1957 by the Air Force because of concerns about 1) developing a 20-ft diameter propeller, and 2) potential structural problems with the C-132 four-engine aircraft caused by a sudden failure of widely spaced 15,000-hp engines [2, p. 24].

JT10 AFTERBURNING TURBOFAN

Perry Pratt asked Dick Coar to put together a turbofan demonstrator engine, the JT10, from PT5 and J75 hardware. In September 1955, Dick started with hardware from existing programs. His engineers removed the reduction gear and added the first few stages from the J75 low compressor to the front of the T57 low compressor with its first three stages removed. The PT5 four-stage, low-pressure turbine was kept in place even though it had more than enough

Fig. 6 PT5 flight test in a Douglas transport (courtesy of U.S. Air Force).

Fig. 7 PT5 reduction gear (courtesy of Pratt & Whitney Museum).

capacity to drive this cobbled up turbofan. Coar's recommendation was as follows [3]:

> The basic turbofan design should be suitable for flight, but the design of the afterburner fuel injection, mixing, diffusing, and burning sections should ... allow ready modification and substitution of alternative designs.... It is essential that all ... the basic parts ... have a satisfactory background of test experience.

This improvised turbofan (Fig. 8) was called the JT10 for want of a more suitable designation. It came right after the JT8 (J52) and the JT9 (J91) in the days before it became customary to refer to a turbofan derivative of the JT3 as the JT3D (and similarly the turbofan derivative of the JT8 as the JT8D).

Pratt & Whitney's design office in Boston was made up of personnel from Chance Vought who did not want to move to Texas. Dick Mulready, working for Dick Coar, was coordinating the turbofan design. In 1956, both Dick Coar and Dick Mulready (Fig. 9) were pulled off the JT10/T57 program to get into a very secret reconnaissance aircraft engine program, one which was to have far-reaching consequences and challenges for the company (which we discuss in the next chapter). Phil Hopper was the Assistant Project Engineer who took over the JT10 turbofan program from Mulready.

Phil Hopper conducted afterburning tests and accumulated over 270 test hr. Two experimental engines were built (Fig. 10). The conclusions drawn

Fig. 8 JT10 afterburning turbofan, right front view (courtesy of Pratt & Whitney Museum).

from this experimentation was that 1) a twin-spool turbofan (front-fan configuration) was straightforward, and 2) fan duct-stream burning and conventional engine-stream afterburning were feasible. The afterburning was in the engine stream, and the cooler bypass air was an annulus surrounding the exhaust from the engine stream (Fig. 11). However, there was no application for such an engine at the time.

JT3D/TF33—TURBOFAN FEVER FUELS INNOVATION

Since 1955, the company had been supplying aircraft manufacturers with information on turbofans for their aircraft studies. By the middle of 1956,

Fig. 9 a) Dick Coar, Project Engineer for the T57, and b) Dick Mulready, T57 turbofan designer.

TRANSITION TO TURBOFANS 297

Fig. 10   JT10 experimental engine, left rear side view (courtesy of Pratt & Whitney Museum).

Pratt & Whitney had a reasonable feel for the advantages of turbofans. Rolls Royce's Conway, which was the first bypass engine in commercial service, had a bypass ratio of 0.6. The JT3D had a bypass ratio of 1.4.

COMPARING TURBOPROPS, TURBOFANS, AND TURBOJETS

A bypass ratio of 0.6 means the amount of flow that bypasses the burner and turbine is 60 percent of the flow that goes through the burner and turbine. Similarly, a bypass ratio of 1.4 means that the airflow that bypasses the burner and turbine is 1.4 times (140 percent) of the flow that goes through the burner and turbine.

A "back of the envelope" way of comparing a turbofan and a turbojet is to compare a turbojet (bypass ratio of zero) with a turboprop plus propeller

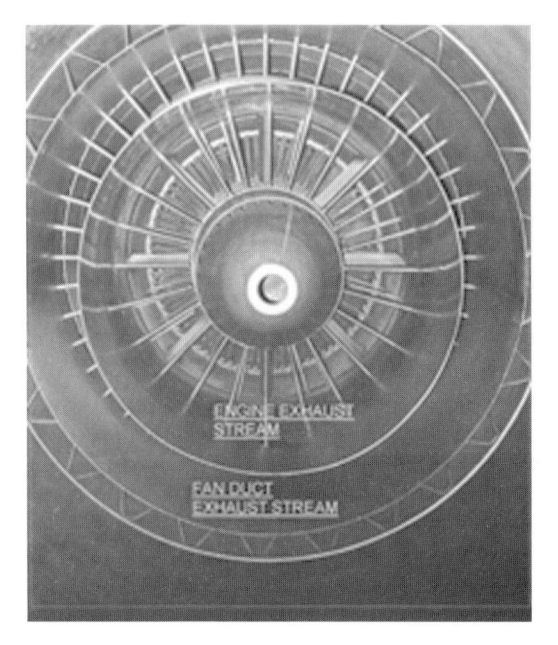

Fig. 11   View looking into the JT10's tailpipe (courtesy of Pratt & Whitney Museum).

(bypass ratio of more than 20). They both operate at the same burner temperature and could be sized to produce the same level of thrust. The turboprop has a lower level of specific fuel consumption than a turbojet. A turbofan fits between those two engines, with a specific fuel consumption lower than that of a turbojet but still higher than a turboprop at moderate flight speeds. Specific fuel consumption is defined as the pounds of fuel per hour per pounds of thrust.

It is possible to produce the same level of thrust by accelerating a small amount of airflow by a great amount (the turbojet route) or by accelerating a huge amount of airflow by a lesser amount (the turboprop way). The turbofan cycle produces lower jet velocities compared to a turbojet and therefore has lower jet noise levels than a turbojet.

PRATT & WHITNEY COMES TO EMBRACE TURBOFANS

Wright Parkins was not enthusiastic about turbofans at first. He was aware of the theoretical benefits, but was more concerned about the effort and financial commitment necessary to bring a new turbofan engine into commercial use. The proven path was to develop the engine in a military application and then convert it to a commercial engine. In contrast to the commercial area, there was no interest in subsonic turbofans on the part of the military establishment at that time.

The source of some of Parkins' annoyance with turbofans was the United Aircraft Research Laboratory (UARL), which did engine performance studies for Navy applications under contract. Some of their studies indicated huge advantages in some aircraft applications for turbofans compared to the kind of turbojets under development at Pratt & Whitney. Since UARL and Pratt & Whitney were part of the same United Aircraft Corporation, the outside world often confused the two organizations and usually confused the types of missions and engine performance assumptions behind the UARL aircraft studies. Parkins was bombarded by government and commercial customers as a result of this confusion, asking why the company was not developing turbofans. The UARL studies showed turbofans were so much better than those P&W J57 and J75 engines. This kind of "loose information" was high-energy fuel to send Wright Parkins into orbit.

One can understand Parkins' attitude. He had done a superb job since 1949 in bringing the J57 along. It was in the B-52 and in a string of Air Force fighters. It was in the KC-135 and would be making its debut in 1958 in commercial aviation with the Boeing 707 and the Douglas DC-8. Of course, the J52, T34, and the J91 developments were moving along too. Then on top of all those worries, Parkins was asked about developing a new turbofan for commercial use when the industry (Boeing, Douglas, Convair, and the airlines) did not have any experience yet with high-performance turbojets.

A letter written in October 1956 illustrates the commercial air industry's interest. William Patterson, President of United Air Lines, wrote to Parkins about turbofans [4]:

> I have heard and read a great deal about the bypass jet engine for commercial use and, as you know, a few of the foreign airlines ordering DC-8s and 707s have specified that type of engine in preference to your company's J57 or J75 jet engines.
>
> Could you ... give me ... the principle of the bypass engine and why Pratt & Whitney Aircraft have not made such an engine type available for these jet transports when so much is claimed for the bypass engine in fuel economy and noise?

Patterson was not the only person to ask Parkins that nettlesome question: Why doesn't Pratt & Whitney supply a turbofan? Parkins quickly replied [5]:

> The question might just as well be asked, "Why don't you provide a ducted fan for military use?" ... The only answer that we can give is that ... not being able to see a need for such an engine now, we have no plans for going forward with one. The ... production of a modern engine costs too many millions of dollars for either the military or the commercial operators to support without clear-cut knowledge of its advantages over other engine types that now exist and without knowing ... how and where it will be used.
>
> The ducted fan has been given a lot of consideration.... Several determined attempts to find a military home for it have been made but they have all fallen by the wayside for the reason that ... the simple straightforward jet or the turboprop-powered airplane can do better.
>
> In conclusion we should like to say that since there are no essential differences in operating costs and since the straight jet type of engine has an enormous amount of operational service experience while the ducted fan has none, the straight jet seems to be the only logical choice for jet-powered commercial aircraft currently under consideration. We see nothing to be gained by using in commercial operation an engine type which has no service background and no economic advantage even when an optimum (but still untried) bypass ratio is employed.

Parkins, as demonstrated in his response to Patterson, was not inclined to ask the organization to invest hundreds of millions of dollars in a new turbofan when United Aircraft was just beginning to get returns on its commercial engine investment.

COMPETITION DRIVES INNOVATION

As far as Parkins was concerned, the airlines were operating quite well with their JT3- and JT4-powered aircraft. Perhaps when he made his oft-quoted statement, "Fans are no damn good!" he was thinking of the comparison of the R-R Conway engine with the JT4. To Parkins, the R-R Conway turbofan had no advantage over the JT4. The Conway had a bypass ratio of about 0.6 with a twin spool configuration of a six-stage low compressor driven by a two-stage low turbine and a nine-stage high compressor driven by a single-stage high-pressure turbine.

The Conway as a bypass engine was really not a serious threat to Pratt & Whitney's market share of the Boeing and Douglas jet aircraft. Relative to a turbojet; however, a new turbofan of the proper size for the 707s and DC-8s would provide specific fuel consumption and noise advantages. American Airlines used the General Electric aft-fan threat to get Pratt & Whitney's attention focused on turbofans.

General Electric was offering an aft-fan version of its commercial J79 turbojet (CJ805-23) and promising about a 12-percent reduction in specific fuel consumption. The situation was brought into sharp focus when Frank Kolk of American Airlines, backed by his boss, C. R. Smith, stated that they would not be ordering Boeing 707s with Pratt & Whitney engines in the future. They preferred General Electric's turbofans, not the J57 and J75 engines. It seems incomprehensible to me that Kolk and Smith were willing to accept an engine with very little airline experience, in contrast to the J57 and J75 engines with a wealth of proven dependability in military and in commercial service. At the same time, Convair had approached Delta Airlines and was offering an attractive deal for Convair 990s with General Electric turbofans. Delta appeared to be moving toward acceptance.

While I was writing this book, Bill Sens (Fig. 12), head of the Advanced Engine Performance Group at Pratt & Whitney, recalled company activities (Fig. 13):

> Frank Kolk had been agitating for a turbofan for commercial aircraft for a long time. Pratt & Whitney management was reluctant to take on a turbofan development ... because they had a huge investment in the

Fig. 12 Bill Sens (photo by author).

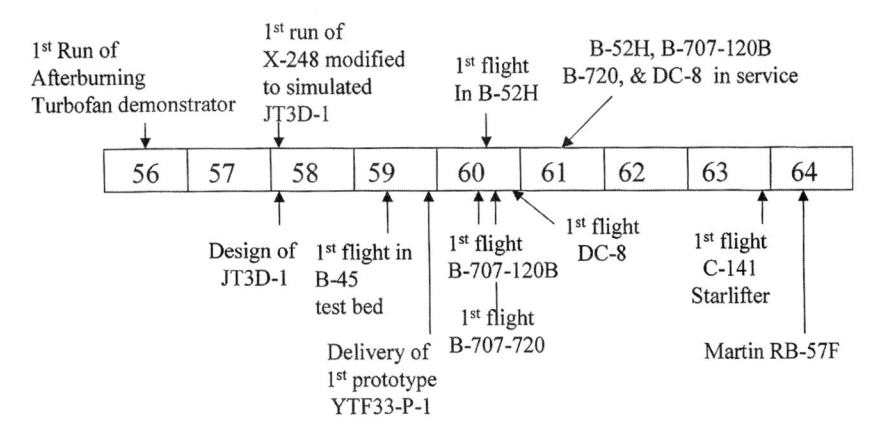

Fig. 13 Time line for the JT3D program.

commercial JT3s and JT4s for which they were expecting a reasonable financial return. Frank saw the CJ805-23 as a means of getting us off our butt. Maynard Pennell stopped at Pratt & Whitney after visiting Frank and reported that American was ready to order 707s with CJ805-23s unless we came up with an attractive alternative.

Perry flew to New York and camped on American's doorstep. Meanwhile an intensive design effort was underway in East Hartford to define a turbofan version of the JT3 best meeting the requirements of the Boeing 707. In about ten days a group of us flew down to New York with a preliminary design ... based on the use of short ducts to conduct the fan air to nozzles located midway aft on the nacelle. This was done to avoid rearrangement of the accessory drives. Subsequent tests of models of the short ducts and nozzles and wind tunnel tests of nacelle models confirmed the good performance of this approach. Perry was enthusiastic....

Bill Sens took his book of turbofan cycle studies that showed the effect of bypass ratio on the thrust specific fuel consumption (TSFC) and estimated how much lower in TSFC a turbofan based upon the JT3 turbojet would be.

> American bought the engine and the JT3D was redesigned and the guaranteed fuel consumption met. Subsequently the JT3D cruise fuel consumption was reduced by an additional three percent. Flight testing in the 707 and the B-52 confirmed the low drag of the JT3D short duct configuration.

The JT3D became a slightly scaled down version of the first two stages of the J91 (JT9) transonic compressor. The low compressor was that of the JT3 without the first three stages. The high spool was the JT3 high spool (7-stage compressor, burner and single-stage turbine). The low turbine was the JT3 low turbine plus another stage.

But the competition was hard at work too. Designer J. DeSantis recalled [6]:

> The story heard in Engineering ... was that Perry Pratt learned that General Electric was developing a fan jet [and] announced during the conversation [with American Airlines] that Pratt & Whitney would have a fan jet on the test stand within two weeks....
>
> Following the meeting, Mr. Pratt called Pratt & Whitney and informed them that the engine would be on the stand and running at the appointed time. This information was relayed to design personnel on Friday afternoon (the same day as the telephone call) and they were instructed to come in on Saturday morning. I was one of a group of approximately six designers told to report for work that day.
>
> Needless to say, the design of that first fanjet was completed. Parts were fabricated directly from the design layouts. Prints were hand carried to C. Wincze, then Chief of Experimental Facilities Design, and the first fan jet was running in the test stand at Willgoos on the required date.

BUILDING THE ENGINE

In 1958, this experimental engine, the X-248 pictured in Fig. 14, demonstrated the potential of a turbofan conversion of the JT3 turbojet. It consisted of the first two stages off the J91 compressor joined to the PT5 low compressor with the PT5's first three stages removed. The rest of this new structure was the same as the rest of the PT5, a four-stage low-pressure turbine. Note that the first three stages of the T57 low compressor were removed and that space was taken up with the first two stages of the larger diameter J91 compressor to put together a turbofan engine for demonstration.

How exactly did the external arrangement of the JT3D come into being? Perry Pratt explained [7]:

> Alden Smith came up with the idea of a short duct so that the fuel control and other accessories could be got at readily. Boeing's engineers told us that the drag on the short duct would kill us. We'd been through a long wind-tunnel program and we disagreed. The airlines went with us.

Fig. 14 Experimental engine X-248, known as the "two-week turbofan" (courtesy of Pratt & Whitney Archives).

Boeing's engineering was split down the middle on the short-duct issue but the airlines demanded it and got it....

Sens deserves great credit on our fan developments.... We had done the computer and paper work and we [knew] that we had to ... get up a full-scale model. We had the J57, the J75, and the J91. So the proposal was to take off the first two stages of the J57 low-pressure compressor and replace them with the first two stages of the low-pressure compressor of the J91, and then add an afterburner, including a duct burner. The design of the duct was a pretty big task. The point is, we learned we could bring such a powerplant through. Not easy. You had to know your stuff.

CLOSING THE DEAL

Parkins, who was on vacation while panic developed, looked at the turbofan data after returning to work, as Stan Taylor told me while I was writing this book:

It took [Parkins] a little while to calm down.... It was ... late Monday afternoon before he looked at the data and by Tuesday morning he was completely converted to the turbofan school and called Wright Field's Power Plant Laboratory chief to set up an appointment ... on Wednesday to tell him about the technical breakthrough that we had made in our testing program on the possibility of a turbofan cycle.

Later that afternoon I was asked to prepare a presentation to be given at Wright field the next morning.... In those days we made presentations on flip charts that came in large pads and we crayoned out presentations on those sheets. So I worked away at that and was told to be at the airport very early the next morning to go in the company Convair airplane to Wright Field.

Well, I got out to the airport and the only other people who were going on this airplane were Wright Parkins and Dick Baseler, who was project engineer on the J57. The reason I was going, I think, was because I was the Military Requirements Engineer at that time and my job consisted in large part of working with the Requirements Groups in all the Air Force and Navy Commands that were dreaming up possible new airplane projects.... I made a lot of presentations. So I thought that I would be the one who would make the presentation.

On the airplane ... the three of us worked over the presentation and scrapped many pages and made many more pages and the crux of our sales program was that there would be a great many parts in the J57 that would be usable in the turbofan version. In fact, the rough estimate was that 50 percent of the parts in the J57 could be used in the turbofan version....

It seemed that ... estimate was pretty low because so much of the engine could be retained that we changed the number on the flight chart to

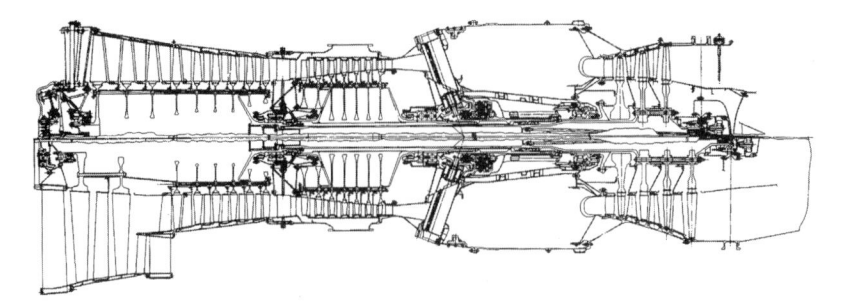

Fig. 15 Comparison of JT3 turbojet (top) and JT3 turbofan (bottom) showing the structural similarity between the two engines (courtesy of Pratt & Whitney Archives).

60 percent commonality between the J57 and the turbofan version of the J57.... As we were on our approach to Wright Field ... Wright Parkins said, "We'd better make that 75 percent." So Dick Baseler and I worked very hard on the table that was in the airplane as it was landing. We did get the charts made and we had them all rolled up and put in the container.

When we got to the Power Plant Laboratory ... Dick Baseler asked [me], "Who's giving this presentation—you or me?" I said, "I don't know. Usually Engineering gives the presentations at Wright Field and we in Military Requirements do the presentations at all the other places." So we ... went into the conference room where the people were gathered. Parkins asked, "Where are the charts? Set them up." So we set them up on a stand provided by Wright Field. Parkins ... proceeded to give one of the most brilliant presentations that you could imagine about this wonderful technical breakthrough that Pratt & Whitney had afforded to the Air Force. If they would like to contract with us for further investigation, we'd be thrilled to do that. And that's how the TF33 program got started. We also quickly got permission from them to talk with commercial customers about a possible commercial version of the same thing.

About the same time I was able to get the engine information on the TF33 engine to Tom Nelson. He was my closest friend, who happened to be the director of sales for Boeing, Wichita—with the responsibility for coming up with new versions of the B-52 to extend the life of the B-52 production program.

He quickly decided that he could promote an extended range version of the B-52 with the TF33/JT3D engines, which at that time were expected to afford a 12-percent reduction in fuel consumption at the altitudes and speeds of which the B-52 cruises. It later turned out, incidentally, that the B-52 range was increased by almost 20 percent for the JT3D.

At Delta Airlines in the meantime, Convair and General Electric were making very tempting offers. Art Ford, who was Delta's Vice President of Maintenance, approached Bob Fitzgerald, a senior executive in Commercial Engine Marketing at Pratt & Whitney to ask whether the company was

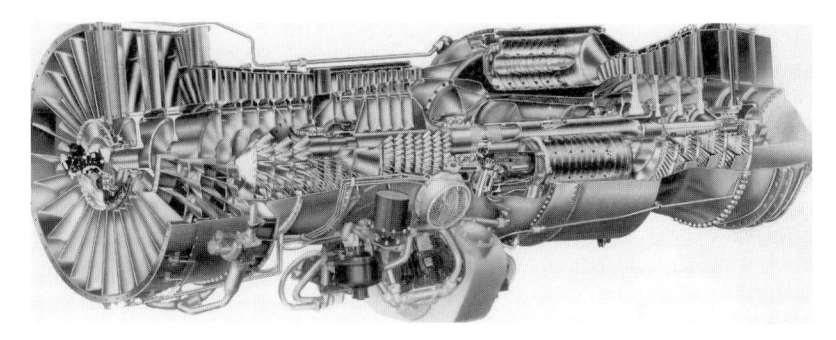

Fig. 16   Cutaway of JT3D (courtesy of Pratt & Whitney Archives).

interested in helping Delta upgrade its fleet. The JT3D was just coming into existence due to the pressure from American Airlines. Boeing was saying that American Airlines was going to order future B-707s with the General Electric turbofan if Pratt & Whitney had no turbofan available.

Bob told me while I was writing this book that Art Ford took him aside and said, "Look, we're going for the CV-990. The only thing that can stop this is if somehow we can get the JT3D in the DC-8s we have." Taylor and Fitzgerald came up with a plan to make it easier for Delta to upgrade its Douglas DC-8 fleet. The plan would retrofit the DC-8s with new JT3Ds (Figs. 15–17) and take the present JT3 engines (on the order of a couple of dozen) back in trade. This would appeal to the finance-oriented Delta management. But Pratt & Whitney management had to approve it—such an arrangement had never been done before. But Fitzgerald sold the idea to Art Smith and then Len Mallet, who was president of Pratt & Whitney at that time. The result of the marketing plan was to sell JT3Ds first and then use the 36 traded-in JT3C engines for land-based gas turbines, where Pratt & Whitney later made money in industrial applications, and the competitor was kept away from the castle gates.

SPECIFICATIONS

5413 engines were produced (Table 1), plus about 3000 upgraded from JT3C engines (Pratt & Whitney sold kits to convert JT3C turbojets into

Fig. 17   JT3D with side nozzles for fan discharge (courtesy of Pratt & Whitney Archives).

TABLE 1 JT3D/TF33 SPECIFICATIONS AND APPLICATIONS

Model	Take-off thrust (lb)	Military or max. cont. thrust (lb)	Normal thrust (lb)	Weight (lb)	Diameter (in.)	Length (in.)	Aircraft
JT3D-1	17,000	14,500	14,500	4130	53.1	136.3	B-720B, B-707-120B, DC-8
JT3D-1-MC6	17,000	14,500	14,500	4540	53.1	145.5	B-707-120B
JT3D-1-MC7	17,000	14,500	14,500	4165	53.0	167.7	B-720B
JT3D-2	17,000	16,500	14,500	3900	53.0	136.3	B-52H
JT3D-2A	17,000	16,500	14,500	4065	53.0	158.5	MATS Cargo
JT3D-3	18,000	16,400	16,400	4170	53.1	136.3	DC-8, B-707-120B
JT3D-3A	18,000	16,400	16,400	4170	53.1	137.4	C-135B, Command Post
JT3D-3B	18,000	16,400	16,400	4300	53.1	136.31	B-707B, VC-135B, DC-8-50, 61, 62
JT3D-7	19,000	17,200	17,200	4300	53.1	136.3	Boeing, Douglas
JT3D-7A	19,000	17,200	17,200	4300	53.1	136.3	AWACS
JT3D-8A	21,000	19,000	18,000	4650	54.1	142.3	C-141
JT3D-8B	23,000 wet[a]	20,800	19,500	4855	54.4		C-141
JT3D-8E	21,000						Douglas
JT3D-9	20,750	17,550	15,810	4700	62.5	135.0	Quiet engine. 1 fan stage for Boeing, Douglas
YTF33-P-1	16,000	15,000	12,500	4350	52.9	145.5	XB-52H
TF33-P-3	17,000		14,500	3905	53.1	136.3	B-52H
TF33-P-5	18,000		16,400	4275	53.1	137.4	C-135B, WC-135B, RC-135E
XTF33-P-7	21,000	19,000	18,000		54.1	142.3	Ground test
TF33-P-7	21,000	19,000	18,000	4605	54.1	142.3	C-141A
TF33-P-9	18,000		16,400	4340	53.1	137.4	Military
TF33-P-11A	16,500				53.1	136.3	Conversions

[a]Water injection cools the inlet air, which results in a greater mass flow of air and hence increased thrust.

JT3Ds). The JT3D consists of a two-stage fan (practically right off the J91) replacing the three front stages of the J57 low-pressure compressor (Figs. 15–17). Then this hardware is followed by the J57 high spool plus a redesigned low turbine with another stage. The serendipitous outcome of this engine was 1) more thrust, 2) lower specific fuel consumption, and 3) about a 10-db lower jet noise level.

APPLICATIONS

Turbofan fever spread so quickly that in 1961 the B-52H and two commercial aircraft were turbofan-powered.

MILITARY. Within three years, two other major military applications of the TF33 turbofan were in use. In late 1963, the Lockheed Starlifter (military C-141 transport) made its first flight with TF33 (military version of JT3D) propulsion (Fig. 18). Later in 1964, the TF33 turbofan gave added power to the Martin RB-57F reconnaissance aircraft.

The story of the B-52 seemed to end with the G-model. Tom Nelson and Frank Sutton at the Wichita branch of Boeing had the vision of increased B-52 production with turbofans. They generated enough interest to involve Cliff Simpson of the Air Force Aero Propulsion Laboratory and to ensure that the engine conversion made sense. General Irvine, Deputy Commander of Air Force Systems Command, saw the turbofan as a way of justifying another

Fig. 18 **C-141 Starlifter Air Force transport (courtesy of U.S. Air Force).**

buy of the B-52s because of the additional benefits in fuel economy. Simpson tells the story [8]:

> In the 57–58 time period, we and Pratt & Whitney converted the J57 (JT3C) to the TF33 (JT3D) for the B-52H. In 1959, Headquarters SAC, including Guy Townsend ... approved the use of this engine. When General Power (Commander SAC) found out about the approval and change he stripped backsides with vigor. You could hear Guy scream "help" without benefit of Ma Bell. Two of us, Lieutenant Colonel Hammet and I, were sent to the Omaha bunker to explain the correctness of our choice ... General Power debated his points man to man and we did convince him on the basis of a fifty-cent bet that he would be pleased. Come to think of it, I am still due the fifty cents. By changing engines we bought 450 miles of range, which was promptly eliminated by the addition of over seven tons of electronics which also used an additional 500 miles of basic range.

Another success for the TF33 turbofan was the Martin RB-57F reconnaissance aircraft. The Canberra, built by BAC (English Electric), was the United Kingdom's first jet bomber. It was powered by Rolls-Royce Avon engines when it made its first flight in 1949. Over 1300 of these aircraft were built. Another 400+ were built under license by Martin in the United States and these used the Bristol-Siddeley Sapphire (J65). About 20 of them were modified as high-altitude reconnaissance aircraft powered by the J57. Another 20 or so were modified by General Dynamics to the RB-57F version (Fig. 19) with the TF33-P-11 and the J60-P-9.

Fig. 19 RB-57F powered by TF33 engines (courtesy of U.S. Air Force).

TRANSITION TO TURBOFANS 309

COMMERCIAL. As mentioned earlier, the JTD3 was used in two commercial competing aircraft: the Boeing 707 and the Douglas DC-8. Pam Am started the rush to fly new jet-powered aircraft by ordering both models in October 1955, thus prompting over a dozen other airline companies to place orders before the planes became available in 1958. The DC-8 was first flown by United and Delta in September 1959. Figure 20a shows the normal configuration of the JT3D installation in the Douglas DC-8-50 and Fig. 20b shows

Fig. 20 The JT3D installed in the Douglas DC-8-50, a) normal configuration, and b) reverse thrust (courtesy of Pratt & Whitney Archives).

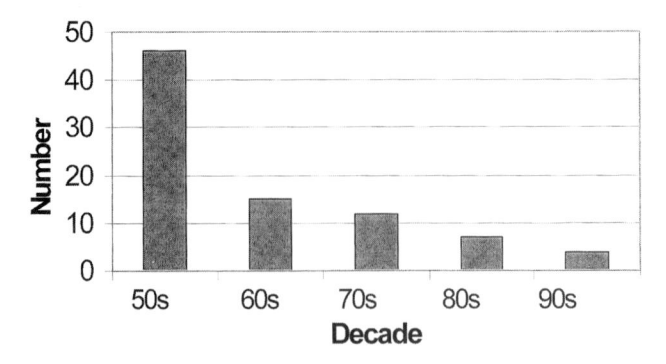

Fig. 21 The number of government-sponsored aircraft programs was greatest right after WWII, but then declined.

the engine in reverse thrust. During reverse thrust, the rear section of the nacelle moves backward, activating internal blocker doors, and exposing a cascade that directs the core engine stream towards the forward direction. The fan stream also uses internal blocker doors and a cascade that directs the fan stream forward. The Boeing 707 became the preferred aircraft, selling three times as many aircraft as the Douglas DC-8.

Three Programs Herald the Future

In the "Fabulous Fifties" Pratt & Whitney engines powered the B-52, A4D, F-100, F-101, F-102, F-105, F-106, U-2, KC-135, A-12, SR-71, F-107—just to name a handful [9]. There were 46 government-sponsored programs ongoing during this time (Fig. 21), making it an engineer's paradise. Pratt & Whitney was supplying engine data to the airframers who were working on these exciting programs. Three engine projects (Fig. 22) that were very significant regarding Pratt & Whitney's future and physical expansion were the Suntan 304, the RL10, and the J58. I'll end this chapter with a discussion of the Suntan project, while the RL10 and the J58 are discussed in Chapter 10. (In Chapter 11 we'll review two other concurrent projects, JTF10A and the JT8D.)

Suntan Project

Start Up

In early 1956, a group of Pratt & Whitney employees, including myself, were invited to a meeting with zero advance information. We gathered together in a small conference room and nervously asked each other, "What's up?" Could it be lay-offs or pay cuts? Gil Cole, head of Design, entered the room and told us in his most intimidating voice that we were going to be participating in a very secret project. He emphasized the need for extremely

Before

J91 Nuclear-powered Turbojet

Scaled down to J58

Navy Attack Aircraft Engine

The Suntan Hydrogen Engine
For Reconnaissance aircraft

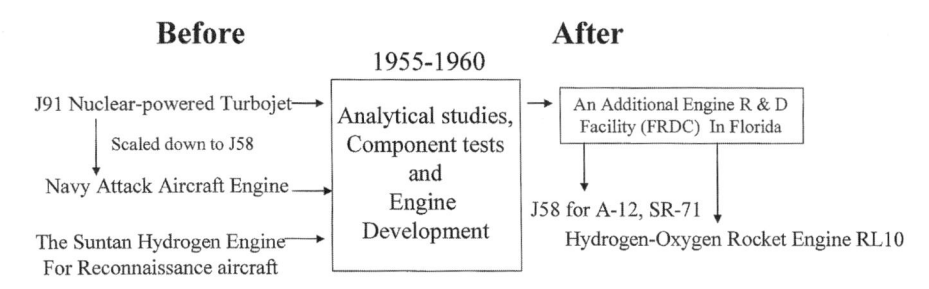

1955-1960

Analytical studies,
Component tests
and
Engine
Development

After

An Additional Engine R & D
Facility (FRDC) In Florida

J58 for A-12, SR-71

Hydrogen-Oxygen Rocket Engine RL10

Fig. 22 Three programs bring new business opportunities.

tight security, way beyond what we had up until that time. The implication was real Federal jail time if there were any leaks. We were not permitted to discuss what we were doing with anybody! This top-secret program was referred to as the "Suntan" or the "304" project (Fig. 23).

Where did the number 304 come from? We were concerned about using the word *Suntan* (the Department of Defense's designation) for fear that it would stimulate the curiosity of people not connected with the project. Of course, after the security briefing we were very concerned about this possibility. I think it was Dick Coar (Fig. 9a) who used the 304 terminology, three digits he pulled off a work order number or something. Our experimental engines were identified by three digits such as 292, 178, and so forth.

Secrecy among Pratt & Whitney, the Air Force, and Lockheed (and other airframers) remained the watchword throughout the 304 time line. During the design of the experimental engines, the 304 Project Engineer Dick Mulready (Fig. 9b), an Air Force officer, and I made a trip to Lockheed and two other airframe companies on the West Coast. The highlight of the trip for me was to see Lockheed's secret research and development Skunk Works in operation. Skunk Work's founder Kelly Johnson was not in the plant during our visit but we met with his second in command, Ben Rich. I noticed that Mulready and the Air Force officer did not carry any papers concerning Suntan. I had all the Suntan engine material in my briefcase that, if misplaced, could send me to

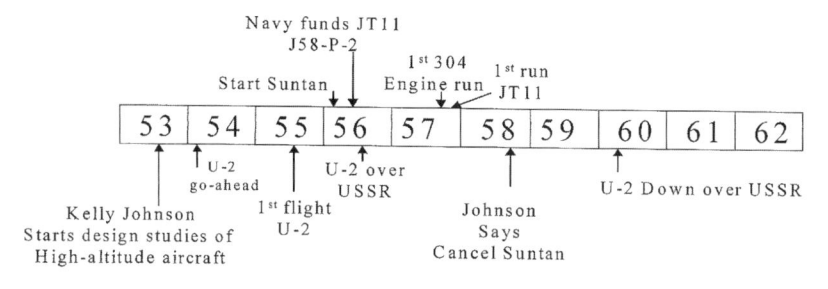

Fig. 23 Time line for the 304/Suntan project.

Leavenworth for life! It slowly dawned on me why I was along on the trip. My traveling companions remained relaxed while I was a nervous wreck.

DESIGN DEVELOPMENT

Kelly Johnson had started studying reconnaissance aircraft configurations in mid-1953. He reasoned that a specialized high-performance, high-altitude aircraft would be able to fly over any territory without being hit by anti-aircraft missiles or being attacked by fighter aircraft. In early 1954, his U-2 development program began and the first flight was in mid-1955. About a year later his aircraft was flying over the Soviet Union, much to the frustration of the Kremlin. It was a fantastic aircraft that could fly anywhere, and with the reconnaissance photographic equipment developed by Dr. Land of Polaroid it could bring back photographs of what was happening on the ground.

The U-2 used jet fuel (JP), which was a hydrocarbon fuel. A fuel such as hydrogen with its 50,000 BTU/lb (compared to about 20,000 BTU/lb JP) implied a much greater range capability because its specific fuel consumption would be on the order of 40 percent of the specific fuel consumption of JP-fueled engine. What was not immediately considered was liquid hydrogen is −423°F and comes in two varieties, para and ortho. When experts explained to the Suntan team that the parallel and anti-parallel spins of the nucleus in each atom in a hydrogen molecule distinguish between para and ortho types of hydrogen, I knew we were getting into something deep!

Bill Sens, project engineer of the Advanced Engine Performance Group (Fig. 12), sent me an e-mail in 2005 that gave me the background on the use of hydrogen fuel:

> One day in October 1955 John Chamberlain, who was then in the Combustion Group, came up to me and said, "You know if you compress liquid hydrogen to a high pressure and then heat it, you can get a hell of a lot of energy out of it by expanding it." In the next week I came up with the 304 cycle as a means of using this energy. In-house studies showed that this powerplant cycle offered a 10,000- to 15,000-ft altitude advantage relative to a hydrogen-powered turbojet flying at the same airspeed. We were under contract to build and test the 304 engine by May 1956. It turns out we were not the first ones to think of this powerplant cycle. A patent search showed a Frenchman [had] a patent about 70 years before. Also I believe Air Research worked on it for a while in this country.

Sens gave us a briefing on what we were expected to do. The first task was to put together a table of pressures, temperatures, and fluid flows at various locations in the engine and to estimate the performance of the engine for Lockheed's airplane studies. The engine reminded me of the

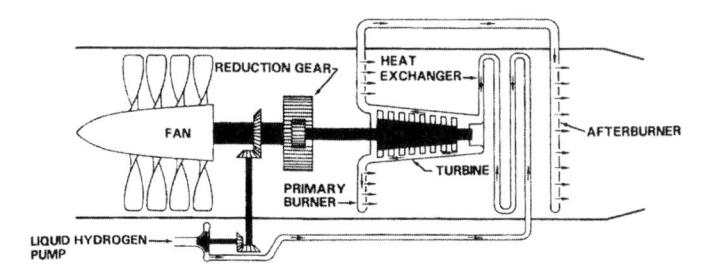

Fig. 24   Engine cycle for the 304 (courtesy of Pratt & Whitney Archives).

1941 Caproni-Campini propulsion system. That airplane first flew for ten minutes in August 1940 and later in the more famous flight of November 1941 from Milan to Rome. The engine consisted of a fan, which was driven by a piston engine, and an afterburner. The Suntan engine consisted of a fan, which was driven by a hydrogen turbine, a primary burner to heat the fan discharge air, a heat exchanger, and an afterburner (Fig. 24).

From Fig. 24, we can see that air enters the nacelle (at a Mach 2.5 flight speed at 90,000-ft altitude), slows down to the fan, and is pumped up to about a 2:1 pressure ratio. This air is then heated in the primary burner by the combustion of hydrogen. The air continues rearward through a heat exchanger in which cold hydrogen from the tank is heated by the hot gas. Then the hot gas is further heated in the afterburner before going through the jet nozzle (Fig. 25). Combustion engineers liked to design afterburners that used hydrogen because of the fuel's rapid burning and its ease of lighting.

Armed with the latest in computational technology (the formidable K&E Log-Log Duplex Deci-Trig 10-in. slide rule), Marks handbook, Keenan & Kaye air tables, and the atmospheric tables, I set off one night to make rough calculations of the 304 engine performance in early 1956. Figure 26 shows some of the concerns and the properties of the fuel at various locations in the engine. You can see where legendary cartoonist and tinkerer Rube Goldberg got his inspiration for complexity.

Fig. 25   The 304 engine with an afterburner (courtesy of Pratt & Whitney Archives).

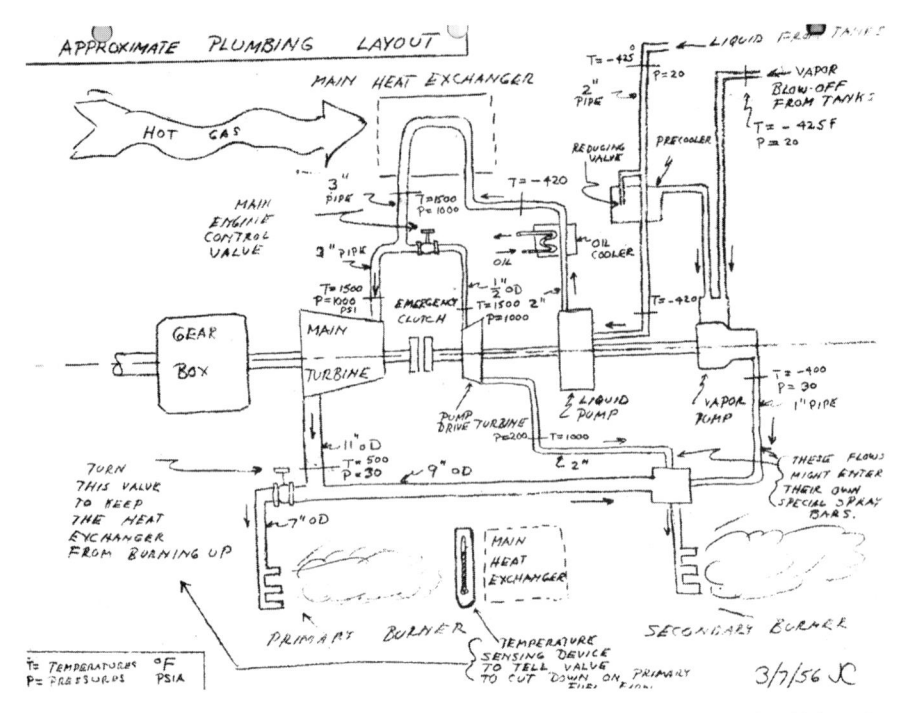

Fig. 26 Overall schematic for fuel, illustrating the difference between the 304 engine cycle and the ordinary gas turbine Brayton cycle (courtesy of Pratt & Whitney Archives).

I estimated that the hydrogen expansion turbine would need 16 stages, but could not believe the number at Fst. In aircraft gas turbines, a single-stage turbine could easily drive a four- or five-stage compressor. I remembered that the turbine power is proportional to the product of fuel flow (pounds per second) and the enthalpy drop (British thermal units per pound). The fuel flow was very small relative to the airflow but the enthalpy drop in the hydrogen turbine was huge. The work (enthalpy change) per stage in a turbine is proportional to the square of the mean wheel speed. I was a little more optimistic on the wheel speed than the designer of the 304 engine. I came up with 16 stages and the official designer put in 18 stages. Figure 27 shows the six stages of the 18-stage hydrogen turbine. This does not look like a Pratt & Whitney turbine, but as if it came from a steam turbine company.

The air to hydrogen heat exchanger was located downstream of the primary burner (Fig. 28). The heat exchanger, according to Suntan Project Engineer Dick Mulready, was "one of the most fantastic parts ever installed in a Pratt & Whitney engine." It was 6 ft in diameter and consisted of 4.5 miles of 3/16-in. diameter tubing. The tubes were made of Hasteloy R® and the manifolds of Waspaloy®—both very difficult materials in manufacturing, and 2400 holes were drilled in each manifold.

Fig. 27 The first six stages of the 18-stage hydrogen turbine (courtesy of Pratt & Whitney Archives).

TESTING IN A NEW LOCATION

In response to the Air Force need for secrecy and also to Pratt & Whitney's need for a facility where neighbors would not complain about noise from test engines at night, United Aircraft started to look for real estate in locations remote from East Hartford. After an extensive search for land and a discovery that engineer recruitment was most successful when Florida was the location, United Aircraft purchased land in the West Palm Beach area (see the end of Chapter 10). The 304 engine development program moved to Florida, while those of us in East Hartford went on to other things (see Chapter 11).

The 304 engine ran on a test stand at the Florida Research & Development Center (FRDC) in the latter part of 1957 (Fig. 29). The inlet diameter of the fan was about the size of the JT3D. To outward appearances, the engine looked like other Pratt & Whitney test engines (Fig. 30). The sea-level thrust was about 4700 lb [10].

Fig. 28 Heat exchanger (courtesy of Pratt & Whitney Archives).

Fig. 29 The 304 engine on test (courtesy of Pratt & Whitney Archives).

SUNTAN FADES

Kelly Johnson and Ben Rich were disappointed with the results of the Suntan aircraft. The lift/drag ratio of the airplane (the L-400) was about 16 percent lower than the goal and the engine fuel consumption was higher than what they had hoped. To achieve their mission objectives of range they would need an aircraft larger than a B-52 [11]!

Johnson explained the problem at a meeting with Assistant Air Force Secretary James Douglas, General Clarence Irvine, and Perry Pratt. He said the airplane with hydrogen was like a flying thermos bottle because of the need for vacuum insulation. There was no way Johnson could stuff hydrogen fuel in every little space in the aircraft because of the need to have a thermos bottle-type container for the cryogenic fuel, which was close to absolute zero. He mentioned that in the past an airplane could grow in take-off weight and yet the range could be achieved by packing hydrocarbon fuel in every available little volume in the wings and fuselage—something that could not be done with cryogenic fuel.

Fig. 30 The heavily instrumented 304 engine getting ready for testing (courtesy of Pratt & Whitney Archives).

The Assistant Secretary of the Air Force and General Irvine then asked Perry Pratt if there was any further potential in the 304 engine for fuel consumption improvement. Perry offered that perhaps there was another 3–4 percent improvement in a five-year time frame [12]. At that moment the curtain came down on the hydrogen-fueled aircraft. It was not just the huge size of the aircraft that killed the program, but also the practical logistics in having hydrogen available at critical locations throughout the world.

Conclusion

What started off as an analytical investigation of high-speed, high-altitude aircraft ended up starting a hydrogen-fueled air breathing engine program, Suntan, only to have it canceled after a couple of years. The outcome from these activities was totally unexpected, which we explore in Chapter 10. The technology developed for Suntan created an innovation in hydrogen-oxygen liquid rockets, the RL10. Suntan was the first program at the new plant in Florida, the Florida Research & Development Center. This new facility is where the J58 was redesigned and developed for continuous Mach 3+ operation in a high-altitude reconnaissance aircraft immune from ground to air missiles, the Blackbird. Pratt & Whitney was in the rocket business!

References

[1] Whittle, F., "The Birth of the Jet Engine in Britain," *Celebration of the Golden Anniversary of Jet Powered Flight 1939–1989*, AIAA (Dayton-Cincinnati Section), Dayton, OH, Aug. 23, 1989 (article provided by Smithsonian Institution).
[2] Mulready, D., *Advanced Engine Development at Pratt & Whitney: The Inside Story of Eight Special Projects 1946–1971*, SAE International, Warrendale, PA, Feb. 2001.
[3] Coar, R. J., "Engineering Order 701990," Pratt & Whitney Archives, Sept. 29, 1955.
[4] Patterson, W. A., "Letter to Wright Parkins," Parkins Files, Pratt & Whitney Archives, Oct. 22, 1956.
[5] Parkins, W., "Letter to William Patterson, United Air Lines," Parkins files, Pratt & Whitney Archives, Oct. 26, 1956.
[6] DeSantis, J., "Memo to D. O'Neil: The First JT3D," Fisher files, Pratt & Whitney Archives, Oct. 29, 1970.
[7] Fisher, P., "Notes on Conversations with Perry Pratt," Fisher files, Pratt & Whitney Archives, 1965–1966.
[8] Simpson, E. C., *The Last Great Act of Defiance: The Memoirs of Ernest C. Simpson, Aero Propulsion Pioneer*, edited by J. J. St. Peter, Aero Propulsion Laboratory, Wright-Patterson AFB, OH, 1987.
[9] *Aviation Week & Space Technology*, Mc-Graw Hill, Columbus, OH, March 13, 2000.
[10] Sloop, J. L., *Liquid Hydrogen as a Propulsion Fuel, 1945–1959*, NASA, 1978.
[11] Rich, B. R., *The Skunk Works*, Little, Brown & Company, Boston, MA, 1995, p. 176.
[12] Johnson, C. L., *Kelly—More Than My Share of It*, Smithsonian Institution, Washington, D.C., 1985, p. 138.

Chapter 10

HIGHER AND FASTER

INTRODUCTION

Although the Suntan program did not result in any applications, it was the technological genesis of the RL10 program that led to the J58 program, discussed herein. This chapter ends with a discussion about where these two programs took place, a then-remote area where the noisy new jet engines could be tested without disturbing anyone—the newly founded Florida Research & Development Center.

RL10 PROGRAM

While the concentration here is on the aviation side of Pratt & Whitney rather than on industrial power engines, rockets, and fuel cells, it is worth mentioning the RL10 hydrogen-oxygen rocket engine design that was essentially inspired by the Suntan or 304 program. An extensive discussion of RL10 can be found in Dick Mulready's excellent book about eight Pratt & Whitney projects [1].

The engineers working on the 304 had almost three years of experience with liquid hydrogen by the end of the program. In October 1958, the Air Force Special Projects Office required a 15,000-lb thrust hydrogen/oxygen rocket. The engine designation became LR115 (P&W RL10). The conventional approach to liquid-fueled rocket design had an auxiliary combustion chamber off to one side to produce high-temperature, high-pressure gas for a turbine to drive the fuel and oxidant pumps. As you can see from Fig. 1, the sequence of events in the RL10 is much simpler.

Starting at the hydrogen pump, hydrogen flows along the convergent–divergent nozzle wall, where it picks up heat while cooling the structure, and into a gaseous hydrogen turbine, which provides the power to drive the oxidant and fuel pumps. The RL10 cycle was also referred to as the *bootstrap cycle*. The beauty of this cycle is that it uses waste heat to power the fuel and oxidant pumps. Mulready gives much credit for this cycle to Bill Sens and Bob Atherton [1, p. 58]. The construction of the thrust chamber and nozzle with 360 tubes seems almost impossible. Some of the engineers connected with the 304 heat exchanger, which was more complicated than the RL10

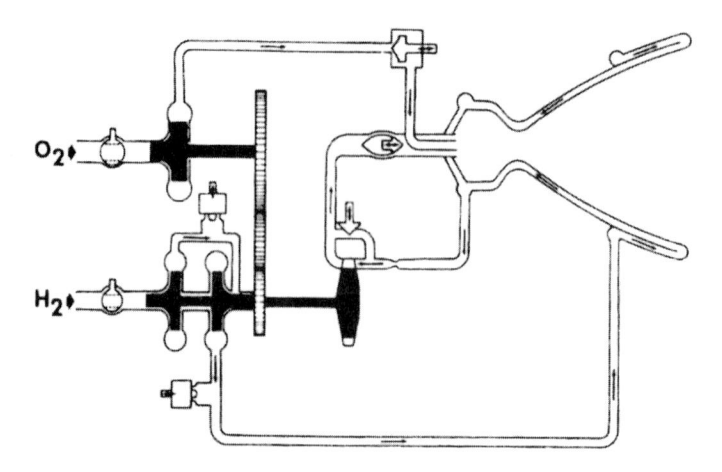

Fig. 1 Flow schematic of the RL10 rocket engine (courtesy of Pratt & Whitney Archives).

structure, obviously have left their signatures on the RL10 (Fig. 2). The combustion temperature is about 6000° F at 300 lb/in^2.

Frank McAbee worked on the RL10 project and remembers the first test of the RL10 [2]:

> I [was] the test engineer responsible for assembling and testing FX-121, the first RL-10 rocket engine. Hydrogen and oxygen had never been used as rocket fuels before, so there was a lot to learn about handling them in large quantities. The "bootstrap" thermodynamic cycle that was the basis

Fig. 2 Fabrication of the RL10 (courtesy of Pratt & Whitney Archives).

Fig. 3 The completed RL10 (courtesy of Pratt & Whitney Archives).

for the engine design came from the Suntan project.... The engine incorporated many materials that had not previously been used in a cryogenic environment.... It was fun to come to work every day.

I remember the first time we tested the RL-10. It was ... in the middle of the night, black as ink out in the swamp but still a starry sky. When the engine lit off it was like doomsday had arrived. A tremendous sheet of flame shot out through the diffuser and there was more noise than I had ever heard. I could feel the pressure pulsations in my chest and they actually affected my breathing. The engine ran, and ran, and ran. That first test was for a duration of 426 sec, which doesn't sound like a lot, but was the longest run duration of any rocket engine at that time. We set a new record that night. I'll never forget it.

The service record is outstanding for the RL10 (pictured in Fig. 3). In November 1963, two RL10s successfully powered the first flight of the Centaur, an upper-stage space vehicle. In January 1964, six RL10s successfully powered the first Saturn S-IV space vehicle, which carried the largest payload ever shot into space. Over a 40-year time period, the engine has been fired 550 times in space with only three failures [1].

J58—GROWING MACH CAPABILITY

PROGRAM STARTUP

The J91 Mach 3 turbojet program was technically successful, but there was no application. Then in mid-1956, during the intense activities of the 304 program (Fig. 4), the Navy had an application: a 30,000-lb thrust

322   JACK CONNORS

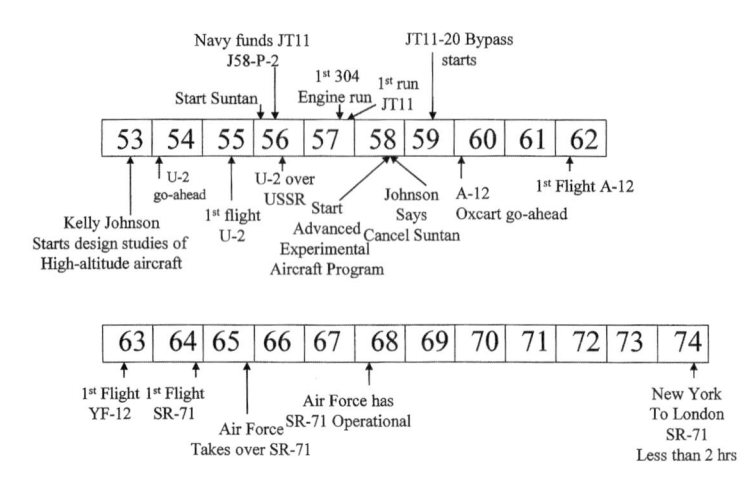

Fig. 4   Time line for the J58 program and the preceding Suntan program.

afterburning turbojet for a shipboard attack aircraft capable of making a dash at Mach 3. The Navy application never came into being, but the Air Force used a version of the J58 in an aircraft of unsurpassed design: the Blackbird (Lockheed SR-71).

The J58 program was moved to the Florida Research & Development Center in 1958 and East Hartford Engineering was gradually phasing out its effort. Both the aircraft and engine programs were so secret that the number of people in Lockheed and Pratt & Whitney as well as the number of people in the government involved were few. Even so, on the senior technical level there were three principal engineers from each organization involved: Kelly Johnson (Lockheed), Bill Brown (Pratt & Whitney) and Cliff Simpson (Air Force). Bill Brown (Fig. 5) recalled that the engine and airplane people very quickly evolved into one team; if a performance variance in the aircraft came

Fig. 5   Bill Brown (left) and Bill Gorton (Florida Research & Development Center General Manager) examine the J58 (courtesy of Pratt & Whitney Archives).

Fig. 6 J58 Program Manager Jack McDermott (courtesy of Pratt & Whitney Archives).

up, then the team made up of the three organizations worked out the best resolution together. Bill Brown made the following assessment [3]:

> On this program, the Government fully recognized that many of the problems ... could be solved ... by a joint engineering effort and the contracts were written to allow this activity without penalties. As a result, an extremely close working relationship between the engineering groups was developed and flourished until the SR-71 became fully operational. This method of operation led to prompt solutions of many problems which under a more cumbersome management system could have severely impeded the program.... The result was an operating system incorporating a magnum step in the state-of-the-art at an earlier time and less cost to the Government than would otherwise have been possible.

The J58 Program Manager was Jack McDermott (Fig. 6), whose career went back to another great engine, the R-2800 (Chapters 3 and 4). In the 1960s, Simpson told me that McDermott was a first-rate engine program manager. McDermott has a special place in the history of propulsion, thanks to the engine he developed.

Engine Design

Recall from Chapter 9 that Kelly Johnson recommended the cancellation of the hydrogen-fueled airplane and engine when he considered airplane size and the magnitude of the worldwide logistics problem with cryogenic hydrogen. Jet fuel can be stored in many places in an airplane in conventional fuel tanks but hydrogen has to be stored in tanks that are thermos bottles. Johnson asked Perry Pratt what level of fuel consumption could be attainable in an afterburning turbojet at Mach 3+ flight speeds at 80,000-ft altitude. Perry Pratt's estimate (most likely after a quick phone call back to Bill Sens in Connecticut) was enough to launch the Blackbird program with a redesigned J58.

In those days, *Mach 3 capability* meant that the engine could operate at that flight speed for less than 15 minutes in a dash situation. The first engine run of the J58 was in late 1957. Both non-afterburning and afterburning versions of the J58 were being promoted for a number of different military applications among the airframers: the Convair F-106, the North American F-108, the Convair B-58C, and the North American A3J. In all applications, the high Mach number condition was considered a short-time situation, not a long-time cruise condition. Studies included the following models:

1) JT11-1 at 26,000-lb afterburning thrust (Mach 3 dash capability)
2) JT11-5A at 32,800-lb afterburning thrust (Mach 3+ capability)
3) JT11-7 at 32,800-lb afterburning thrust (Mach 4 capability)

McDermott moved to Florida with the program and started all over again with essentially an all-new engine design to meet the Lockheed requirements for Mach 3 cruise capability. He started off with the JT11 military version of the J58 engine (Fig. 7) when the engine was targeted for a Mach 3 dash capability. The JT11 was a three-quarter scale of the J91 (discussed in Chapter 8), and its resemblance to the earlier engine is obvious.

In 1959, the J58-P-2 engine was completely redesigned to operate continuously at Mach 3.2 up to 100,000 ft. The J58 as used in the SR-71 program was like the original Navy J58-P2 only in the compressor and turbine aerodynamics and the engine centerline. Just about every part had to be redesigned to handle the hostile operating environment of continuous operation at Mach 3. The fuel and oil temperature limits are nominal values. Bill Brown said that these numbers at times were considerably higher [3]. Obviously special fuel and lubrication liquids had to be developed for this special application. The companies who created these special liquids were Ashland, Shell, and Monsanto.

At take-off, an aircraft gas turbine is close to its maximum airflow and pressure ratio. When the aircraft is at subsonic cruise, the compressor corrected airflow is very close to the sea-level take-off values. However, as the aircraft moves into the supersonic region the corrected flow keeps decreasing. This was not a problem with the J57 and J75 applications, which were limited to Mach 2 maximum speed. However, once the engine was expected to

Fig. 7 Cutaway of the J58-P-2 configuration (courtesy of Pratt & Whitney Archives).

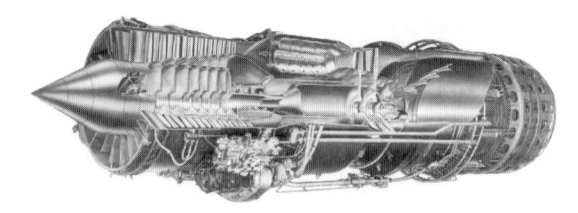

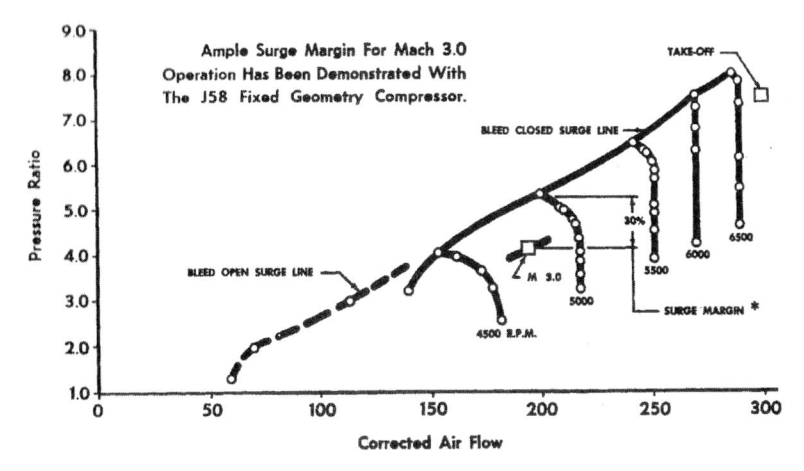

Fig. 8 J58 compressor map showing the take-off and Mach 3 operating points (courtesy of Pratt & Whitney Archives).

provide thrust above Mach 3, the corrected flow and pressure ratio of the compressor dropped to about 65 percent of the design airflow, because the corrected temperature at Mach 3+ is less than 50 percent of its design value.

In addition to the engine's variation in corrected flow with Mach numbers, there is also a greater decrease in a fixed inlet's corrected flow with increasing Mach numbers. When the engine and inlet flow capacities are matched at the Mach 3 condition, for example, the inlet has much more flow capacity at Mach 1 compared to the engine's flow capacity. At the Mach 1 condition with a fixed inlet, there must be a great amount of spillage around the engine, which causes a considerable amount of drag. By bringing this bypass air into the nacelle and dumping it into the engine exhaust as tertiary air the total drag can be decreased. However, the use of a variable inlet (with translating spike) minimizes the need for bypass air during the subsonic flight regime.

The thermal efficiency of the cycle is a function of corrected temperature, the component efficiencies, and the overall pressure ratio (the ram pressure ratio and the compressor pressure ratio). The turbine exit total pressure may not be much higher than the total pressure entering the compressor (Fig. 8). The turbomachinery in that case is like the pre-WWII industrial gas turbines that had hardly any net output because of low-compressor pressure ratio and low corrected temperature (turbine inlet temperature in degrees R divided by the inlet total temperature in degrees R). It would be better at Mach 3+ to have a pure ramjet.

Bypass Bleed Cycle. In April 1959, Bob Abernethy thought of a way of increasing the thrust of the J58 at maximum Mach number. He suggested bleeding air out of the fourth compressor stage to unstall the early stages and unchoke the rear stages of the compressor. In the process, the improvement in compressor airfoil performance would pull more flow through the inlet. When the bleed air is dumped back into the turbine exhaust, the higher airflow from the inlet should result in a greater thrust, as long as mixing losses from the turbine exhaust and bleed air do not become excessive. In other words, the *bypass bleed cycle* takes air from the fourth stage compressor and ducts it downstream of the turbine where it joins the core-engine air and then goes into the afterburner (Fig. 9).

Bob Abernethy described this theory in a memo to George Armbruster, a Project Engineer on the J58, and recommended a modification to the J58 configuration to accommodate the bleed air. If all worked out, not only could there be more airflow but also there could be more surge margin, which implies a higher altitude capability. Bill Brown gave the memo to Bill Sens in East Hartford for comments. Sens responded positively, "... the by-pass bleed engine has attractive features ... as far as application to the advanced J58 is concerned. I think it offers sufficient potential to justify the detailed study that would be required to evaluate it versus development of the basic turbojet" [4].

After the smoke cleared, Abernethy was right—20+ percent more thrust! This improvement permitted the engine to act like a turbo-ramjet at maximum flight speed without the weight and complexity of a turbo-ramjet. A pure turbo-ramjet, of course, would bypass all of the inlet air around the engine at the high flight speed and would operate as a pure ramjet. The turbo-ramjet,

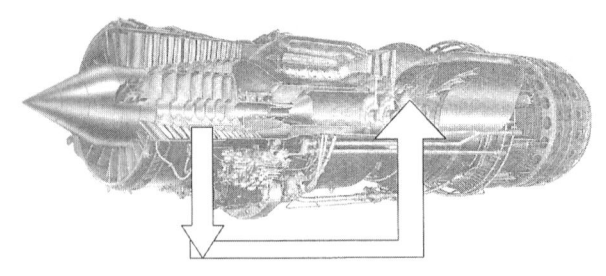

Air from 4th compressor stage

Thrust is proportional to (Airflow) times (Jet velocity minus airplane speed).

Airflow = Up
Jet velocity = Up
Thrust = Up

Fig. 9 The bypass bleed cycle (adapted from a Pratt & Whitney Archives cutaway of a J58).

Fig. 10 J58 before the bypass bleed cycle (courtesy of Pratt & Whitney Archives).

however, becomes more mechanically complex and heavier than what Abernethy suggested. For one thing, it would have blocker doors to close off the turbojet compressor inlet and would also have a good size annular passage around the engine for the ramjet air, which would discharge into a common afterburner. Figure 10 shows the engine before the bypass bleed cycle has been developed. Compare this with Fig. 11, which shows the redesigned J58 with afterburner showing the ducting that brings air from the middle of the compressor and dumps it into the afterburner.

JET NOZZLE. The J58 jet nozzle consists of a blow-in door ejector with free-floating flaps in the divergent part of the nozzle. The flight speed range of the Blackbirds was so great that the jet nozzle had to take on various positions at each flight speed. At maximum speed, the performance required a convergent–divergent jet nozzle. This same nozzle at high subsonic speeds would have too high a base drag because the exit area of the nozzle is too large for the

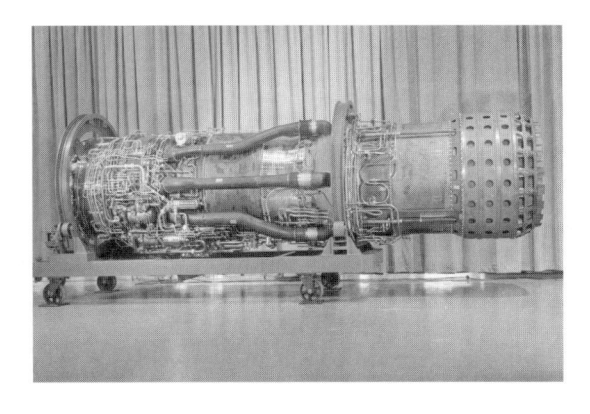

Fig. 11 The redesigned J58 with afterburner shows the ducting (courtesy of Pratt & Whitney Archives).

reduced nozzle pressure ratio (compared to maximum speed). The blow-in doors are open in the lower end of the flight spectrum and are closed during the higher speeds (Fig. 12). In the late 1950s, Bill Sens and Stu Hamilton invented the blow-in door ejector jet nozzle that earned them the prestigious United Aircraft Corporation's Mead Engineering Award.

There are three functions of the nozzle system. At low flight speeds the blow-in doors are open to bring in external air that fills in the area that would normally contribute to base drag. These doors close at the higher speeds because of the higher pressures in the nozzle relative to atmospheric pressure. The ejector action of the primary flow from the afterburner sucks the secondary air for cooling the engine and nacelle interior at all flight speeds. The divergent part of the jet nozzle has free-floating flaps that fold inward at low speeds and extend outward at high speeds. Figure 13 shows models of the blow-in door ejector tested in the United Aircraft wind tunnel.

There was some trouble with the free-floating flaps on the ejector and its Lockheed gearbox; George Armbruster relates this experience with Kelly Johnson [5]:

> When I was out there working with Kelly ... he was having trouble designing his gearbox and he said, "You don't know anything about designing flaps and things.... We've been designing them on airplanes for years and we're going to build the ejector and you're going to build the gearbox." As far as I was concerned that was a good decision and so that's how it came about.

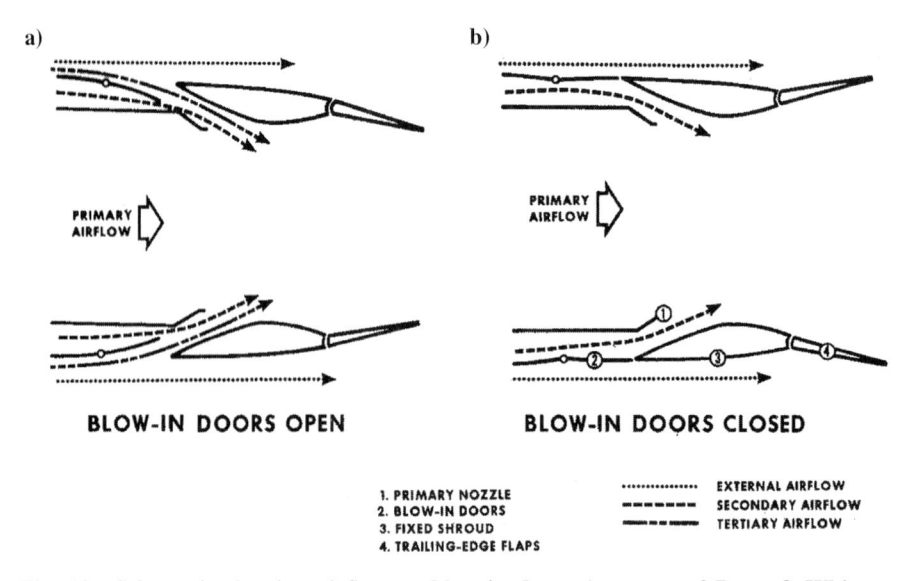

Fig. 12 Schematic showing airflow on blow-in doors (courtesy of Pratt & Whitney Archives).

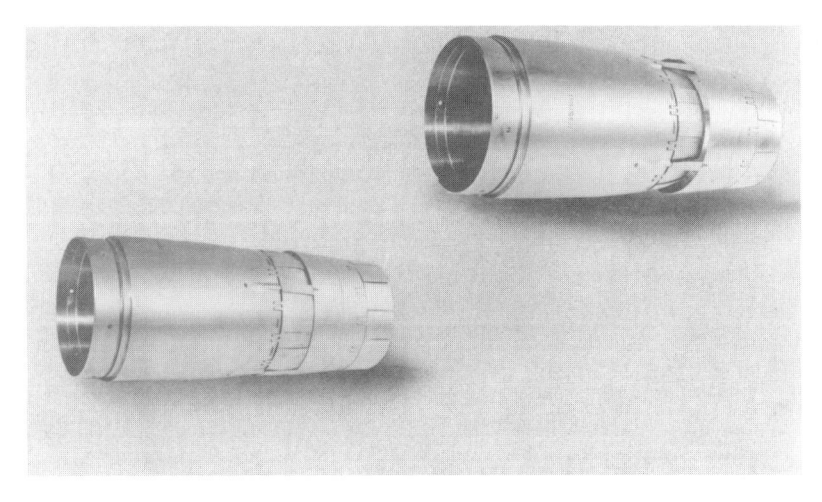

Fig. 13 Models of the blow-in door ejector (courtesy of Pratt & Whitney Archives).

RECOLLECTIONS OF THE PROGRAM MANAGER

The design requirements for the J58 to power an aircraft to cruise continuously at Mach 3+ at 80,000 ft seemed way beyond the reach of available technology. In 2008 I asked Jack McDermott (Fig. 6) for his recollections of the J58 program in the Florida Research & Development Center (FRDC). He provided a vivid picture of the day-by-day issues that arose while developing the engine:

> The redesign was laid down in September 1959 to power the Lockheed A-12 surveillance airplane. It would become the first FRDC turbine engine to power an aircraft. It was neither an Air Force nor Navy engine and to disguise the customer we gave it the company designation of JT11-20. However, most people called it the J58.
>
> Actually it was a major redesign of the J58 to give it the ability to power cruise at Mach 3.2 and up to 100,000-ft altitude.... It was a big design and development challenge for FRDC.... We got some critically needed help from Connecticut in 1961 when Bill Gorton was assigned as our General Manager. He brought key East Hartford people with him.
>
> Gorton['s] ... doubts about the engine were brought home to me [during] our attempts to stop the pipe failures we had been having, which we attributed to the pressure "ripple" coming off the hydraulic pump. A friend in East Hartford told me that the airliner manufacturers used a "Pulsetrap" to smooth their hydraulic pump "ripple." I decided to try it and mentioned it at one of Bill's three times-a-week meetings. He seemed ... unimpressed. So, I relegated it to a low-priority test engine.

A few weeks later, I reported that one of our pipes was still failing and he startled me by telling me to put the "Pulsetrap" into the Bill of Materials ... I was very uneasy about releasing something to Production with so little experience on both it and its manufacturer. So I had the test engineers move the "Pulsetrap" to an endurance engine and run a 60-hr PFRT [Preliminary Flight Rating Test] over the weekend. Monday morning when I reported to Bill that we had run the test without a problem, he said he didn't think the engine was capable of a 60-hr endurance test. It was certainly disappointing to discover his lack of confidence in the engine, but I had to admit he had reasons. We were cracking first-stage compressor blades and first-stage turbine blades frequently....

Our government customer was interested in results, not micromanagement. Also, severe secrecy spared us the usual interference of politics and the gloom and doom with which Washington newspapers usually report on military development programs. Of course the secrecy wasn't all pleasant. We had no bragging rights with people who weren't cleared....

In his 1964 re-election campaign, President Lyndon Johnson revealed the existence of the Blackbirds. We had been told in advance of his intention and warned that as far as we were concerned security rules had not changed, no matter what he said. When the next morning's newspaper carried the President's announcement, my 15-year-old son became very inquisitive. I had to dodge all of his questions instead of taking a bow.

The customer warned us to observe secrecy in our telephone conversations.... This got me thinking about another potential problem related to our proximity to Cuba. Hijacking of commercial airlines and forcing them to land at Havana was a common practice in those days. We made frequent flights from Miami to Los Angeles. If I detoured to Havana, my briefcase might be searched. Hoping that I would be allowed a trip to the toilet before we landed, it would be handy if my secret documents were paper that dissolved in water. Our Purchasing Department found paper that ... did disintegrate in water. We used it on all our airline flights ...

Through it all, our morale was great and our relations with the Skunk Works were better than I had experienced with airframers on any other program. Norm Cotter had a group of his Performance people at Burbank for months. He also used our United Aircraft Research Lab to model test in the wind tunnel to help Lockheed with the design of their ejector. We received a daily wire on the scrambler keeping us informed of all flight test results. The flight test program was unusually long and difficult because of the need to develop the complex and sophisticated inlet required for Mach 3 flight.

Of course, working with Kelly Johnson present[ed] some interesting problems. Many people called him a genius for his many outstanding accomplishments.... I'm sure it helped Lockheed's Skunk Works win a lot of contracts. However, when Kelly had to reveal an unforeseen problem to a customer he seemed to [mention] somebody else's problem at the same time.

He caught us flat-footed at a joint meeting with the Air Force after they took over management of the program by claiming that the engine's idle thrust was so high that his pilots were wearing out brakes holding back the airplane while taxiing. We spend weeks coming up with a redesign to increase the overboard air bleed enough to avoid compressor stall at a lower idle speed. When we submitted the change ... for approval, the Air Force rejected it because the Skunk Works said they needed the existing idle speed to run some of their accessories during taxi.

As for the two engine problems mentioned earlier, the compressor blade problem ceased when we stopped using Zyglo fluid for inspection of the titanium compressor blades. The turbine blade failures ceased when we changed the material to Inco 718 recommended by Joe Moore [who had been] hired ... to run our Materials Development Laboratory. My first meeting with him occurred when he came to tell me we were using the wrong material (SM200). He had a good story about how he and an International Nickel metallurgist had improved the alloy while Joe was at General Electric. I was skeptical because I didn't know Joe and our own people disagreed. He ended by casually mentioning that the density of 718 was 10 percent less than that of SM200. Wow! A 10 percent reduction in the centrifugal load on that blade, which had been failing in the root, sounded like just what we needed. We tested 718 in half of our experimental engines and never failed a blade. To save weight and cost, we wrote the Engineering Change to 718 to include the turbine vanes as well as the first-stage blade. It was the only Engineering Change I ever signed that improved durability with reduced weight and cost....

Joe ... brought with him fellow metallurgist Roy Athey and Corinne Stevens, a wizard on the electron microscope.... They were responsible for a number of important developments, including Gatorizing discs made from power and rapid solidification.

Although the compressor blade failures were easy to avoid after we discovered the cause, one blade did fail in flight before corrective action took effect ... the day before I arrived at the flight test area on a routine visit. I was in the hangar looking at the engine with the broken blade sticking through the inlet cast when Bill Park, Skunk works Chief Test Pilot, joined me and said, "Boy, this is a hell of an engine." I replied, "This one is a little worse for wear." His response was, "Yeah, I know. I was flying it and I shut the engine down, but I'm sure if I needed it to go around, I could have started it and it would have been fine." That left me speechless, but I realized the pilots did like our engine. They had lost a number of friends in the F-104 flight test program due to engine failure. We looked very good comparatively. No aircraft or pilot was lost in the Blackbird program due to an engine failure. We took some chances that made Bruce Torell [FRDC Executive Vice President] uneasy but our close working relations with the Skunk Works allowed us to keep close watch on our "bets" during the flight test program....

The J75 engines were used to power the first flights. During the development work there were verbal jabs from Kelly but in the end he was complimentary. In his autobiography, *More Than My Share of It All*, he said, "The powerplant for the Blackbirds is a marvelous development on the part of Pratt & Whitney. It is the only engine of its kind in the world."

Earlier, Ben Rich, number two man at the Skunk Works, was invited to speak at FRDC (renamed the Government Products Division by then), during our celebration of Engineers' Week. He [said that] the J58 was the only jet engine he had ever worked with that never had a compressor stall. This was surely a benefit of the recover bleed air system invented by Bob Abernethy, which bypassed the rear stages of the compressor above Mach 1.5, delivering air from the front stages directly to the afterburner. It provided adequate stall margin over the full operating range from sea-level static conditions to Mach 3.2 as well as increased thrust at high Mach number cruise....

I am extremely thankful to have been in the right place at the right time to participate in this unique program.... By the way, we earned maximum fee under the incentive clause of the contract.

APPLICATIONS

The J58's capability has never been exceeded and its applications are listed in Table 1. Its use in the first operational stealth aircraft, the A-12, was part of aviation history that created new possibilities in reconnaissance capability. I had the pleasure of meeting Colonel Fox Stephens (Robert L. Stephens), the first Air Force officer to fly the YF-12 interceptor and the SR-71 (Fig. 14), when he was a consultant to Motoren & Turbinen Union (MTU). In May 1965 he established a world's speed record in the YF-12 at 2070 mph. I asked Colonel Stephens about flying the SR-71. He said as a pilot he had to pay close attention to what he was doing. He gave me an example of heading north from Mexico to land at Edwards Air Force Base in California. He said if you did not pay close attention to what you were doing, you would have to make a U-turn around Seattle. That's how fast the aircraft was.

TABLE 1 APPLICATIONS FOR THE J58

Model	Application	Comments
J58-P-2	Proposed Navy Fighter	Program canceled in mid-1959.
JT11D-20	Lockheed A-12, SR-71 (Blackbird)	Redesigned in 1959 with bypass feature.
J58-P-4	North American A3J	Non-afterburning version of above engine without the bypass feature in the compressor. Two engines built and tested. Program canceled in 1962.

Fig. 14 The SR-71 (courtesy of U.S. Air Force).

LIQUID AIR CONDENSING ENGINE (LACE)

Before the J58's bleed bypass cycle was perfected at FRDC, it was the subject of another project. In 1958, Pratt & Whitney was awarded a contract from Wright Field to study recoverable air-breathing boosters for the first-stage of orbiting or near-orbiting space vehicles. At the same time, Wright Field awarded a contract to Convair-San Diego to use engine information in Convair's vehicle studies. Gene Montany was the program manager, assisted by Ted Slaiby (responsible for engine design information) and Dick Staubach (in charge of the airframe design and performance). Montany was also in charge of the Aircraft Performance Group under Bill Sens. Montany gave me a two-page summary of LACE engine/vehicle studies in 2001, which he kindly allowed me to reproduce here:

> Our studies concentrated on turbojet and turbo-ramjet engines for take-off and acceleration to various speeds and altitudes for upper stage launch. We did very comprehensive and detailed design studies, which optimized airframe and engine studies for parametrics of launch altitudes and speeds, take-off field lengths, payload size, and orbital inclination among others.... The Convair program manager spent a lot of time at Pratt & Whitney but we saw very little of his vehicle studies. In fact, during our weekly or bi-weekly oral reports at Wright Field we were not allowed to attend the Convair presentations!
>
> Our studies provided the customer with ... studies outlining the many launch scenarios and associated costs available with air-breathing launch vehicles that could be launched from practically any military and most commercial airfields. By virtue of their ability to fly dog legs, vehicles could allow upper-stage flights over virtually any target on the first pass. The resulting security and precision could not be duplicated with rocket launches from ... fixed bases. Convair's studies were eventually revealed to us. They were limiting their studies to single-stage-to-orbit vehicles utilizing a liquid air condensing engine (LACE) under study at Marquardt.

A typical case of pie-in-the-sky trumping solid, attainable, and realistic approaches!

The LACE engine was conceived as a hydrogen-burning air-breathing/rocket engine. The liquid hydrogen would be pumped through a heat exchanger on its way to the air-breathing engine's combustion chamber that would use the heat-sink of the fuel to condense some of the incoming air and store it in the emptying liquid hydrogen fuel tanks. Then the air-breathing engine would be converted to a rocket engine using the hydrogen as fuel and the stored liquid air as an oxidizer. The rocket engine would push the entire vehicle to low-earth orbit where the payload would be deposited and the entire vehicle would re-enter the atmosphere to be used again.

The possibility of a single-stage-to-orbit recoverable launch vehicle was obviously intoxicating the AF [Air Force] people in spite of tremendous technical problems and little or no payload capability. Pratt & Whitney was provided the Marquardt data so we could use our extensive experience with hydrogen and heat exchangers to improve the engine. In going this ill-fated route, I believe the AF ceded the recoverable launch vehicle to the rocketeers of NASA. Thus the shuttle was born and the launch costs never reached the low target levels desired of launch vehicles, and the AF relies on expendable rocket vehicles launched from vulnerable fixed launch sites, and without the significant capability to launch into virtually any orbit.

A few years later, Pratt & Whitney was awarded a contract to study rocket and air-breathing engines as propulsion systems for long-range ASM [air-to-surface missile] and AAM [air-to-air missile] missiles. We were directed to do the missile airframe design work without a companion airframer! I was again Program Manager and Ted Slaiby was again responsible for engine work. Bill Gaffin was loaned from CANEL (the nuclear engine project) for the vehicle design work. Since we had to do the missile designs we were assigned two senior designers—Walt Ledwith and Bruno Fiore.

We soon found that the designers were not free to innovate in engine and vehicle designs because the Pratt & Whitney *Engine Design Manual* restricted them.... We did, however, prevail on the more progressive minds in management and the designers were unshackled. New and advanced approaches to vehicle and engine designs were invented and two designers were enthralled by their new-found ability to innovate and to add new procedures to the design manual. Bruno eventually left Pratt & Whitney but Walt was transferred to Florida where his new-found skills were put to great use over the ensuing years.

FLORIDA RESEARCH & DEVELOPMENT CENTER

PRATT & WHITNEY EXPANDS

This chapter, along with Chapter 9, discusses several of the programs begun in the 1950s that took many resources. One important resource was physical space—something that the company was lacking in the highly

Fig. 15 This aerial view shows the congestion surrounding the East Hartford facility (courtesy of Pratt & Whitney Archives).

developed area of Hartford, Connecticut (Fig. 15). In 1956, H. Mansfield Horner became Chairman of the Board of United Aircraft Corporation and Bill Gwinn became its president. Gwinn liked the Florida location and negotiations progressed satisfactorily so that Pratt & Whitney obtained 7000 acres in Palm Beach County with no neighbors in sight (Fig. 16).

This plot of land really was at the edge of nowhere at the time, which was handy. Chuck Roelke became General Manager of the Florida center and had two highly classified major programs on his books, the Suntan program (which gradually morphed into the RL10 program) and the J58. The first burden was to move a hot program (Suntan) to Florida without missing a beat. The second burden was to build a completely new engineering facility from scratch in the middle of a huge swamp. The third burden was manpower. Only a limited number of East Hartford engineers could be shipped to Florida because the main plant also had many programs going on and was tight on manpower. As a consequence, Roelke had to hire engineers who were not accustomed to the Pratt & Whitney system.

In November 1956, Roelke leased office space in downtown West Palm Beach to start the engineering operations. Out at the site, the Air Force authorized the building of a hydrogen plant (known as "Mama Bear"), which was completed in the fall of 1957. Two senior members of the Training School in East Hartford went to Florida to set up a training facility in Riviera Beach in mid-1957 to teach methods for precision work to new hires for the shop. The

Fig. 16 Two views of the Florida facility, a) looking north, and b) looking south (courtesy of Pratt & Whitney Archives).

facility was dedicated in May 1958—just about two years from the start of the Suntan program in East Hartford. Another larger Air Force hydrogen generating plant (known as "Papa Bear") was built and put on-line in 1959.

Bill Gorton (Fig. 5), from the East Hartford plant, became the new General Manager at FRDC in July 1961. He brought with him more than 20 years of engine development experience. Achievements came into focus when the J58 passed its PFRT in 1962 opening the way to the A-12 flight test. The Florida center had grown from about 800 personnel in 1956 to over 5000 by the beginning of 1963. An important part of this facility was a new materials research and development department.

METALLURGY

At this point it is worth mentioning a branch of engineering that has not been mentioned yet, metallurgy. The major obstacle to the widespread use of gas turbines was the limitation of turbine materials that could stand up under temperatures higher than 1200° F. Materials development became one of the major efforts of Pratt & Whitney in gas turbines, including titanium.

Titanium made high-bypass turbofans possible because of its great strength to weight ratio compared to steel; it also made lighter turbojets. The company bought $1.5 million worth of titanium, only to find out that it was contaminated with an impurity. Rudy Thielman, a consultant to Pratt & Whitney in the early days of the J42, the J47, and the J57, was an expert on titanium. Rudy had it vacuum-melted, and created the highest purity of material then known.

In the early days of the J57 (see Chapter 7), there was great concern about the life of ball and roller bearings. Ed Schneider, from the Design Department, and I contacted Fafnir, New Departure, and Rockwell as well as bearings expert Burt Jones, who was conducting analytical studies of the forces acting on bearings in aircraft gas turbines. The greatest improvement was in vacuum-melting of the best bearing materials, as well as the mechanical engineering work done at Pratt & Whitney by Dana Waring and Dick Chevchenko. Dana and Chev concentrated on design details to use the oil most effectively not only for lubrication but also for cooling.

Pratt & Whitney had recognized the need for better materials during the piston engine era. As the jet era dawned, however, the need for improved materials was even more critical. When the United Aircraft Scientific Advisory Board (a group of prominent scientists and engineers headed up by Professor Soderberg) recommended that Pratt & Whitney set up its own materials research and development organization, United Aircraft made a substantial investment in people and facilities. In East Hartford, there was a Materials Research Laboratory set up under Dr. Bud Shank from MIT.

One striking example of what came out of this facility was the development of directionally solidified metal for turbine blades and a single-crystal turbine blade that improved the reliability of higher turbine temperature

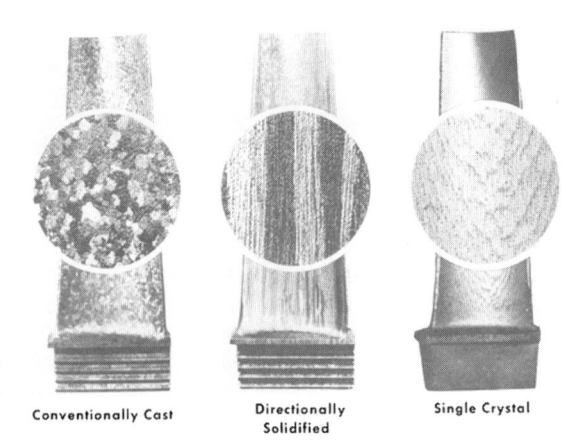

Fig. 17 Three blades of greatly different structures (courtesy of Pratt & Whitney Archives).

Conventionally Cast   Directionally Solidified   Single Crystal

operation. Figure 17 shows conventionally cast blade with individual grains visible in the enlargement. The stress failure here was usually along the grain boundaries. The next advancement was making the individual grains grow along the span direction for increased strength. The ultimate advancement was to make the blade out of a single crystal.

In Florida, a materials research and development organization was established under the direction of Joe Moore, mentioned earlier in this chapter. Moore developed a plastic forging system so that high-strength parts could be made more easily. His name for the process was a whiff from the local wildlife—*Gatorizing*®! In addition, Joe Moore and his group pioneered the use of powder metallurgy in the manufacturing of machine turbine parts. Moore described this process [2]:

> We ... took some material and ... tested the tensile properties in the 1700 to 2200 °F range—stress rupture and tensile. We were absolutely amazed with what happened. As soon as the guy loaded the specimen it just stretched like chewing gum. We became very excited over this because these alloys, like IN100, are very difficult to forge. Here they were flowing like putty.
>
> So, if you can get the material in that kind of condition and if you can pull like that in tension ... you can come up with forgings that are very close to the final shape with a reduction in machining.
>
> So we set up a little forging rig. We put the material in a super-plastic condition and heated up the whole system. While it was at temperature we were able to make a small pancake with beautiful surfaces. [Moore also pioneered the development of powdered alloys with the innovation of RSR.]
>
> RSR stands for Rapid Solidification Rate and you make powder under inert conditions.... Our boys came up with a fantastic machine for making powder and rapidly solidifying it ... the metal would be poured in a vacuum chamber from the crucible into a nozzle—all into a spinning

disk. The rotation of the disk throws the metal out in 360° all around and then we had a blast of Argon (a circular curtain around the spinning disk) in order to increase the speed of solidification. What we were doing was, as we always had done since the J58 development, we were reaching for a stronger turbine blade material using RSR.

REFLECTIONS

In 2005, the RL-10 was designated as a National Historic Mechanical Engineering Landmark by the American Society of Mechanical Engineers (ASME). Frank McAbee, a test engineer on RL10, describes what it was like to work on the project at FRDC [2]:

> Our group of ... engineers, along with our families, had been transferred from Hartford down to Pratt's new Florida facility.... The communities near the plant had hardly been developed ... a whole new housing development was being built in Lake Park.... Everything was new at work as well—the technology, the challenge, and especially the relationship with our government customer. We ... weren't put off by the fact that the things we were working on had never been done before. There was tremendous camaraderie within our group and our families.

Dick Mulready also described the esprit de corps of the initial group of engineers who were transferred to the Florida facility [1]:

> It was a very happy time ... when people loved to come to work and they were so devoted to making this equipment work that they would spend all kinds of extra hours at it. It was hard to explain how people felt at all levels. Everybody was just happy with the job.

During my career at Pratt & Whitney I visited the Florida facility many times and had the pleasure of working for Dick Mulready in East Hartford on the Suntan project (before it moved to Florida). He was a gung-ho type of guy with a great sense of humor. His words sum up how many of us feel.

REFERENCES

[1] Mulready, D., *Advanced Engine Development at Pratt & Whitney: The Inside Story of Eight Special Projects, 1946–1971*, SAE International, Warrendale, PA, Feb. 2001.
[2] *Thunder in the Sun: A Look at 40 Years in Florida, 1958–1998*, video produced by Communications Department, Pratt & Whitney, 1999.
[3] Brown, W. H., "J58/SR-71 Propulsion Integration or The Great Adventure into The Technological Unknown," *Lockheed Horizon*, No. 9, Winter 1981/82, Lockheed Martin, pp. 7–13.
[4] Sens, W. H., "Memo to W. H. Brown Regarding a Bypass Bleed Engine," Pratt & Whitney Archives, May 6, 1959.
[5] Armbruster, G., "Interview by Harvey Lippincott," Pratt & Whitney Archives, March 13, 1984.

Chapter 11

GOING COMMERCIAL

INTRODUCTION

In Pratt & Whitney's earlier days, engines were developed under government contract for military use. During the "Fabulous Fifties," however, commercial customers were pushing for technological advances and would go to a competitor if they did not get a fast enough response from Pratt & Whitney. That meant the company had to put its reputation and resources on the line to secure a sale without a contract. In this chapter we'll review two projects managed in East Hartford, one being the company's first engine developed for commercial use without a predetermined military application, the JT8D. First we'll review the JTF10A program, a huge effort launched before the company knew that the JT8D would end up as a concurrent project.

ORIGIN OF THE TF30 (JTF10A)

The Douglas DC-8 made its first flight in May 1958. Shortly after the DC-8 program got into full swing, Douglas began to think about its next aircraft program—a small four-engine short-range jet-powered aircraft. Douglas had information on the J52 (Chapter 8) because its military department was studying the use of the J52 in the A4D aircraft to replace the J65 engine. The JT8 (J52) was a possibility but "turbofan-mania" was in the air.

The four-engine configuration seemed to be the trend in transports—DC-4, DC-6, DC-7, and DC-8. In early 1959 Donald Douglas Jr. approached "Jack" H. Mansfield Horner (who replaced Rentschler as United Aircraft Chief Executive Officer [CEO]) about a joint venture on a small jet transport, tentatively named DC-9. The agreement seemed to be that 1) Pratt & Whitney would launch the engine program right away, and 2) Douglas would conduct a marketing program for the aircraft. Since it took at least a year longer to develop the engine compared to the airframe, it made sense to start the engine right away. Then, depending upon the market reaction, Pratt & Whitney and Douglas would either continue or stop and neither party would owe the other party anything. Development Engineer Dick Smith wrote the Engineering Order in April 1959 that put the life into the JTF10A program as a commercial venture (Fig. 1). Pratt & Whitney went ahead with the

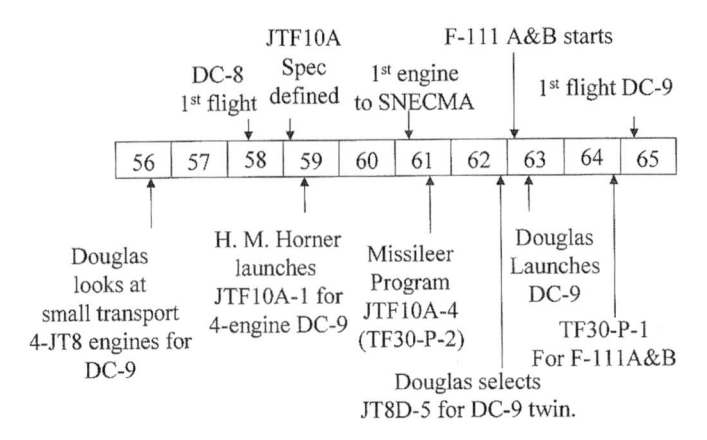

Fig. 1 Time line for the JTF10A.

design of a turbofan half the size of the JT3D and supported Douglas in its studies and sales work.

Douglas conducted a sales effort with potential airline customers over the subsequent several years. Douglas described the aircraft as carrying between 68 first class and 87 tourist class passengers. Donald Douglas, Jr. said his company could be prepared to deliver aircraft early in 1963 for short- and medium-range applications. In the meantime Pratt & Whitney's engine development continued.

The JTF10A caught the eye of a French company. Leonard C. Mallet, General Manager of Pratt & Whitney, and Henri Desbrueres, president and director-general of Societe Nationale d'Etude & Construction de Moteurs d'Aviation (SNECMA) signed an agreement in 1960 for the French company to manufacture and sell many of Pratt & Whitney's gas turbine engines as well as the complete line of piston engines. This agreement gave Pratt & Whitney a 10.9-percent ownership of SNECMA and the right to be represented on the French company's board.

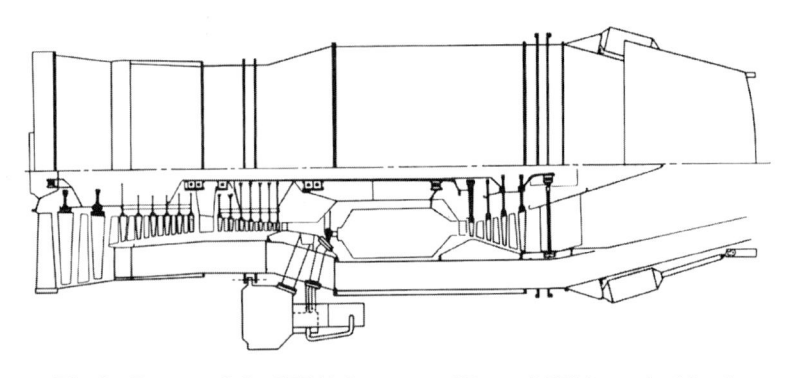

Fig. 2 Layout of the JTF10 (courtesy of Pratt & Whitney Archives).

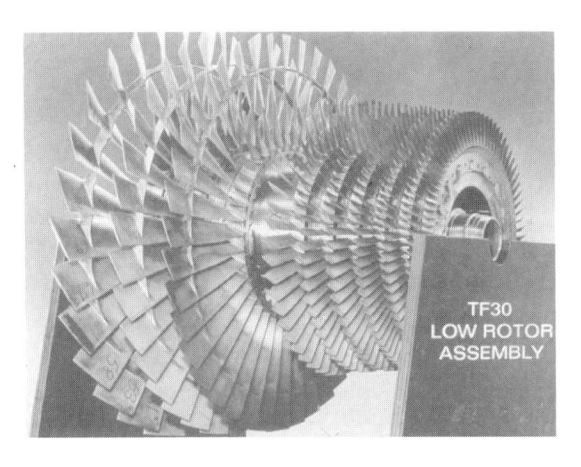

Fig. 3 The TF30 low-pressure fan/compressor combination (courtesy of Pratt & Whitney Archives).

Engine Description

The engine was given the designation JTF10A. There are two fan stages and six low-pressure compressor stages driven by a three-stage low-pressure turbine (Fig. 2). The high spool has seven compressor stages driven by a single-stage high-pressure turbine. The fan stages have shrouds at about two-thirds of the span to minimize blade vibration (Fig. 3). This is the same configuration as the JT3D but with only half the JT3D airflow (Fig. 4). There was one major difference between the JTF10A and the JT3D. The former had a long duct while the JT3D had a short duct. The reason for the difference was to make room for a military version in a fuselage installation and a mixed-flow afterburner. Specifications as well as applications are given in Table 1.

Military Applications

The half-scale engine from the JT3D started off as a commercial venture but ended up with no commercial applications and limited military

Fig. 4 The original JTF10A-1: literally a half-size version of the JT3D (courtesy of Pratt & Whitney Museum).

TABLE 1 JTF10A/TF30 Specifications and Applications

Model	Take-off thrust (lb)	Military or max. cont. thrust (lb)	Normal thrust (lb)	Weight (lb)	Diameter (in.)	Length (in.)	Aircraft
XTF30-P-1	18,500	10,750			48	235.9	BuWeps Ground test Engine
YTF30-P-1	18,500	10,750		3934	48	235.9	F-111A / F-111B
TF30-P-1	18,500	10,750		3869	48	235.5	F-111A
TF30-P-1A	18,500	10,750		3899	48	235.5	F-111B
XTF30-P-2	10,000	9200		2050	41.6	156.7	Missileer Douglas XF6D-1
TF30-P-3	18,500	10,750	9800	4058	49	241.7	F-111A,C,K / YF-111A / F-111E
TF30-P-5	19,750	11,770		4423	50.7	241.4	FB-111
YTF30-P-6		11,350	9100	2786	42.1	128.1	A-7A
TF30-P-6		11,350	9100	2716	42.1	128.1	A-7A
TF30-P-6A		11,350	9100	2716	42.1	128.1	A-7A
TF30-P-6C		11,350	9100		42.1	128.1	A-7A
TF30-P-7	20,350	12,350	10,800	4121	50.7	241.4	FB-111A
TF30-P-8	12,200	12,200	9600	2526	42.1	128.1	A-7B, A-7E

GOING COMMERCIAL

TF30-P-9	19,600		12,000	4070	49.0	241.7	F-111D
TF30-P-12	20,250	10,750	12,290	4027	43.5	241.4	FB-111B
TF30-P-14	15,000	9235	11,780	3141	42.1	179.9	A-7D
TF30-P-16	14,500	8750	10,950	3316	42.1	206.4	A-7D
TF30-P-18			15,000	2916	42.1	130.7	A-7E
YTF30-P-100				3980	48.9	241.7	F-111F
TF30-P-100	25,100	13,170	14,560	3985	48.9	241.7	F-111F
TF30-P-408			13,400	2597	42.1	128.1	A-7B
JTF10A-1	8250	6310	7280	2110			DC-9 Proposal: 4-engines
JTF10A-3	10,000						SNECMA
JTF10A-8		9100	11,350	2716	42.1	128.1	A-7A
JTF10A-9		9600	12,200	2445	42.1	128.1	A-7A
JTF10A-10	11,350			2716			SNECMA
JTF10A-15			15,000	2926	42.1	130.7	A-7E
JTF10A-16A			13,400	2597	42.1	128.1	A-7A
JTF10A-20	18,500	8500	10,750		48.0	235.5	F-111A
JTF10A-22	18,500	8500	10,750	3725	42.8	213.1	Mirage 3V-02 Dassault
JTF10A-23	18,500	8500	10,750	3052	42.8	123.4	Miraage F2 & 3G
JTF10A-26	14,500			3316	42.1	179.9	A-7A
JTF10A-27D	20,350		12,350	4120	50.7	241.4	FB-111A
JTF10A-31	15,000		11,780	3141	42.1	179.9	A-7D
JTF10A-32C				3985	48.9	241.7	F-111F
JTF10A-35	19,750		11,770	4423	50.7	241.4	FB-111
JTF10A-36	20,840		12,430	4070	49.0	241.7	F-111D

Fig. 5   TF30-P-8 non-afterburning engine (courtesy of Pratt & Whitney Archives).

applications, with 3739 TF30 engines and 18 JTF10 engines produced (Figs. 5 and 6).

In 1961 the U.S. Navy had a need for an aircraft that carried missiles (air-to-ground and air-to-air) for fleet protection to the outer perimeter of the fleet's operations. The Navy selected the JTF10A-1 (TF30-P-2) for the Missileer in late 1961 because engines were already in development. Unfortunately, the program died a natural death leaving the JTF10A-1 once more without an application. Then, all of a sudden, the Tactical Fighter Experimental (TFX) program opportunity appeared where the Air Force and the Navy would use the same aircraft and engines—a philosophy promoted by Secretary of Defense Robert McNamara. He believed in the commonality principle to get more bang for the buck. Pratt & Whitney had experimental engines running while the other engine competitors had only proposals. So, Pratt & Whitney's gamble years earlier on a strictly commercial venture at last appeared to have a payoff in a military application.

There ended up being three military applications of the TF30 engine: the F-111, the A-7, and the F-14. The Department of Defense started its F-111 A

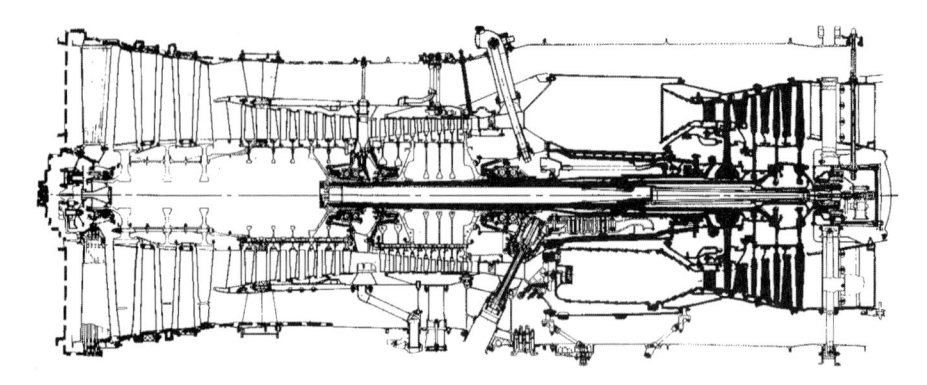

Fig. 6   Layout of the advanced TF30, with three fan stages and only six low-compressor stages (courtesy of Pratt & Whitney Archives).

and B aircraft program in early 1963. By the end of 1964 the TF30-P-1 became the engine for the Air Force F-111 and the Grumman F-14.

In 1960 the Department of Defense tried to meet the needs of two services for a tactical aircraft. The Air Force was looking for a replacement for the F-105 and the Navy was trying to replace its F-4 aircraft. The proposed solution was the TFX program. The Air Force selected the team of General Dynamics and Grumman, which proposed the P&W TF30 afterburning engine. General Dynamics worked on the swing-wing F-111A for the Air Force and Grumman concentrated on the Navy's F-111B aircraft. Both aircraft featured the variable-sweep wing—which was swept back for supersonic flight and straight out for good subsonic range, take-off, and landing. The Air Force model had its first flight in 1964. Aircraft production began in 1967— eventually amounting to more than 550 aircraft (Fig. 7). The initial flight-testing brought into focus inlet-engine compatibility problems that were initially daunting but were eventually overcome. The later aircraft, F-104 and F-105, used variable geometry in their inlets, which helped to provide a less distorted total pressure profile at the engine inlet. The F-111B for the Navy was canceled in 1968. Subsequently the Navy sponsored the TF30-powered Grumman F-14 in the Navy's VFX program [1].

In 1964 the Navy awarded a contract to Vought to build an aircraft to replace the Douglas A-4 Skyhawk. The configuration of the new aircraft (the A-7 Corsair II) was very similar to the J57-powered F-8 Crusader. The first flight was in 1965 and deliveries to the Navy started the following year. Approximately 450 aircraft were TF30-powered (Fig. 8). The Air Force had its own version of the A-7. The Air Force wanted the TF30-powered A-7 but Pratt & Whitney was not able to produce the engines to meet the Air Force aircraft schedule. I believe Art Smith offered a license to Allison to produce the TF30s but Allison preferred to produce the Rolls-Royce Spey. The Spey became the A-7 engine. The Air Force and foreign customers ordered about 600 of the Spey-powered aircraft.

The F-111B was supposed to be the Navy version of the one-size-fits-all philosophy promoted by Robert McNamara, Secretary of Defense under President Lyndon Johnson. This gradually developed into a no-way-are-we-going-to-use-the-same-aircraft situation. Grumman, which had been working

Fig. 7 The F-111 at Pratt & Whitney (courtesy of Pratt & Whitney Archives).

Fig. 8 The Vought A-7 Corsair with a non-afterburning TF30 (courtesy of U.S. Navy).

on the Navy F-111B as part of the General Dynamics team, resolved the Navy requirement with the F-14 Tomcat (Fig. 9) after the F-111B program was canceled in 1968. Grumman submitted a proposal for a high-performance swing-wing aircraft in 1969 that the Navy accepted. The first flight of the F-14 took place in December 1970. The aircraft, eventually powered by P&W TF30-P-412A, entered service in 1972. The production program ended in 1987 after 514 aircraft had been manufactured. In its time it had the reputation of being "one of the finest warplanes in the world" [2, p. 479].

JT8D—FEET FIRST INTO COMMERCIAL SERVICE

ORIGIN OF THE JT8D

About $3 million and nine months had been spent on the JTF10A-1 program when a new opportunity was discovered. Development of the JT8D engine began in 1960 as opportunity to sell Boeing on putting a version of the J52 into its intermediate-range aircraft, the B-727, when they had already

Fig. 9 The F-14 Tomcat (courtesy of U.S. Navy).

decided on Rolls-Royce's engine. This happened during of time of growing organizational complexity. For example, even in the Pratt & Whitney groups closest to the customers, Marketing and Installation Engineering, it was a challenge to get customer problems and recommended actions from those in the trenches up to those in the Officers Club.

The Pratt & Whitney Advanced Planning Group had just been formed under the direction of Gene Montany with a location near the Marketing Department. Wright Parkins, who was out of Pratt & Whitney and into the United Aircraft Corporate structure called a sudden meeting regarding what he saw as a blossoming engine opportunity at Boeing. On the spot a team was created to respond: Jack Craig (Marketing), Charlie Brame (Installation Engineering), and Gene Montany (Advanced Planning). Jack Craig was a behind-the-scenes type of guy and undoubtedly had discussions with his friends in the Boeing Engineering Department about the 727 engines prior to the meeting with Parkins. He contacted his friends at Boeing and learned that Boeing had selected the Rolls-Royce engine and was not interested in waiting around for a Pratt & Whitney engine.

In the Boeing organization there were more analytical engineers working on engine studies than Pratt & Whitney had in its Advanced Engine Performance Group under Bill Sens. The Boeing engine group made studies of engines for the 727 that included the Rolls-Royce Spey and the JT8D engines (Pratt & Whitney Engineering had been supplying Boeing powerplant data for its studies of a three-engine transport with a variety of potential engine configurations). Their conclusion was that the Spey should be selected. It should not be surprising that their report reached this conclusion because an engine group in an aircraft company would probably not have the experience to realize what it takes for an engine to meet performance and durability specifications (any more than an engineer in an engine company would understand all it takes to build a successful aircraft). A striking difference between the engines was engine weight. An engine heavier than the so-called ideal engine would eat away at the 727 performance capability.

Boeing's speedy decision is understandable when you consider that the airframe manufacturer is anxious to make a sale. The lower the price of an aircraft, with any given payload capacity, the greater the chances of making the sale. When an engine company comes along and tries to point out that the engine should be bigger than the airframer is talking about, the airframer does not want to hear it. The engine company offering the lowest weight matched with an attractive performance has the best chance of winning the sale. Apparently Boeing had what it believed to be a winning combination using three Rolls-Royce Spey engines manufactured by Allison.

At this point, Pratt & Whitney, while recognizing the opportunity to challenge a competitor, did not have senior management approval. Stan Taylor (Fig. 10), in Marketing at the time, told me while I was writing this book

Fig. 10 Stan Taylor (courtesy Pratt & Whitney Archives).

how the lines of communication were shortened to make sure topside got the message completely unfiltered:

> In 1960 Boeing was promoting the design of a three-engine airplane with ... a fairly tortuous inlet arrangement for the middle engine, which was under the tail. Pratt & Whitney's management, both Engineering and Financial, was very leery of this because of the difficulties that had been encountered with the inlet on the F-101 airplane built by McDonnell with the J57.
>
> So even though there was a great hue and cry both at United Air Lines and Eastern Airlines asking us to offer a turbofan version of the J52, there was really no apparent action being taken to do that. Fred Fuetsch, who worked with United, and Frank DuLyn, who worked with Eastern, said that these two airlines were going to buy this airplane and they were going to buy it with Rolls-Royce engines!
>
> Rolls-Royce was very actively promoting an engine for that airplane. So Frank and I sat in my office mulling this over one day and we said, "We've got to get this message to Jack Horner [United Aircraft CEO] that they were going to buy these airplanes with Rolls-Royce engines because we don't seem to be making any impression upon the management." So we decided that Frank would take on Charlie Froesch and I would take on Jack Herlihy.
>
> Charlie Froesch was the Chief Engineer at Eastern. Frank went down to see him and told Charlie, if he was ever going to get Pratt & Whitney to offer an engine for the new Boeing airplane, it was essential that he call Jack Horner personally and make an appointment to come to see him as soon as possible. At the same time Fred Fuetsch and I made an appointment with Jack Herlihy, vice-president of United Air Lines Engineering in San Francisco and his right-hand man, Bill Menser, to go out there in a couple of days.

Going Commercial 351

    Well, Jack Herlihy and Bill Menser responded with a phone call to Jack Horner, and Charlie Froesch came up to East Hartford and had a meeting with Jack Horner as well as the Pratt & Whitney Technical and Financial Management and it was decided right then and there in that day that Pratt & Whitney would in fact offer an engine for this airplane.

    A delegation, in a couple of days, went down to Eastern Airlines and made a presentation and we were off and running with a big promotional program with both Boeing, United and, of course, Eastern to put a turbofan version of the J52 in the 727 airplane. So that's how it got started.

As Stan explains so clearly, the next move had to be with the airlines that were ready to buy the Boeing 727. Don Jordan (Fig. 11), Chief Powerplant Engineer with Chance Vought, came to Pratt & Whitney when Vought moved to Texas and was soon embroiled in the JT8D engine campaign with Eastern Airlines because of his extensive experience with engine installations and his grasp of engine technology. Jordan tells us the final scene, after Jack Horner got the message:

    Our JT8D was something in the ballpark of 1000-lb heavier than the Spey because it was a bigger engine with more thrust. Of course that did not cut it with the airframer because he is looking for the smallest engine to do the job....

    I went to Bill Gorton and said "we've got to try to take at about 750 lb out of this engine and we've got to assure Boeing that we do not see the center engine inlet as a problem." The S-duct did not bother me because I had the experience from Vought with engine inlets. Bill's reaction was one of disbelief that we could get close to the weight of the Spey but he said by all means see what you can do. Well, the designers went to work with a fury and the best they could do was to take out about 500 lb. Even

Fig. 11 Don Jordan (courtesy Pratt & Whitney Archives).

this amount meant that the inlet case had to be made out of aluminum—which was not the favorite material for an inlet case that was a major support for the rotors.

Next, H. Mansfield Horner made an appointment with Eddie Rickenbacker of Eastern, to discuss what Pratt & Whitney could do to power the Boeing 727. Horner brought me for the technical presentation and Bill Gwinn and Frank DuLyn from Marketing. The meeting was at the Waldorf Astoria in New York. Rickenbacker had his chief engine man, Charlie Froesch, there to hear Pratt's technical message.

Horner asked me to give the technical presentation on the JT8D, which included not only a description of the engine and also a list of things we have learned from the J57 and J75 service experience—the kinds of experiences on bearings, seals, accessories, and compressor and turbine blades. I described the problems we ran into and how we got out of them. Also, I had a copy of the Boeing report and went down the line commenting on item by item. Charlie seemed to agree with what I was discussing.

Then Horner talked to Rickenbacker directly with the message that we are going to make this an airline engine with the same degree of performance and reliability that our other engines have. Also, Horner said he was behind this engine and would make the resources available to make this engine at least as good as other Pratt engines. That's all Eddie Rickenbacker had to hear.

Then Rickenbacker went to Boeing and told him he wanted the 727 with the Pratt & Whitney engine. You can imagine the subsequent turmoil at Boeing.

Boeing management was furious. I understand there was a subsequent phone call from Boeing's management to Horner advising Pratt & Whitney in strong, forceful tonality to keep its engineers away from Boeing's customers, that is, the airlines. But wasn't it just over little over a quarter century earlier that Boeing and Pratt & Whitney partners in the United Aircraft and Transport Company? Times had changed, but Pratt & Whitney still got the job.

DEVELOPMENT BEGINS

Gordon Beckwith, who became the Senior Project Engineer on the JT8D, gave a presentation to a distinguished group of Pratt & Whitney and United Aircraft executives in January 1961 only nine months after program go-ahead. (Beckwith was critical to the development of this engine, and we wrap up this chapter with a few words about him.) The list of attendees included H. M. Horner, Bill Gwinn, Len Mallet, Wright Parkins, Art Smith, Barney Schmickrath, Dick Coar, Charles Kearns, and Erle Martin. Beckwith pointed out that the JT8D would be the company's first engine to jump into commercial service without any military service experience. The result of this presentation was that Horner, Chairman of United Aircraft, provided

GOING COMMERCIAL 353

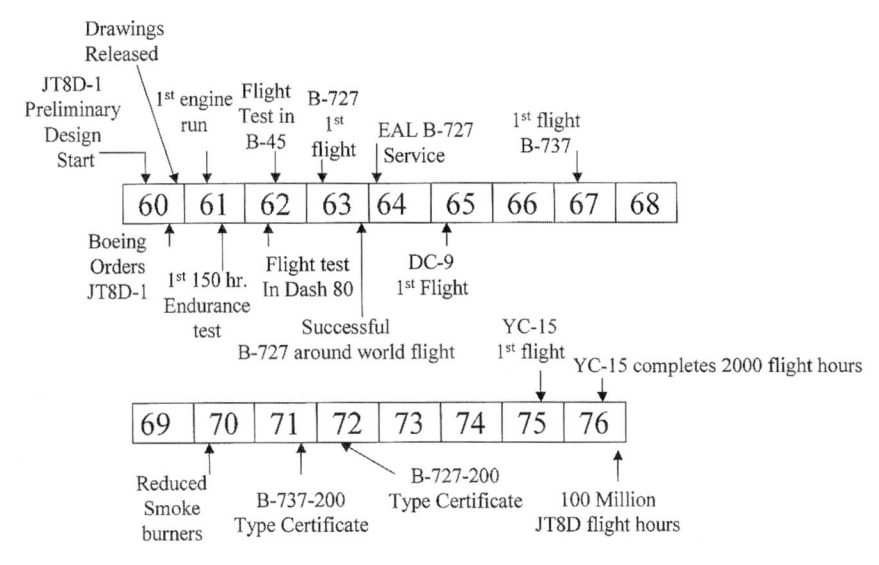

Fig. 12 Time line for the JT8D.

$75 million budget for a five-year time period (Fig. 12). Beckwith managed to have the engine certified about midway through that period and was within his budget at that point.

ENGINE DESCRIPTION

An engine in commercial service accumulates engine operating hours at the rate of 3000–4000 hr/yr. On the other hand, an engine in military applications accumulates about 300–500 engine-operating hr/yr. Low-cycle fatigue concerned Boeing. Although the JT8D low spool was new, the high spool came from the military J52 (JT8) with a good reputation (Fig. 13).

The fan section consists of two stages as in the JT3D and JTF10A engines. The low compressor, which is on the same low shaft as the fan, has four stages and is driven by the three-stage low-pressure turbine. The high-pressure compressor is made up of seven stages and is driven by a single-stage high-pressure turbine. The combustion chamber is a can-annular configuration with nine cans. The fan stream and the core engine exhaust into a common tailpipe and jet nozzle.

Figure 14 shows the comparison of the JT8D to the JT3D (Chapter 9). The JT8D was designed from scratch to become what Beckwith named "the airline man's engine" for its ease of maintenance. The JT3D had some limitations, as it was built upon an existing JT3 engine designed in 1949. Both engines share the same fan transonic aerodynamics. However, the compressor design of the JT8 (Fig. 15) was much more advanced as is apparent from the fact that the JT8D achieves an 18:1 overall pressure ratio from 13 stages (2 fan, 4 low compressor, and 7 high compressor) while the JT3D has

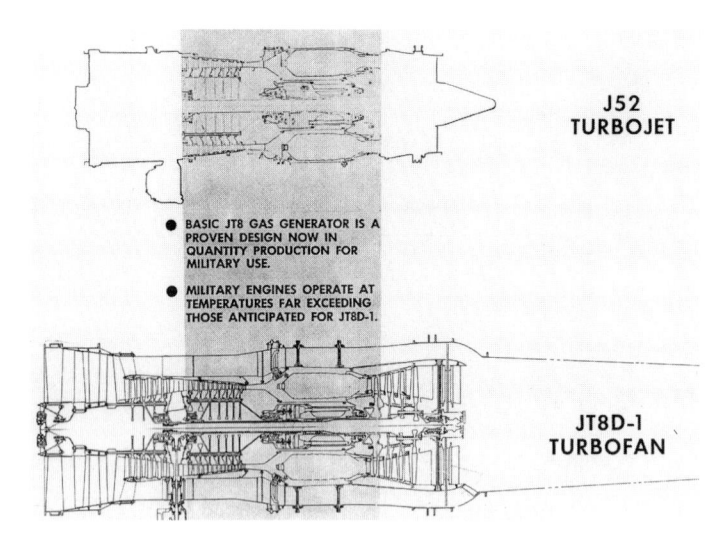

Fig. 13   Origin of the JT8D engine (courtesy Pratt & Whitney Archives).

15 stages (2 fan, 6 low compressor, and 7 high compressor) for an overall pressure ratio of 12:1.

About three months into development, the aluminum inlet case was replaced with a titanium structure. Pratt & Whitney Engineering felt confident that the JT8D would be a great engine and it turned out their expectations were realized: The JT8D was Pratt & Whitney's most vigorous program involving more than 100 airlines (Table 2), with about 14,000 standard JT8Ds and 2900 JT8D-200s produced. The added engine weight was forgotten because of the aircraft's need for more thrust than was required in the original thrust specification—a trend that engine manufacturers run into only too frequently.

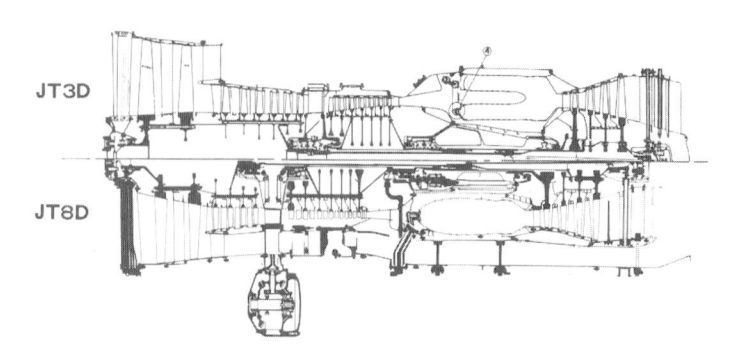

Fig. 14   Comparison of JT3D and JT8D turbofans (courtesy of Pratt & Whitney Archives).

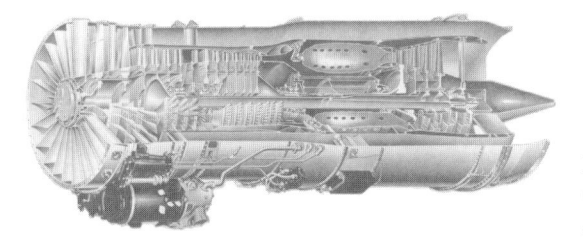

Fig. 15 Cutaway of the JT8D (courtesy of Pratt & Whitney Archives).

COMMERCIAL APPLICATIONS

There were three major commercial applications: the DC-9, the B-727, and the B-737. Douglas and Boeing competed with vigor to dominate the U.S. market. In addition to the U.S. airframers, there were two other applications in Europe: the Mercure and the Caravelle Super B (Fig. 16). The Douglas DC-9 with two fuselage-mounted JT8Ds went into commercial service with Delta Airlines at the end of 1965. The DC-9-50 is about 20 ft longer in fuselage length. It used the JT8D-17 model engine.

Engine design started in April 1960 and the B-727 made its first flight in early 1963—just three short years. Toward the end of 1963 Boeing took a 727 around the world on an ambitious marketing adventure. But it was an adventure just to get the engine in flight. The center engine received its airflow from the S-duct. Company engineers were concerned with the degree of inlet distortion (non-uniform total pressure profile) that the engine would have to ingest (Fig. 17). On the first flight of the B-727, the center engine encountered stall. Jack Steiner, 727 Program Manager, did not hesitate to call Beckwith late at night to inform him in forceful tonality that the center engine stalled on take-off. Beckwith, slowly coming out his nightly slumber responded, "Your damn inlet stalled!" The solution to this situation was the installation of vortex generators strategically located in the S-duct.

In 1959 Lord Douglas was chairman of British European Airways. He was looking at the three-engine Trident and the Boeing 727, which looked somewhat alike except for the size (727 was larger). He suggested that the two companies get together on a joint development [3]. Hawker Siddeley sent a group of engineers to Seattle and Boeing sent a group to England to look

Fig. 16 The French Super Caravelle (courtesy of Pratt & Whitney Archives).

TABLE 2 JT8D Engine Specifications and Applications

Model	Take-off thrust (lb)	Military or max cont. thrust (lb)	Normal thrust (lb)	Weight (lb)	Diameter (in.)	Length (in.)	Aircraft
JT8D-1	14,000	12,600	11,400	3155	44.9	123.6	B-727-100 DC-9-10, 0F * 30 SUD Caravelle
JT8D-1A	14,000	12,600	11,400	3155	44.9	123.6	Same as –1 but with higher EGT limits
JT8D-2		14,500	13,000	3200			B-52 & B-47 re-engine studies
JT8D-3	14,000 wet to 90°F	12,600	11,400	3025	44.9	123.6	B-727 for A.A.L.
JT8D-5	12,250 to 90°F	12,250	10,250	3155	44.9	123.6	DC-9
JT8D-7	14,000 to 84°F	12,600	11,400	3155	44.9	123.6	727, 737, DC-9, Caravelle
JT8D-7A	14,000 to 84°F	12,600	11,400	3155	44.9	123.6	Same as –7 but with smokeless burner
JT8D-9	14,500 to 84°F	12,600	11,400	3217	44.9	123.6	727-200 737. DC-9-30 & 40 Caravelle
JT8D-11	15,000 to 84°F	12,600	11,400	3309	44.9	123.6	B-727-200 DC-9-30,40
JT8D-15	15,500 to 84°F	13,750	12,400	3309	44.9	123.6	Adv. B-737 Dassault Mercure
JT8D-17	16,000 to 84°F	14,200	12,800	3330	44.9	123.6	Adv. 727, 737
JT8D-22	26,000 A/B	14,750	12,600	4260	42.5	236.0	Saab AJ37 Viggen
JT8D-23	27,600 A/B						Uprated –22 for SAAB fighter studies

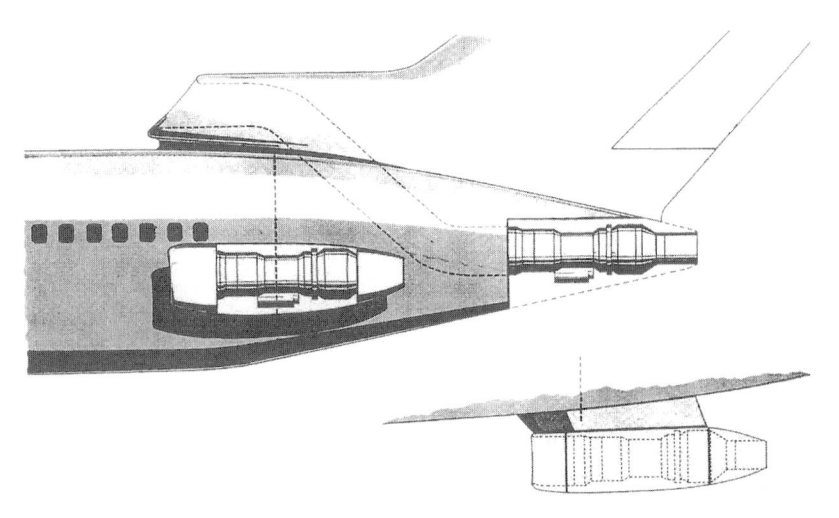

Fig. 17 The installation of the center engine was of great concern (courtesy of Pratt & Whitney Archives).

over the Trident. Boeing declined the joint venture. However, what both engineering groups had in common was the belief that the S-duct for the middle engine would not be a severe problem. This 727 program went ahead in 1960 with the JT8D. The aircraft went into commercial service at Eastern Airlines and United Air Lines in February 1964, and 727-200 became the most advanced and the most popular model of the 727. There were over 1800 B-727s built. More than two-thirds of them were the 727-200 model (Fig. 18), which was about 20 ft longer than the initial model.

Boeing came back again with another aircraft, the short-range B-737, which had its first flight in early 1967. The B-737 rounded out a great family of aircraft for Boeing—short range, intermediate range, and long range. Boeing announced its go-ahead in February of 1965. It would use the same engine that was in the very successful B-727. The more popular model became the 737-200, which entered service with United in 1968 (Fig. 19). The B-737 grew into the B-737-300 model in 1984 with high-bypass fans

Fig. 18 The B-727-200 (courtesy of Pratt & Whitney Archives).

Fig. 19 JT8D installed in a United B-737 (courtesy of Pratt & Whitney Archives).

(CFM56) replacing the JT8D engines. The use of the same fuselage diameter as the B-707 (for six-abreast tourist seating) in the 727 and the 737 saved considerable tooling costs. This 737 has become the world's most popular aircraft with about 3000 in service.

GOVERNMENT REGULATIONS ON ENGINE EXHAUST SMOKE

The federal government issued regulations for the elimination of smoke from aircraft engines. In the 1970s Gordon Beckwith told me it took about 100 redesigns of the burner and numerous flight tests to get to the final configuration. Of course the reaction was great for military applications of jet engines because the enemy could no longer see the aircraft coming from a long trail of smoke. On the other hand, the commercial pilots commented

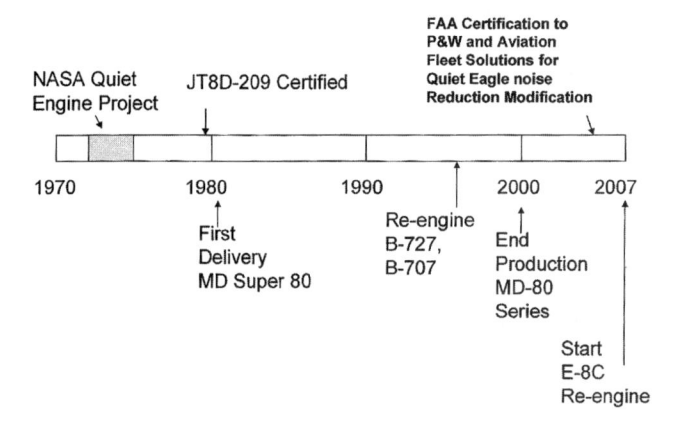

Fig. 20 Time line for the JT8D-200.

GOING COMMERCIAL 359

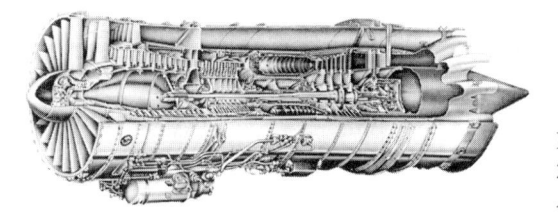

Fig. 21 Cutaway of the JT8D-209 (courtesy of Pratt & Whitney Archives).

that smoke trails were good from a safety viewpoint because they were able to track the air traffic more easily when there was smoke. You can't make everybody happy.

JT8D IS REFANNED AS JT8D-200

Pratt & Whitney participated with NASA in the Quiet Engine Project from 1972–1975 (Fig. 20). The JT8 core was used with a new larger fan resulting in a higher bypass ratio for the JT8D-109. The flight-testing of the engine in a modified DC-9-30 was promising enough to warrant the design of a higher bypass ratio fan. The basic JT8D had a bypass ratio of about 1.0 and the newer design, the JT8-209 (Fig. 21), had a bypass ratio of 1.78. The JT8D-200 (Fig. 22 and Table 3), like the earlier JT8Ds, has an annular duct surrounding the core engine; however, this engine adds a mixer to reduce the noise level from the core-engine exhaust.

One Dash 209 engine first flew as one of the four engines in a Douglas YC-15 in March 1977 for a total of about 50 hr. The Dash 209 was certified in June 1979, and it is the engine in the Super 80 DC-9 with about a 12-percent reduction in specific fuel consumption compared to the JT8D-17. Table 4 compares the refanned JT8D-200 to the original JT8D. Both engines have an

Fig. 22 JT8D-200 (courtesy of Pratt & Whitney Archives).

TABLE 3   JT8D-200 ENGINE SPECIFICATIONS AND APPLICATIONS

Model	Take-off thrust (lb)	Weight (lb)	Diameter (in.)	Length (in.)	Aircraft
JT8D-209	18,500	4410	54.0	168.6	MD-81
JT8D-217	20,850	4430	54.0	168.6	MD-82/87/88
JT8D-219	21,700	4634	54.0	168.6	MD-83

automatic 1000-lb increase in thrust should the other engine in the DC-9 lose thrust on take-off.

REDUCING NOISE

The standard JT8D uses a *hush kit* to bring its noise generation into line with regulations. Noise-level theory says that noise level is proportional to the eighth power of jet velocity. Therefore, the quickest way to reduce jet noise level is to mix the slower fan bypass air with the higher jet velocity of the turbine discharge air. About 77 percent of the JT8D-powered fleets have converted to these kits for noise reduction. The JT8D exhaust mixer, shown in Fig. 23, "averages" through mixing the exhaust velocity of the bypass duct stream and that from the low turbine discharge.

The engine has been FAA-certified in the QuietEagle configuration to meet the FAR Part 36 Stage 4 noise standards and the International Civil Aviation Organization's noise standards. The standard Eagle system reduces the noise level by about 6 decibels. The engine modification for the noise reduction system is estimated to take about 160 man-hours and can be installed with the engines on the aircraft. The system was developed by Pratt & Whitney and Aviation Fleet Solutions. This system does not cause any loss of engine performance.

TABLE 4   COMPARISON OF JT8D-17 AND JT8D-209

Engine	JT8D-17	JT8D-209
Thrust, sea-level static (lb)	16,400	18,500
Bypass ratio	1.0	1.78
Overall pressure ratio	17.5	17.4
Fan stages	2	1
Low-compressor stages	4	6
High-compressor stages	7	7
High-turbine stages	1	1
Low-turbine stages	3	3
Exhaust nozzle	Fixed common exhaust	Fixed common exhaust with mixer

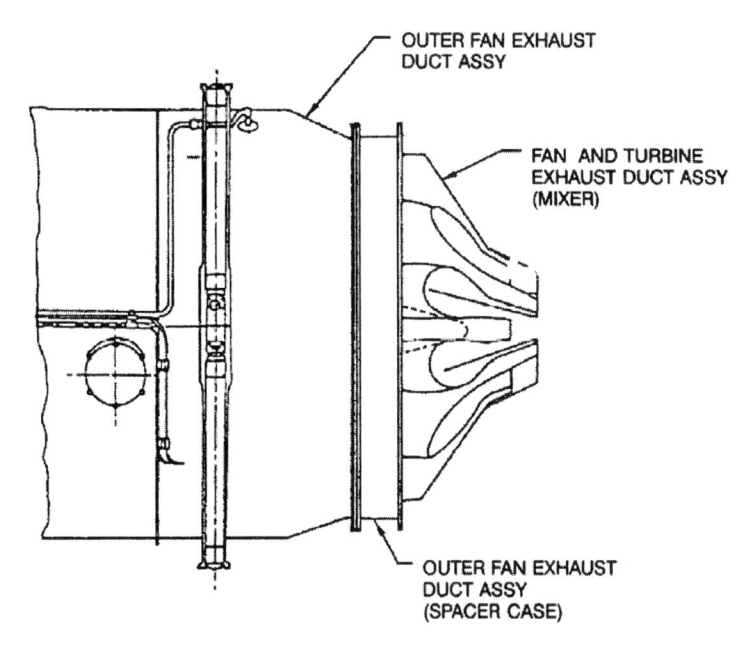

Fig. 23 The JT8D exhaust mixer, which "averages" through mixing the exhaust velocity of the bypass duct stream and that from the low turbine discharge (courtesy of Pratt & Whitney Archives).

INFLIGHT SHUTDOWN RATE

Figure 24 shows the trend in improving the inflight shutdown rate, which set a new standard for engine reliability. The arithmetic of an inflight shutdown of 0.02 implies the need to shutdown an engine in flight will happen about once in 17 years! This is a remarkable statistic when one considers the number of moving parts in an engine. The reliability of aircraft engines today is such that the traveling public does not give a thought to flying across the Atlantic in a two-engine aircraft.

MILITARY APPLICATIONS

The Saab-37 Viggen was the only military application of an afterburning JT8D. Saab engineers did some design and development work on the afterburning version of the engine with help from Pratt & Whitney. Saab tried to market the engine to the NATO nations, Japan, Australia, and India, but was not successful—not because of aircraft performance but restrictions from their own government [2, p. 182]. Saab built about 330 aircraft. The RM 8 had a sea-level maximum thrust of 26,500 lb.

In the United States, outside of the military use of commercial aircraft adapted for special military applications, such as executive transports or air-ambulances, there were no extensive military applications for the JT8D.

There was one gallant try though for the Advanced Medium STOL Transport (AMST). McDonnell Douglas, along with Boeing in 1972, won contracts to demonstrate the operation of a STOL transport aircraft. Douglas proposed the JT8D for its YC-15 demonstrator aircraft (Fig. 25). Its first flight was in mid-1975. The program died in 1979 when funding was cut. There was no production fallout from this successful demonstrator program. It did serve, however, as a great promotion for the CFM56 later, prior to the C-17 program. The CFM56 was mounted on the left outer position and given a test flight to altitude with the comforting feeling of security that goes with three JT8D-17 engines operating on the same aircraft.

On August 28, 1976, a YC-15 flew to the east coast and then on the 31st flew to Mildenhall Air Base for the Farnborough Air Show, a 12-hr non-stop flight. After the air show it demonstrated its capability in the United Kingdom and later in Germany. When the aircraft returned to Long Beach on September 29 it could boast of another 50+ flight hr. Later in November it visited about a half dozen airbases in the United States to provide flight demonstrations to guest pilots on handling the aircraft as well showing its cargo carrying capability. After all this marketing effort the aircraft had put on about 80+ flight hr (without engine problems) since it left Long Beach in August [4].

Boeing took over the McDonnell-Douglas aircraft programs in 1997. No further orders for the aircraft developed, so the MD-80 series of 1192 aircraft went out of production in 2000. The production line was re-activated in 2007 to produce JT8D-200s for the U.S. Air Force Joint STARS program in the E-8C aircraft. This replaces the old TF33 engines on the Boeing 707 aircraft. The JT8D-200 engines (designated as the JT8D-210 model) gives the aircraft

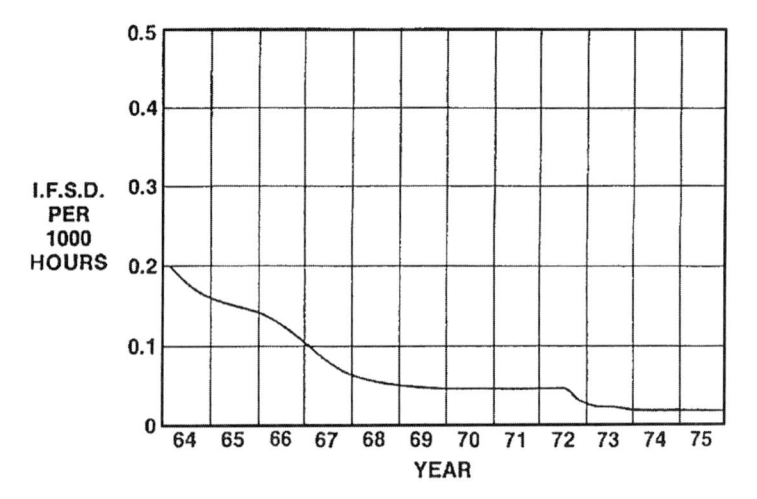

Fig. 24 The inflight shutdown rate for the JT8D (courtesy of Pratt & Whitney Archives).

Fig. 25 The Douglas AMST demonstrator aircraft was powered by four JT8D-17 engines (courtesy of Pratt & Whitney Archives).

greater range, payload capability, and lower maintenance cost than the older engines. The 17 aircraft in this Northrop Grumman program are referred to as the Joint Surveillance Target Attack Radar System (STARS).

JT8D-POWERED B-727-200 HIJACKED

Sometimes an airline pilot has a problem greater than any concern about engine reliability. American Airlines Captain Al Mitchell, with more than 20,000 flight hr with Pratt & Whitney engines, was bringing his 727-200 towards Chicago from the east over Flint, Michigan when the stewardess unexpectedly opened the cockpit door. Right behind the stewardess was a hijacker who threatened to detonate the two bombs he was carrying if Al did not turn the 727 around and take him to Brazil. Al, a former United Technologies employee, gives us the rest of the story in his words:

> It was obvious that we could not fly to Brazil in the 727. The principles of aeronautical engineering guaranteed that the 727 with its current load of fuel and payload could not reach such a long-range destination. The first thing we had to do is to refuel and work out a plan of action to save the aircraft and all of its occupants (except perhaps the hijacker) from any harm.
> The situation grew grimmer by the minute. Chicago was alerted and the full security team was now in operation—police, airport security, the FBI, sharpshooters, fire trucks, and ambulances—all in the ready position. I brought the plane down and then while on the runway I talked to the American Airlines people who had to respond to the demands of the hijacker. We got the passengers off the aircraft safely (but without their baggage!) and then admitted a hippie lawyer, who was going to help the hijacker negotiate with American Airlines.
> The hippie lawyer advised the hijacker to get to Ireland because then he could avoid extradition and from there could arrange a flight to Brazil from Ireland. Canada and Mexico were not options according to the lawyer. So now I made arrangements with American management to fly this hijacker and his lawyer to JFK where a Boeing 707 would be ready (with another flight crew) to take the kooks to Shannon Airport in Ireland.

I focused on the immediate objective of getting fuel at Chicago—with the JT8Ds still running (ordinarily a no-no, of course!). In fact the engines were on idle for about 8 hr while on the ground in Chicago. Then I maneuvered the airplane toward the runway for take-off. The Tower told me I could not take off until there was an FOD inspection on the runway. I, of course, had more important considerations at the moment and runway debris was not one of them. So I took off and one hour plus 58 minutes later (an unofficial speed record for airlines) I set the beautiful 727-200 down gently on the JFK runway and pulled along side of the 707 to get the hijacker and his quack lawyer off this plane. I was "passing the baton" on to the unfortunate crew of the 707. All's well that ends well—lives of the crew and passengers and American's expensive aeronautical asset, the 727-200.

What is interesting to an engineer about this story, aside from its practical example of resolving a potential disastrous situation successfully, is that the JT8Ds were on idle for 8 hr and then were pushed up to maximum cruise power for a couple of more hours. Because of the situation, refueling and take-off required standard operating procedures (SOPs) were ignored. However, as expected and as the situation demanded, the engines were up to the job.

40 YEARS IN SERVICE

The JT8D and its derivative the JT8D-200 powered about 25 percent of the world's aircraft by 1999 (Fig. 26). The following is a summary of the JT8D history after 40 years of service:

- JT8D (standard and refanned) is first aircraft engine to reach 0.5 billion hr.
- 25 percent of the world fleet is JT8D-powered.

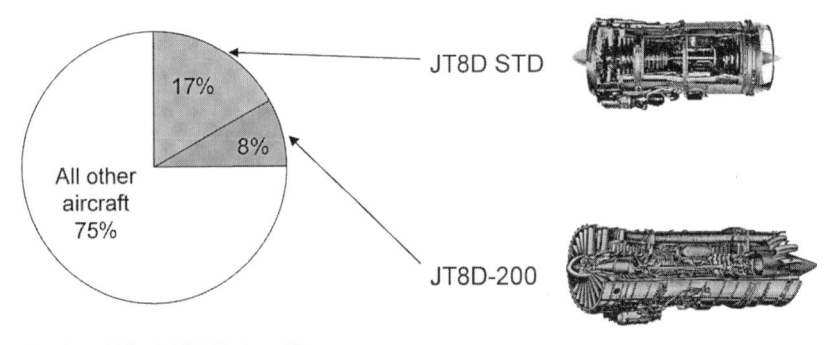

Total world fleet 16,645 aircraft*
  • Over 14,700 JT8Ds Produced
  • More Than 400 Customers
* Includes Government and regional jets

Fig. 26 Achievement of the JT8D (courtesy of Pratt & Whitney Archives).

Fig. 27   Gordon Beckwith.

- JT8D powers the MD-80 (1181), DC-9 (783), B-727 (1306), and B-737 (923).
- Over 14,700 JT8Ds produced and 73 percent are still flying.
- JT8D engines have flown the equivalent of 10 million trips around the world.

It would be impossible to sum up of the history of the JT8D without shining the spotlight on the original project engineer, Gordon Beckwith (Fig. 27). He later went onto become Assistant Chief Engineer and then Western Region Marketing Manager, whose clients included airframers Boeing and Douglas and airlines American, United, Braniff, Continental and Pacific Southwest.

On the occasion of the 40th anniversary of the program, the JT8D Development Team presented Beckwith with an award of appreciation inscribed with the words "his perseverance, and energetic and Inspirational Leadership, resulted in one of the most successful programs in Pratt & Whitney history." Dick Baseler, who was Beckwith's boss during the JT8D development, wrote Gordon in 1984:

> You and I, and perhaps a few others some of whom were not able to be there, remember well the "blood, sweat and tears" of getting the original 8D launched and sold and the unquestioned success of the program is certainly primarily due to your efforts and expertise as Program Manager (or whatever we called you in those days) and to your continued support of that and other programs throughout your career at the aircraft.

Yet when I discussed the JT8D history with him in 2005, Beckwith wanted me to know that the success of the engine was due to the efforts of hundreds of Pratt & Whitney engineers involved in the design, development, production, and product support, as well as the efforts of airline engineers and maintenance personnel who contributed valuable input to create the "Airline Man's Engine."

REFERENCES

[1] St. Peter, J., *The History of Aircraft Gas Turbine Engine Development in the United States: A Tradition of Excellence*, ASME, New York, 1999, p. 310.
[2] Donald, D., *The Complete Encyclopedia of World Aircraft*, Barnes & Noble, Inc., Lyndhurst, NJ, 1997.
[3] Serling, R. J., *Legend and Legacy: The Story of Boeing and Its People*, St. Martin's Press, New York, 1992, p. 182.
[4] Norton, B., *STOL Progenitors: The Technology Path to a Large STOL Aircraft and the C-17A*, Library of Flight, AIAA, Reston, VA, 2002, p. 100.

Chapter 12

CHALLENGES AND NEW TURBOFANS

INTRODUCTION

The early 1960s brought on some significant pressures to Pratt & Whitney's gas turbine engine business. The first was an effort to seek markets in space technology, and we discuss the company's first space technology product, a fuel cell for NASA in this chapter. A second pressure was the changing nature of the aircraft engine business, where lower-level personnel in the government were having more input about the selection of an engine than before. A third pressure was the escalation of the military action in Vietnam, which demanded an instant response to shortages of spare parts. The escalation also brought on the need for a more modern fighter aircraft, which turned out to be the F-15 built by McDonnell. We will explore these pressures as we discuss Pratt & Whitney's effort to win the Air Force contract.

SUPPORTING ACTIVITIES FOR CHANGING TIMES

Pratt & Whitney had always conducted engine installation and aircraft performance studies to get a better understanding of the airframers' challenges. In the piston engine era, one engine had many potential applications and an engine company could look forward to a lucrative production run. The same situation, to a lesser extent, was true when aircraft gas turbines came into use after WWII. But when the space age arrived, it was not clear that there would be such a thing as a long production run. Bill Gorton at the Florida Research & Development Center asked Gene Montany's Advanced Planning Group in East Hartford for help sorting out the business opportunities in the space programs. Montany, with some help from Joe Sabatella, described their studies in the fields of space propulsion, space power, and high-speed aircraft propulsion in the early 1960s, with a paper titled *SPOPS, SPOPS II, and SHEPPS*, which Montany gave to me in April 2004. The paper is reprinted here with his permission.

SPACE PROPULSION

In the early sixties when Advanced Planning was being initiated at the management level at Pratt & Whitney under my direction, I received a call from Bill Gorton, then General Manager at FRDC. Bill was concerned that opportunities in space propulsion systems would not provide the kind of production programs consistent with the "Pratt & Whitney Model".... Bill asked me to come to FRDC to set up a study group to address this concern ... I offered him the option of my staying in East Hartford [to run the] planning, and setting up a study group to address his concerns. Bill accepted this compromise and obtained both permission and money from Len Mallet [Pratt & Whitney President]. I formed the new study group in June 1963, comprising Stan Greenberg, Harry Robinson, and Ed Quinn from Engineering under the direction of Joe Sabatella; and Frank Tomlinson, Norm Turkowitz, Larry Moran, and Frank Greene from Installation Engineering under the direction of Charlie Brame.

The first effort was focused on space propulsion. The title of the project was Space Propulsion Opportunity Planning Study (SPOPS). The study was completed in 1965, with final reports issued in February and August, confirming the accuracy of Bill's gut feelings about the small production potential.

In addition to ... technical characteristics of space propulsion systems over the next 15–20 years, the team studied other aspects of space operations and technical advances ... including guidance systems and on-board systems. Three possible space operations models were postulated and the differing requirements for space propulsion systems were analyzed.... The data ... [allowed] very quick interpretation of new ... requirements to be quickly analyzed.... This assured the timeliness of the study results ... [which] were constrained by ... our economic models to hold expenditures to a constant percentage of GNP.

At the time of the study, the RL-10 and the North American Aviation J-2 engines existed. This left the field for new engines ... severely limited. The opportunities ... seemed to be limited to two ...: a new larger RL-10 type of 40,000-lb thrust or so, and a new high-pressure engine to be associated with a new undefined manned space transportation system. R&D opportunities were continuing for nuclear and other chemical and non-chemical systems with no clear requirements....

Space Power

Even with the sobering results of the SPOPS study, management found the study useful for deploying company resources. In fact, they were so enthused [that] they extended the budget and called for the formation of a study group to do a similar study on the opportunities for space power systems with the title of Space Power Opportunity Planning Study (SPOPS II). As project leader again, I immediately formed the new group and final reports we issued in February and August of 1967.

In this study Joe Sabatella led his team of Dave Kleinhen, Ed Quinn, and Dale Schenk from Engineering. Charlie Brame's team from Installation Engineering consisted of Rusty Fancher and Larry Moran. Jack Connors had a team from Advanced Power Systems made up of John Gerstle, W. Kennedy, Bob Lombardi, and C. Skorski. Advanced Planning assigned Stan Greenberg and Ron d'Arcy to the study effort.

The basis for this study was the SPOPS mission model updated. This model showed 60 programs ... which would require one or more power systems.... One hundred twenty (120) applications were defined, not including launch vehicle requirements. Detailed specifications for each for these applications were derived.

Each of the power systems were arrayed against the candidate systems determined to be ... attractive for each application.... The projected capabilities of each system ... were defined. Potential systems studied included batteries, fuel cells, photovoltaic, thermoelectric, thermionic, magneto-hydrodynamics, Brayton cycle, Rankine cycle, solar, and nuclear reactor.

The advanced manned re-entry class of vehicles will have the most prolific launch rates among manned spacecraft.... These vehicles will typically have power levels of several kilowatts with peak up to 20 kilowatts and mission durations of one week or so.... To capture this important market, fuel cells should be promoted as the best means of supplying all electrical power. It is essential to develop fuel cell systems which provide low specific weight and closely approach the ruggedness, reliability, and flexibility of batteries. Fuel cells should also be promoted as a means of providing potable water, cooling and/or attitude control by making effective use of the exhaust products.

The essential, actionable, conclusion of the study was stated in the final report as follows:

> The advanced manned re-entry class of vehicles will have the most prolific launch rates among manned spacecraft during the late 1970s and beyond. These vehicles will typically have power levels of several kilowatts with peak up to 20 kilowatts and mission durations of one week or so. In order to capture this important market, fuel cells should be promoted as the best means of supplying all electrical power. It is essential to develop fuel cell systems, which provide low specific weight and closely approach the ruggedness, reliability, and flexibility of batteries. Fuel cells should also be promoted as a means of providing potable water, cooling, and/or attitude control by making effective use of the exhaust products.

HIGH SPEED AIRCRAFT PROPULSION

This study ... by Joe Sabatella was of particular importance ... throughout the aircraft and propulsion community ... because of its pioneering work in conceiving and developing the concept of regression analysis to bring the workload of multi-parametric studies down to manageable size. I always like to characterize this as "statistical inference tool" as it allowed one to develop a full matrix of performance numbers while requiring a precise calculation of only a small number of points. Joe Sabatella describes, in the following pages, the Supersonic/Hypersonic Engine Performance and Planning Study (SHEPPS).

In early 1968, Pratt & Whitney's Advanced Planning Department kicked off the SHEPPS study under the leadership of Gene Montany. At that time, the British/French Concorde and Boeing/GE B-2707 were being actively developed as prototypes of a commercial supersonic transport aircraft (SST) and Pratt & Whitney was left out of a potentially large area of the future air transport market.

SHEPPS was undertaken to investigate the spectrum of Mach 2.7-12 airbreathing engine opportunities. Most of the analytical work was performed by personnel in the Advanced Propulsion Systems and Vehicle Systems Analysis sections of the New Products Development Group in Engineering.

SHEPPS was divided into two phases. Phase I dealt with pure cruise transport, reconnaissance and interceptor aircraft in the Mach 3-12 regime, and provided a valuable background for the Phase II part of the study. Phase II of the study received substantially greater emphasis and covered Mach 2.7-4 transport aircraft.

A primary objective of this study was to determine the most advantageous cruise speeds and fuel types for transport aircraft. Another key objective was the identification of optimum engines and cycle parameters for SSTs. In the process, a considerable amount of insight was gained into areas such as vehicle configurations, sonic boom, airline economic considerations, engine noise, and operational benefits of SSTs.

The study of Mach 3–4 SSTs involved many parameters pertaining to engine and vehicle characteristics as well as economic aspects. Thus there were thousands of combinations to consider. Computer programs and statistical methods were utilized to reduce the workload to manageable proportions without unduly compromising the level of sophistication of the individual analyses. One of the key advantages of this approach is that many alternatives could be compared on a consistent basis in order to identify their relative advantages and disadvantages. One of the methodologies (regression analysis) developed and used in SHEPPS became a valuable tool in many future studies throughout the aerospace industry.

In addition to Mach number, the study included variations in mission profiles (e.g. loiter time, range, etc.), fuselage shaping (for sonic boom), variable vs fixed sweep aircraft, JP vs methane fuel, engine type (turbojets/turbofans, pre-cooled turbojets, turboramjets) and engine cycle (bypass ratio, overall pressure ratio, and turbine inlet temperature). Aircraft TOGW [take-off gross weight], DOC [Direct Operating Cost], and ROI [Return on Investment] were used as figures of merit. A computer simulation program was used to determine aircraft utilization as a function of aircraft speed and range.

A summary of the study results is presented here along with a description of the regression analysis methodology used in the engine cycle analysis.

STUDY SUMMARY AND CONCLUSIONS

- SSTs with cruise speeds of Mach 2.7–3 can make almost twice as many flights per year as subsonic jets. But, since mission profiles include an acceleration and deceleration phase, the block time of SSTs does not improve significantly at speeds higher than Mach 3 for practical airline routes.
- There does not appear to be a practical way of significantly reducing sonic boom through either aircraft shaping or flying higher than altitudes for best fuel consumption.
- Although methane fuel offered small TOGW benefits, the operational disadvantages probably offset it.
- The benefit of methane increased as Mach number increased. Methane pre-cooled turbojets and turboramjets had no advantage over conventional methane turbojet cycles.
- The JP turbojet engine cycle showed a small advantage in the baseline mission, while the turbofan had the advantage as loiter time increased. Engine noise would tend to favor the turbofan.

REGRESSION ANALYSIS METHODOLOGY

To cope with the large quantity of computations involved in the study, regression analysis was employed to reduce the time and cost of data computation. Statistical analysis methodologies (regression analysis) had been widely used in engineering disciplines such as experimental design layout and materials development, but to our knowledge, this was the first time it was employed in an analytical study to optimize engine cycles. In SHEPPS, engine bypass ratio, overall pressure ratio, and turbine inlet temperature were chosen as cycle variables at each of three Mach numbers. Conventional analysis would require 3 times 81 = 243 engine cycles to be evaluated, but the use of regression analysis reduced the number to 3 times 11 = 33 engine cycles for a very significant cost reduction.

In the mid-1970s, McDonnell Douglas, Boeing, and the U.S. Air Force had also developed an interest in the application of regression analysis to engine cycle optimization in order to reduce the cost of engine data generation and to provide an interactive means of cycle optimization as mission and/or aircraft parameters changed. The experience gained using this methodology in SHEPPS allowed Pratt & Whitney to win the "Regression Simulation of Turbine Engine Performance" (RSTEP study contract from the Air Force Aero Propulsion Laboratory in 1977). The purpose of this study was to further develop the methodology and improve accuracy.

The RSTEP study led to significant improvements in the regression analysis methodology with techniques such as transformation of variables, variable role reversal, and pattern selection. Pratt & Whitney's performance in the RSTEP study led to its being selected as the lead team (with aircraft data support from several aircraft companies) in other USAF sponsored studies that eventually led to the F119 engine cycle.

Since SHEPPS, regression analysis methodology has been used by Pratt & Whitney in numerous studies, both in-house and under contract, rocket engine optimization was another significant application. Under the NASA-sponsored Space Transportation Booster Engine (STBE) study, regression analysis was used in two ways: 1) in characterizing the trajectory characteristics in terms of ideal velocity as a function of engine and vehicle parameters, and 2) engine variables such as area ratio, chamber pressure, etc.

New Possibilities

Montany's paper gives us an idea about what possibilities lay ahead for the company. The Engineering Department provided lectures on space subjects with films put out by the Jet Propulsion Laboratory in California. All of a sudden, the new future appeared to be in space technology. While engineers were fascinated by aerospace, hard-nosed businessmen were concerned about the real market potential. On a flight to Chicago one day, I happened to sit next to a man who was with the Chicago Bridge & Iron Company. He was talking about how low tech his efforts were and how his company would probably not make much money in the space business, and neither would the high tech hopes of engineers be fulfilled. Years later I read that the Chicago

Fig. 1 A nighttime scene at the Hartford Graduate Center (RPI Extension) in which engineers fight to stay awake (courtesy of Pratt & Whitney Archives).

Fig. 2 Bill Podolny (courtesy of Pratt & Whitney Archives).

Bridge & Iron Company cleaned up in the space age by constructing huge facilities at launch sites. Back then all of us in aviation were just feeling our way regarding the new possibilities.

The United Aircraft president, H. Mansfield Horner, established a branch of Rensselaer Polytechnic Institute in the Hartford area, in East Windsor. Dr. Warren Stoker was the Dean who managed the graduate school. He recruited some personnel from Pratt & Whitney. I became an Adjunct Associate Professor of Mechanical Engineering in the new facility teaching a course called "Gas Turbines & Jet Cycles." From 1958 to 1962, I had a class size of about 25 students. This was a great experience for me because I had to scramble to keep more than one class ahead of the students (Fig. 1).

DEVELOPING SPACE TECHNOLOGY

PRATT & WHITNEY'S EXPANDING ROLE IN ENERGY CONVERSION

Walt Doll, as head of the Technical & Research Group, was asked to seek out new markets for Pratt & Whitney. The company was in the business of converting chemical energy in hydrocarbon fuel into shaft power and thrust power, so it made sense to seek other energy conversion opportunities. Walt assigned the mission to Bill Podolny (Fig. 2), who was the innovative spark plug in the compressor group. He was uniquely qualified because not only was he a brilliant engineer, he also had a firm grasp of the business aspects of new ventures.

Podolny looked into a long list of potential applications for future business. He explored solid-state physics, thermoelectrics, thermionics, and fuel cells, among other possibilities. In the early 1960s, he mentioned to me the possibility of a laser printer (long before Hewlett Packard brought one to

Fig. 3 Apollo space vehicle's fuel cell (courtesy of International Fuel Cells).

market). He zeroed in on fuel cells, which offered the potential of high efficiency in the production of direct current electrical power.

Podolny convinced the company to get into fuel cells. His first selection was technology from the research of Englishman Francis T. Bacon, who was a descendent of Sir Francis Bacon (a scientist himself). This fuel cell used a high-temperature alkaline electrolyte. Gaseous oxygen and hydrogen flowed to separate electrodes to produce the direct current electrical power and the products of the combination became water. In 1961, Bill demonstrated for NASA a six-cell fuel cell, and drank the water that resulted from the hydrogen and oxygen reaction. Not only could the fuel cell produce electrical power, it also could provide potable water for the astronauts. This demonstration won him a 250-watt fuel cell contract with NASA. Then in 1962, North American Aviation, Inc., Space and Information Division, selected Pratt & Whitney to develop and produce a 1.4-kilowatt fuel cell to provide electricity and water on the Apollo missions (Fig. 3). As a result of Podolny's efforts, International Fuel Cells (a United Technologies Company) is a world leader today in the production of fuel cell powerplants.

Apollo Fuel Cell Project Engineer Don Brendal remembers the fuel cell development well:

> The moon mission could not be carried out without the fuel cell. Batteries ... were too heavy.... North American's concern was [that]

a full-fledged fuel cell had never been developed nor put into production ... and ... expressed great concern that we might not "cut the mustard" and followed our program diligently and relentlessly. I don't believe any of us on the fuel cell program at Pratt & Whitney ever gave a thought that we might not meet all of our program objectives and do our part to go to the moon and back. Solving engineering problems was our bread and butter.

Of course the development did have many problems as would be expected with something as new as the fuel cell ... one problem had us buffaloed for some time. A fuel cell would be running normally on test and then start to diminish in power output with no known cause, just as a dry cell battery in a flashlight toward the end of its life.... Our engineers found that by lifting one side of the fuel cell with a two by four and letting it drop immediately restored it to full power. Of course the cycle would repeat itself. This led us to the cause of the problem which was "dendrites" growing in the electrolyte between anode and cathode causing electrical shorts. Jarring the fuel cell would dislodge them and eliminate shorts for a time. With this knowledge we developed a fix.

The electrolyte used in the fuel cell was extremely corrosive. We used to joke that if a mechanic dropped his pliers in a vat of it they would be dissolved before he could fish them out. We needed a flat o-ring type seal about 10 in. in diameter between anode and cathode to seal in the electrolyte. Teflon was the only suitable material but we had problems with it extruding out of position. DuPont was called in and they ... went to work in their labs and developed a Teflon that did the job and without charge to the program.

So problems arose and were solved. But to North American (and NASA) the problems must have appeared to be never ending. So they enlisted Bell Laboratories, a very prestigious research and development organization, to come to Pratt Whitney and evaluate the fuel cell program. Several engineers and scientists spent ... days with us. We were elated with their conclusion: "We see no cause for concern and we do not have a single thing to suggest that Pratt & Whitney is not already pursuing." So the development continued. We celebrated when for the first time a fuel cell completed a 50-hr endurance test without incident. We continued on with our development and completed a very rigorous and formal 150-hr certification test. This major milestone cleared us to deliver production fuel cells.

Our first production unit was ready for delivery on the contractual date. North American did not want us to ship it because, as we understood it, we were the only subcontractor to be on schedule.... The great concern for the success of the fuel cell program was replaced with praise. They admonished other subs to be like Pratt & Whitney.

The fuel cells performed admirably on the 18 Apollo missions. On the ill-fated Apollo 13, the media for a time erroneously reported that the fuel cells had failed when in actuality the oxygen tank had ruptured thus depriving the fuel cells of this necessary element.

Following Apollo, United Technologies continued to provide more advanced fuel cells for all of the NASA manned space programs and have accumulated over 90,000 hr operation in space.

Pratt & Whitney moved a large number of engineers over to the fuel cell operations. I was one of those and remember the five years that I spent there as a great learning experience. It changed my focus from concentrating on a specific engineering detail to one of standing back to see the big picture. The assignment also involved meeting a large number of potential customers in the military, in the space vehicle business, and at NASA.

Adventures in Marketing

I remember one occasion where I wanted to compare the potential of fuel cells in space with solar cells. One of the engineers at Lockheed Missiles & Space invited me to come out to see him so that the two of us could map out areas where solar cells and fuel cells were applicable in a variety of space applications. I worked through the Pratt & Whitney system to get the appointment set up. That meant a representative from the Marketing Department arranged the meeting for me. On the appointed day the marketing man decided to go along with me. As we walked into the reception area at the Lockheed building for the meeting, I noticed a huge crowd, and asked the receptionist at the sign-in desk what was going on. She fingered through her desk calendar and noted that there was a big presentation this morning by Pratt & Whitney on fuel cells. That was puzzling to me because nobody from the fuel cell operation was scheduled to be at Lockheed that week.

As if on cue, the local Pratt & Whitney customer service representative showed up and sheepishly confessed to a mistake. A memo had gone out to many Lockheed employees, giving the impression Pratt & Whitney was there for a mega fuel cell extravaganza. It was about five minutes to show time when I told customer service representative that this meeting logically belonged to the marketing man, whose field was fuel cells; in the meantime, I would be going to be in a very different meeting. As I glanced at the marketing man who was turning pale, I realized that suddenly the problem was now mine. I drew upon my ancestral background in Ireland—not far from the fabulous Blarney Castle—and years of practice in extemporaneous speaking in Toastmasters International to use persuasive fluency of speech to salvage the situation. I told the group I would not hold them up very long but would give them a brief rundown of what was going on at Pratt & Whitney regarding fuel cells. The whole operation was over in less than 10 minutes, and much to my surprise, the audience appreciated that brevity.

Developing New Engine Technology

Evolution of Sales and Marketing

After WWII, advances in gas turbine and jet propulsion technology propelled the aerospace industry forward. New and increasingly complex possibilities of engine configurations as well as types of aircraft became so confusing that senior management at Pratt & Whitney and its competitors had to rely on lower-level experts for guidance in evaluating business opportunities. As a result, the marketing and sales operations became larger and more complex as teams of people were needed to review potential customers' needs, prepare proposals, and pitch them. Once a proposal was accepted, communications between the company and the customer were complicated by the sheer amount of development effort, time, and labor necessary to create, manufacture, and then test a new engine. We will review two such efforts in this section, the supersonic transport engine program and the F-100 engine proposal. First, let's review the standard procedure for developing and testing an engine. Imagine that you are not in the aviation industry, but instead a member of the public. As such engine development may seem like a mysterious as well as complicated process.

Procedure for Engine Development

Let's designate the Air Force as the customer for a new engine. The Air Force and the airframer specify what the aircraft has to do in terms of mission profile. Then the airframer outlines a profile of what the engine has to do in terms of performance at each of the segments of the aircraft's mission, which is just a first pass at what might be truly required. It is not until the aircraft is test flown for the first time that a pilot discovers what the aircraft can really do. Later on, when the Air Force starts its flying, the military learns how best to use the capabilities of the aircraft. Flight data is gathered throughout testing and feedback directs any redesign (Fig. 4). The net result of that body of

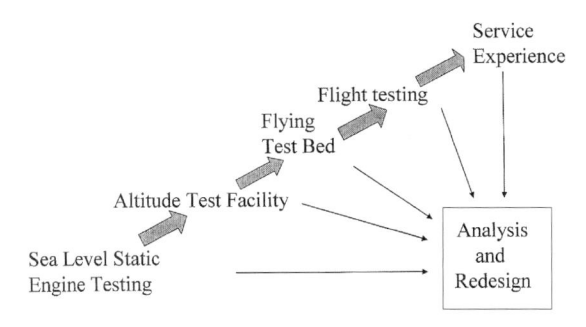

Fig. 4 The design and development process for aircraft engines.

new knowledge could well mean that engine specifications should have been different than what was first estimated.

The development process for aircraft engines starts with sea-level static testing, which can bring into focus engine parts that need to be improved. The next step in the process is to take the engine into an altitude test facility, such as the Arnold Engineering Development Center (AEDC), to evaluate engine performance at altitude conditions. Sometimes the next step of installing the test engine in a flying test bed is used, especially when the engine is more apt to be used in a nacelle type of installation. The real test, of course, is in the aircraft and it is here that problems that did not show up in sea level or at AEDC are likely to quickly become apparent. Engine development is a lengthy process of refining the design in response to problems that are identified in the whole spectrum of the testing process.

The engine is usually made up of eight components, including the controls (Fig. 5). Some of these components interact with each other in ways that cannot always be anticipated in advance of the flight test; for example, movement of the variable nozzle or sudden changes in afterburner temperature feed directly back to the turbomachinery that cause the engine to operate in a regime different from the normal operating regime. An aircraft engine, which can be considered as a dual spool gyroscope, can be moved around in six different ways (Fig. 6). As the engine is moved around during aircraft maneuvers, the spools act as gyroscopes that resist the change by exerting forces on the bearing mounts. This action could cause a rotor to rub the stationary case in some circumstances.

One could design a stationary gas turbine and conduct a test program to determine if any design changes would be necessary to correct performance or durability problems by mounting the engine on an immovable, solid base. On the other hand, if you moved the base through six different types of motion, then one could imagine the test program would require considerably more design iterations compared to the stationary gas turbine. To do so, the engine would have to be mounted to a structure that has more flexibility than the solid concrete base. That condition in the aircraft opens up possibilities for some sympathetic vibrations between the aircraft and the engine under certain circumstances.

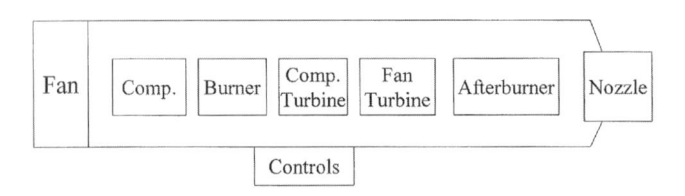

Fig. 5 These eight components in an engine interact with each other.

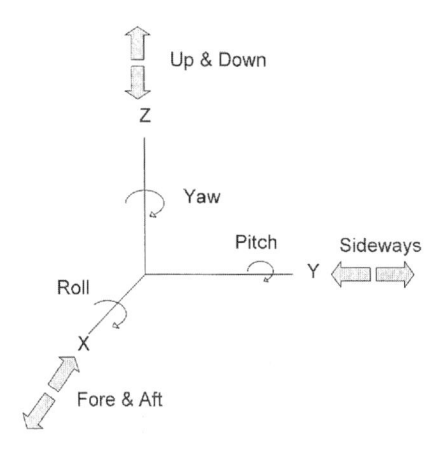

Fig. 6 Aircraft can move the engine in six different ways (the arrows illustrating rotation about XYZ axes can go either way).

There are three fundamental goals for the engine design of a fighter aircraft—high thrust to weight, durability, and stability of operation in all flight conditions. One achieves high thrust to weight by pushing present design limits on stress and aerodynamics. Higher thrust per pound of air comes with higher turbine temperatures. This requires better cooling methods for the areas exposed to higher temperatures. Weight can be reduced by cutting down on the number of stages in the fan, compressor, and turbines. Higher aerodynamic loading on the airfoils could cut down on the number of stages. Running higher rotational speeds is another way of getting more output per turbine stage. Higher strength materials and more advanced manufacturing technology can permit higher stress levels. Running higher stress levels make greater endurance more difficult. One of the reasons parts must be replaced is creep or failure due to stress to rupture. Another reason for parts to be replaced is low cycle fatigue, a failure that first came to light in commercial gas turbine service. It is one level of challenge to fly on the straight and level. The real challenge is to fly during combat maneuvering. This latter could provide extensive inlet distortion of the total pressure profile and lead to engine surge and potential engine shutdown. The task of balancing the goals of high thrust to weight, durability, and operational stability is a daunting engineering challenge. We'll hear about such challenges as we review two programs that Pratt & Whitney bid on.

SUPERSONIC TRANSPORT ENGINE PROGRAM

In January 1964, the Federal Aviation Agency (FAA) published summary statements from the airframe and engine bidders for the U.S. Supersonic Transport Program. The engine bidders were General Electric, Wright Aeronautical, and Pratt & Whitney. General Electric and Wright Aeronautical proposed afterburning turbojet engines while Pratt & Whitney proposed a

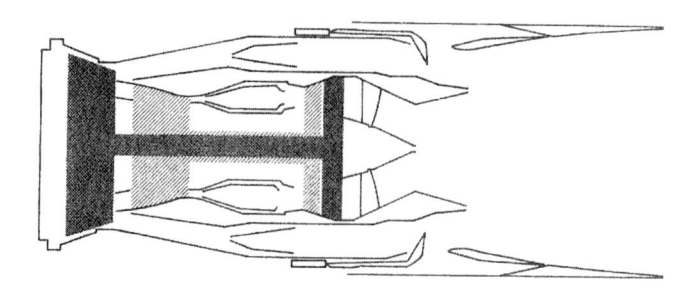

Fig. 7   JTF17 schematic (courtesy of Pratt & Whitney Archives).

duct-burning turbofan. General Electric and Pratt & Whitney went on in the demonstrator phase with their engines. By June, Pratt & Whitney's Florida facility received a $50 million contract from the FAA for construction and test of an engine suitable for a supersonic transport.

In 1965, Boeing and Lockheed executives Maynard Pennell and R. A. Bailey, gave glowing forecasts of at least 500 supersonic transports that the airlines would order by 1985 [1]. Pratt & Whitney was at the ready by designing and building a duct-burning turbofan. The schematic drawing in Fig. 7 shows the two spools and the burner in the duct. This engine, JTF17, was first tested on March 31, 1966.

JTF17 Engine Description

The take-off thrust of the JTF17 was 61,000 lb. The engine had no inlet guide vanes in the fan section. This made it possible to overhang the fan and reduce the number of bearings supporting the rotors to four. The fan consisted of two stages, which were driven by a two-stage low-pressure turbine. The high spool had a six-stage high compressor powered by a single-stage high-pressure turbine. The exhaust nozzle was similar to the TF30 and J58 blow-in door ejector nozzles with free floating flaps with one exception.

Fig. 8   Model of JTF17 showing the burner in the duct around the core engine (courtesy of Pratt & Whitney Archives).

CHALLENGES AND NEW TURBOFANS

Fig. 9 The JTF17 supersonic transport engine (courtesy of Pratt & Whitney Archives).

The JTF17 also had reverse thrust capability, a requirement for its commercial application. The duct-burning turbofan, compared to an afterburning turbojet for the SST, offered a greater subsonic cruise range because of its better specific fuel consumption. In addition, the duct-burning JTF17 shown in Figs. 8 and 9 offered lower noise signatures than an afterburning turbojet.

FAA Decides

Bill Gorton outlined the technical challenges in developing an engine for commercial service at a meeting of the Society of Automotive Engineers in April 1966. Gorton pointed out what was required in an engine to make the grade in commercial service (Fig. 10). One has to put in as much development effort after certification as before certification. In the case of the SST engine, 80 to 90 percent of the flight time would be at maximum turbine temperature as opposed to 1 to 2 percent for subsonic commercial engines [2].

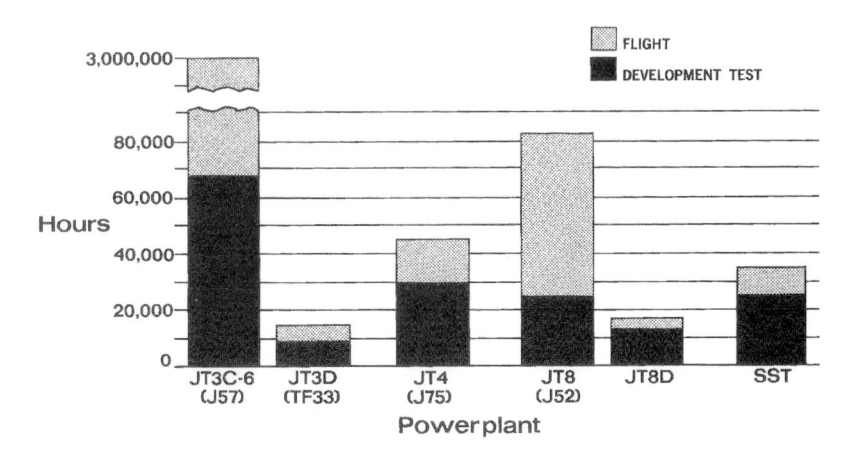

Fig. 10 Historical summary of development testing and flight-testing experience of a commercial engine before entering airline service (courtesy of Pratt & Whitney Archives).

Gorton, of course, was also responsible for the J58 engine development and knew what was required for development of a reliable Mach 2.5 to Mach 3.0 engine. Engine development is something only a small fraction of aviation's population understands. A couple of months later, the FAA announced that it was awarding the engine development contract to General Electric.

Dependability in a commercial engine demands that more than 3000 parts operate without failure for at least three years on the wing. It is impossible to predict which parts will fail from the design layout. That is why the development group keeps piling up test hours looking for a weak spot.

F100 — THE ULTIMATE MILITARY ENGINE

In late 1969 the Air Force wanted a more modern fighter aircraft, a twin-engine airplane, and selected McDonnell to build the airframe. The aircraft needed an engine that enabled Air Force pilots to fly at faster speeds and higher altitudes than the first afterburning turbofan, the TF30, could handle. The Air Force's proposal competition for such a powerful engine gave Pratt & Whitney the opportunity of creating the company's finest military engine, the F100, in an environment that seemed at times as public as Macy's windows at Christmas time. Pratt & Whitney was able to rise to the challenge by having already assembled a team that could communicate and respond effectively in both the evolving military structure and aviation market place.

A CHALLENGING CHANGE OF ASSIGNMENT

In the spring of 1967, I was working with fuel cells, trying to interest the Army Corps of Engineers and the Signal Corps in small research programs for quiet power generators using hydrocarbon fuel. Don Jordan asked me if I would move back into the aviation effort at Pratt & Whitney. He suggested that I see Dick Baseler, Vice President of Engineering. Baseler announced to me that I was being transferred back into the aviation gas turbine activities at Pratt & Whitney starting next Monday. Then he proceeded to tell me what this new assignment was all about. Pratt & Whitney had lost a couple of important contracts to General Electric—the Air Force C5 Transport engine and the Supersonic Transport engine. A Pratt & Whitney representative at Wright Field attributed the failures to our inability to communicate effectively with our military customers. Baseler was under great pressure from the management to "do something about the problem" with the Air Force. The problem also included the Navy; in the Supersonic Transport competition, the powerplant evaluation team included experts from both the Air Force and the Navy.

So where does Connors fit into this? Dick Baseler wanted me to represent the Engineering Department, together with a representative from Marketing, to improve communications with the Air Force and Navy. That was quite a vague challenge and I was not sure that it was possible to achieve such a noble objective. I told Dick of my concerns in taking on an impossible job with a high probability of failure. Baseler understood my feelings and agreed that the chances of success were not great—but he would not hold that against me. I had very little choice but to take on this new assignment and do my best. Baseler said I would report to my former boss Bill Sens, but he wanted me to keep him up to date on what was going on at Wright Field. He also wanted me to understand that he would help me in any way that he could.

I wrote a job description to control all visits to the Air Force Aero Propulsion Laboratory and the Navy facility at Trenton. The reason for this restriction was that contacts between Pratt & Whitney and the Aero Propulsion Laboratory were sporadic and interpreted by the Aero Propulsion Laboratory people in a negative way.

KNOWING THE CUSTOMER

My counterpart in this new assignment was Bill Weaver from Marketing, who had been a representative at Rohr, Grumman, and Boeing (Fig. 11). He had made many friends at those companies and these connections would be of great use to us later. Weaver spoke in straightforward sentences that expressed exactly what he had on his mind with little concern for collateral emotional damage. He also impressed me as a "gung-ho" type of guy who made things happen. Weaver made a call to a friend of his at Boeing, Walt Swan, who provided insight about working with the Air Force.

Weaver and I quickly agreed on a plan, which was to find out what the customer wanted, and work with our organization to give the customer what he wanted, within the best interests of the company. In this case the "customer" consisted of a group of engineers with various specialties related to gas turbines. Our objective was to give each customer special attention to make sure he

Fig. 11 Bill Weaver.

had the information he needed from Pratt & Whitney. We would control the flow of company information to the customer so that propriety information would be protected. We did not employ a sophisticated smooth-talking salesmen's approach, but instead a simple approach to improve communications.

We met with Cliff Simpson, who was head of the Air-Breathing Engine Division in the Aero Propulsion Laboratory, which went well. Simpson told us to work with his technical expert, Howard Schumacher.

MEETING THE CUSTOMER

Colonel Herb Lyon, who was the new head of the Aero Propulsion Laboratory, had a meeting at Pratt & Whitney, and Bill and I spent a few moments with him. It turned out that I had met the Colonel a couple of years earlier. He remembered me as the guy who talked about fuel cells and responded to his questions readily. The Colonel impressed us as a straight shooter. We told him we were going to do our best to improve communications between the Aero Propulsion Laboratory and Pratt & Whitney. The starting point was to find out what the laboratory wanted, and then Bill and I would try to match that with what Pratt & Whitney could provide. We parted on fairly good terms.

Our first meeting with Schumacher was productive. He had seen a number of people from Pratt & Whitney, but he said he would work with us and would introduce us to his staff of engineers. Weaver expressed our off-the-cuff impression of Howard as, "Howie is a diamond in the rough and we can work with him." Bill and I always looked at our association with Cliff, Howie, and the fellows at the Aero Propulsion Laboratory as an enjoyable and productive experience.

WORKING WITH THE CUSTOMER

Weaver and I were not visiting the Air Force as salesmen. We were there to help the laboratory personnel do that part of their job that involved Pratt & Whitney. Over the subsequent six months we got to know each of the engine component specialists in the laboratory. What I did was to line up corresponding specialists at Pratt & Whitney's Engineering Department and take them to the laboratory to do whatever was needed. In many cases, trust had already been established between individual engineers at the two organizations. It became clear that Bill and I were of practical help to the laboratory. At the same time, we would check in with the Colonel and Cliff to keep them up to date.

Bill and I were on the job for only about four or five months when we got a surprise from an unexpected quarter. Someone in Pratt & Whitney's Dayton office recommended that Bill and I be taken off the job. Perhaps he felt we were not keeping him well enough informed, or he was just not happy with our less-than-formal way of working with "his" customers. At that moment, our friend Hugh Gosselin, a Marketing Manager, came to our rescue. He conducted his own assessment of the so-called problem by contacting his friends in the Air Force and concluding there was no problem.

F100 ENGINE COMPETITION

When the first rumors of the TFX (Tactical Fighter Experimental) program came up, Weaver and I talked it over with Simpson, who gave us the background. He pointed out that in 1962, the Department of Defense, trying to save costs, declared that the Air Force and the Navy could use the same airplane. This was a compromise of the Air Force's need to replace the F-105 and the Navy's interest in an F-4 replacement. The F-111A would become the Air Force airplane and the F-111B would go to the Navy. Pratt & Whitney created the TF30 engine for both airplanes. General Dynamics would build the Air Force model and Grumman would build the Navy model. But the Navy and the Air Force could not see eye to eye on sharing a common airplane, and the Navy decided to go with Grumman with the F-14, which was also powered by the TF30 (discussed in the last chapter). The technology developed for the TF30, Pratt & Whitney's first afterburning turbofan provided the genesis for the F100 engine (Fig. 12).

The Air Force was looking for an air-superiority fighter, the F-15, which called for a state-of-the-art engine (Fig. 13). This time, the Department of Defense granted that the Air Force (FX) and Navy (VFX) airplanes could be different, but to save resources, there would be essentially only one engine for both applications. The Initial Engine Development Program (IEDP)

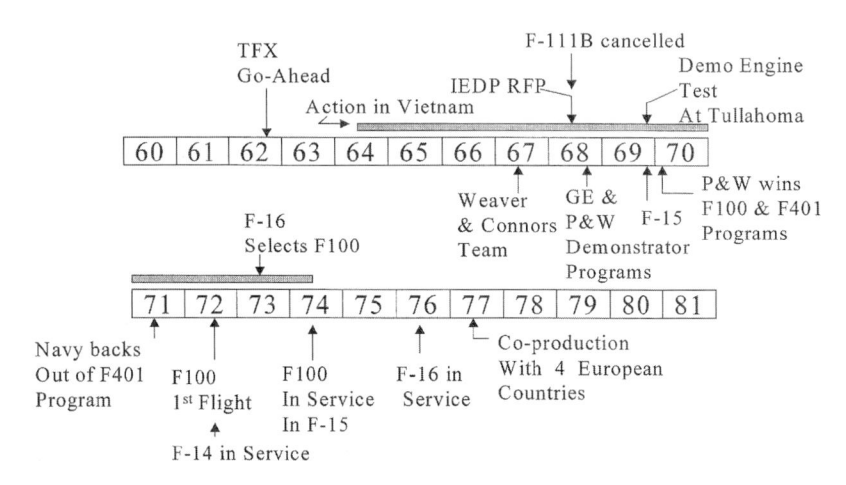

Fig. 12   Time line for the F100 program.

Fig. 13 a) The F100 alongside the F-15 Eagle (courtesy of Pratt & Whitney Archives), and b) the F-15 in flight (courtesy of U.S. Air Force).

Request for Proposal (RFP) came out in mid-1968. The military's proposal defined commonality as using the same core (the high spool), which is the more expensive engine spool from a development effort viewpoint. The Air Force and Navy engines could have different low spools. This was a cost-saving gesture in view of the escalating situation in Vietnam that was gobbling up military resources at a fierce pace.

The Air Force oversaw the engine development. The common core would be in the Air Force F100 engine with one low spool and the Navy's F401 engine with a different low spool to provide a greater thrust level. The winner of the IEDP would develop two separate engines. The Navy's VFX became the F-14, and shortly thereafter the Navy bowed out of the F401 engine program. This departure left the Air Force with a bigger share of the development cost, putting a strain on the Air Force's budget.

When the IEDP RFP appeared in 1968, Allison, General Electric, and Pratt & Whitney were participating in advanced technology component programs with the Air Force Aero Propulsion Laboratory at Wright Field. The new RFP called for an 18-month demonstrator program aimed toward the air superiority fighter application, the F-15. Two contractors would be selected, and each would get about $50 million to build and test two demonstrator engines, one of which would be tested at the Air Force's Arnold Engineering Development Laboratory (AEDC) in Tullahoma. This demonstrator program reflected the "fly before you buy" philosophy that had been used in other Department of Defense programs. Allison declined to bid, leaving General Electric and Pratt & Whitney to compete with each other.

Airframers and engine companies were now gearing up for the long, drawn-out competition. Back in East Hartford, Pratt & Whitney was preparing for a vigorous proposal effort for the huge engine contract that was expected from this F-15 fighter program, which was expected to have more potential than the B-1 Bomber program. The management decided to give full responsibility for the effort to the Florida Division in West Palm Beach.

Weaver and I (in the East Hartford Division) were not sure how we fit into this new situation. We had developed a good working relationship with the personnel in the Aero Propulsion Laboratory over the past year, but the engineers in the Florida Division had not had a chance to develop the same type of relationship. Therefore, we saw our role as helping Florida personnel to work effectively with the laboratory.

We found out that there would be a big meeting in Florida to discuss the upcoming competition. We called up the appropriate fellow employees in Florida to tell them we thought we should be at such a meeting because we had much to offer and wanted to help them. Weaver and I were concerned that if Pratt & Whitney lost the competition, we would be the most likely candidates to take the heat. We went to Florida and gave the group what we knew was going on regarding the upcoming program and told them we would help them with the Laboratory in any way we could.

FIRST FRDC MEETING AT THE AERO PROPULSION LABORATORY

Our contacts in the laboratory told us about a forthcoming meeting between Cliff Simpson and a high-level group from Florida headed up by Bill Brown, an example of him finding out something from his customer instead of from his own (large and complex) organization. Weaver and I asked Simpson if it would be a problem for him if we showed up at the meeting. He assured us that it was no problem. The meeting ended with Bill Brown's assurances that the laboratory would have no problem working with the Florida branch of Pratt & Whitney. Bill Weaver and I could see that our jobs had just gotten more complicated because now we had to start earning the trust of the Florida organization, and some of the work critical to component substantiation was being done in East Hartford.

F100 ENGINE PROGRAM IS LAUNCHED IN FLORIDA

Bill Gorton, General Manager of the Florida Research & Development Center (FRDC), appointed Gordon Titcomb as the Program Manager for this "must win" program. Gordon had previously been the Program Manager for the supersonic transport demonstrator program. FRDC had its own group of component specialists. Weaver and I tried to line them up with their counterparts at the Aero Propulsion Laboratory. As far as we could see, the communications between the two organizations were proceeding fairly well.

WORKING WITH THE INTERNAL CUSTOMER

Our next big encounter with the Florida bureaucracy was in the preparation for a quarterly review meeting with the Air Force at the Florida location. Weaver and I had talked to the laboratory people about the review. They requested breakout meetings with component specialists, some time off for them to digest briefing material, and time for the Air Force to have its own private meeting to discuss what they learned during the review. They said they would like to have a brief recap or wrap-up from the Program Manager at the end, in addition to a question and answer session toward the end of the final day.

Weaver and I wondered how to get this message across to the fellows in Florida. We knew that if we started out with, "Here's what you guys have to do," we would get no place other than the exit door. We had to devise a strategy to work with our Florida counterparts, which we did by thinking of them as our *internal customer*. Once again I had to fall back on my ancestral heritage (which Weaver shared) and bring in a little blarney. We had a 20-slide presentation in cartoon form to show how ridiculous the typical Pratt & Whitney quarterly review meeting is, in which we run the customers ragged by first blasting them in a crowded conference room filled with smokers, rushing them through lunch, and then running them through every facility we have. Then I had a couple of slides on what we felt the Air Force wanted but was hesitant to request.

Our message to FRDC was simple and not at all technical. We reminded them that the laboratory people attended more contractor briefings than one would imagine. We needed to stand out by reformatting the typical quarterly review meeting by tailoring it to the preferences of the government attendees. The format we suggested was to start the first day with a general overview for all attendees, which would tell them what would happen in the two-day meeting; then get into the presentations aimed for all attendees; and after lunch let the specialists in the government talk to their counterparts at Pratt & Whitney. This would make it possible for engineers to talk with engineers, finance people with finance people, lawyers with lawyers, and program management people with management people. For the second day we recommended that the specialists' meetings continue in the morning; after lunch, let the government attendees have their own private meeting; and then conclude with the F100 program manager giving a summary and the government people's spokesperson would make his or her remarks. After Weaver and I got through with our humorous presentation, Bill Gorton said, "Okay. Let's do it that way!"

The other suggestion Weaver and I made was to let us get to know the Air Force personnel in the potential jury pool, those who would make the assessment, evaluations, or recommendations on the technical, financial, and management parts of the Pratt & Whitney proposal. The selection would be under the Aeronautical Systems Division (ASD), with assistance from the laboratories. Our recommended priority was what the customer wanted to know, rather than on what we assumed the customer should know. In other words, take care of the customer's concerns first, and then add whatever message you think should be expressed.

INITIAL ENGINE DEVELOPMENT (IED) PROGRAM

In the meantime, the F100 team in Florida, under the direction of Gordon Titcomb, was working around the clock. The FRDC team designed what they thought the customer requested and got the first demonstrator engine to test in Florida on July 30, 1969 [3]. The second engine was delivered to the Arnold Engineering Development Center in Tullahoma (AEDC) in November 1969 for the final testing of the engine.

Senior Test Engineer Roger Bursey (Fig. 14) described the last minute activities [3, p. 53]:

> We were trying to get everybody lined up to get home for Christmas and immediately return afterwards because we were working through the holidays. The schedule was that tight. Two days before Christmas we had released all our people except seven of us, and AEDC says, "There's an air test period available. Do you want it?" We said, "By all means." On the 24th with our mini-crew, we ran that engine and got another test point out of the way. In the process, we had a malfunction in the fuel control that we didn't understand.
>
> Marshall Seymour, on-site Pratt & Whitney representative, had arranged with Gordon Titcomb, Program Manager, to send the Sabreliner to pick us up around six o'clock that night at the AEDC airstrip. It was around three o'clock in the afternoon when we were finally able to get into the test cell. It had to be purged of nitrogen. Our controls engineer says, "Hey, we can't do anything on this here. We're going to have to bring the control home." So Jerry Spangler and Bob Jackel proceeded with wrenches.

Fig. 14 Roger Bursey (courtesy of Pratt & Whitney Archives).

> As you know, keeping people waiting around on Christmas Eve is pretty sensitive, but Jerry and Bob finally got the fuel control off and dumped the fuel out. Marshall Seymour saw the Sabreliner come in and took us over to the airport. We convinced the pilot that the fuel control had to go home with us. He said, "There's no place for it. In fact, there's one too many passengers. It's a six-place airplane—someone's going to have sit on the head." We ... put a light guy back there. Then the fuel control, all wrapped up in plastic, was placed in the airplane and after the door was closed we were able to set it on the floor.
>
> The pilot took off and got to altitude. Then the smell of fuel. This thing (the fuel control) was dumping fuel out. The pilot and all of us got a little nervous. So the No Smoking Light was on the whole way and fuel kept running out on to the floor of the Sabreliner.
>
> Here we are making a mess of the Corporation's airplane. Bob and Jerry took paper towels, T-shirts, anything they could to sop up the fuel; but the fumes were burning our eyes to high heaven.
>
> We arrived at Palm Beach International Airport that night, Christmas Eve, around nine o'clock. Jerry Spangler loaded the fuel control into the trunk of Bob Jackel's car, and the team went home for Christmas. The day after, we were out at FRDC working on the fuel control, and we were able to correct the deficiency, and take the control back to AEDC as the crews were getting back on line up there. We tested, if my memory serves me right, the 27th or 28th of December. So it only took us three days to solve the problem. Normally, with something like that, you'd lose two weeks to a month—encountering a difficulty, correcting it, and getting back on-line.
>
> Looking back on these events, Titcomb believes the Air Force chose the F100 not only because it was a superior design and included advanced materials, but also because FRDC had demonstrated its total commitment to do whatever was necessary to build and test a successful engine.

Bursey recalled that the FRDC had never used the Air Force Tullahoma facility before the F100 program. During the F100 competition with General Electric; however, the Air Force required Pratt & Whitney to run its engine test at the Arnold Engineering Development Center (AEDC) facilities. The concern was that FRDC engineers did not know the Air Force personnel at the facility and our competitor had developed an effective relationship with them. In 2007 Bursey told me how the Pratt & Whitney team prepared the way for the demonstrator engine test.

> As FX202 (the demonstrator engine) was being instrumented and assembled, as Senior Test Engineer, I made many trips to AEDC to interface and coordinate descriptions of the engine, operating parameters, [and] instrumentation.... Test plans, expected results, and reams of paper became an absolute requirement. All of these requirements were met and delivered to the AEDC test managers on time.

Marshall Seymour and I followed the truck carrying FX202 from FRDC to AEDC. During the stop in Orlando, Marshall obtained two plastic toy inflatable alligators, inflated them, and tossed them into the nozzle of the engine. It was not long at AEDC when the shipping covers were taken off the engine when the alligators were discovered.... Security people took photographs of the engine, and it was not long when word got around that the engine had alligator contamination. The joint Pratt & Whitney/AEDC team went on to accomplish 100 percent of the test objectives.

The total commitment of the test team ... was evident during the return of the test engine FX202 to Florida. AEDC had placed in the engine nozzle a Pratt & Whitney blue box, which contained a bottle of Jack Daniels wrapped with performance curves in two plots.

The Air Force selected Pratt & Whitney to develop the F100 engine in early 1970. One of the reasons the military selected the company was its knowledge of inlet–engine compatibility technology and experience—lessons learned from the F-111/TF30 program. Over 7000 F100 engines were manufactured and are still purchased today.

ENGINE DESCRIPTION

The F100 is a two-spool turbofan with a three-stage fan driven by a two-stage low-pressure turbine (Figs. 15 and 16). The high spool consists of a 10-stage compressor, a burner, and a two-stage high-pressure turbine. The sea-level take-off thrust was 24,000 lb with a weight of 3000 lb. The ratio of thrust to weight was 8. One of the details my team got into was the effect of inlet airflow distortion on the performance of the engine. The laboratory personnel and Pratt & Whitney component specialists did a great job trying to quantify the effects of inlet distortion on performance and more important on the tendency for an engine to stall.

Fig. 15 Cutaway of the F100 engine (courtesy of Pratt & Whitney Archives).

Fig. 16 The F100 (courtesy of Pratt & Whitney Archives).

F-16 OPPORTUNITY ARISES

A few years after Pratt & Whitney was awarded the contract for the F-15's F100 engine, and it was in development, the Air Force had a need for a highly maneuverable light-weight fighter (LWF), and contracted with two finalists, Northrop and General Dynamics, in early 1972. The Northrop aircraft became the YF-17 and the General Dynamics aircraft was the YF-16. The Air Force's evaluation in 1974–1975 selected the F-16. Figure 17 shows the installation in an F-16. I remember Larry Clarkson, FRDC Program Manager on the F-16 engine proposal, describing the opportunity. It looked like a tremendous challenge because of the need for production-sharing overseas, and working with the Air Force to limit the transfer of technology. His deputy was Bob Shanower, from the United Technologies Washington Office.

One of the keys to winning the business was the willingness to outsource the production of the engine, as the sale of the airframe and engine was contingent upon how much of the aircraft would be manufactured overseas. On

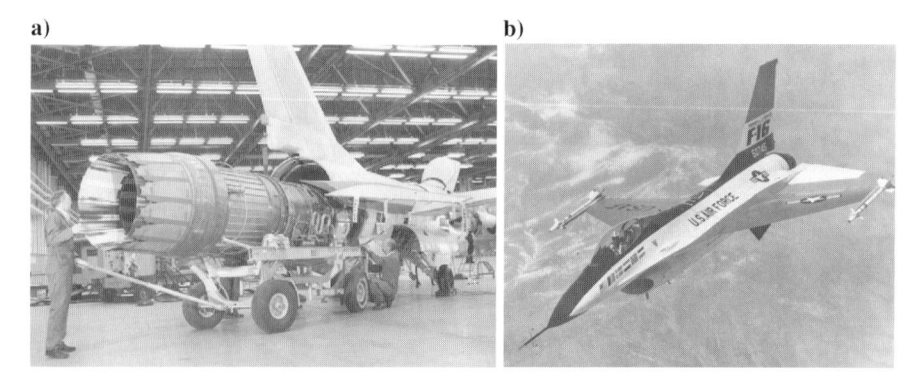

Fig. 17 a) The F100 being installed, and b) the F-16 in flight (Courtesy of Pratt & Whitney).

CHALLENGES AND NEW TURBOFANS   393

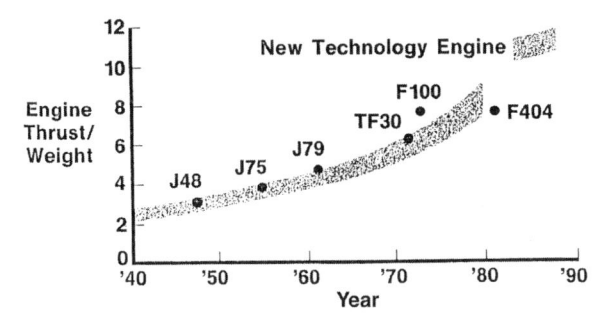

Fig. 18 The F100 engine was a big step in thrust/weight beyond other fighter engines (courtesy of Pratt & Whitney).

May 1977, Pratt & Whitney signed a contract with four NATO nations (Denmark, Norway, Belgium, and the Netherlands) for the co-production of the engine. The share of engine production was 40 percent by the Europeans and 60 percent by Pratt & Whitney. Fabrique Nationale in Belgium assembled the first F100 and tested the engine in July 1978. One year later the first consortium F-16 was built and delivered.

ENGINE DEVELOPMENT SUMMARIZED

When the F-111 came along (Chapter 11), its mission required an afterburning turbofan for improved fuel consumption in subsonic flight (non-afterburning) with the capability of supersonic dash. Figure 18 shows how much progress in thrust/weight was made from J48, the J75, and the J79 turbojet engines to the afterburning turbofanned TF30 and the F100. Development of the TF30 brought into sharp focus the challenge of inlet-engine compatibility and the effects of inlet distortion on the ability of an engine to operate under severe inlet flow distortion. A great deal of experience was gained during the development of the F-15 aircraft and F100 engine when test pilots and later Air Force pilots explored new capabilities for this high performance aircraft. These new capabilities introduced greater inlet distortion and higher structural loads that were not anticipated in the original specifications.

THE GREAT ENGINE WAR

No engine has generated as much publicity as the F100 engine program.[*] It became the star of "The Great Engine War" of the early 1980s, a subject of media frenzy. The dramatis personae comprised politicians, television news

---

[*]For an in-depth analysis, see David M. Kennedy's "The Great Engine War," Case Study C16-85-629, Kennedy School of Government, Harvard Univ., Cambridge, MA, 1985.

gurus, reporters, lawyers, accountants, and engineers. Frank McAbee, who became president of the Government Products Division (GPD), explains what went on in an arena monitored by the public [4]:

> Traditionally when you won a contract, you had that business in your business base ... and you knew that you were going to have production volume in two or three years to support you. That particular tradition [changed] very rapidly in the early 80s. When I became president of GPD it had almost reached the boiling point.
> Those of us who worked in Washington were well aware of the fact that the F100, which had just been introduced into service, while it was performing beautifully, was having some problems.... Opponents to the military program in Congress could use [it] as an example of how the military really didn't know how to run a program....
> We knew that the military was seriously considering giving some of our business to GE [General Electric]. We knew that GE had developed their F110 engine in competition with the F100 earlier on.... They pursued a ... vigorous lobbying campaign with both the Congress and the military ... to show that our problems were bigger than we could handle. Any issues that arose, where there was any type of question as to Pratt & Whitney's responsiveness to the value to the military problem, the competition obviously took advantage of that and played that up. So it was a very tense time. The engine was working fine. Actually it was working too well. It had such a high thrust to weight ratio—so much better than any previous engine—the pilots ... were flying the airplane well outside the envelope.... As a result ... the engine was being pushed beyond some of its design limits and it was having some premature failures.
> So we had that to contend with that. The Air Force ... had decided that the way to get our attention ... was to threaten to give part of the business to GE. Our management did not respond well to that.... It was our engine. We would fix the problem.... Just give us money and go away ... the Air Force decided they were going to run a competition and thus began the Great Engine War. It was Pratt & Whitney against GE. And the proposal we submitted was based on, to a great degree, us getting all the business. And that really wasn't what the customer wanted, and as a result, GE was given 65 percent of the contract the first year. That was ... a tremendous shock to us and to everybody in the plant.

With the competition lobbying Congress and the military, the F100 program was bound to attract the media's, and therefore, the public's attention. Pushing the flight envelop was exciting, but attention was focused on the engineering problems. Terms used by engineering professionals may take on a sinister meaning when the public hears them; for example, when a term such as "design deficiency" is used, it sounds as if Company X has done something wrong and that company should be punished. To a design engineer; however, the term simply means that we have learned something about a particular engine part and now we can make such and such a change to improve it. Earlier in this chapter we reviewed the steps in developing an engine. That may seem like basic knowledge to you, but for the public hearing about challenges encountered during engine design and development they would not be familiar with the development process and not know about the effort involved in bringing an engine to market. Hence, the "Great Engine War" came to pass.

THE BATTLE WON

With such negative attention, you might think that the F-100 was not a success. When all smoke from "The Great Engine War" lifted, it turned out that all Air Force F-15s and about two thirds of F-16 Falcons worldwide are powered by F100s. We will trace further evolution of the F100 from the 1980s to the present day in Chapter 14. But first we will continue reviewing the second generation of turbofans in the next chapter.

REFERENCES

[1] Bulban, E. L. "Possible Market for 500 SSTs Forecast," *Aviation Week & Space Technology*, May 24, 1965.
[2] Gorton, W. L. "Supersonic Transports—Propulsion Systems, Application of Advanced Technology," SAE Paper 660297, April 26, 1966.
[3] "Hal," Government Products Division (GPD) Files, Pratt & Whitney Archives.
[4] *Thunder in the Sun: A Look at 40 Years in Florida, 1958–1998*, video produced by Communications Department, Pratt & Whitney, 1999.

Chapter 13

HIGH-BYPASS FANS

INTRODUCTION

Pratt & Whitney's 40th anniversary in 1965 was a year of exclamations. Since the beginning of commercial jet aviation, Pratt & Whitney had delivered to airlines more than 5300 commercial turbojets and turbofans. The JT3/J57 engine that had started the commercial jet age with the Boeing 707 and the DC-8, came to the end of its 21,000+ military and commercial production run. The last J57 was for the French Navy's F-8E Crusader. The JT3Ds and TF33s amounted to about 8000 engines, some of which had been converted JT3/J57 engines. Since the beginning of the commercial jet era, foreign and domestic airlines had ordered over 600 aircraft with company engines. The Boeing 727s were flying, followed by Douglas DC-9s, and soon to be followed by Boeing 737s, all powered by JT8Ds. Commercial engine duration records were being set in Time Between Overhauls (TBOs) with the JT3D at 6100 hr, the JT8D at 3600 hr, and the J75/JT4 at 6800 hr (at that time, the highest of any engine under FAA regulation) [1]. The best of the Pratt & Whitney piston engines (R-2800) had a TBO of about 2500 hr. The world's first afterburning turbofan, TF30, completed its qualification test in 1965. Its non-afterburning configuration powered the LTV A-7's first flight. The afterburning TF30 powered the F-111. A French version of the TF30 (TF106) was being developed for the Dassault Mirage III-V. The J58-powered YF-12A set new speed and altitude records in May of 1965.

On the one hand, Pratt & Whitney had all this good news of which to be proud, but on the other hand, in mid-year it discovered it did not win the C-5 engine competition. It is important to review the demonstrator engine for this competition as it provides the basis for Pratt & Whitney's second generation of turbofans, the first of which was the JT9D for commercial use (Fig. 1).

LIGHT WEIGHT GAS GENERATOR (LWGG) PROGRAM

The roots of the technology for the Lockheed C-5 military transport engine program at Pratt & Whitney go back to the early 1960s and the Air Force Light Weight Gas Generator (LWGG) program at Wright Field. Cliff Simpson

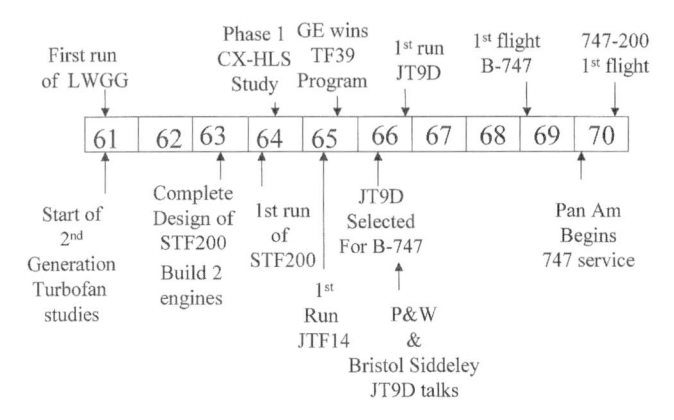

Fig. 1 Time line leading up to and including the JT9D program.

was the Air Force sponsor of the program to push aerodynamic loading of compressor and turbine blades and to explore higher turbine inlet temperatures. The LWGG technology program started in March of 1959 and the demonstrator engine first ran in 1961 (Fig. 2). The technology program, under the direction of the Aero Propulsion Laboratory at Wright Field, continued under the name of Advanced Turbine Engine Gas Generator (ATEGG) program.

In the early 1960s, both General Electric and Pratt & Whitney had been involved in various lift engine propulsion schemes. Airframers had a voracious appetite for study engines from the engine manufacturers in support of their various studies of VSTOL and STOL aircraft. Then the U.S. government and the Federal Republic of Germany discussed collaborating on a lift-cruise type of military aircraft, the JTF16 lift-cruise engine shown in Fig. 3. This program shot up like a Fourth of July rocket, only to fade from view in terms of a marketing opportunity for U.S. engines.

Fig. 2 The Light Weight Gas Generator program demonstrator engine (courtesy of Pratt & Whitney Museum).

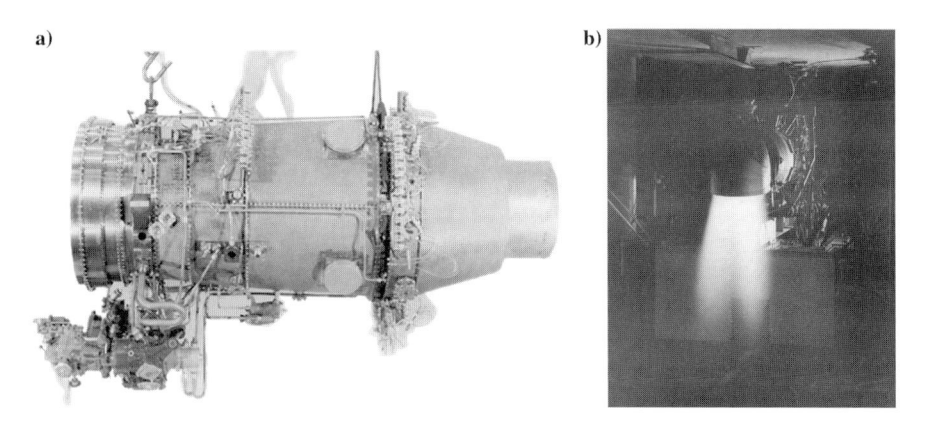

Fig. 3 a) The JTF16 lift cruise engine, and b) demonstrating the thrust vectoring to 90 deg (courtesy of Pratt & Whitney Archives).

The TF30 and the JTF16 generated about the same amount of thrust; however, the JTF16 was considerably shorter, as Fig. 4 illustrates. It was shorter because there were fewer number of stages of turbomachinery in the continued pursuit of increasing thrust/weight ratio.

C-5 ENGINE PROGRAM

In the early 1960s, the Air Force was evaluating its heavy lift capability in transports. It had launched the C-141 program (with JT3D/TF33 turbofans), but the fuselage was not large enough to accommodate some of the essential fighting equipment that the ground forces needed. The operational studies sponsored by the Air Force converged on a mix of 130 C-141 transports and 130 larger aircraft with fuselages wide enough for the equipment the Army

Fig. 4 Comparison of non-afterburning TF30 (top) and JTF16 (bottom) (courtesy of Pratt & Whitney Archives).

2-STAGE FAN   9-STAGE HPC   2-STAGE HPT   3-STAGE LPT

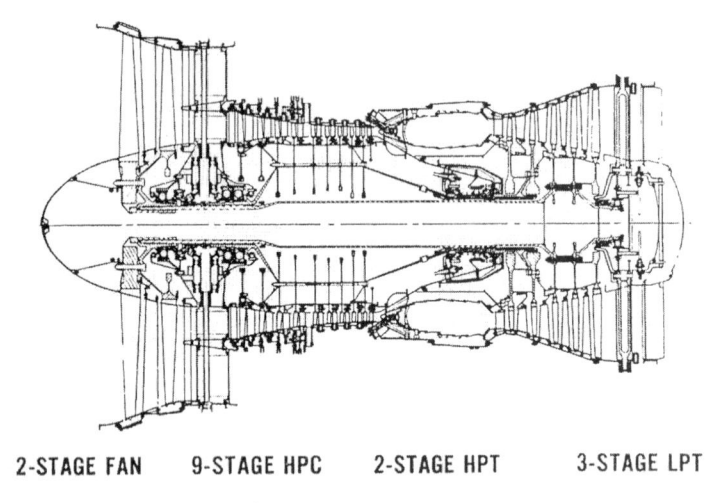

Fig. 5 Line drawing of the STF200C (courtesy of Pratt & Whitney Archives).

needed for its support. The latter aircraft had the designation of CX-HLS for Cargo Experimental–Heavy Logistics Systems [2].

The Zero Phase program for airframers and engine companies started in April 1964 with a Request For Proposal (RFP). Phase 1 started that summer, with Boeing, Douglas, and Lockheed as the airframers, while the engine competitors were General Electric and Pratt & Whitney.

Pratt & Whitney, anticipating the need for a more efficient turbofan than the JT3D, had started the detail design of its STF200 demonstrator engine back in January 1963, financed by the company. In a 31,000-lb thrust size, Pratt & Whitney applied technology from the Air Force-sponsored Light Weight Gas Generator Program and from other component research programs. The STF200C, with a 20:1 overall pressure ratio, first ran on April 1964. It had a bypass ratio of 2 with a two-stage fan driven by a three-stage

Fig. 6 The STF200C (courtesy of Pratt & Whitney Archives).

TABLE 1   STF200 VARIATIONS

Engine	Thrust (lb)	Bypass ratio
STF200C	31,000	2
STF200D	34,000	3
STF200F	39,200	4
JTF14	41,000	3.5

low turbine. The high spool had a nine-stage compressor driven by a two-stage high turbine. The STF200 demonstrator engine family had subsequent variations in bypass ratio and thrust level (Figs. 5 and 6). There were three variations in the STF200 series (Table 1). However, the engine proposed for the C-5 program was the JTF14.

There is a considerable difference in design between the first-generation turbofans (JT3D, JT8D, and TF30) and the higher bypass STF200 turbofan. The mechanical structure is simpler because there are only two bearings per spool and only three bearing support structures. In addition, the number of turbomachinery stages in the JT3D was 19 (for 12:1 overall pressure ratio) compared to 17 for the STF200 (with 20:1 overall pressure ratio). Another feature of the turbomachinery was the extensive use of high aspect ratio airfoils, which helped to shorten the engine. Figure 7 shows that, when scaled to

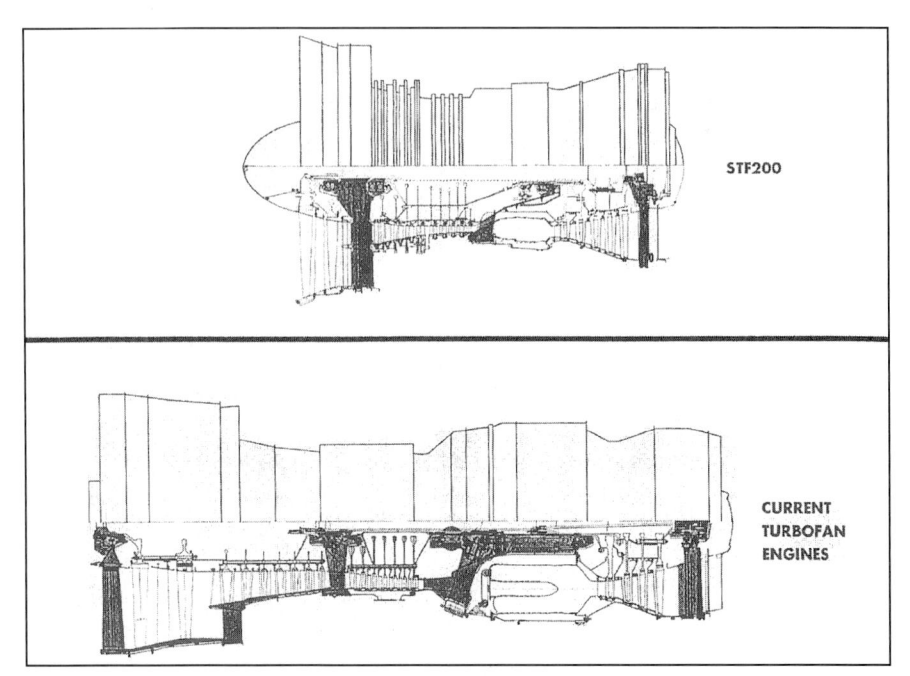

Fig. 7   The STF200 (top) compared to the JT3D (bottom) (courtesy of Pratt & Whitney Archives).

the same level of thrust, the STF200 represented a considerable shortening of the engine structure compared to the then-current turbofan JT3D.

The engine proposed for the C-5 was the JTF14E with a bypass ratio of 3.5 and 41,000-lb thrust (Fig. 8). It had a single-stage fan driven by a four-stage low turbine. The fan had no inlet guide vanes and looked very much like modern high-bypass commercial fans. The high spool featured a 10-stage compressor driven by a two-stage high turbine. The weight of the engine was about 7000 lb, which resulted in a thrust to weight of close to 6.

In the early 1960s, General Electric had been involved in a number of propulsion schemes involving the use of J85 turbojets whose exhaust drove a form of a Whittle-type aft fan (his Thrust Augmentor 3, aft fan with tip turbine) with the turbine blades on the tip of fan blades. The other type of Whittle aft fan (Whittle's Thrust Augmentor 2), for which he had filed a patent application in 1936, was the CJ-805 configuration with the fan blades on the tips of the turbine blades [3]. The CJ-805 turbojet powered the Convair 880 and the CJ-805-23B aft fan powered the Convair 990.

The downstream turbine-fan combination engine was equivalent in its thermodynamic cycle to a very high-bypass turbofan (in the 10–12 range). Lockheed made studies that showed the very high-bypass cycle looked promising in terms of range and payload. This got the attention of the Air Force, and when General Electric reduced that high-bypass cycle to practice, it focused on the more conventional configuration for an 8:1 bypass ratio engine. It subsequently became a winner for the C-5 competition. Martin C. Hemsworth of General Electric said on August 23, 1989 in discussing turbofans at the AIAA Golden Anniversary of Jet Powered Flight conference in Dayton, Ohio, "The CJ805-23 is noted here because it was the first commercial

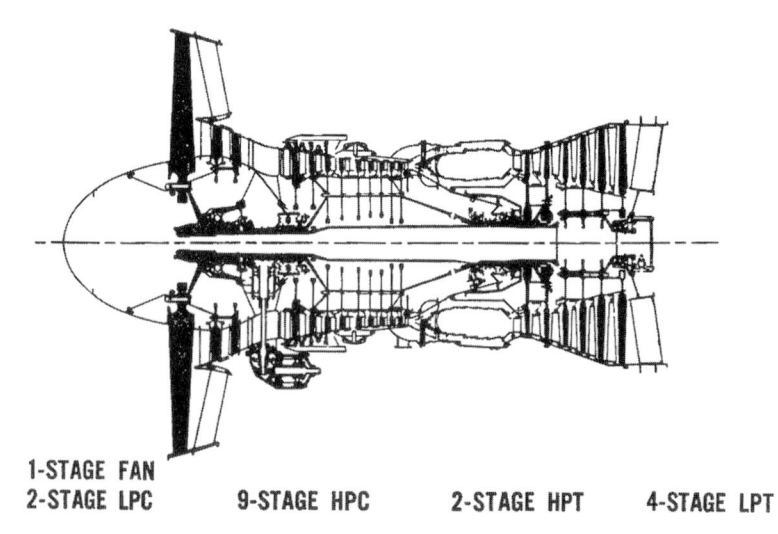

1-STAGE FAN
2-STAGE LPC   9-STAGE HPC   2-STAGE HPT   4-STAGE LPT

Fig. 8 Schematic layout of the JTF14 demonstrator engine (courtesy of Pratt & Whitney Archives).

HIGH-BYPASS FANS

Fig. 9 The JTF14 demonstrator engine (courtesy of Pratt & Whitney Archives).

fan engine in the United States which was a significant advance in transport engine powerplants. Unfortunately, as one of our people said at the time, 'We converted the heathen but the competitor sold the bibles.'"

The JTF14 low spool had a single-stage fan with two low-compressor stages driven by a four-stage low-pressure turbine (Fig. 9). The high spool was the same as the STF200 high spool (nine compressor stages powered by a two-stage high turbine). The major change from the STF200 was a single-stage fan and a low compressor, plus another low-turbine stage. The higher bypass ratio of 3.5 required more low-turbine power. The single-stage fan at the same limiting tip speed resulted in a lower rotational speed. This condition meant that the overall pressure ratio at the root decreased. That caused a need for several low-compressor stages to make up for the pressure rise across the root of the STF200's two-stage fan.

The result of the STF200 program was great improvements in the specific fuel consumption of turbofans beyond that of the JT3D, or first, generation of turbofans. As mentioned earlier, the impetus for high thrust to weight core engines came from the Air Force's Aero Propulsion Laboratory. The objectives of the research programs were higher overall pressure ratios, higher turbine inlet temperatures, higher rotational speeds, and the use of newer materials to permit higher stress levels. In the end, an engine designer could use the technology to create higher performance military fighter and transport engines.

PATH TO THE JT9D PROGRAM

The STF200 had already been run when the RFP for the CX-HLS program came out. In the spring of 1965 (before the C-5 award), the faint flickering light of the possibility of a large commercial airplane could be seen at Boeing. At a meeting in New York at that time, Boeing's Clancy Wilde and Bruce Connelly were discussing what might be considered for the next step in commercial aviation with Juan Trippe and John Borger of Pan American Airways. The discussion centered around a large aircraft with a 4000-mile range, like the new Douglas Super DC-8 series. Boeing engineers were not enthusiastic about stretching the B-707 any more. Lindbergh had talked to Trippe earlier about how the C-5 engines looked interesting and maybe someone ought to design an airplane around an engine like those. Juan Trippe mentioned his interest in a 400-passenger airplane to Boeing [4, p. 284].

Trippe had previously checked in with Lockheed after its win of the C-5 to see if it was planning a commercial airliner version of the huge aircraft. The answer was no. He also discussed with Douglas the possibility of a new larger commercial aircraft, and there the response was that stretching the DC-8 was the way to go. That did not appeal to Trippe. On December 22, 1965, Juan Trippe and William Allen of Boeing signed a letter of intent for the B-747 at 550,000-lb weight [5]. Then in April 1966, Pan American signed a purchase contract with Boeing for 25 aircraft at 680,000-lb GTOW (23 passenger and two all-cargo). The engine thrust was spelled out as 41,000 lb, which was a considerable increase over the 33,000-lb engine thrust for the aircraft in the letter of intent [6].

Pan Am's interest activated a circuit at Boeing, and soon Jack Steiner (father of the B-727) was directing aircraft design studies. This was a whole new challenge in contrast to what had been considered a dead end with a further stretch of the B-707. Several months later, in October 1965, Boeing learned that it did not win the C-5 competition and its attention now focused on the 400-passenger commercial airplane. There were three engine sources for Boeing to evaluate: Pratt & Whitney, General Electric, and Rolls-Royce. A rule of thumb in new aircraft engine and airframe development at that time was as follows: it takes about four years to develop an airframe; it takes about five years to develop the engine for the airframe; and therefore, the engine must be launched about a year before the airframe.

Boeing's T. Wilson assigned Ken Holtby to recommend an engine supplier after a careful evaluation. Rolls-Royce was judged to be further behind Pratt & Whitney and General Electric. General Electric was concentrating its effort on its C-5 engine development (TF39) and declined the opportunity, as Jack Parker commented that the time available for development (three years) was too short. He was guided by the rule of thumb and was correct in his assessment [7, p. 40]. Pratt & Whitney appeared to have sufficient work behind it from the C-5 engine competitive programs to be a viable candidate. Boeing selected Pratt & Whitney in the early part of 1966, perhaps by default but perhaps also by its ability to provide information almost instantaneously as we hear from Stan Taylor in the next section, highlighting an example of how direct communication between engineers and senior management can produce results.

JT9D Program for Boeing's 747 Begins

The net result of the C-5 program for Pratt & Whitney was the creation of the JT9D-1 turbofan for the first wide-body commercial aircraft (B-747). In 2005 Stan Taylor described to me the start of the engine program:

> I was Commercial Sales Manager in 1965. I went down with Rene Poucel ... the person that I had assigned to cover Pan American, to see John Borger, Pan American Chief Engineer. We'd been following Borger's interest in Boeing's studies for the successor to the 707 for some time and we'd kept him up to date on our STF200 high-bypass ratio demonstrator engine that we were working on in a NASA program under Bill Witherspoon in Engineering.
>
> We'd also given the STF200 information to Boeing without offering an engine for their wide-body airplane study but GE had. The reason we hadn't offered an engine was that Boeing had two teams working—one on a growth version of the 707 and another on a high-bypass ratio engine powered wide-body airplane. And it appeared that Boeing management as well as United Aircraft management favored the growth version of the 707.... Borger told me ... that he had pretty well concluded that Pan American should buy the high by-pass ratio engine powered wide-bodied airplane for which GE had offered an engine.
>
> Well, the next morning back in Hartford I was writing up my report of that visit when the phone rang and it was Eastern Airlines, who desperately needed another JT8D spare engine immediately. To get a JT8D or any other engine delivered prior to the formal engine delivery schedule required action by top management. So I went ... to see Art Smith, who at that time was Assistant General Manager of Pratt & Whitney Aircraft, reporting to Len Mallett, who was the General Manager ... in that period the titles went from manager to vice-president and president and I am not sure which level applied to them [at the time].
>
> Art was a very good person for picking the brains of anyone who came in to see him. When I told him about the fact that Eastern Airlines had called and they needed to have an engine quickly and I would appreciate his taking the appropriate action to see that this was done, he said that he'd do that. Then he asked me what's been going on lately, what have you been doing? So I told him about my visit the day before with Borger of Pan American during which Borger had told me that he was on his way out to Boeing on Monday of the following week to start working closely with Boeing on finalizing the studies on the high-bypass ratio engine for this airplane and I wish we were able to do something pretty soon.
>
> Art took this all in and he didn't say anything particularly and then he said, "You know it requires the signature of an officer of the Corporation to change the engine schedule to get this engine for Eastern Airlines and I have to issue an excess ordering authority which you will have to take back to the engine-scheduling people and get them to take

care of that and I could just have it approved by Len Mallet when he comes back to the office tomorrow, but," he said, "let's see, Len Mallet isn't here and Bill Gwinn isn't here." He said, "Let's go up and see Jack Horner." So, we started up the stairs and he said to me, "While you're in there with me, why don't you tell Jack Horner about your talk with Borger yesterday."

So, I did. It was easy for me to talk to Jack Horner because he'd been the one who had sent me ten years earlier to the Harvard Business School for a year and a half to get an MBA at company expense. So, I gave him the story that I'd given Art and I ended up saying "Don't you think we need an engine for this wide-body Boeing airplane!" and he said "By all means!"

Art and I got out of Jack's office as fast as possible and went downstairs. As we got back to Art's office he turned to me and said, "You know what to do now, don't you?" I said, "Yes, sir!" That's the last time Art and I talked to each other about that venture.

I went back to my office and called Barney Schmickrath, who was the top man on the engineering scene that day, and I told him that Jack Horner had said that we should offer an engine for the Boeing wide-bodied, high-bypass engine powered airplane. Barney said, "Wonderful." I said, "We've got to get something down there right away because Borger has told me that he would like to take study information from us that he could take out to Boeing on the following Monday." Barney said, "Fine. We'll have it for you in two days." And sure enough, two days later he had the study dimensions, weight, and curves giving the performance of a high-bypass ratio engine.

In the meantime, I had called Borger and asked him what size engine he wanted and he responded, "33,000-lb thrust." So on Friday, we got the information to Borger to take to Boeing and we also sent a messenger out to Seattle with the same. And that is how the JT9D concept got started.

But complications soon set in.

WEIGHT GAIN

Pan Am's dream continued with an accompanying increase in weight from 550,000 to 710,000 lb (Fig. 10). As the aircraft weight increased, the engine would have to be redesigned and the program would have to be stretched. Pratt & Whitney started with the 41,000-lb thrust for the 747 at 650,000-lb take-off weight. The development schedule was to go to 43,500 lb thrust three years after the initial engine certification rating. John Borger (Pan Am Engineering Vice President) and Art Smith (Pratt & Whitney Executive Vice President) recommended slipping the program by six to nine months to focus the design on 43,500-lb thrust for the 710,000-lb take-off weight. Boeing management did not want to delay the program because they

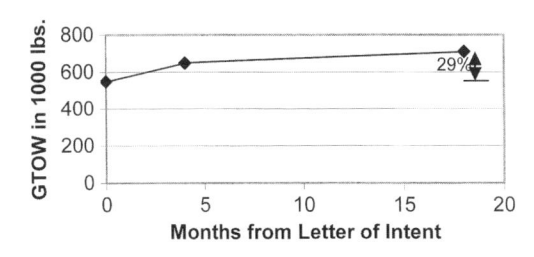

Fig. 10 Growth in maximum take-off weight of the Boeing 747.

felt it would cost too much. They demanded that Pratt & Whitney develop the increase in thrust from the present engine design—in essence giving Boeing the advanced rating of 43,500 lb from the present design at aircraft certification time instead of three years later. John Newhouse reported that Pratt & Whitney had sunk so much money into the JT9D that they went along with it, noting that "by pushing to keep the program on schedule in the face of excellent reasons to slow it down, Boeing's management made worse a situation that came very close to destroying the company" [8, p. 165]. Why did Pratt & Whitney agree at an inadequate time for a normal development schedule? Certainly the collective wisdom of Luke Hobbs and Perry Pratt most likely would have discouraged this venture. But this was a gamble analogous to the need to make the J57 successful in the B-52 because the 747, like the J57 in the B-52, was the only game in town for Pratt & Whitney.

BUSINESS RISK

When Boeing launched the 707, it had the B-52 and the KC-135 programs in hand. The situation with the B-747 was completely different: this time Boeing was taking its greatest business risk in its history when it signed up for the 747 deliveries to a very demanding customer, Pan American Airways. Perhaps Boeing's Board of Directors looked at its commitments to buying land and erecting the largest airplane building in the world and then decided "engine problems" are small potatoes by comparison even though "the engine is the heart of the airplane." Boeing's senior financial officer at that time was Hal Haynes, who put together the financing for the 747 program. The situation looked very bleak. Boeing was out of sources of money to help it get over a slump in cash flow until the program started to pay back on the investment. United Aircraft (later to become United Technologies), Northrop, and Rohr, helped Boeing with the equivalent of a $150 million loan in the form of an agreement to defer payments [4, p. 333]. United Aircraft also made a direct loan to Boeing around that critical time of about the same magnitude, as Gene Montany, former Vice President of Technology and Strategic Planning, at Pratt & Whitney told me.

Boeing climbed out of its financial predicament due to several factors, pointed out to me by Montany. The Arab Oil Embargo in 1973 brought on the U.S. government's fuel allocation program for the airlines. This situation, in addition to the movement of airlines toward the hub-and-spoke concept for more efficient operations, made the B-727 a great aircraft for increased revenues for the airlines. In other words, it was more profitable to operate full 727s than half-full wide-body transports, which on paper had lower seat-mile costs. The airlines could then provide more frequent flights, which gave their customers more choices. Thus, increased B-727 use paid for the B-747.

PROGRAM MANAGEMENT

The primary application of the JT9D was in the Boeing 747 Jumbo Jet. Bob Toft, JT9D Engineering Project Manager (Fig. 11), and his team deserve great credit for their extraordinary efforts in bringing the initial JT9D-3 and subsequent JT9D models along. A complication inside Pratt & Whitney during the JT9D-3 development in the late 1960s was the need to provide military engine parts during the Vietnam operations. The JT9D development team had to compete with the U.S. military needs in the experimental shop and on the production line. Toft's effort on the JT9D-3 started with the engine design on January 12, 1966, and the certification of the engine came less than 40 months later on April 25, 1969.

Toft joined Pratt & Whitney in June 1942 as a test engineer on the company's largest piston engine, R-4360. His subsequent career took him

Fig. 11   Bob Toft.

through engine controls on the J42 and J48 centrifugal turbojets, managing the engineering of J57 afterburning engines, development of the in-flight reverser for the JT3D turbofan, managing the TF33 program for the Lockheed C-141 Star Lifter, and conducting the company's early experimental work on high-bypass turbofans. He became engineering program manager for the JT9D turbofan in 1966.

The JT9D was a complex program with a very tight schedule for meeting the demands of the Boeing Company and Pan American Airways within, of course, budget. Pratt & Whitney had used charts to keep track of progress on previous programs but the JT9D program needed something more detailed to keep track of the tasks and the cost of each task. Toft's principal project engineers were Bob Rosati (deputy program manager), George Woodger (hot section), Bill Witherspoon (high compressor), Cliff Horne (fan and low compressor), Joe Phillips (program management) and Ted Slaiby (technology). Toft also had the help of Andy Anderson, head of the JT9D Cost Group, in setting up a Work Breakdown Structure (WBS) system to keep track of the accumulated costs of each principal task in the development of the engine. The following is a description of the WBS [9]:

> The project work breakdown structure is defined as a systematic breakdown of the overall JT9D project which proceeds downward from the definition of the program objectives through successive areas to the lowest level of detail required for effective program management. It enables assignment of responsibilities, orderly monitoring of progress and provides a basis for uniform planning and program visibility. The JT9D structure establishes the basis for:
>
> - Defining the work to be performed in successively greater detail.
> - Summarizing actual status and forecasting progress of the program
> - for progressively higher levels of management.
> - Estimating labor and material for detailed activities.
> - Constructing schedules for various activities.
>
> As engineers you are primarily concerned with technical achievement; yet your ultimate success depends upon your ability to get the job done within certain financial constraints.

JT9D ENGINE DESCRIPTION

The JT9D is referred to as a second-generation commercial engine. It took its structure from the STF200/JTF14 program, and added features to enhance its use in commercial airline operations (Fig. 12). Its bypass ratio was 5 in contrast to the first-generation's JT3D ratio of 1.4 (Fig. 13 and Table 2). Its two spools had only four bearings—a thrust bearing and a roller bearing

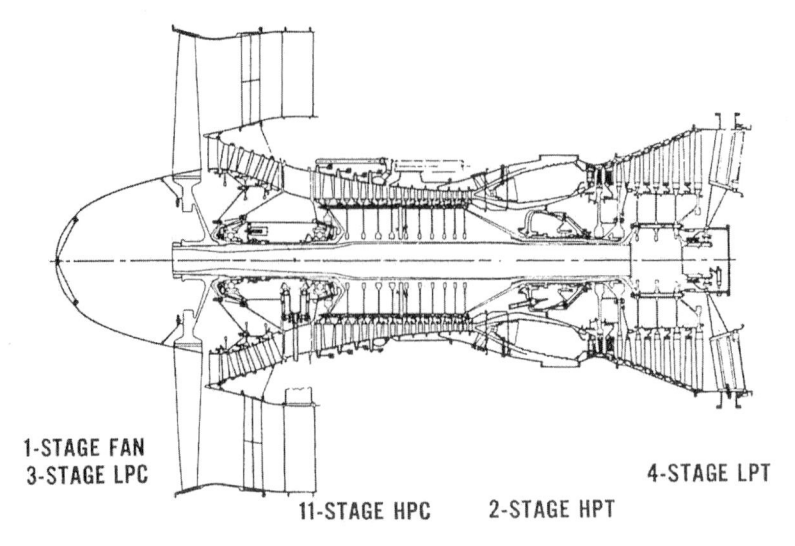

Fig. 12 Layout of the JT9D (courtesy of Pratt & Whitney Archives).

on each spool. The number of bearing supports had been reduced to three compared to the JT3D's four supports. What made the JT9D different from earlier JT3D and JT8D turbofans was the fact that the JT9D was a new engine without previous military experience. In contrast, the JT3D used the high spool from the J57 and the JT8D had the high-spool experience from the Navy J52 turbojet engine.

The elimination of the inlet guide vanes on the fan simultaneously eliminated the problems associated with inlet guide vanes deicing. One of the

Fig. 13 JT9D (left) and JT3D (right) compared (courtesy of Pratt & Whitney Archives).

TABLE 2   COMPARISON OF FIRST GENERATION AND SECOND GENERATION TURBOFANS

	First generation	Second generation
Engine	JT3D	JT9D
Bypass ratio	1.4	5.0
Bearing support structures	4	3
Number of bearings	7	4
Fan inlet guide vane	1	None
Variable stators in HPC[a]	None	In first 4 stages of HPC
Intershaft bearing	Yes	No
Overall pressure ratio	12	24
Burner type	Cannular	Annular
Turbine inlet temperature	Base	+400F
Thrust/weight	Base	+20%
Relative TSFC	Base	−20%

[a] High-pressure compressor.

problems on JT3D was associated with the intershaft bearing and seals. These problems, of course, were not on the JT9D because the intershaft bearings and seals were eliminated (Fig. 14). The rotating elements in the modules were pre-balanced so that the engine could be taken apart and reassembled without the need for rebalancing. In addition, there were a number of borescope inspection positions, strategically located, to make on-the-wing inspection more convenient. Thus, the JT9D modular concept improved maintainability (Fig. 15).

The noise patterns of a high-bypass engine were somewhat different from the first generation turbofan, where the major source of noise was in the high velocity exhaust jet from the core. The high-bypass engine had a considerably lower jet velocity, which resulted in such a reduction in noise that the fan noise

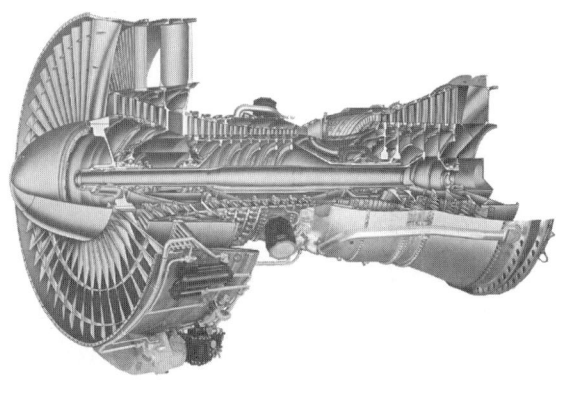

Fig. 14   Cutaway of the JT9D (courtesy of Pratt & Whitney Archives).

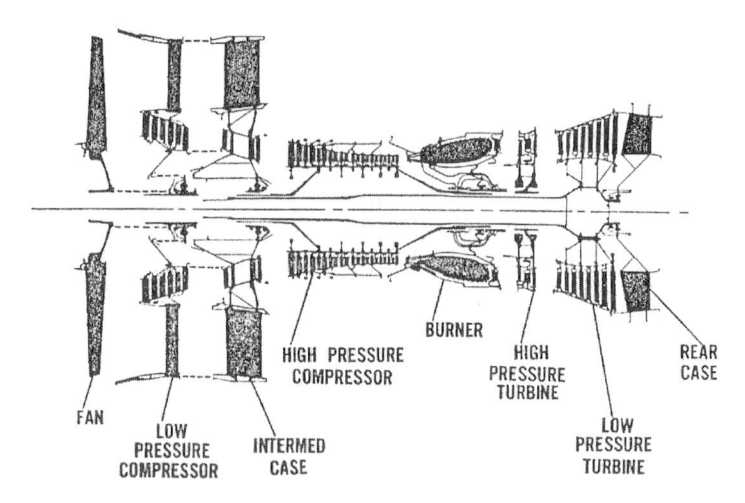

Fig. 15 The JT9D major modular assembly (courtesy of Pratt & Whitney Archives).

generation became the louder source. The sources of noise generation in the fan came from the wakes of the rotating fan blades sweeping the stationary airfoils downstream, the combination tone noise generated by the forward movement of sound waves from the rotating fan, and the noise from the turbulent flow through the fan. The approaches to sound suppression were flow path padding with sound absorption material and wider spacing of the rotating fan and its downstream stators, all of which implied weight increases.

Pratt & Whitney leased a B-52 as a flying test bed for the JT9D. Its smoke-free exhaust makes it easy to identify relative to the older J57 engines. Figure 16 shows the JT9D on the right wing (from the pilot's viewpoint). The first flight of the B-747 took place on February 9, 1969. The aircraft was so fast that the chase plane had to be an F-86. In spite of its accelerated schedule for more thrust, the 747 in its first 16 months of airline service accommodated 10 million passengers and accumulated 24 billion passenger miles. Table 3 lists the various applications, along with the engine's specifications.

Fig. 16 The B-52 powered by the JT9D and six J57 engines (courtesy of Pratt & Whitney Archives).

TABLE 3 JT9D Engine Specifications and Applications

Model	Take-off thrust (lb)	Max cont. thrust (lb)	Max cruise thrust (lb)	Weight (lb)	Diameter (in.)	Length (in.)	Aircraft
JT9D-1GT	40,500			9200	95.6	154.2	B-747 Ground test
JT9D-3 Proto	42,000				95.6	154.2	B-747 Flight test
JT9-3	43,500	35,400	33,100	8470	95.6	154.2	B-747 Flight test
JT9D-3 Block II	45,000 wet 43,500	36,400	33,100	8470	95.6	154.2	B-747-100
JY9D-7	47,000 wet 45,500	39,650	35,500	8850	95.6	154.2	B-747-100 B-747-200
JT9D-7A	47,670 wet 46,150	40,080	37,275	8850	95.6	154.2	B-747-100 B-747-200 B-747SR B-747SP
JT9D-7F	50,000 wet 48,000	40,200		8850	95.6	154.2	B-747-200 B-747SP
JT9D-7J	50,000	40,200		8850	95.6	154.2	B-747-200 B-747SP
JT9D-7Q	53,000	46,000		9295	95.6	154.2	B-747-200
JT9D-7R4A	44,300	42,800		8675	97.0	132.7	B-767-200
JT9D-7R4D	48,000	45,400		8675	97.0	132.7	B-767-200
JT9D-15	47,900 wet 46,300	40,400	38,000	8450	95.6	154.2	DC-10-40
JT9D-20	49,400 wet 46,300	40,400	38,000	8450	95.6	154.2	DC-10-40
JT9D-59A	53,000	46,000		9140	97.7	153.6	DC-10-40 A-300
JT9D-70A	53,000	46,000		9155	97.7	153.6	B-747-200

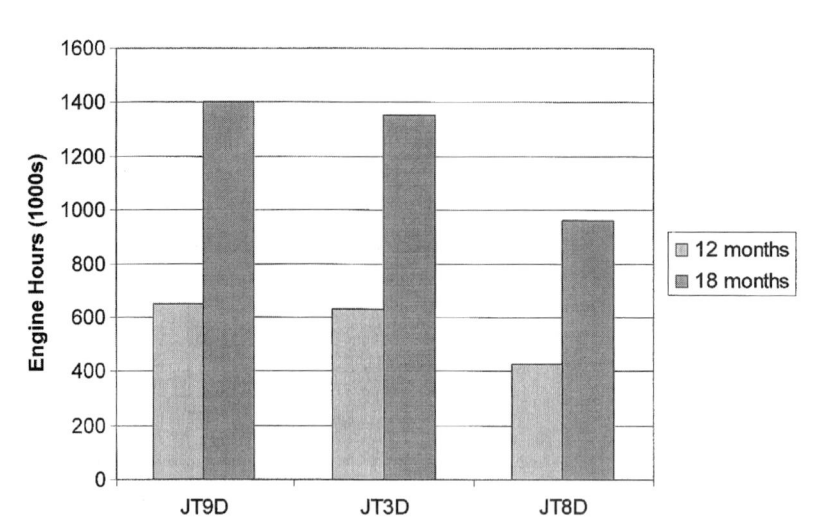

Fig. 17 The JT9D accumulated more engine-flight hours in its first 12 months and 18 months than the JT3D and JT8D engines.

ENGINE-OPERATING HOURS

There is nothing like commercial engine service to accumulate engine-operating hours. In military applications an engine piles up about 300–500 hr/yr. However, an airline puts 3000–4000 hr on an engine in a year (Fig. 17). The challenges Pratt & Whitney faced with the JT9D were complicated because of three demanding circumstances. First, there was not enough time for normal engine development. Second, the JT9D piled up engine-flight hours in service faster than previous turbofans. And third, commercial service brought to light too quickly the need for rapid redesign of certain parts. Retrofitting airline engine spare parts inventories quickly was a horrendous challenge because of the need to meet military demands for spare engine parts during the hostilities in Vietnam.

SYNOPSIS OF THE FIRST 16 MONTHS IN THE AIR

United Aircraft Chairman, William Gwinn, made the following comments at a symposium sponsored by *Time Magazine* in 1971, when the JT9D had been in service for 16 months.

> Boeing, like ourselves, a disappointed competitor in the C-5 program, was the first airframe company to commit itself to build a wide-bodied jetliner, based on an initial order for 25 747s from Pan American World Airways in April 1966. In placing the first order for the 747, Pan Am again demonstrated its long-practiced philosophy of pioneering the introduction of new aircraft offering dramatic improvements in comfort, service, and economy. Pan Am had exhibited similar leadership in

helping launch the commercial jet age by placing the initial order for the 707 back in 1955.

Boeing chose the JT9D as the powerplant for the 747, specifying a take-off thrust of 41,000 lb. Production models at this thrust were scheduled for delivery beginning in December 1968. In addition, we offered ... to provide a new, growth model of 44,000-lb take-off thrust for delivery three years after the 747s entry into scheduled airline service, or, in other words, January 1973.

However, it subsequently became apparent that Boeing could not meet customer requirements without substantially increasing the plane's weight. This weight increase demanded more than the 41,000-lb of take-off thrust planned for the initial production engines. In short, a more powerful engine was essential to a successful aircraft program.

We had no choice, therefore, but to support Boeing. In lieu of the original 41,000-lb thrust engine, a 43,500-lb thrust powerplant was now required. Because of commitments made by Boeing to the airlines, this increased thrust ... was needed at approximately the same time as the original engine. This had the effect of compressing our development schedule for the growth model by more than three years. The delivery of a higher thrust engine at an earlier date than initially planned was a difficult undertaking, but it was essential to the success of the 747 program. Despite the accelerated schedule, the first engine was delivered on time in April 1969. This increased thrust was made necessary because the takeoff weight of the 747 had increased from the 625,000 lb planned when the JT9D was selected by Boeing in 1966, to 710,000 lb, an increase of more than 40 tons.

The 747 was introduced into passenger service at a very rapid rate beginning in January 1970, much faster than that for the predecessor 707. Nearly 100 747s were flying by the end of the first year. After scheduled service began, technical problems were encountered that caused engines to be removed from service for repair. Despite widespread publicity about engine troubles, the number of times engine had to be removed for either inspection or repair to date compares favorably with the JT3D experience at the same stage of its life.

In considering the 16 months since the JT9D entered airline service, one fact stands out: More engines accumulated more hours of operation in that period than any other new engine in history, and did it with a perfect safety record.

EVOLUTION OF THE JT9D FAMILY

The progression in engine models from the first production engine (JT9D-3A) up through the JT9D-59A and JT9D-70A evolved throughout the 1970s. The JT9D-20 powered the Douglas DC-10-40. The JT9D-59A powered the Airbus A300. The JT9D-70A powered the Boeing 747-200. The group of engines from the JT9D-7J through the JT9D-7Q2 powered the

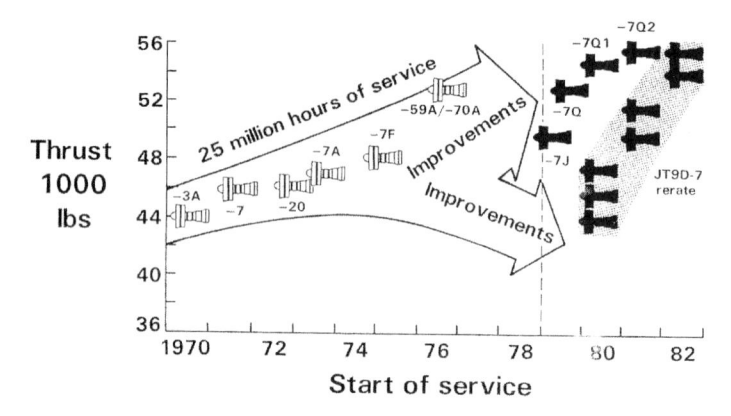

Fig. 18 Succession of engine models to meet increasing thrust requirements (courtesy of Pratt & Whitney Archives).

B-747 and had improved performance. The rerate engines were engines derated from their full potential thrust and made lighter because of their lower operating temperatures as a result of the derating.

The rerated engine series used parts appropriate to the reduced thrust requirements and performance improvement. The following are some of the changes:
1) Lower aspect ratio fan blades
2) Added another low-compressor stage
3) Improved materials in disks to reduce weight
4) Active clearance control in the high-pressure turbine
5) Single crystal blades in the high-pressure turbine first stage
6) Titanium blades in the last turbine stage
7) Carbon seals in the bearing compartments to reduce air leakage

Figure 18, created in 1978, shows the succession of engine models to meet the increasing thrust requirements. Figure 19a shows the variation in thrust specific fuel consumption for the rerated series of JT9D-7R4 engine models. The lower end of the curve shows an improvement in thrust specific fuel consumption of about 6 percent for the B-767 and A310 applications. Figure 19b shows the variation in thrust size.

Boeing designed the nacelle for the JT9D-7A in the B-747. It was a tightly wrapped nacelle for minimum drag and the engine accessories were located on the core engine. The nacelles for the JT9D-20 and the JT9D-59A were called the Common Nacelle System (CNS), and were larger because the accessories were mounted on the fan case for easier maintenance. When the larger JT9D-70A engine in the CNS was installed on the B-747, its installed performance was not up to expectations.

A change was made to the location of the engine accessories on the core engine, and the Boeing-made tightly wrapped nacelle provided a performance improvement on the B-747. The certification of this combination was done in

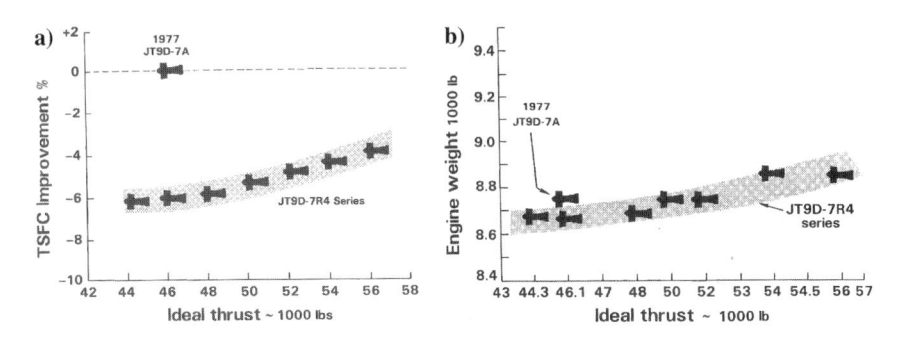

Fig. 19 a) Thrust specific fuel consumption, and b) the variation in 7R4 series engines with thrust size (courtesy of Pratt & Whitney Archives).

1978, using a Northwest Airlines B-747. The engine model for this nacelle was the JT9D-7Q, which resulted in a three-percent increase in NAMs (nautical air miles). The total installation weight reduction in airplane weight (four engines with reduced weight nacelles) was about 6000 lb.

In addition to the performance and installed weight improvements, Pratt & Whitney worked with the airlines and the airframers to make it easier to access the engines.

ADVENTURES IN MARKETING

COMPETITIONS FOR BOTH THE A310 AND THE B-767

Airbus was created in 1970 as a French and German joint venture, that later included the United Kingdom and Spain, to design and manufacture commercial aircraft. Since that time, it has become one of the world's significant producers of commercial aircraft, with about 4000 aircraft in service by 2006 throughout the world. In early 1978, there appeared to be two new potential engine opportunities on the horizon with Airbus and Boeing in the 200-passenger size aircraft. Airbus was marketing its A310, which was a twin-engine installation similar to its A300. The one from Boeing was an "either or" situation for the 200-passenger size known as the B-7X7. One possibility was a twin-engine configuration like the A310 and the other was a trijet with a B-727 type of configuration. United Airlines was considering Boeing's twin and the Airbus A310. On the other hand it seemed that American Airlines was comparing the Boeing trijet with the Airbus A310. From Pratt &Whitney's viewpoint, the Boeing trijet would have a PW2000, while the Boeing twin would need an engine smaller than the current JT9D. It seemed likely that Boeing would offer only one aircraft rather than two 200-passenger aircraft.

Pratt & Whitney's marketing organization assessed the Boeing twin aircraft as one that would be extremely difficult to win because United Airlines would be the most likely launch customer. First of all, the thrust requirement from Boeing was 40,000 lb. United had DC-10s in its fleet powered by CF6-6D engines rated at about 40,000-lb thrust. This engine would be a natural for what Boeing described as its 767 aircraft, a one-stop transcontinental aircraft. Boeing talked about 40,000 lb as all the thrust United would need in an engine, and United already had the engine in the inventory. United was familiar with the engine operation not only in its DC-10 fleet, but also was experienced in maintaining the engine on the flight line and in the shop. Obviously there had to be great savings in selecting the General Electric CF6-6D engine. General Electric would be in a position to offer a lower price on its engine as it came down the production learning curve.

The only engine Pratt could propose quickly would be a JT9D, which was too big in thrust and too heavy, according to airplane performance studies. The general feeling at Pratt & Whitney was, "If United accepts Boeing's twin and launches the program before American, then we don't have a chance because the CF6 is already in United's fleet and is the right size If American goes first, then we have a reasonable chance in the Boeing trijet with our PW2000." Thus, the nightly prayers at that time in East Hartford among the aviation cognoscenti were for a timely American Airline launch of the Boeing trijet.

In 1978 I was working for Bob Rosati, senior vice president of the Program Management Office (Fig. 20). I don't know how it happened but suddenly I became the marketing manager in charge of the proposal response to United Airlines' RFP for an engine to power the B-767 twin. I got the assignment not because management perceived me as a brilliant marketer (a perception only my mother held) but rather because I suspect nobody else would be associated with such an obviously hazardous and career-tarnishing opportunity. On the bright side, however, I was given reasonable freedom to select a team for the proposal effort.

Fig. 20   Several of the key players on the B-767 twin proposal: a) Bob Rosati, b) Bill Andersen, and c) Don Rudolph (courtesy of Pratt & Whitney Archives).

I enjoyed working for Rosati because he was a serious student of history and had an uncanny insight into human behavior. He introduced me to the concept of *heat seekers*. The Sidewinder missile had a heat sensor in its nose and its control system directed the missile towards a source of heat—usually jet engine exhaust. Rosati explained how certain types of people are heat seekers, those who become associated with winning projects and as a result garner a not insignificant amount of management attention—without investing much effort in the successful project. I did not really appreciate his point until after the 767 marketing campaign was over. My favorite Rosati-ism was "The combat veterans never get to march in the victory parade." The proposal team, after about five months of 70-hr weeks, would have been too tired to march anyway.

I was fortunate to get a team of the best people for this assumed *Mission Impossible*. The man who understood and knew the United Airlines personnel was Bill Andersen (Fig. 20). The best man to get things done inside Pratt & Whitney was Don Rudolph (Fig. 20). Another capable engineer to help Rudolph with the inside work was Howie Latimer. Our secretaries were Dolores Pinwar and Terry Wagner: Not only were they the best but they also could have earned a Mother Teresa Patience Award as they typed and retyped many versions of transmittal letters as each rung in the chain of command made their inevitable changes. In order to appreciate the efforts of these dedicated women, one must recall that they labored over typewriters without the relative ease of word processing software!

There is a lot to be said for working on a project where everyone avoids you. Nobody bothered us. It was as if there was a sign in our work area that read "Quarantine: Losing Project. Possible side effects are headache, nausea, and diarrhea." Even top management left us alone, not exactly a bad circumstance. Our happy little team had its own private area. As I think back, our operation was more like a scene from the television show *Mash* rather than a highly disciplined, sedate organization in the orderly pursuit of marketing excellence.

We needed the help of the Engineering Department. So we scheduled a general meeting in one of the conference rooms in the Engineering building for any who could help us on the marketing campaign. One of the senior people stood up, claimed we did not have a chance in the world, and said he was not wasting his time in this meeting. Then he swaggered out of the conference room, leaving an astonishing wake of silence. I believe at that moment one of our team, most likely Don Rudolph, declared, "Well, if that's what he thinks, we're not going to invite him to the victory celebration party!" The meeting broke up with the understanding that Engineering would definitely help us.

Bob Rosati already had the Engineering Department working on taking weight out of the JT9D-7A. Andersen, Rudolph, Latimer, and I had a very old-fashioned marketing plan. We would find out what the customer needed and give it to him in accordance with the best interests of Pratt & Whitney.

One of the problems facing a vice president of a maintenance department in an airline is the uncertainty of his monthly maintenance expenditures. He has a yearly budget, but sometimes he can blow that in a few months if some serious problem comes up unexpectedly. We thought about offering a guaranteed maintenance plan so that the airline maintenance vice president would not have to worry about being way off-budget and looking bad in the monthly reviews.

Our team approached the Pratt & Whitney Product Support Department at the lower level and got a study underway to see what we would have to charge for a 10-year guaranteed maintenance plan for United. The vice president of Product Support had lunch in the executive dining room and, as was the custom, the company president might sit down at the table and start the conversation with, "What are you guys up to these days?" One day the Product Support vice president innocently mentioned that Connors had asked for a 10-year maintenance guarantee for United Airlines. We were not offering a 10-year maintenance guarantee to the airline. We were studying the situation to see how much we would have to charge—all subject, of course, to the president's approval. Unfortunately, Dave Hines, Commercial Products President, took it to mean that Connors was making a direct offer to United without even discussing it with him. Dave sent me megabites of high-temperature information at 186,292 mps over the internal phone line, ordering me to discuss things like that with him before making commitments. End of conversation—period! I never got the chance to explain. Yes, sometimes an innocent man is sent to the gallows.

Bill Andersen, our Marketing Account Manager for United, knew the people at United and understood their needs. Bill helped us to put together a package that would appeal to the airline's management by satisfying their most pressing needs. Our proposed engine satisfied Hobbs' recommendation of having at least a 20-percent thrust growth capability (because no matter what the customer says to the contrary he is going to demand greater thrust in the near future). Our offer to United was an engine that admittedly was heavier than the competition for a one-stop transcontinental aircraft. However, if United wanted to get 15 to 20 percent more thrust for a non-stop transcontinental B-767, nothing more was required than a small adjustment in the fuel control. On the other hand, the competition would either run higher turbine temperatures or go to a bigger engine, neither of which ideas would appeal to United.

In July 1978, we had a meeting with United in San Francisco that (as far as I can recall) included Dick Tabery (Vice President of Maintenance) and Bob Collins (Vice President of Operations) from United with Rosati, Andersen, and me from Pratt & Whitney. Rosati made a low-key, conversational-style presentation using only chalk on a blackboard. His main points were that the

HIGH-BYPASS FANS

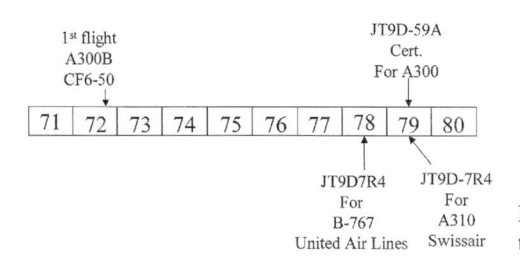

```
                            JT9D-59A
   1st flight               Cert.
   A300B                    For A300
   CF6-50                      │
      │                        ↓
┌──┬──┬──┬──┬──┬──┬──┬──┬──┬──┐
│71│72│73│74│75│76│77│78│79│80│
└──┴──┴──┴──┴──┴──┴──┴──┴──┴──┘
                      ↑    ↑
                   JT9D7R4  JT9D-7R4
                     For      For
                    B-767    A310
                 United Air Lines  Swissair
```

Fig. 21 Time line for second generation turbofans.

JT9D-7R4 had the capability of making the aircraft non-stop transcontinental whenever United wanted it at no extra cost, and in the meantime, the engine in the B-767 (one-stop transcontinental) would be operating at lower than present normal turbine inlet temperatures for the JT9D in the B-747. The increased engine hot-section life implied in the second point was not lost on the United attendees.

In July 1978, many people at Pratt & Whitney were surprised that the so-called heavy JT9D launched the B-767 with United. Figure 21 shows the time line for the JT9D-7R4 for the Boeing 767 and Airbus A310; the JT9D-59A gets into the A300 program after about seven years after the aircraft's first flight. The next scene was my introduction to the heat seekers who were at the victory celebration at the Hartford Golf Club. Those who had protested that this campaign was a waste of effort were the ones bending CEO Harry Gray's ear on the esoteric subject of "how we won." The evening was a great physical and mental relief to the non-marching combat veterans—strategically positioned at the open bar.

INTERNATIONAL MARKETING PROPOSALS

After the United Airlines campaign, I was transferred to International Marketing. That old adage about "no rest for the weary" must have come into play when I found myself as Director of Marketing–Europe. That title sounded best in France when the Chief of Staff (Chef du Cabinet) at Air France introduced me to management as *Monsieur "Ko-Nohr," Le Directeur de Vente Internationale.* My few moments of fame in Paris quickly faded when I soon became immersed in one proposal after another for the next few years.

Perhaps it was the luck of the Irish at work, but I inherited a first-class team in my staff covering Europe. The first tough campaign was trying to influence Swissair in a JT9D-powered B-767 or A-310. This was like the 767 campaign, a launching situation because nobody had bought the A310 yet. The luck I mentioned was having people like Barry Lucas, Dick Inman, and Karl Domeisen working the challenge at Swissair. In subsequent campaigns, Bill Woodburn, Bob Cook, Jack Finnegan, George Peterson, and Bill Patterson helped to advance my career. It was not long before it was announced that Swissair signed up for the A310 powered by the JT9D and would be the first in Europe to add the A310 to its fleet (Fig. 21). Airbus arranged its production positions so that Swissair (with Pratt & Whitney engines) and Lufthansa (with General Electric engines) would receive their A310 aircraft around the same time.

Soon my team was creating a new proposal about every six weeks, a strenuous activity. A proposal ascends the review hierarchy, consisting of the departments of Engineering, Legal, Finance, Marketing, as well as the offices of the executive vice president and the president of Commercial Products. After the proposal manager survives this bureaucratic jungle, he wafts it by the chairman of the corporation, Harry Gray, who needs to be kept aware of potential effects on the cash flow situation.

One time, under a very tight schedule, I sent a copy of a proposal to Mr. Gray's executive assistant with the request to hold onto it until the Pratt & Whitney president, Dave Hines, had approved it. The corporate office was located across the Connecticut River in Hartford (about 30 minutes away). While waiting for an appointment with Hines late that afternoon, I got a call from Gray's assistant who was happy to tell me he approved the proposal. Between that moment and the subsequent 60 or so minutes outside of Hines' office I developed a severe case of the dreaded "I'm about to be fired syndrome!" You could not imagine the mental relief I experienced when Dave Hines gave me his approval and then told me to be sure to show it to Harry.

My boss at that time was Jim Kennedy, Vice President of International Marketing (Fig. 22). He was a natural born salesman who had a remarkable

Fig. 22 Jim Kennedy.

HIGH-BYPASS FANS

Fig. 23 A happy moment with the 1970 Collier Award; seated from left to right: Art Smith (President of United Aircraft), Bill Gwinn (Chairman of United Aircraft), and Barney Schmickrath (Engineering Manager); standing from left to right are Dick Baseler (Engineering Vice President), Bob Rosati (Deputy JT9D Program Manager), Bob Toft (JT9D Program Manager), Don Nigro (Manufacturing Vice President), and Gus DeCamilis (Manager, Manufacturing) (courtesy of Pratt & Whitney Archives).

intuition in assessing risks and taking action quickly. I got a kick out of General Electric's Robert Garvin, who lamented in his book that he wished General Electric had on-the-spot authority to make decisions as the Pratt & Whitney rep did [7, p. 236]. Garvin was referring to Jim Kennedy in China where Kennedy would make commitments when the marketing opportunity was hot, without the time-consuming task of seeking management approval. Kennedy taught me that it was easier to explain to management why it cost so much to win the competition than to explain why we lost. And win we did: Fig. 23 shows management gathered to celebrate JT9D's win of the 1970 Collier Award.

ADVENTURES IN INTERNATIONAL MARKETING

One of our amusing exploits (in retrospect) was with a sales campaign at Britannia in 1979. The day before we arrived, General Electric's Brian Rowe, a former U.K. citizen, made his sales pitch for Britannia to purchase the aircraft with his company's engines. Next the Pratt & Whitney marketing team of Kennedy, Connors, and Finnegan shows up. It was not one of our most successful marketing forays.

The tour of duty in domestic marketing was in sharp contrast to marketing in Europe. In the United States, one had to convince the various organizations in the airline of the soundness of one's proposal, covering technical, financial, and how things will work out as proposed. However, in Europe, governments own the airlines. The first challenge is to sell to the airline and the bigger challenge is to sell to the government. In the first challenge one talks with engineers, financial specialists, lawyers, and management people. The technical and financial merits of the proposed business deal are of primary concern to the airline customer. In the second challenge, one must deal with politicians. The technical merits tend to fade from view compared to how many jobs can be created as a result of one's business proposition.

After I had been covering Europe for a while and had a chance to meet many people in the various airlines, I got a call from Pratt & Whitney's Military Marketing Department in Florida for help with a potential military sale in a small European country. This particular country's airline was supplying aircraft engine experts to the military people to help in the engine evaluation process. After our meeting with the evaluation team, one of the airline men told me we were not bringing enough muscle into the campaign. He said your competition's chief salesman is the Premier of France, who talks about the "French" engine, which of course was really a General Electric engine. In addition, this "French" salesman can cater to the political concerns of the prime minister of the small European country. This French influence is discussed in more detail by Newhouse [8, p. 39], where Mr. Giscard d'Estaing in March 1980 visited a number of Arab countries and spoke of political matters in a way that made it extremely difficult for Boeing to sell its 767 against the French-led A310. The underlying message appeared to be that the French and the Arab world were together against the Americans, who were on the side of Israel.

Despite the complications of international sales reviewed in this chapter and an evolving domestic marketplace discussed in the previous chapter, Pratt & Whitney continued to refine its second generation turbofans into the engines that are developed and sold today. While Pratt & Whitney's attention was focused on the development of the JT9D in the Boeing 747, General Electric was looking at ways of using its TF39 high spool (as Pratt & Whitney did earlier with its J52 high spool) toward a highly successful commercial turbofan. Also, Rolls-Royce was preparing to enter the forthcoming wide-body aircraft engine market. Pratt & Whitney found the competition in commercial engines was ratcheting up. The story of the further evolution of high-bypass commercial engines continues in the next chapter including how the Department of Defense saved money by buying a commercial engine for its C-17 military transport.

References

[1] *The Power Plant*, Vol. 22, No. 22, Pratt & Whitney, December 17, 1965.
[2] *Aviation Week & Space Technology*, May 11, 1964, p. 22.

[3] Whittle, F., "The Birth of the Jet Engine in Britain," *Celebration of the Golden Anniversary of Jet Powered Flight 1939–1989*, AIAA (Dayton-Cincinnati Section), Dayton, OH, Aug. 23, 1989 (article provided by Smithsonian Institution).
[4] Serling, R. J., *Legend & Legacy, The Story of Boeing and Its People*, St. Martin's Press, New York, 1992.
[5] Bender, M., and Altschul, S., *The Chosen Instrument: Pan Am, Juan Trippe, The Rise and Fall of an American Entrepreneur*, Simon & Schuster, New York, 1982, p. 503.
[6] "Pan American World Airways Press Release," Pan American, April 1966.
[7] Garvin, R. V., *Starting Something Big: The Commercial Emergence of GE Aircraft Engines*, Library of Flight, AIAA, Reston, VA, 1999.
[8] Newhouse, J., *The Sporty Game: The High-Risk Competitive Business of Making and Selling Commercial Airliners*, Alfred Knopf, New York, 1982.
[9] Toft, B., "JT9D Project Work Breakdown Structure Memo," Pratt & Whitney Archives.

Chapter 14

THE MODERN ERA

INTRODUCTION

In this chapter we will see the whole spectrum of high-bypass ratio engines. A century ago the Wright brothers produced one hundred pounds of thrust with two propellers driven by a 12-hp engine. Today, a 100,000-lb thrust level from a single engine is possible. Furthermore, the degree of engine reliability is so high that should there be an engine shutdown on a twin-engine aircraft, a 3-hr flight on one engine to a suitable airport is not considered an unacceptable risk—perhaps an unimagined possibility 100 years ago. We will also review the Integrated Product Development (IPD) process, which evolved during the development of the F119 engine.

JT10D PATH TO THE PW-2037

When I first met Walt Swan at Boeing in 1967, he told me, "You guys ought to develop a 25,000-lb thrust engine for the B-727 replacement. That opportunity will be coming along." At that time, it was difficult to detect much interest in such an opportunity because of the lack of an aircraft just dying for such an engine. An engine company does not usually start a two-billion dollar development program without a clear-cut opportunity for recovering its investment. The path to the JT10D goes back to the Air Force's Advanced Turbine Engine (ATE) program in the early 1970s (Fig. 1). I was program manager of the ATE program up through running the JT10D demonstrator engine in 1974. In the late 1960s the Air Force had issued an operational requirement for military transports. After a number of studies and coordination meetings between the U.S. Army and U.S. Air Force, the latter issued a Required Operation Capability (ROC) document in mid-1970. This document led, near the end of 1972, to the Advanced Medium STOL Transport (AMST) program for the demonstrator aircraft. Boeing and McDonnell Douglas were the winners and were to produce two demonstrator aircraft for each contractor. The Advanced Turbine Engine (ATE) program for a new high technology turbofan engine to be used in the AMST was part of the AMST program.

In early 1971, the French engine company SNECMA was anxious to enter the commercial engine market with an American partner to design, develop,

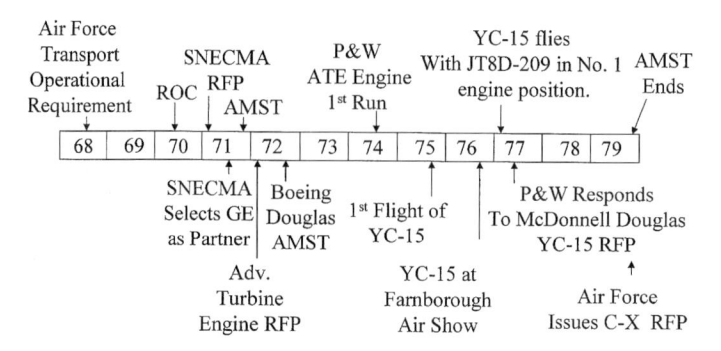

Fig. 1 Time line of events leading to the JT10D.

and produce a "10-tonne engine" (22,000 lb). Pratt & Whitney had a long history of cooperation with SNECMA; in fact, Pratt & Whitney was about a 10-percent owner of the French company and even held a position on its board of directors. René Ravaud, head of the engineering operations, was a longtime friend of Pratt & Whitney and was looking forward to developing such an engine. SNECMA held extensive meetings with both General Electric and Pratt & Whitney. Both companies submitted proposals for the development and production of the 22,000-lb thrust engine.

General Electric made a bold offer to provide the high spool from the Air Force B-1 engine as its 50-percent contribution to the joint venture. This move would provide the French with the latest high technology and would very likely enhance the marketability of the engine. On the other hand, the Pratt & Whitney proposal, based upon current commercial engine technology, was less attractive. This was a fortunate situation for General Electric because it did not require the financial commitment for the development of a new high spool. It was already developed at Air Force expense. General Electric's offer of the B-1 engine technology to the French initially incurred the wrath of the Air Force because of the concern about exporting technology; however, in time the concern and wrath subsided.

In early 1972, the Air Force issued an RFP and Pratt & Whitney responded with a 23,000-lb thrust engine (PW-1D) with a bypass ratio of 6 and an overall pressure ratio of 21.5. This was an all-new engine. General Stewart at Wright Field was in charge of the engine and aircraft programs. Just when it looked as if there was a real opportunity for a new engine program, the Department of Defense (DOD), with its melting research & development (R&D) financial resources, asked General Stewart to choose between a new engine and a new airplane because there was not enough funding for both programs. The General chose the new aircraft program, which created the YC-14 (Boeing) and the YC-15 (McDonnell Douglas).

Pratt & Whitney had its ATE engineering team in place and continued on with the design and construction of an experimental engine, which first ran in

August 1974. Project engineer Sel Berson was in charge of the engineering, and he and his team did a great job in reducing an engine concept to working hardware. Berson was ably supported by Frank Latanzio (design) and Nils Carlson (technology). Bill Weaver was in charge of marketing. The program was essentially a technology program to explore approaches to engine technology beyond what was in current engines.

The AMST program was officially launched near the end of 1972. Boeing's technical approach to the YC-14 was to mount two large high-bypass engines on top of the wings and use the engine exhaust to help turn the flow downward through the aircraft flaps on take-off and landing. This technique was called upper surface blowing (USB). Boeing's engine selection was General Electric's largest commercial engine. On the other hand, McDonnell Douglas followed a more conventional approach in its YC-15 (Fig. 2), with a four-engine installation under the wings. The engine exhausts were deflected downward for increased lift on landing and take-off by the extended flaps. This type of lift augmentation was called externally blown flaps (EBF), which required titanium flaps to withstand the hot exhausts from the JT8D-17 turbofans. The JT8D-17 engines used fluted exhaust nozzles to mix ambient air with the hot exhaust.

Marvin Marks, Executive Vice President at McDonnell Douglas, was the enthusiastic motive force behind the YC-15, which made its first flight in August 1975. In the following year, he sent one of his two demonstrator aircraft to the Farnborough Air Show and on a subsequent trip around Europe. He showed considerable interest in the JT10D, which was an outgrowth of the ATE Demonstrator program. During one of his trips to Pratt & Whitney in 1976, he hoped to obtain from management a firm outlook for the future

Fig. 2 The YC-15 demonstrating STOL capability (courtesy of Pratt & Whitney Archives).

of the JT10D. Such a commitment was not in the cards because of the lack of a commercial application on the horizon. Marks was disappointed because he realized the full potential of the YC-15 could be achieved best with a high-bypass turbofan in the 25,000-lb thrust category.

The YC-15 demonstrated the installation of the JT8D-209 in the number one position in early 1977. In the second quarter of 1977, Pratt & Whitney responded to a McDonnell Douglas RFP for JT8D-17 and JT8D-209 engines in the C-15. Later in 1977, McDonnell Douglas submitted its proposal to the Air Force for a C-15 powered initially by the JT8D-209 and later the CFM56. The flight test program continued with the JT8D-17, and then was cancelled in late 1979. I understand the AMST program objectives were achieved by both the YC-14 and YC-15 aircraft programs. The C-17 program came along later in the year and the Air Force RFP for the C-17 was due in very early 1981.

Ed Granville, the youngest of the Granville brothers, was a man of exceptional ability in charge of the Experimental Assembly Floor where the ATE engine was built (Fig. 3). Even on the list of Engineering Programs, ATE was just a pigeonhole in the budget where money could be stored in a holding tank until needed for more important operations. Ed saved me on many occasions when the official priority list would not even include ATE. He knew that the JT9D and the TF30 requirements were of top priority, but he also believed in moving the advanced engines along. When I would ask him about his schedule for the ATE, he would give me a wink and a smile as he said, "Don't worry. We'll get it done!" And he did.

Fig. 3 Ed Granville (right) explains to the author how an experimental engine with the lowest of priorities can still be built (courtesy of Pratt & Whitney Archives).

When Granville retired in 1976, I was the master of ceremonies at his retirement party. I asked my friend Bill Russell (Vice President in United Aircraft International and later Pratt & Whitney International Marketing) about getting aviation pioneer General James Harold "Jimmy" Doolittle to write a nice greeting. Russell gave me a phone number for someone in New York City. I told the gentleman who answered the phone what I would like to ask Doolittle. His enthusiastic response, as best as I can recall after about 28 years, was something like this: "Here's Jimmy's phone number and address, and his secretary's phone number. Tell Jimmy I said he should get this letter out right away." What I learned a few hours later was I was talking to Dick Knobloch, one of Doolittle's B-25 pilots on the famous mission over Tokyo in 1942.

In General Doolittle's greeting to Ed Granville, he said, "This note is to thank you for all you have done to advance aviation and to extend you every good wish for happiness in whatever you decide to do from now on. All the best." Then he signed the letter with, "As ever, Jim." I had the letter framed; Bruce Torell, Pratt & Whitney President, presented it to him at the retirement party; and Ed was happy to mount it on his wall. What I treasure is Doolittle's cover letter to me, which started with "Dear Jack" and signed "As ever, Jim."

JT10D Moves On

After the demise of the ATE program, the program continued as a technology demonstrator program. When Harry Gray succeeded Art Smith as CEO at United Aircraft, he gradually became familiar with the programs going on at Pratt & Whitney. I believe it was in early 1972 when Gray reviewed the JT10D program. Gene Montany, in charge of Advanced Planning at the time, helped me with the presentation by giving me a couple of charts that showed the tremendous market for a 25,000-lb thrust engine. As I remember, the numbers turned out to be conservative. The hesitancy to proceed with a real development program was a lack of Boeing, McDonnell Douglas, or Lockheed commercial aircraft being offered to the airlines. Gray understood the need for foreign partners in this risky venture for two reasons: 1) to help share the financial risk, and 2) to make the engine more marketable in areas where foreign engine companies had political influence. A look at sales at that time showed that the percentage of engine business overseas was getting larger than the domestic business.

The JT10D activity over the next few years was concentrated on providing the airframe manufacturers with study engine data for their commercial aircraft (Table 1). The fuel crisis of 1973 and 1974 put the emphasis on fuel economy. Boeing was looking at the whole spectrum of aircraft configurations to meet the anticipated needs of airlines. Gradually the 7X7 program evolved into two products, the B-767 and the B-757.

TABLE 1 DATA FROM JT10D STUDIES

	JT10D-2	JT10D-X	JT10D-3
Potential certification date	1980	1979	1983
Take-off thrust (lb)	24,600	19,700	28,600
Bypass ratio	5.67	2.89	5.4
Overall pressure ratio	24.0	27.1–29.2	27.3
Fan tip diameter (in.)	68.3	53.0	68.3

Dick Mulready, from the Florida Research & Development Center, the RL10, and the liquid hydrogen products, returned to East Hartford and took over the JT10D as a joint venture with foreign partners. In the spring of 1973, Bruce Torell sent Dick Mulready and Gene Montany to Europe to put together an international collaboration for the design, development, and production of an engine in the 25,000-lb thrust class. By the time of the Paris Air Show that year, they had lined up Motoren & Turbinen Union (MTU) in Germany, as well as Fiat and Alfa Romeo in Italy. Later on, Rolls-Royce joined the venture, and soon the whole collaboration was designing advanced versions of the JT10D from 25,000- to 30,000-lb thrust.

Most of us did not see either what was unfolding before us in the 22,000-lb thrust class or the subtle movements of the French government. Neither the CFM56 nor the JT10D were considered certified engines in the early 1970s. Pratt & Whitney would not proceed with or without foreign partners unless there was a relatively certain market for aircraft powered by 25,000-lb thrust engines. On the other hand, the French government ordered a few Boeing surveillance-type aircraft with the requirement for a "French engine." Slowly the CFM56 attained certified status, and in addition the costs of certifying the engine on the airplane seemed to be no problem, as it would be for the JT10D.

Another re-engine opportunity arose when about 20 Douglas stretched DC-8s looked for a high-bypass engine in the 22,000-lb thrust class. There was no competition for the CFM56 at that point. The next opportunity was for the B-737-300. By that time, the CFM56 had such a head start that nothing could catch it in spite of several Boeing executives' solemn vows never to have a single engine source on another Boeing aircraft. The B-737 turned out to set world production records for commercial aircraft. Montany mentioned to me that the U.S. government's deregulation policy in the 1970s helped to move the airline operations in the direction of hub and spoke type. This made the B-737 type of aircraft, launched in 1965, very much in demand to bring passengers into and out of the hub.

As early as 1970, Montany, whose group concentrated on long-range planning for the Pratt & Whitney Division, urged management to consider the 25,000–30,000-lb thrust engine. Montany told me that he wrote a confidential memo to the president of Pratt & Whitney in October 1971, "Since the 30,000-lb thrust turbofan engine may dwarf even the JT8D in number of applications and production volume, its importance to Pratt & Whitney Aircraft is obvious. Because no single application appears to be large enough to support the development of an engine, the first engine on the scene could very well capture the entire commercial market, and the military market as well." The great success of the CFM56 has given Montany's prediction legitimacy, but unfortunately success to the wrong company—General Electric.

A significant focus of the JT10D program was an exercise in designing a gas turbine for as low a manufacturing cost as was feasible. We had a member of the Production Department participate in the design process. Figure 4 shows how the JT9D provided about a 15-percent reduction in specific fuel consumption over the JT8D and JT3D engines. This improvement came, however, at the expense of an engine with many more parts. The JT10D exercises showed ways to cut down on the number of airfoils by using wider chords. For example, on either a compressor or turbine rotor, if we double the chord of the airfoils, then the number of airfoils would go down by a factor of two. The JT10D studies also showed ways of reducing fuel consumption by higher overall pressure ratio and higher bypass ratio, as well as other ways of refining the uses of cooling air and the application of a full authority digital electronic control (FADEC). Figure 5 is a vivid comparison of the use of wider chord airfoils in the JT10D engine relative to the airfoils in the JT9D. The longer chord airfoils, for the same overall pressure ratio in the case of the

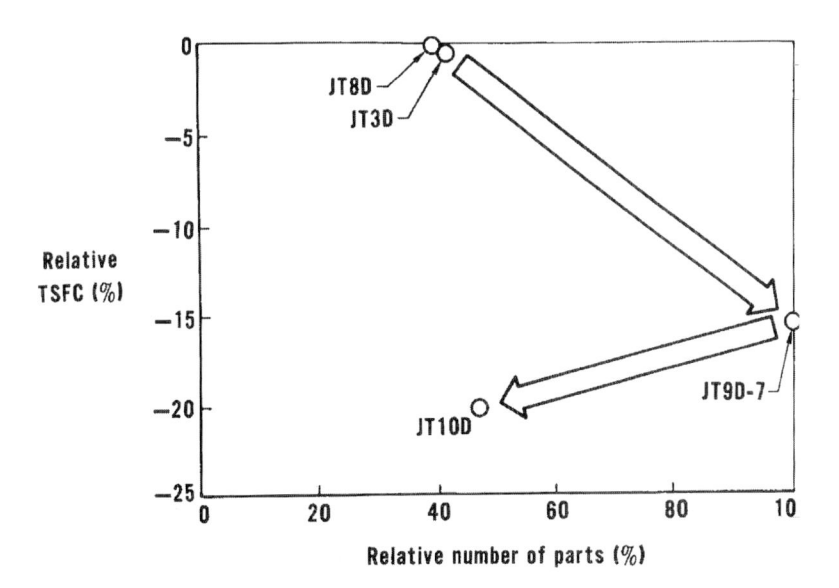

Fig. 4 Comparison of number of engine parts in various gas turbines (courtesy of Pratt & Whitney Archives).

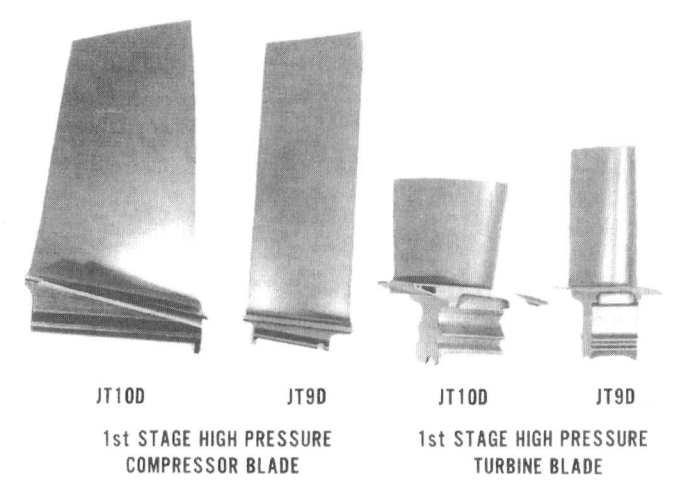

JT10D JT9D JT10D JT9D
1st STAGE HIGH PRESSURE   1st STAGE HIGH PRESSURE
COMPRESSOR BLADE   TURBINE BLADE

Fig. 5 Comparison of the JT9D and JT10D blades (courtesy of Pratt & Whitney Archives).

compressors, result in a longer engine between the engine inlet flange and the last stage of the low-pressure turbine.

EVOLUTION OF THE B-757 AND PW2037

In January 1981, Robert J. Carlson, president of the Pratt & Whitney Aircraft Group (in the United Technologies Corporation) introduced a more descriptive naming system for new engines. It was a simple system that reminded one of the manufacturer's name and it gave an indication of the thrust size. Because the JT10D was in the category as being a new engine, it qualified for the renaming process and was referred to as the PW2000 engine. This new name put the engine in the 20,000–30,000-lb thrust category. While the engine was waiting for a home and full development status, other events were happening in the aviation marketplace.

The B-7N7 started out as a replacement for the very successful B-727. Boeing's concept, after many design studies, came out as an aircraft with about the same length and fuselage diameter as the 727. The most obvious visual differences were in the twin-engine (under wing) configuration and a conventional empennage (no T-tail and no rear-mounted engines). It is interesting to reflect that after all the studies, the most attractive configuration of the commercial aircraft is that of the Boeing 247 and Douglas DC-3 of the early 1930s. The legendary Ed Wells of Boeing got his start in engineering on the B-247 aircraft, often referred to as the first modern commercial transport in the United States. He retired in 1972 but was still active as a consultant. He is the one who is credited in 1976 with the B-767 configuration as a twin rather than a trijet. His words still ring true today as they did in the days when

THE MODERN ERA

the trend in aircraft design was from the Ford and Fokker trimotors to the twin-engine configuration [1]: "You're spending money on that three-engine project and paying lip service to a twin. You've got it backwards—you should be spending money on the twin and paying nothing more than lip service to a trimotor."

The appeal of the trimotor was in its ability to continue flying after an engine failure. One of the design requirements for a twin-engine aircraft was that it had to be able to fly on one engine after an engine failure. The result of this requirement was the aircraft was over-powered under normal operating conditions and therefore had a steeper climb out path compared to the trimotor—a characteristic that pilots liked. Another aspect of the requirement was that during cruise conditions the engines were throttled back, compared to the engines in the trimotor. With time the reliability of aircraft engines improved so much that a twin-engine aircraft in the jet era even made sense for long-range travel over water. Today many twin-engine aircraft can fly over long stretches of water as long as they are within 3 hr of an airport. Extended twin-engine operations (ETOPS) involves the airframe, the engine, and specific airline engine maintenance practices.

While the JT10D was being studied in the 7X7 and the 7N7 aircraft configuration, all at once the ground rules were changed. There were two resounding world events in the 1970s that had a profound effect on the future of aviation in the United States, 1) the fuel crisis of 1973, and 2) the Airline Deregulation Bill in 1979. After the fuel crisis, the price of fuel rose so much (about a factor of 10 to make fuel about half the cost of operations) that fuel burn per passenger became extremely important to airlines. What the airlines meant by a modern replacement of the B-727 was one with a much lower fuel consumption than the 727-200. Pratt & Whitney's solution to this request was to offer an engine for the B-757 with the lowest specific fuel consumption that advanced technology could develop—the PW2037. The new engine designation system tells you the size was 37,000 lb of thrust. The immediate response from the airlines to the fuel crisis was to change their airplane mix towards full B-727s and away from less-than-full-wide-body aircraft. The further sales of the B-727 provided the needed financial foundation to help the Boeing Company climb out of its financial hard times.

Rolls-Royce and Pratt & Whitney parted company when the B-7N7 opportunity transformed into the B-757. Boeing was estimating a market of about 1400 aircraft. Of the three potential engine companies, Rolls-Royce was quick to jump at the opportunity and provided a derivative engine to launch the Boeing program, energized by British Airways and Eastern Airlines commitments. Rolls-Royce offered Pratt & Whitney a chance to participate in their derivative engine program, but Pratt & Whitney declined. General Electric also declined to enter the competition for this particular engine opportunity. This 757 program was launched a month after the 767 program in 1978 and, as one might suspect, there was a certain amount of similarity between the two programs at Boeing. Both aircraft had the same two-person flight deck so that airlines with both aircraft would have pilots familiar with both aircraft operations.

TIME LINE

The JT10D technology in the 1970s was aiming for a 25,000-lb thrust requirement. Studies were concentrating on configurations to minimize the weight and the specific fuel consumption. Suddenly the requirement was there with the B-757-200, not exactly a huge market but Pratt & Whitney looked at it as "the only game in town" for a new engine. The company launched the PW2037 in December 1979 (Fig. 6) to try to capture as much of a marketing opportunity as possible with an engine that was as fuel efficient as the latest technology would permit. Two foreign companies, MTU (at 11 percent) and Fiat (at 4 percent), joined Pratt & Whitney in the venture.

As mentioned, Rolls-Royce was first into the marketplace with its derivative engine for British Airways and Eastern Airlines. It was not until October 1984 that the PW2037 entered service on a Delta 757-200. One of the other potential applications was in the military C-17 program that was given the go-ahead in the early 1980s. The PW2037 was selected as the engine in August 1981. The economic situation in the early 1980s was at a low point in the United States as far as military budgets were concerned. It was not until about 10 years later, in September 1991, that the first flight occurred with the C-17 powered by four PW2037s. The military version of the engine is called F117-PW-100 and is rated now at 41,700 lb of thrust.

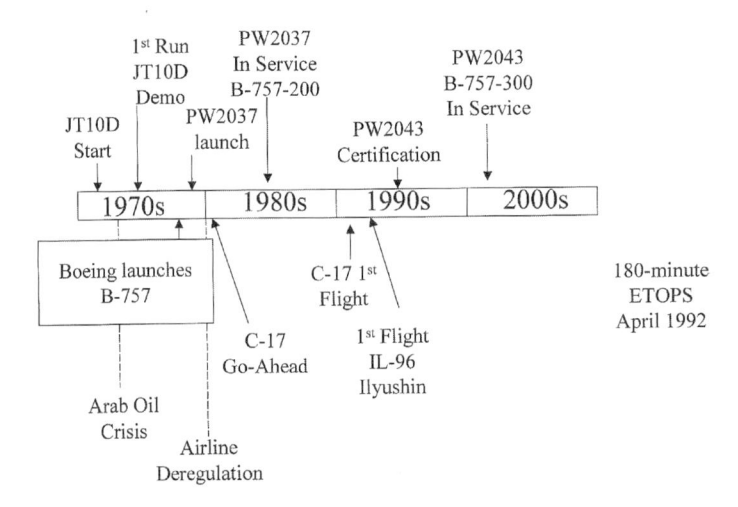

Fig. 6   The sequence of events associated with the PW2037.

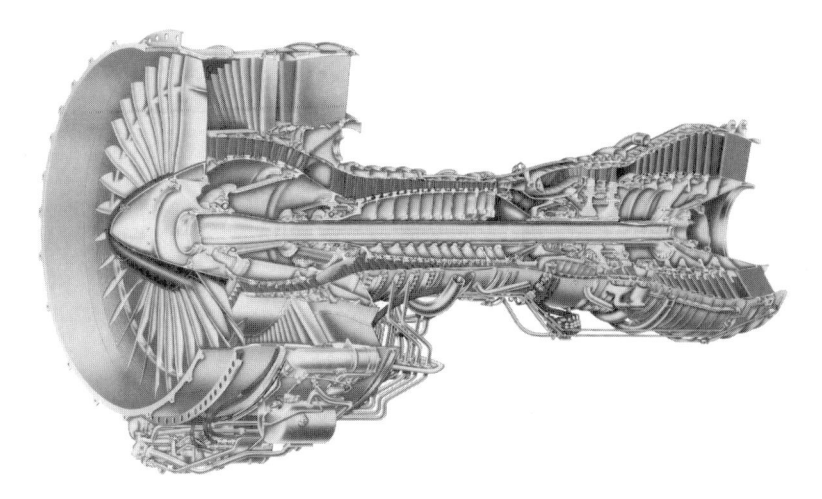

Fig. 7 Cutaway of PW2037 (courtesy of Pratt & Whitney Archives).

ENGINE DESCRIPTION

The PW2037 is a bypass-six turbofan with four low-pressure compressor stages, 12 high-pressure compressor stages, an annular burner, two high-pressure turbine stages, and five low-pressure turbine stages (Fig. 7). It was in this engine that all of the technology advances from the JT10D program now were used in the lowest specific fuel consumption engine that Pratt & Whitney was able to design. MTU took responsibility for the low-pressure turbine and Fiat built the gearbox and some external accessories. The following are general characteristics for engine models PW2037, PW2040, and PW2043:

Bypass ratio: 6 : 1
Overall pressure ratio: 27.6–31.3
Take-off thrust (lb): 38,400–43,734 (flat-rated to 96 F)
Fan pressure ratio: 1.63 : 1
Fan tip diameter (in.): 78.5
Length, flange to flange (in.)141.4
Weight (lb): 7300

This was Pratt & Whitney's first commercial engine to have a FADEC, visible in Fig. 8, which was mounted on the left side of the fan case (at the 2 o'clock position) to keep the engine cooler. Up to this time, the furthest anyone would trust "an electronic fuel control" was in the supervisory position, where the hydromechanical part of the control system was a backup should something go wrong with the electronics. By the time of the PW2037 go-ahead, however, demonstrated reliability gave the company full confidence in a control that could simplify the operation of the engine.

Another innovation in aerodynamics was in the compressor airfoils, which controlled the pressure rise on the airfoil's suction surface to prevent boundary layer separation, which is why they are called "controlled diffusion airfoils." This permitted a greater pressure rise per compressor stage and reduced compressor weight. The use of single-crystal turbine airfoils in the high-pressure turbine provided a higher temperature capability, which meant more power per stage.

Another performance improvement in the turbine was related to clearances between the turbine blade tips and the stationary case. This clearance changes as the engine powers the aircraft from take-off through climb to the cruise condition where the aircraft has settled down to a state of relative mechanical stability. As the engine case over the turbine heats up, it expands, and at the cruise condition the clearance over the turbine blades is greatest. The performance improves when this clearance is reduced. If the turbine case is cooled, then the clearance goes down and the engine performance improves. That is what happens in the high and low turbines. The cases close down on the rotating machinery at the mechanically stable condition of cruise.

On the take-off roll, with subsequent rotation of the aircraft prior to climb, the gigantic gyroscope represented by the whirling fan tries to bend the low shaft. During the rotation maneuver it is extremely important to have the shaft and cases stiff enough to minimize any such deflection caused by the gyroscopic action of the fan. Therefore, the PW2037 designer added another

Fig. 8 The PW2037, left 3/4 view (courtesy of Pratt & Whitney Archives).

Fig. 9   The C-17 (courtesy of U.S. Air Force).

bearing on the low shaft a short distance aft of the fan thrust bearing and upstream of the high spool forward bearing.

APPLICATIONS

One of the byproducts of the new airfoil profiles in the compressors is a more erosion-resistant airfoil because of a larger than usual leading edge thickness. There is also provision to blow off sand or similar particulates into the fan discharge area before it goes into the high compressor. The use of a commercial engine in the C-17 (Fig. 9) saved the Air Force the need to fund the design and development of a new engine.

Pratt & Whitney pursued a potential opportunity in the Russian Ilyushin four-engine commercial wide-body aircraft in the early 1990s. The first flight of the Ilyushin-96 aircraft with the PW2037 took place in April 1993. This 300-passenger aircraft opportunity unfortunately did not develop into a market. A more powerful version of the 2000 series engines came along and was certified at 43,000 lb of thrust (PW2043) in March 1995. This engine powered the B-757-300, which entered commercial service in July 2002. The major users of Pratt & Whitney-powered B-757s are Delta, Northwest, and United Parcel Service.

INTERNATIONAL AERO ENGINES (IAE) AND THE V2500

In 1981, I was Program Manager of Current Engines (JT8D, JT3D, and older commercial engines). We had about 12,000 engines in the air. There were constant lively communications with airframers, airlines, engineers, and procurement people. Daily we responded to customers' requests for quicker delivery of an engine, more thrust, or concessions on engine performance when airframers made proposals to airlines. I was fortunate to have Howie Latimer and Joe Raffin working for me, two capable engineers who knew how to get things done in a large, complex organization.

In late 1981, my secretary asked me if I wanted a smoking or non-smoking room in London for the following week. I was puzzled and I told her I had no plans to go to London. She did not know why the trip came up, so I called the source, Larry Clarkson, who said, "Oh yes, I should have told you earlier. You and I are off on a new venture—an international collaboration with Rolls-Royce and other companies." Clarkson had been on a special assignment, and now was tapped to put together this new collaboration, after which he went on to become president of the Commercial Products Division (CPD).

Rolls-Royce had been working with three Japanese engine companies on a "10-tonne" engine. At the same time, Pratt & Whitney had been working with MTU and Fiat on the JT10D engine program. Now Clarkson was trying to put together a group of engine companies to participate in this so-called 10-tonne engine market opportunity. This would be Pratt & Whitney's second attempt at collaboration with Rolls-Royce. The first collaboration was interrupted in 1977 when Rolls-Royce and Pratt & Whitney each jumped at the B-757 opportunity.

I had enjoyed working with the engineers from Rolls-Royce and some of the same people would be involved again. This looked like a great change of pace from a program of current engines where each day a new challenge had to be faced and resolved. For the next year I had a chance to talk to engineers from Rolls-Royce, three Japanese companies (IHI, KHI, and Mitsubishi),

Fig. 10 Members of the collaboration who worked out the details for the formation of IAE; Bob Rosati—seated second from the right, Connors and Stanwood—standing far left (courtesy of Pratt & Whitney Archives).

THE MODERN ERA 441

TABLE 2 V2500 SPECIFICATIONS AND APPLICATIONS

Model	Thrust or range (lb)	Weight (lb)	Diameter (in.)	Length (in.)	Applications
V2500-A1	25,000	5210	67.5	126	A320-200
V2500-A5	22,000 to 33,000	5230	67.5	126	A319, A320 A321-100 A321-200
V-2500-D5	25,000 to 30,000	5610	67.5	126	MD-90-30 MD-90-30ER MD-90-50

MTU, and Fiat. Many of the meetings took place in East Hartford. Bob Stanwood, who worked for me, made my job much easier with his experience in working with customers and his background on nacelles.

The collaboration came together in March 1983, when those representing the many companies finished up the agreement paperwork. The International Aero Engines (IAE) was in business about six months later, under the direction of Bob Rosati, the first IAE president and J.M.S. Keen (Rolls-Royce) as Executive Vice President. The IAE milestone (Fig. 10) occurred when I had spent 35 years at Pratt & Whitney and at age 60 I retired. It was a good decision because in the past 26 years I have had a chance to visit the cities I once rushed around while on company business.

The IAE collaboration has evolved into the following division of responsibilities defined by the V2500 engine modules:

1. Pratt & Whitney: Combustor and high-pressure turbine
2. Rolls-Royce: High-pressure compressor
3. Japanese Aero Engines Corporation (JAEC): Fan and low-pressure compressor
4. MTU: Low-pressure turbine

ENGINE DESCRIPTION

The V2500 is a two-spool high-bypass ratio engine. The low spool consists of a single-stage wide-chord fan, a four-stage low-pressure compressor and a five-stage low-pressure turbine. The high spool has a 10-stage high-pressure compressor and a two-stage high-pressure turbine. Table 2 lists the engine's specifications and applications.

TIME LINE

It took about six years from the initial agreement to the first engine in commercial service (Fig. 11). This was a considerable accomplishment as the critics held out no hope that seven companies could get together and produce

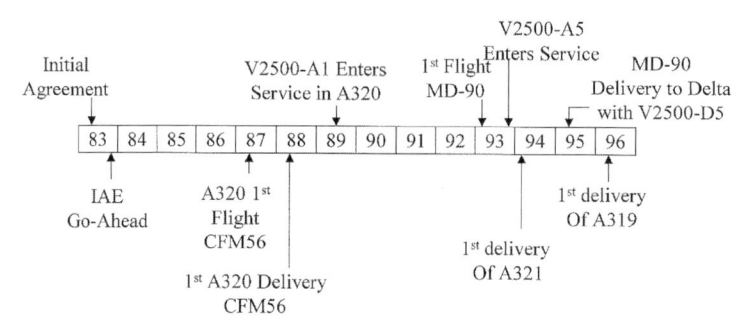

Fig. 11　The sequence of events associated with the V2500.

an engine that the airlines would buy. Well, the critics were wrong. IAE managed to break into the Airbus A320 series aircraft, even though it got there just a little over a year later than the first aircraft powered by the CFM56.

APPLICATIONS

The V2500 has two principal applications in narrow-body aircraft, the Airbus Industrie A320 family and the Boeing MD-90. These are short-range and medium-range aircraft covering the passenger range in the Airbus series (A319, A320, and A321) from 124 to 186 passengers. The MD-90 is in the 155-passenger size.

PW4000 SERIES OF HIGH-BYPASS ENGINES

In the early 1970s, Pratt & Whitney had small engines (the JT8D), medium engines (JT3D), and large engines (JT9D). By the late 1970s, it was time to look over the menu and determine how it should be revised. During the 1970s, a considerable amount of research had been done, and much of it was funneled into the JT10D high-bypass engine studies in the areas of performance improvement, weight reduction, and reduced cost of manufacturing. By the end of the 1970s, the company launched the PW2037, which could be classified as a medium-sized engine. In the large engine category, it was clear that the JT9D was near the end of its life, and a new family of large engines with newer technology would be needed.

In December 1982, the company launched the PW4000 engine series. The first development engine was the PW4052, with plans for a whole family of engines approaching 100,000-lb thrust. In the days of the B-707, powered by four 10,000-lb thrust engines, it was theoretically feasible to build a twin configuration with two 20,000-lb thrust turbojet engines. However, at that time there was a question of whether gas turbines would ever develop the reliability of piston engines. Therefore, the financial risk was too great because the market was not that well-defined.

The Modern Era

Since then, however, the reliability of gas turbines has been phenomenal. Twin-engine aircraft have been built up to over 500,000-lb gross weight. Pratt & Whitney's 4000 series has an in-flight shutdown statistic of 0.013/1000 hr. In other words, one can expect such an event once in 77,000 hr. A commercial airliner accumulates about 3000 to 4000 hr/yr. Therefore, one could expect an engine shutdown once in about 20 years! One hundred thousand pounds of thrust engines are feasible today. Therefore, aircraft the size of the DC-10 and MD-11 could theoretically be powered by two 100,000-lb thrust turbofans.

The challenge facing Pratt & Whitney was to replace the JT9D with a series of engines from 50,000 lb of thrust all the way up to 100,000 lb of thrust—without driving the company into bankruptcy. The overall plan was to develop a large core engine (high-compressor–burner–high-turbine) that would operate with three different low spools, each with a different fan size. At the same time the focus was on dependability, high performance, and thrust growth at low risk.

The PW4000 family of three fan sizes builds heavily on commonality in technology, hardware, and tooling. The common core-engine approach made sense because in high-bypass engines at least 60 percent of the engine development effort is tied up in the core engine. Figure 12 sums up why this family of engines appealed to customers.

ENGINE DESCRIPTION

All high-bypass engines of Pratt & Whitney and of IAE seem to look the same. These high-bypass engines look very different from the JT3D and JT8D engines in layout view, because the fan blades are long compared to those in

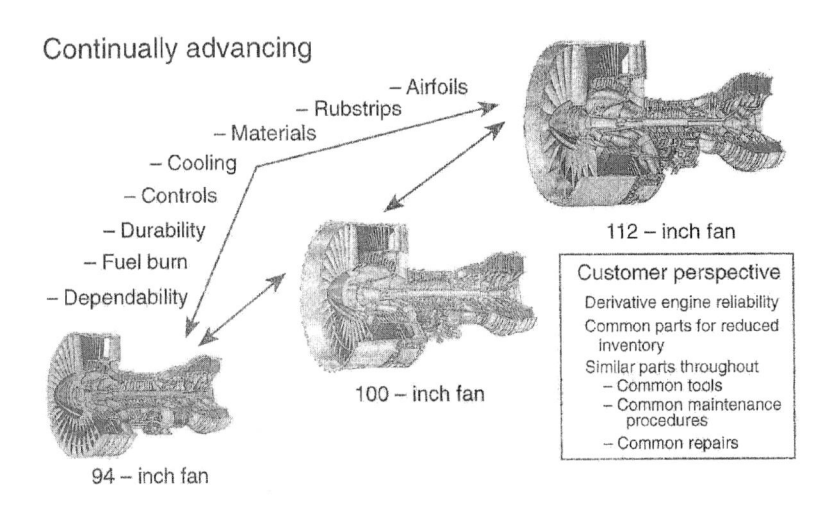

Fig. 12   The three PW4000 engines (courtesy of Pratt & Whitney Archives).

low-bypass engines. If you think of the fan tip as being limited in approach Mach number, then as the bypass ratio increases from the JT8D/JT3D values, the rotational speed must slow down to keep the tip speed below a critical limit. How much pressure rise one can get from a low compressor (at fan rotational speed) and how much work output one can expect from the low-pressure turbine (also at fan rotational speed) depends upon the tangential wheel speed (the product of rpm and radius). Therefore, the designer bulges out the location of the low compressor and the low turbine to get more performance. The alternative is to add many more compressor or turbine stages which makes a longer, heavier engine.

At first glance it is difficult to distinguish the PW2037 when it is standing alone from the 94-in. PW4000 (Fig. 13), even though there is about 12 in. in maximum diameter difference. This confusion is not surprising as both engines share the same technology. This same illusion occurred when one tried at a glance to identify a stand-alone J57 or a J75. It does not take long, however, to identify the engine after you study the externals. The 100-in. fanned engine has a sheet of Kevlar (for fan blade containment) wrapped around the fan

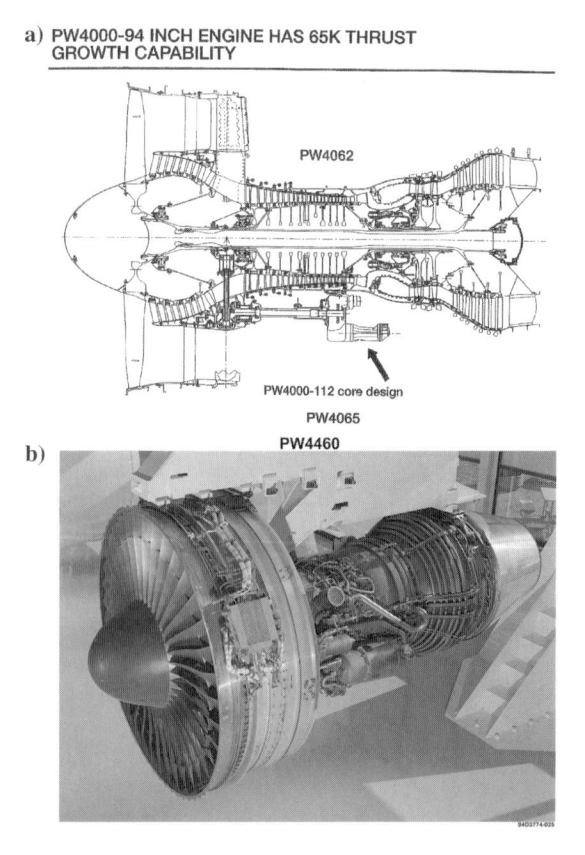

Fig. 13 The PW4000 with 94-in. fan (courtesy of Pratt & Whitney Archives).

The Modern Era

a) **PW4000 100-INCH FAN PROPULSION SYSTEM**

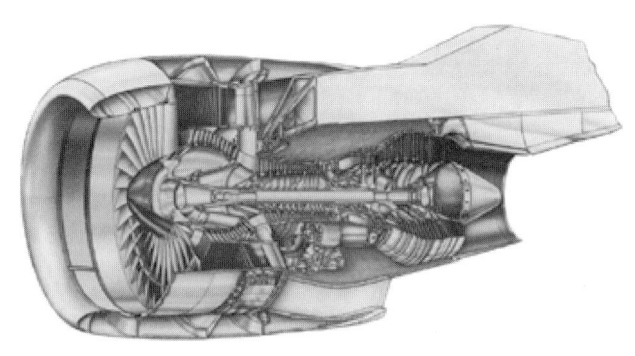

b)

Fig. 14 The PW4000 with 100-in. fan (courtesy of Pratt & Whitney Archives).

case, visible in Fig. 14. The 112-in. fanned engine (Fig. 15) is a bit different from the previous large fan engines: it does not have part-span shrouds and the fan blades have wider chords. Table 3 lists the specifications and applications for the PW4000 family of engines.

Applications

For decades, the PW4000 family of high-bypass ratio engines has powered quite a few commercial passenger wide-bodied airliners (Fig. 16). The 94-in. fanned engine has powered two Airbus aircraft. The A300 was a short- to medium-range aircraft, launched in 1972 as the world's first twin-engine widebody. The A310 was a medium to long-range aircraft first launched in 1978. This version of the PW4000 also powered the McDonnell Douglas MD-11

Fig. 15 a) The PW4000 with 112-in. fan, b) which powered the B-777 (courtesy of Pratt & Whitney Archives).

three-engine medium- to long-range aircraft, with two engines mounted on underwing pylons and a third engine at the base of the vertical stabilizer. This engine also powered three Boeing planes: the B-747 (dubbed the *jumbo jet*), the B-767 (the first wide-body twin-engine jet produced by Boeing), and the B-777 (the world's largest twin-engine jet, known as the *Triple Seven*). Another version of the B-777 used the 112-in. fanned engine. The 100-in. fanned engine powered the Airbus 330, another medium-to-long-range aircraft.

THE MODERN ERA

TABLE 3 THE PW4000 FAMILY OF ENGINE SPECIFICATIONS AND APPLICATIONS

PW4000 engine model	Nominal thrust range (lb)	Weight (lb)	Maximum diameter (in.)	Length (in.)	Applications
94-in. fan	50,000 to 64,000	9400	97	154	A300, A310 MD-11, B-767 B-747, B-777
100-in. fan	68,000	12,480	105	163	A330
112-in. fan	84,000 to 98,000	15,000 to 16,260	119	192	B-777

ENGINES FOLLOWING THE PW4000

The JT9D was Pratt & Whitney's first high-bypass ratio engine for wide-body commercial aircraft. Subsequently, the PW4000 series engines provided a greater range of thrust and better performance with fewer engine parts. The PW6000, with still fewer engine parts, was the next step after the JT8D engine series for narrow-body aircraft.

Airlines buy equipment to fly domestic and international routes that meet noise and emission regulations. The airlines want engines that meet the various airport regulatory requirements as well, at the lowest cost. Part of the

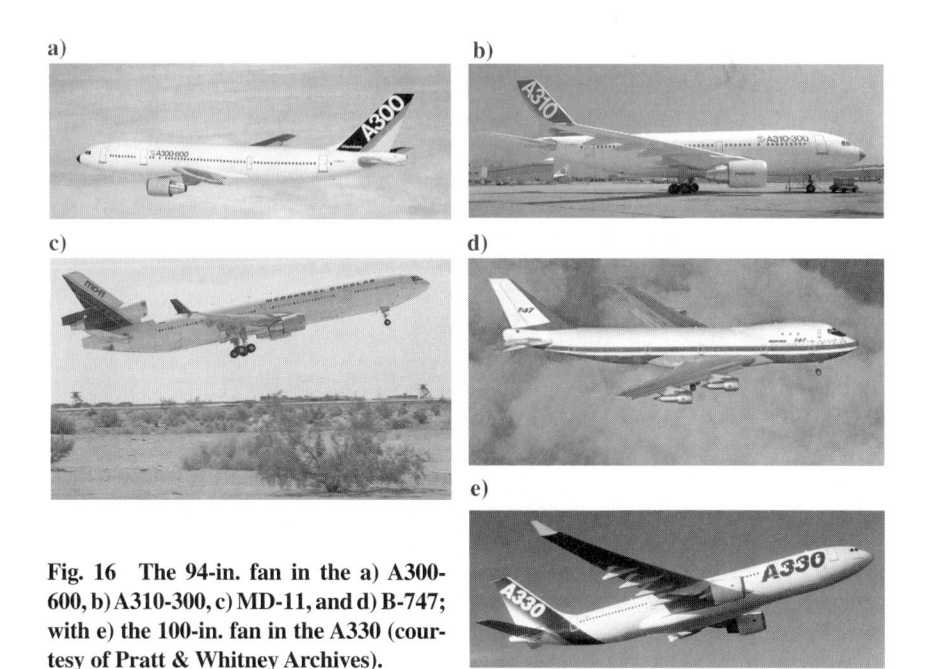

Fig. 16 The 94-in. fan in the a) A300-600, b) A310-300, c) MD-11, and d) B-747; with e) the 100-in. fan in the A330 (courtesy of Pratt & Whitney Archives).

airline costs come from engines that pull the airplanes. The engine manufacturer responds to the airline's needs by designing an engine that minimizes the total of costs associated with the engine:

1. *Initial cost*: The engine manufacturer tries to keep the cost down by reducing the number of parts in the engine and by using more cost-effective manufacturing processes
2. *Maintenance cost*: This cost is reduced by building an engine with fewer parts that have longer lives and thus easier online maintenance actions, done through simplifying the engine configuration. In addition, the life-limited parts, such as disks, are designed for a sufficient number of cycles (take-offs and landings) to stay on the wing for about six years. The positioning of the external components (line replaceable units, LRUs) is such that any of them can be removed and replaced in less than 30 minutes. This is an important feature for the short-range aircraft that make several stops during a typical flight. The LRU could be changed during the unloading and the boarding of a flight.
3. *Fuel cost*: This cost is reduced by improved performance. Higher overall pressure ratio and bypass ratio with higher turbine inlet-temperatures are factors that reduce specific fuel consumption.

Almost all of the factors that result in lower cost are mutually dependent. If a designer decreases the number of airfoils (at a certain chord size) on a compressor or turbine disk, there is a point at which the performance falls off rapidly. So, at that point, while the number of parts decreased (good), the engine performance deteriorated (unacceptable). The gradual refinement in turbomachinery design practices enabled Pratt & Whitney to push the limits on airfoil design to minimize the engine parts without sacrificing engine performance.

PW6000

The first commercial engine after the PW4000 series was the PW6000, which focused on the lower end of the thrust spectrum, 18,000 to 24,000 lb of thrust (Table 4). The PW6000 program started in September 1998 (Fig. 17). The aircraft that could use this engine would be in the 100-passenger size. The first application of the PW6000 was in the Airbus narrow-body A318 aircraft.

TABLE 4 PW6000 SPECIFICATIONS

Engine model	Take-off thrust (lb)	Weight (lb)	Fan tip diameter (in.)	Length (in.)	Bypass ratio
PW6122	22,000	5034	56.5	108	5.0 : 1
PW6124	24,000	5034	56.5	108	4.8

The Modern Era

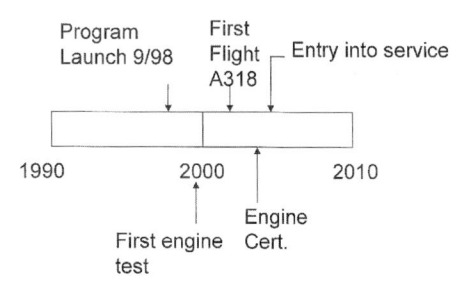

Fig. 17 Time line for the PW6000.

The PW6000 was a joint program conducted by Pratt & Whitney, MTU, and Mitsubishi Heavy Industries (MHI). The most striking characteristic of this engine is its design configuration for reduction of maintenance and acquisition costs. Experience says that about 60 percent of the maintenance costs are tied up with the parts that run at high temperatures (high-pressure compressor, burner, high-pressure turbine, and low-pressure turbine). Therefore, reducing the number of compressor and turbine stages and airfoils results in a real cost saving (Fig. 18). In addition, during the flight profile, the tip clearances of the turbines are controlled to minimize rubbing during aircraft rotation at take-off, and to maximize performance during cruise conditions. Table 5 compares the turbomachinery stages in the PW4000 (94-in. fan) and the PW6000. Right away, we can see that the number of turbomachinery stages has been significantly reduced at the same overall pressure ratio.

In piston engines, increasing the displacement (volume swept out by the piston) increases the power. In gas turbines the equivalent parameter to increase power is *wheel speed*, rpm times the radius. The low compressor

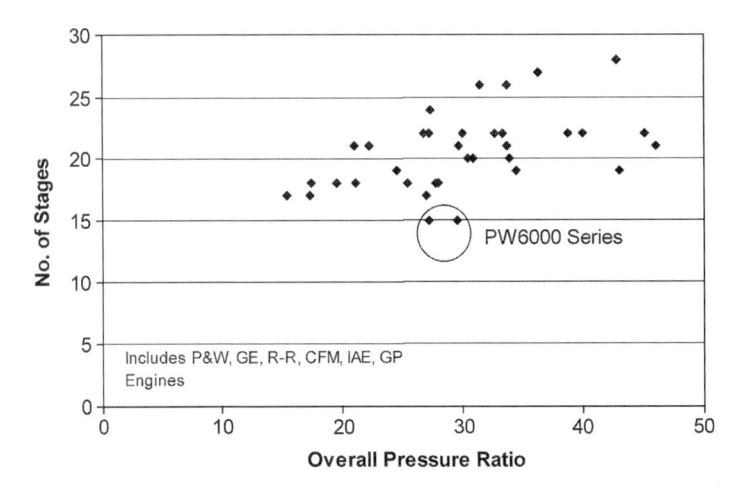

Fig. 18 The PW6000 series minimizes the number of turbomachinery stages in the engine compared to other commercial engines (data from [2]).

Table 5 Turbomachinery Stages of the PW4000 and PW6000

Component	PW4000	PW6000
Fan	1	1
Low compressor	4	4
High compressor	11	6
High turbine	2	1
Low turbine	4	3
Total stages	22	15
Overall pressure ratio	28	28

bulges out in diameter to increase the wheel speed (in feet per second) at the mid-span of the blades and vanes. This higher wheel speed permits the compressor blades to do more work per pound of airflow to obtain a higher pressure rise than what would have been possible at a lower radius. An analogous trend is true in turbines: the greater the wheel speed at a turbine's mean radius, the greater the work output per pound of airflow. The PW6000 airflow path bulges out the low spool in the low compressor and low turbine as in other Pratt & Whitney commercial engines.

Another striking feature is the absence of part-span shrouds on the fan blades to improve fan efficiency. The exhausts from the fan and the core engine go through a mixer to reduce the jet velocity of the core engine in a similar form used earlier by the JT8D-200 engines (Fig. 19). Another difference from older engines produced for commercial aircraft is that the fan rotates clockwise, when viewed from the front (Fig. 20). This contra-rotation minimizes the gyroscopic moments of the rotating machinery so that the engine has less effect on aircraft maneuvers.

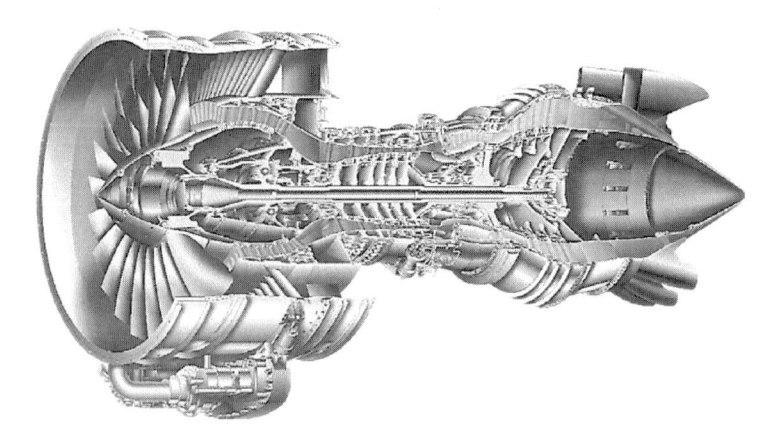

Fig. 19 Cutaway of the PW6000 (courtesy of Pratt & Whitney Archives).

THE MODERN ERA 451

Fig. 20   The PW6000 (courtesy of Pratt & Whitney Archives).

APPLICATIONS

The launching airline for the PW6000-powered A318 was LAN Airlines of Chile (Fig. 21). The A318 is a 107-passenger size aircraft with a fuselage shorter than the A320 and A319 Airbus family. The aircraft has a wingspan of about 112 ft and a length of about 103 ft.

WHAT BECAME OF THE F100?

In Chapter 12 we discussed the F100, Pratt & Whitney's finest military engine that first powered the F-15 in the early 1970s and still powers the Air Force's F-15 and F-16 today. The F100 engine improved into the F100-PW-229

Fig. 21   The A318, powered by the PW6000 (courtesy of Pratt & Whitney).

model, which went into operational service in April 1992. Its thrust rating was about 29,000 lb at a weight of 3740 lb, a thrust/weight of about 7.6. Its bypass ratio is about 0.4 and has an overall pressure ratio of 32 : 1.

In the spring of 2001, the Air Force and Pratt & Whitney conducted a series of accelerated mission tests in the Arnold Engineering Development Center (AEDC) to evaluate a production engine's ruggedness under different mission requirements [3]. These tests at AEDC were really to see how much margin was in the F100-PW-229 to handle a variety of mission profiles.

The engine survived 4500 tactical aircraft cycles. In addition, it performed 2 to 4 times as many as spec requirements for number of afterburner lights, hours of duration, and number of hours at full military thrust. The extensive operational experience with the F100 engine family helped Pratt & Whitney to win the advanced tactical fighter F/A-22 Raptor propulsion contract in April 1991 with the F119-PW-100, as we shall see later in this chapter.

The F100-PW-100/200/220 (rated at 23,770-lb thrust) has had more than 30 years in service, accumulating more than 19 million flight-hr. The F100-PW-229 model (rated at 29,100-lb thrust) has had 17 years of service accumulating 1.3 million flight-hr.

F119—Advances in Engine Development

In 1981, the U.S. Air Force outlined its requirements for an advanced tactical fighter (ATF) to replace the F-15 aircraft in the future. In early 1983, Pratt & Whitney, anticipating such a program, started the design of an engine for the ATF under the designation of PW5000. Later in 1983, the U.S. Air Force awarded seven study contracts to airframers for the ATF and demonstrator engine contracts to General Electric and Pratt & Whitney.

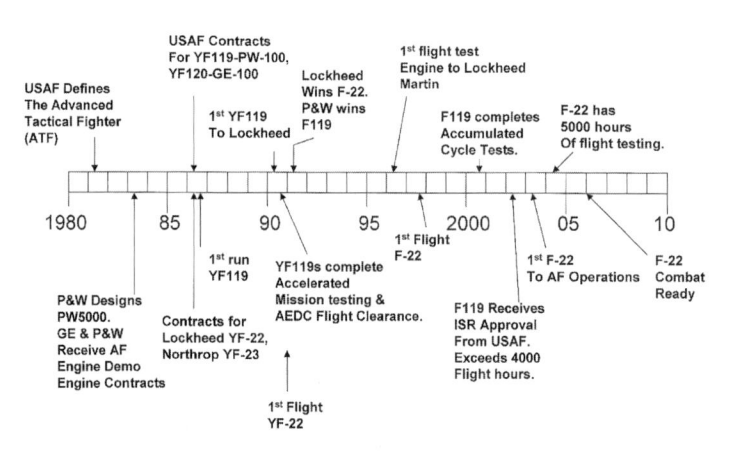

Fig. 22 The F119/F-22 timeline extends from ATF definition to combat readiness over a time span of a quarter century.

The Modern Era

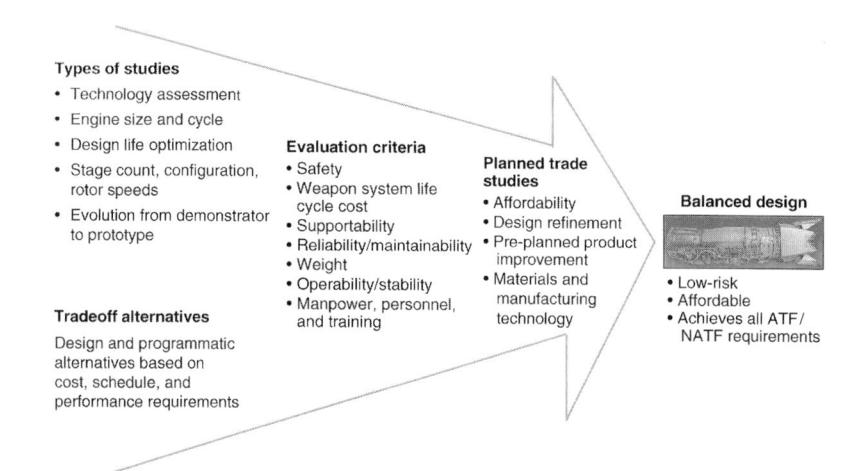

Types of studies
- Technology assessment
- Engine size and cycle
- Design life optimization
- Stage count, configuration, rotor speeds
- Evolution from demonstrator to prototype

Tradeoff alternatives
Design and programmatic alternatives based on cost, schedule, and performance requirements

Evaluation criteria
- Safety
- Weapon system life cycle cost
- Supportability
- Reliability/maintainability
- Weight
- Operability/stability
- Manpower, personnel, and training

Planned trade studies
- Affordability
- Design refinement
- Pre-planned product improvement
- Materials and manufacturing technology

Balanced design
- Low-risk
- Affordable
- Achieves all ATF/NATF requirements

Fig. 23 Trade-off studies led to a low-risk and affordable F119 (courtesy of Pratt & Whitney Archives).

In 1986, the Air Force awarded prototype engine contracts for the YF119-PW-100 engine to Pratt & Whitney and the YF120-GE-100 engine to General Electric (Fig. 22). Later in the year, it selected its two prototype airplanes from the previous ATF phase, the Lockheed YF-22 and the Northrop YF-23. In the meantime, the two engine companies continued with their development of their prototype engines. In 1990, the YF119 successfully completed its accelerated mission tests on one engine, and another prototype engine completed its flight clearance testing at the AEDC. In March 1991, the Lockheed Corporation merged with Martin Marietta to form the Lockheed Martin Corporation. One month later, the Air Force selected Lockheed Martin for the F-22 and Pratt & Whitney for the F119.

Satisfying the ATF's needs while staying within the DOD's budget and schedule was a complex challenge. A detailed program management systems and the systems engineering discipline were used for the aircraft and engine development programs. Not only were there the customary work breakdowns and scheduling techniques, there were also program reviews at specific points to assess risk and propose any alternate courses of action. Another characteristic of this process was the involvement of the customers (Air Force and Lockheed Martin) in the engine program's progress earlier than in previous engine programs. Over 200 trade-off studies were completed (Fig. 23) and considerable attention was paid to every element, which led to a successful flight test program.

Frank Gillette (Fig. 24), father and engineering manager of the F119 program, gives us some background on how the program got started [4]:

> We built this engine on the information we got out of the ATEGG program—the Advanced Technology Engine Gas Generator. In our plant

Fig. 24 Frank Gillette, who has been honored with the 2006 Collier Award (courtesy of Pratt & Whitney Archives).

we called it FX689. The compressor, combustor, and turbine were designed before 1983 and we just picked up the technologies and refined them so that we had a real good common core, which gave us a basis to do everything. The ... objective at that time was a balanced design. What we wanted was performance, reduced cost, more durability and ... most important increased supportability. For performance we wanted the highest specific thrust engine that had ever been made. We ran our first demo engine in October of 1986 and it was a thrill to see this very powerful engine run down in the test stand with a two-dimensional CD nozzle. It was very exciting for us and we were very proud of all the work accomplished by our people.

The airplane wanted supersonic persistence. And that means an airplane can go faster than Mach 1 without using afterburning. And that was the exciting thing. That's the big change that technology in the ATF would have over any other airplane. In other words, the ATF [F-22] can fly the same place that an F-15 can go without using the afterburner. The other big change for this airplane was it was low observable, which means it was a stealth airplane. We've accomplished that over a lot of technology. This can be an awesome airplane. This airplane will be as good as an F-117 stealth fighter but it will go supersonic and it will turn on a dime. And it will put an air-dominance into this United States that we always have security. We don't talk about air security. We talk about air-dominance.

In my closest moment [to a mechanical disaster] I'll have to say during the 15 years as being Engineering Manager of the F119 was an engine called YF4. It was the first engine that was supposed to go into the YF-22. We had run the engine down to pass the qualification to ship it to Edwards and the engine passed the course. Two people were saying to me, "I know we passed the qualification but the engine isn't right". ... They showed me the tracking plots. The vibrations were what we had not seen for seven years. This was on a Thursday and the C-141 was supposed to come on Tuesday of the next week. So we had a lot of work to do to get ready to get packed to go out. And obviously General Electric was already flying

THE MODERN ERA

in YF-22 and we had to fly to be a winner. The guys kept telling me the engine wasn't right. On Friday we decided to take the engine apart and I know a lot of people didn't agree with that but these guys really insisted we should do it. They took the engine apart and we found out that on the number two bearing support we had left all the bolts out. The only thing holding this engine were the snaps ... we replaced the bolts.... We flew it out to Edwards and that engine was the first one to fly with engine number five in the YF-22. So the lesson that I learned is to listen to your people. Without that—this engine on the first rollup—we could have lost the contract.

ENGINE DESCRIPTION

When a jaded Pratt & Whitney retiree takes a first glance at the F119, he has a flashback to Rolls-Royce engines. They rotate clockwise when viewed from the front while American engines rotate counter-clockwise when viewed from the front. Could there be counter rotation in the two spools? Yes. The F119 also has a reduced number of turbomachinery stages, compared to earlier Pratt & Whitney fighter engines. The 35,000-lb thrust F119 has a three-stage fan driven by a single-stage low-pressure turbine. The high spool has a six-stage compressor, annular combustion chamber, and a single-stage turbine (Fig. 25). Table 6 provides a quick comparison of the number of turbomachinery stages from earlier engines.

This comparison of the three engines is even more startling when the overall pressure ratio is taken into account. When the F100 program went ahead in 1970, the state-of-the-art was the TF30, with a total of 19 rotating stages and an overall compressor pressure ratio of about 20. The F100 reduced the number of stages, and increased the overall pressure ratio by a little more than 50 percent to a value of 32. The overall pressure ratio of the F119 is a little less

Fig. 25 The F119 (courtesy of Pratt & Whitney Archives).

TABLE 6   COMPARISON OF THE NUMBER OF
TURBOMACHINERY STAGES

Engine	Number of turbomachinery stages
TF30	19
F100	17
F119	11

than that of the F100 because of the optimum specific thrust at non-afterburning supersonic cruise occurs at an overall pressure ratio lower than 32. The advanced technology in the F119 is apparent from the comparison with the TF30 and F100 engines. Figure 26 shows the higher non-afterburning supersonic cruise thrust of the F119 compared to the F100. The F119's afterburning thrust is higher than the F100 thrust throughout the flight envelope. Later in this chapter, we will discuss how the advances in higher thrust came about.

The F199's turbomachinery stage aerodynamic loadings on the fan, compressor, and the two turbines are higher than the earlier engines. The materials in the high-pressure turbine not only are superalloys in single-crystal form, but also have sophisticated cooling schemes to make them operate reliably. The fan and the compressors have integrally machined blades with the disks (Fig. 27). The term for this combination of blades and disk, when it first was used many years ago, was called a *blisk*.

The exhaust nozzle can direct the flow up or down 20 degrees to provide vectored thrust. Like other military engines, it uses FADEC, which not only manages the engine operation but also provides diagnostic and on-condition

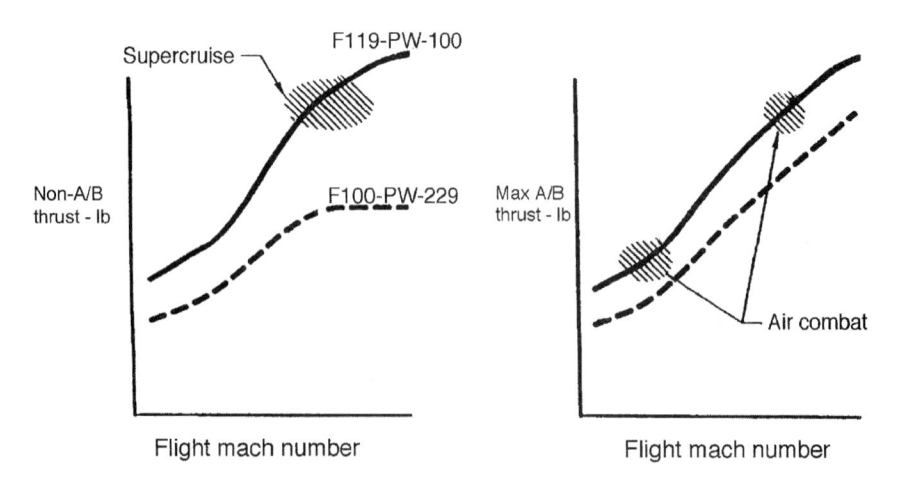

Fig. 26   The thrust of the F119 is compared to the F100 (courtesy of Pratt & Whitney Archives).

35,000 pound thrust class Two-dimensional vectoring nozzle
Fuel efficient design
Equivalent useful life (15-20 years operation)

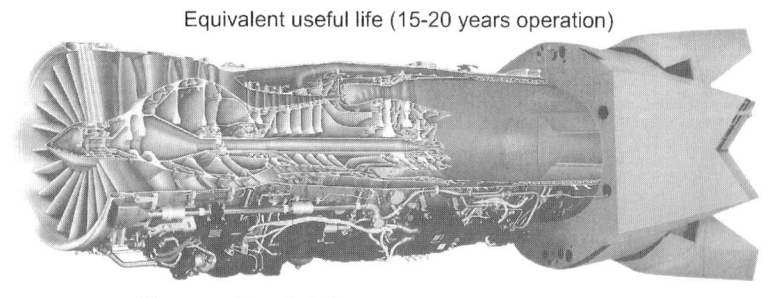

4th generation digital electronic control system   Maintainable design
Low Observable features

Fig. 27 A three-dimensional view of the F119 engine (courtesy of Pratt & Whitney Archives).

monitoring information. The engine is equipped with an afterburner, but can pull the F-22 at supersonic speeds without the use of the afterburner. This reminds one of the J52 in the Hound Dog missile on the B-52, which could go supersonic without the need for an afterburner. This feature, shown in Fig. 28, provides better fuel consumption (and aircraft range) at supersonic speeds.

Paul Metz, Lockheed Martin Chief Test Pilot, made the first flight in the F-22 in 1997 (Fig. 29). Previously designated as the F-22 and then the F/A-22, it was last designated as the F-22A, which implies a fighter aircraft with attack aircraft capability. Called the Raptor, built by Lockheed Martin, it was touted as the world's best fighter due to its stealth, speed, and precision, qualities described earlier in this chapter by Gillette. In 2009 after a battle for funding was well publicized in the media, the American public became aware that the F-22A required more than 30 hr of maintenance for every flight hour (not because of the engine) and cost billions of dollars while never being flown in either Afghanistan or Iraq. On July 22, 2009 the U.S. Senate voted to end F-22A production at 187 aircraft, following the recommendation of

Fig. 28 The F119 with a downward component to the thrust (courtesy of Pratt & Whitney Archives).

Fig. 29   F-22 with Chief Test Pilot Paul Metz (courtesy of Paul Metz).

Defense Secretary Robert Gates. While the F119 lost its aircraft, the engine technology was adopted for other military applications, as we shall see later in this chapter.

INTEGRATED PRODUCT DEVELOPMENT (IPD)

The Integrated Product Development (IPD) process is a multi-disciplined approach to design, where working-level teams have full responsibility for designing, manufacturing, and supporting a product in the field. The IPD teams strive to balance performance, weight, cost, schedule, R&M, and other factors. This approach to developing engines at Pratt & Whitney evolved during the F119 program.

In 2007 Frank Gillette gave me more detailed information on the genesis of the F119 engine program, pointing out the increased emphasis on supportability compared to earlier military and commercial engine programs (I retell Gillette's story with his permission). On June 22, 1983, at the beginning of the Joint Advanced Fighter Engine (JAFE) program, General James P. Mullins, USAF Commander, Air Force Logistics Command, and General Robert T. Marsh, USAF Commander, Air Force Systems Command, published a joint agreement. The bottom line was that improvements in support and readiness were a major DOD objective in both weapon systems and support technology areas. They agreed to increase emphasis on technology areas that can increase mission reliability; reduce dependence on support equipment, spares, and repair facilities; and reduce the need for highly skilled personnel.

In 1984, the Air Force initiated the Reliability and Maintenance (R&M 2000) program that emphasized the need to increase operational effectiveness through improved R&M. The Air Force set five goals to achieve the objectives of R&M 2000 plan:

1. Increase warfighting capability
2. Increase survivability of the combat support structure
3. Decrease mobility requirements per deploying unit
4. Decrease manpower requirements per unit of output
5. Decrease costs

During the development of the F119, the engine design team placed R&M on an equal footing with performance, weight, cost, and schedule. The key in developing a supportable engine is to design R&M in from the start. Five steps were taken in the design process to increase R&M:

1. Integrated product development (IPD)
2. Supportability awareness
3. Supportability reviews and trade studies
4. Early support tool involvement
5. Full scale engine mockup and nacelle

SUPPORTABILITY AWARENESS

Early supportability awareness is a key step for designing R&M features. By understanding the supportability concerns on current operations weapon systems, these teams can apply lessons learned in new designs. The Blue Two program, begun in the early 1980s, sponsored by the Air Force is one method of increasing awareness. The term *Blue Two* is the nickname for the Air Force men and women (blue-suiters with two stripes) who must maintain the weapon systems in the Air Force inventory. By enabling the design teams to visit Air Force and Navy operational facilities and work on actual combat aircraft, they show the design teams what the real world of aircraft maintenance is like. These visits have been conducted at bases all around the world and in all climate conditions. Some key issues stressed for supporting fighter engines was to provide component accessibility (for example, one-level deep LRUs), reducing the number of support tools, lockwireless fittings, performing flight line maintenance tasks in chemical-biological-radiological (CBR) warfare gear, minimizing the quantity and volume of support equipment and improve mobility.

Frank Gillette was able to replace an augmentor flap on an F100 engine by simply reading the technical manual. Gillette had his design teams experience performing maintenance tasks while wearing the heavy, confining CBR gear, which gave them an idea of actual engine maintenance conditions in the field. This experience was comparable to one of my own. Earlier in my career, a Navy officer suggested that I take a trip on a carrier plowing through the North Atlantic in wintertime, and experience the pleasure of conducting a maintenance action on the J57 while the carrier was doing pushups in the rough sea. He rightly said the experience would be a worthwhile addition to my engineering education.

SUPPORTABILITY REVIEWS AND TRADE STUDIES

During the creation of the designs, tools are laid out on the drawings to determine accessibility and the skill required by mechanics to assemble and disassemble the components. In some cases, the engine design was actually changed to accommodate easier engine assembly and disassembly. Another aid used during the IPD process is a standard engine parts display board (Fig. 30) showing the standard parts (nuts, bolts, clips, etc.) to be used in the engine. Such a board is prominently displayed in the design room and results in reducing the number of standard parts in the engine.

EARLY SUPPORT TOOL INVOLVEMENT

In previous engine programs, the major focus had been on engine performance in the airplane. However, there is another aspect to an engine's life on the ground where mechanics do their jobs to keep the aircraft flying. In the F119, the engine externals were centrally located on the bottom half of the engine and designed for easy removal. Figure 31 lists the benefits of this

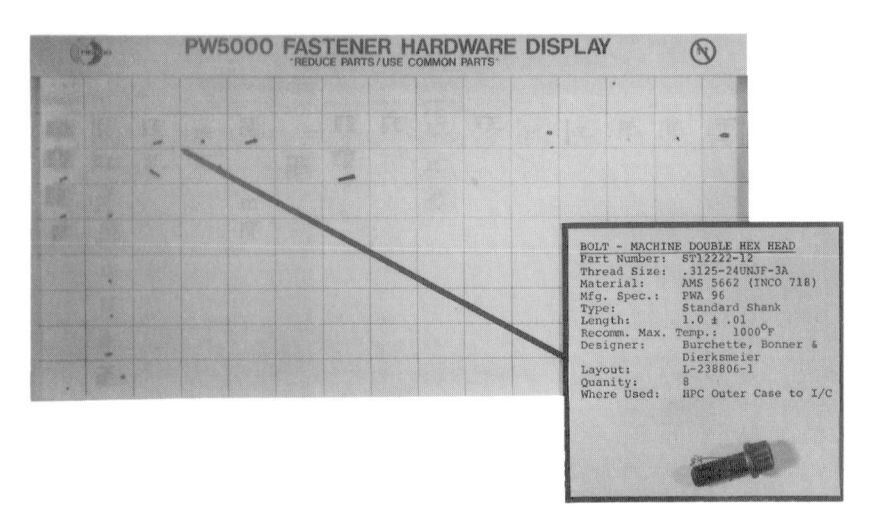

Fig. 30 A display board keeps a running tally on the standard parts (courtesy of Pratt & Whitney Archives).

The Modern Era

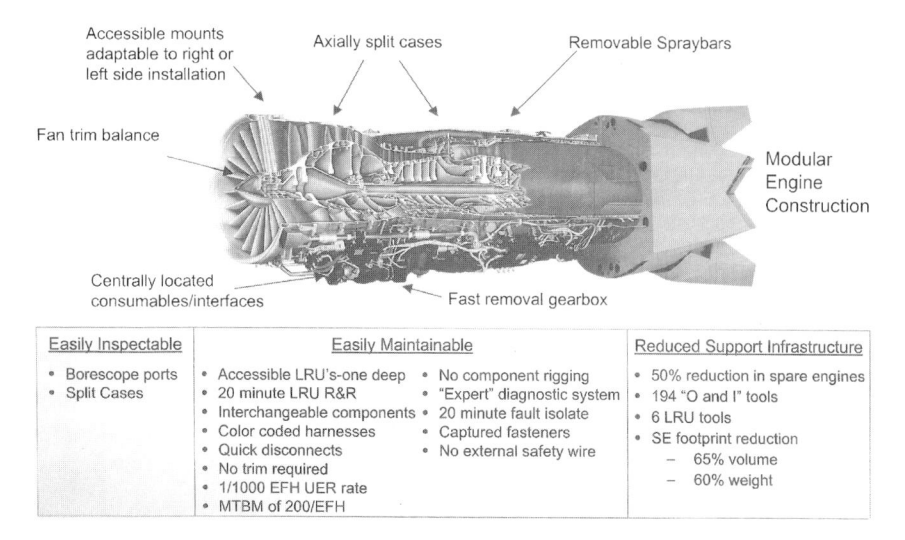

Fig. 31 The F119 engine was designed to be supportable (courtesy of Pratt & Whitney Archives).

Callouts: Accessible mounts adaptable to right or left side installation; Axially split cases; Removable Spraybars; Fan trim balance; Modular Engine Construction; Centrally located consumables/interfaces; Fast removal gearbox.

Easily Inspectable	Easily Maintainable	Reduced Support Infrastructure

• Borescope ports
• Split Cases
	• Accessible LRU's-one deep
	• 20 minute LRU R&R
	• Interchangeable components
	• Color coded harnesses
	• Quick disconnects
	• No trim required
	• 1/1000 EFH UER rate
	• MTBM of 200/EFH
		• No component rigging
		• "Expert" diagnostic system
		• 20 minute fault isolate
		• Captured fasteners
		• No external safety wire
			• 50% reduction in spare engines
			• 194 "O and I" tools
			• 6 LRU tools
			• SE footprint reduction
			    – 65% volume
			    – 60% weight

configuration. There is no safety wire to remove because of the self-locking devices, and the tube joints are of the quick-disconnect type. The LRUs are just *one deep*, which means the mechanic can get at them without removing other components or plumbing lines. The average time to remove and replace a LRU is 20 minutes. Each LRU can be removed using only one of six standard tools. Overall, the F119 had 40-percent fewer parts than previous comparable engines.

The design of an engine dictates the support tools required for assembly and disassembly from the start. Having the tool designers involved early in the design phase minimizes the amount and complexity of the tools, which reduces the airlift requirements. When the engine designer also focuses on the needs of a mechanic, the tools can be improved to make maintenance easier. One example of such a tool is the lifting adapter shown in Fig. 32. This tool was designed to be simpler with light weight and yet meeting all of the other tool requirements.

FULL-SCALE ENGINE MOCKUP

Dimensional drawings cannot always show how external components fit together on an engine or the access the mechanic will have when performing maintenance. Mockups allow an assessment of maintenance tasks and show the skill levels required, and the LRU replacement times. Pratt & Whitney invited Air Force maintenance personnel to participate in external hardware reviews, using the mockup to make it easier for the customer to see and understand (Fig. 33).

New design

- Aluminum alloy material (AMS 4117)
- Meets 5:1 safety factor requirement
- 64% lighter weight
- Tool handled by one person

Fig. 32 F119 support equipment, new lifting adapter tool on the right (courtesy of Pratt & Whitney Archives).

The new IPD philosophy applied to the development of the F119 led to the benefits listed in Fig. 31, as outlined here:

1. All components, harnesses, and plumbing are located on the bottom half of the engine.
2. All LRUs are located one deep with easy access from the aircraft engine bay.
3. Flightline on-condition maintenance is maximized through integrated diagnostics and automated storage of health monitoring and life-usage data.
4. There is no scheduled maintenance.
5. There is no engine trim equipment.
6. Certain LRUs hang in place during assembly.
7. Each LRU may be removed with one standard tool.
8. Average LRU replacement time is less than 20 minutes, using standard U.S. Air Force tools.
9. Average fault detection and isolation time is less than 20 minutes.

Self-locking devices eliminate the use of lockwire

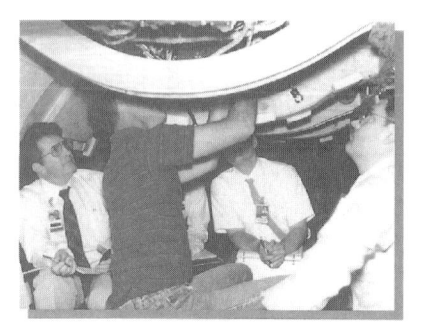

Easily accessible maintenance

Fig. 33 Using a full-scale mockup of the F119 (courtesy of Pratt & Whitney Archives).

The Modern Era

10. Cast manifolds minimize fuel plumbing and mounting.
11. Aircraft main fuel supply disconnects quickly.
12. Color-coded electrical harnesses aid in diagnostic fault detection.
13. Engine configuration allows installation in either bay of a twin-engine aircraft.
14. Positive self-locking devices are used to eliminate the use of lockwire as shown in Fig. 33.
15. Fan and compressor variable vane actuators are interchangeable.
16. Plumbing brackets are composite and mounted to the ducts with captured fasteners.
17. There is a common borescope plugs removal tool.
18. There is a common oil and fuel filter removable wrench.
19. Plumbing termination points at every LRU use flex-lines for the last few inches to prevent damage.

The IPD teams deserve great credit for the designs that have set new standards for R&M. The Air Force sponsored Blue Two visits and the Air Force R&M 2000 initiative have led the way for providing the user with a truly awesome, supportable F-22 weapon system.

THRUST VECTORING

There was another significant activity going on at that time—thrust vectoring, conducted by Roger Bursey. Here is what Bursey has to say [4]:

> Well, there was a lot of discussion back in the 60s by a German aerodynamicist, Dr. Wolfgang Herbst. He felt that new aircraft of the future would have to have a lot of agility in the upper left-hand corner where there was a minimum of aerodynamic power. Some maneuvers would enhance the majority of aircraft in what was known as the post-stall region of the flight envelope. One of the means of achieving this agility was to have the use of the thrust of the engine to augment the aerodynamic surfaces. In order to accomplish that he felt it could be done with a series of paddles that could deflect the thrust. However, that was cumbersome and added weight, added drag. The aerodynamics of that era determined that if we could come up with a system to vector the thrust of the engine, and have it transparent to the pilot, we could accomplish those feats.
>
> With that in mind, our vast technology team here started a series testing with scale models that eventually led to a very successful flight test. I can recall the first flight. It was absolutely rewarding on May 10, 1989 out at Edwards Air Force Base to see the two-dimensional nozzle in flight.
>
> The [Air Force] folks indicated that now we've had the success in taking thrust vectoring into the pitch mode and [asked] "if we can do

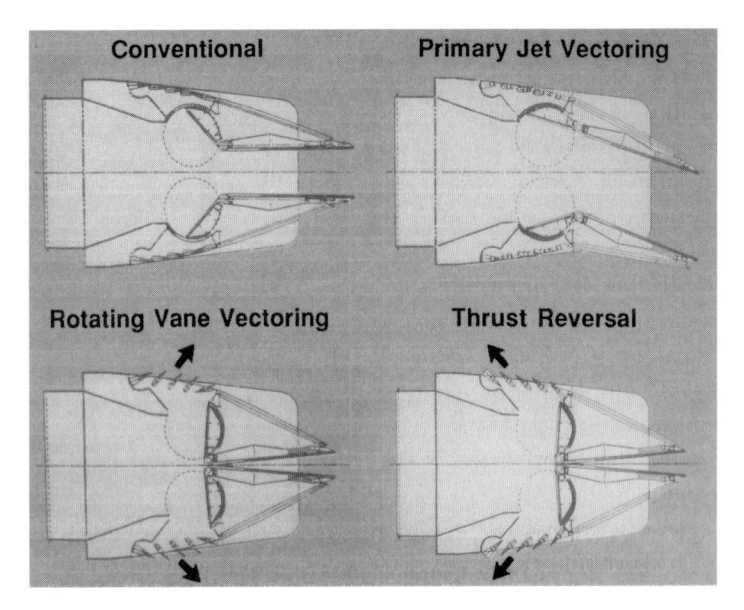

Fig. 34 Using the engine exhaust nozzle to provide thrust vectoring (courtesy of Pratt & Whitney Archives).

that, why can't we do this?".... So we launched a review in 1992 for the ground demonstrator—made up of various components and successfully demonstrating a multi-axis thrust vectoring.

Roger Bursey and his team looked at various concepts of thrust vectoring. Figure 34 shows some of them. Pictured in the upper left is the conventional nozzle. Thrust vectoring can be done in the pitch mode as well, as shown in the upper right and in Fig. 28. On the lower left is another type of vectoring using vanes to direct the thrust from the engine centerline. The engine exhaust nozzle segments rotate to the closed position, forming a blocker door so that the engine exhaust is turned either up or down depending upon whether the top or bottom vanes are open. The thrust can also be directed in reverse, similar to what is used in commercial engines. The engine is not in the afterburning mode during thrust reversing. Once the engine is capable of providing a force off the engine's centerline, then the aircraft's control gets more complicated. The need for directing the aircraft's movements with airframe surfaces and the engine's vectored thrust would require a close integration of the airplane and engine controls systems. Figure 35 shows thrust vectoring at work in the F-15.

F135—How Times Have Changed

When the Wasp (R-1340) air-cooled radial engine made its debut in 1926, it gradually found its way into more than 100 different military and commercial

Fig. 35 The F-15 in a) first flight with F100-PW-220 thrust vectoring in the pitch axis and thrust reversing, May 1989 at the McDonnell facility in St. Louis (Courtesy of U.S. Air Force) and b) fitted with multi-axis thrust vectoring nozzles on two F100-PW-229 engines, at the NASA Dryden facility (courtesy of NASA).

applications. At the beginning of Pratt & Whitney's gas turbine era after WWII, the J57, initially aimed at the B-52, ended up in about a dozen different aircraft. The airframers were aware of those engines and designed aircraft that made effective use of the capabilities of each engine. The F135, on the other hand, came into being during a much more sophisticated period in aviation history, one with complex design objectives.

BEGINNING OF THE PROGRAM

In late 1996, the military contracted with Pratt & Whitney to proceed with the demonstration of an engine that could, with minor modifications, power three different types of aircraft:

1. The F-35A, a conventional take-off and landing aircraft (Air Force)
2. The F-35B, a replacement for the Harrier type of aircraft with short take-off and vertical landing capability (Marines)
3. The F-35C, a carrier variant of the F-35A with greater range and more difficult landing requirements plus arresting gear and wing-folding capability (Navy)

The project had to be an international venture with hardware similarities among the three aircraft, with minimum lifecycle cost relative to previous programs. The significant performance objectives of this joint strike fighter (JSF) were that is be supersonic, have stealth, accomplish short take-off and vertical landing (STOVL), and be an affordable multi-role fighter accommodating operating requirements of U.S. military and its allied partners.

In the F-111 situation, the General Dynamics/Grumman team could not satisfy the special requirements of both the Air Force and the Navy. The "common aircraft" transformed into two different aircraft, the Convair F-111 for the Air Force, and the Grumman F-14 for the Navy. In both the DOD and the aircraft industry, much has changed since the F-111/F-14 programs. Gas turbine technology has greatly improved. There has also been a great advancement in managing complex programs and in working with international partners.

Today, the DOD uses the terms defined below when evaluating aircraft. An engine manufacturer sees its contribution in the following areas relative to earlier propulsion systems:

1. *Survivability*—Minimizing the propulsion system's visible signature, plus keeping the engine running in combat missions by redundancy of critical systems, and health management systems to minimize non-recoverable in-flight shutdowns.
2. *Lethality*—Reduced fuel consumption for greater range and doubling the time between maintenance actions for greater mission readiness.
3. *Supportability*—Reducing the cost of ownership by 50 percent through fewer maintenance man-hours and spare parts in the pipeline.
4. *Affordability*—Reducing total ownership cost by about 50 percent through fewer parts, lower maintenance man-hours, and extended periods between engine maintenance actions.

The two airframer teams competing for the three types of aircraft were Boeing (X-32) and Lockheed Martin (X-35). Both teams demonstrated greater operational capability than could ever have been imagined at the

The Modern Era

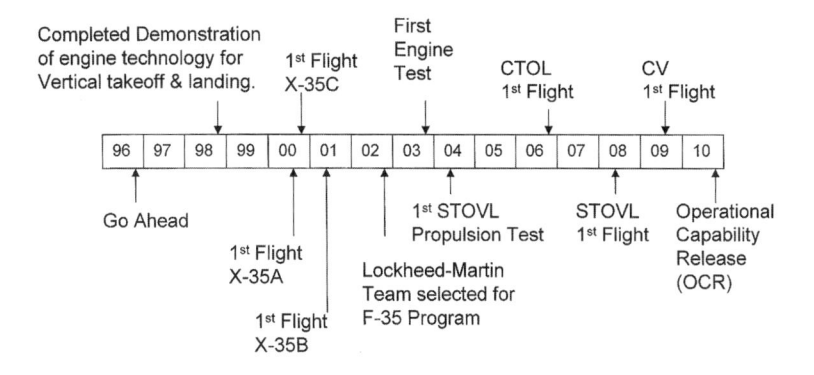

Fig. 36   Milestones in the F135 time line.

Timeline (1996–2010):
- Completed Demonstration of engine technology for Vertical takeoff & landing
- 1st Flight X-35C
- First Engine Test
- CTOL 1st Flight
- CV 1st Flight
- Go Ahead
- 1st Flight X-35A
- 1st Flight X-35B
- 1st STOVL Propulsion Test
- Lockheed-Martin Team selected for F-35 Program
- STOVL 1st Flight
- Operational Capability Release (OCR)

CTOL – Conventional Takeoff & Landing
STOVL – Short Takeoff & Vertical Landing
CV – Carrier Variant

beginning of the gas turbine era. The government selected Lockheed Martin in late 2001 to develop the F-35 to meet the challenging requirements of the three Services (Fig. 36). Lockheed Martin started the first flights of the Air Force X-35 in October 2000, and the Navy X-35C in December of 2000. The Marine X-35B demonstrated its vertical landing capability in mid-2001.

What was not obvious was the magnitude of Pratt & Whitney's and General Electric's roles in supporting three propulsion systems for Boeing's X-32A, X-32B, and X-32C aircraft, and another three propulsion systems for Lockheed Martin's X-35A, X-35B, and X-35C. Pratt & Whitney's approach to the F135 engine configuration was to use the high spool of the F-22's engine (F119), and then tailor the low spool to the specific requirements of each aircraft. The Air Force and Navy use the same basic engine. There is a difference in the aircraft because the Navy version is subject to the stresses of carrier operation in landing and take-off.

Engine Description

The F135 uses the F119-PW-100 core high spool consisting of a six-stage compressor driven by a single-stage turbine. The compressor has integrally bladed rotor disks. The turbine uses cooled single-crystal, superalloy blades. A shrouded cooled two-stage low-pressure turbine drives the fans in each of the applications. This is a unique feature because in the past it has not been easy to put shrouds on cooled turbine blades. The shrouds improve turbine efficiency but present challenges to the effective cooling schemes. The focus on reducing maintenance costs and improving reliability moved the engine design toward 40-percent fewer parts than in previous engines. Furthermore,

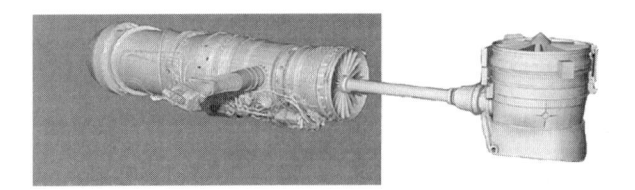

Fig. 37 F135 propulsion system, with the lift fan on the right—the right outrigger arm is visible about a third of the way down the engine length (courtesy of Pratt & Whitney Archives).

the line replaceable components (LPCs) can be removed in about 20 minutes with a set of only six common hand tools.

The requirements for the Marines' application with a vertical landing were the most challenging for the F135. The vertical thrust provided by directing the engine nozzle 90 degrees into the vertical position and sending 25,000 hp from the front of the engine along a six-foot shaft to power a Rolls-Royce lift fan whose exhaust is vertically downward (Fig. 37). In addition, compressor bleed-air duct "outriggers" from each side of the engine extending outward provided aircraft roll control. The propulsion system provides not only the conventional thrust for normal flight but also can permit the aircraft to move in the vertical direction (Fig. 38).

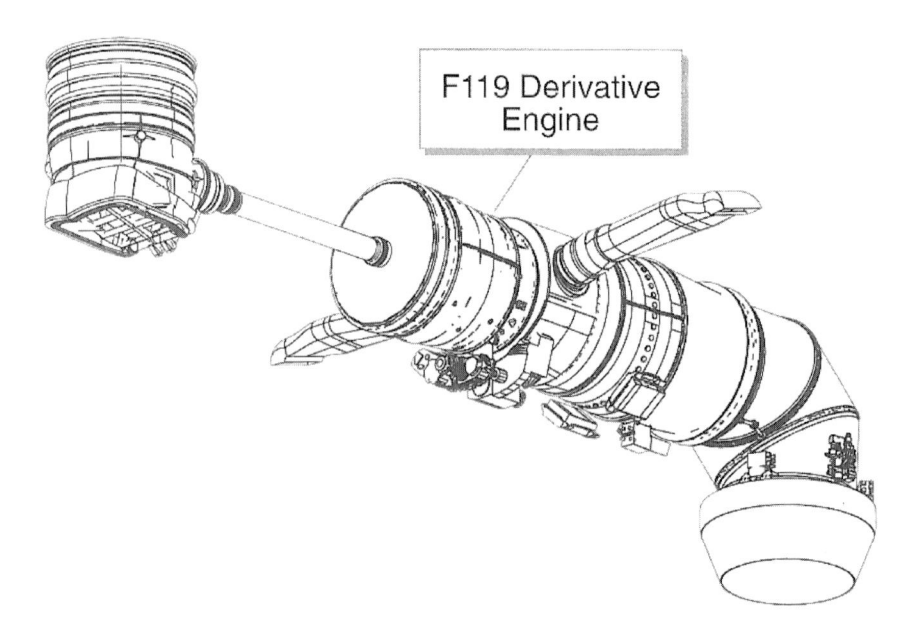

Fig. 38 The multi-tasking engine, shaft power to the upfront fan, thrust vectoring nozzles with compressor air, and engine exhaust thrust vectoring (courtesy of Pratt & Whitney Archives).

The Modern Era

Applications

The first flight test of the F135 occurred on December 5, 2005, and since then over 100 flights have been logged. The total engine market is estimated to be about 5000 engines. The F119 and F135 together will have accumulated more than 800,000 hr before the F-35 goes into service in 2012. On July 28, 2009, Lockheed Martin rolled out the F-35C (CF-1), the U.S. Navy's first stealth fighter, the Lighting II.

F135 Team Wins the Collier Award

The 2001 Collier award went to the team of Pratt & Whitney, Rolls-Royce, Lockheed Martin, Northrop Grumman, and BAE Systems for the Integrated LiftFan™ Propulsion System (ILFPS). At the award ceremony, the reasons why this team was honored were explained as follows [5]:

> [ILFPS] is a revolutionary break-through in propulsion system technology. It redefines the relationship between thrust and vertical lift to achieve a 1.5 : 1 lift-to-thrust ratio, as well as major increases in performance, efficiency, and safety. And it was thoroughly proven as it powered the Lockheed Martin X-35B Joint Strike Fighter (JSF) demonstrator aircraft to a stunning success in the summer of 2001.
>
> Following months and more than 1200 hr of successful ILPS ground testing, flight validation trials were conducted on the X-35B Joint Strike Fighter prototype under demanding hot desert conditions. The propulsion system performed flawlessly, meeting or exceeding all program expectations.
>
> In only 38 days, the ILFPS powered 39 flights and more than 20 hr of testing, including 22 hovers, 17 vertical take-offs, 18 short take-offs, 27 vertical landings, and a total of 116 conversions—95 ground and 21 in flight. Of particular significance, on July 20, 2001, the X-35B performed a short take-off, a level supersonic dash and a vertical landing in a single flight, proving superior versatility for combat missions.
>
> The ILFPS is a major step forward in propulsion system technology. It's this technology, and the teamwork that developed it, that has led to this achievement of excellence.

Pratt & Whitney is the prime contractor and has full responsibility for the F-35 propulsion system. The X-35B is the closest integration of an engine with the airframe, more than any earlier airplane programs. The off-design demands on the engine for vectored thrust, shaft horsepower to the lift fan, and air-bleed for roll stability require very close integration of the engine with the airframe.

PW1000G

The PurePower™ PW1000G is Pratt & Whitney's latest engine for commercial applications, and appears to be a great departure from current high-bypass turbofans. At the time of this writing, few specific details on the engine are available. Its design objectives compared to current engines are reduction in fuel burn by 12 percent, reduction in NOx emissions by about 50 percent, and reduction in noise (relative to Stage 4) by 15–20 decibels.

So far, a 30,000-lb thrust demonstrator engine has undergone sea-level testing and has accumulated over 40-hr flight testing in a Boeing 747 test bed for engine performance and operability. Further flight testing has occurred in a joint Pratt & Whitney–Airbus program for the engine in an A340-600. So far it looks as if the engine has demonstrated its fuel burn objective.

Compared to present high-bypass engines, the keys to the PW1000G's great performance and the reduction in number of engine parts is the use of a much higher bypass ratio and the introduction of a reduction gear between the fan and the core engine's low spool. The gear permits the low-spool turbomachinery to rotate at its optimum rpm, which is somewhat higher than that of the fan (Fig. 39). When the low spool has to rotate at fan rpm (as is the case in present high-bypass engines), the engine designer increases the outer

Fig. 39 A geared turbofan may look like this (courtesy Pratt & Whitney).

THE MODERN ERA 471

TABLE 7 TURBOMACHINERY STAGES FOR THE PW1000G

Thrust range (lb)	Mitsubishi 14,000–17,000	Bombardier 17,000–23,000
Number of stages		
Fan	1	1
Low compressor	2	3
High compressor	8	8
High turbine	2	2
Low turbine	3	3
Total	16	17

diameter of the turbomachinery or uses more stages in the compressor and turbine. The work input per compressor stage (and work output of a turbine stage) is roughly proportional to the square of the tip speed. So, the designer can get the required pressure rise in the low compressor by going to a larger radius or by adding more stages. Going to a larger compressor diameter bulges out the flow path for both the low compressor and low turbine. The path to lower weight and fewer stages of turbomachinery is to bring the diameter inward and run at a higher rotational speed.

George Mead and Andrew Willgoos faced a similar problem in the original Wasp piston engine. The engine rpm was limited by propeller tip speed aerodynamic limitations. The power section (the nine cylinders) could get more horsepower by running at a higher rpm. Willgoos put a reduction gear between the power section and the propeller shaft. In the 1970s, the company studied the use of reduction gears in turbofans with higher bypass ratios than 6 : 1. Previous to that time the most powerful reduction gears were the 15,000-hp class in the T57 turboprop. From 1987 to 1989, Allison and Pratt & Whitney collaborated on a propfan that was demonstrated on a Douglas twin-engine aircraft. This engine required a reduction gear of about 40,000 hp. In 1992 Pratt & Whitney tested a reduction gear system for an advanced ducted propeller system. Subsequently, the company got into the development of a lighter weight reduction gear system and tested it in a demonstrator engine during 2001–2002.

The number of turbomachinery stages in the engine for the Mitsubishi and Bombardier applications are listed in Table 7. One has to conclude that 16 or 17 stages for a high overall pressure ratio is indeed a very impressive technology for the PW1000G.

There are already two applications for the PW1000G. One is in the Mistubishi Regonal Jet with a thrust level in the 15,000–17,000-lb range. The other is in the Bombardier CSeries™ aircraft in the 100–149 passenger size with an engine thrust level in the 17,000–23,000-lb range.

ADVANCES IN ENGINE TECHNOLOGY

Throughout this chapter, we have seen how Pratt & Whitney has advanced the field of engine technology in its development of high-bypass ratio engines.

Pratt & Whitney's earlier research on emissions brought out the TALON X (Technology for Advanced Low Nitrogen Oxide) combustor, which achieved its desired turbine inlet temperature while minimizing the formation of the oxides of nitrogen.

The noise levels of the PW1000G engine are considerably below current high-bypass ratio engines because of lower fan tip speeds and lower exhaust velocities from the fan and core engine exhausts. Because the noise energy level of an engine exhaust is roughly proportional to the 8th power of the jet velocity, any reductions in jet velocity pay off quickly in noise decibel level. The high-bypass ratios in the ballpark of 10 and above reduce the jet velocities so low that noise from the turbomachinery becomes important to quiet down. Sound-absorbing material in the engine inlet and flow path attenuates the turbomachinery-generated noises. It is essential for an airline to have lower noise aircraft in smaller passenger sizes so that they can land at airports with a curfew for aircraft which cannot meet noise regulations. The low-noise aircraft give the airline more flexibility in scheduling their flights.

Over the past 15 years the Pratt & Whitney Military Engine division has advanced the state-of-the-art in engine performance and versatility by reducing the cost of ownership and in managing complex programs to achieve program objectives and schedules. The PW1000G engine program is such a big step in technology from the present high-bypass commercial engines that it reminds one of the Luke Hobbs/Perry Pratt J57 program of about 60 years ago. It is beginning to look as if Pratt & Whitney is pulling it off this time as well!

REFERENCES

[1] Serling, R. J., *Legend & Legacy, The Story of Boeing and Its People*, St. Martin's Press, New York, 1992, p. 388.
[2] *Aviation Week & Space Technology Source Book*, Vol. 152, No. 3, Jan. 17, 2000, p. 122.
[3] Candebo, S. W., "F100, F119 Meet Ground Test Challenge," *Aviation Week & Space Technology*, May 28, 2001, p. 36.
[4] *Thunder in the Sun: A Look at 40 Years in Florida, 1958–1998*, video produced by Communications Department, Pratt & Whitney, 1999.
[5] "Collier Award," brochure, National Aeronautic Association (NAA), Washington, D.C., 2002.

Chapter 15

LOOKING BACK 80 YEARS

AVIATION IN HISTORY

In this final chapter we will look at how far aviation has come since the founding of the Pratt & Whitney Aircraft Company and what role the aircraft engine has played in aviation's progress. Today we take air travel for granted; yet before WWII, air travel was an adventure. When you think about the circumference of the world at about 25,000 miles, it is a fantastic achievement to be able to go to just about any point in the world by commercial airlines within two days (Fig. 1). When you look back on about 4000 years of recorded history, you see that transportation was an important enabler for the advancement of civilizations.

The American historian Stephen E. Ambrose in his book, *Undaunted Courage*, pointed out that up to the early 1800s (about 95 percent of recorded history) the fastest man traveled was at the speed of a horse [1]. The flight across the Atlantic by Alcock and Brown in 1919 was an almost unbelievable achievement. Shortly thereafter, commercial aviation came into being, first in Europe and then later in the United States. Since that time, commercial aviation has made international travel relatively effortless compared to travel before the aviation age.

As aviation progressed after the Wright brothers, WWI intervened, and within a short time the airplane was being used as an aid to military operations. Throughout the 1920s and 1930s the airplane began to take on a number of roles, the sum of which was the capability of deterrence. The many roles of military aviation became reconnaissance; transport; bombing; fighter escort and attack; and platform for battlefield management.

The effectiveness of military aviation was enough to keep the Cold War (right after WWII) from heating up. The mammoth B-36 and the B-52 became effective deterrents to aggression. Today the aircraft has become a flying platform for launching missiles that are directed precisely to a designated target. In military operations, a strong air force takes command of the air, and in the process helps ground troops advance without fear of enemy air attacks. At the same time, the air force can attack the enemy's ground troops and their equipment with devastating accuracy. Aviation has been good for the

- 1519 Magellan – about 1100 days
- 1889 Sailing Ship (Nellie Bly) – 72 days
- 1924 Army Planes – 15.1 days
- 1933 Wiley Post – 7.7 days
- 1946 Lucky Lady II – 3.9 days
- Today's commercial aircraft – less than 2 days

Fig. 1   Transportation milestone.

peaceful development of civilization and has been a positive force for peace because of its powerful deterrence capability.

COLLIER AWARDS AS OVERVIEW OF AVIATION HISTORY

Robert J. Collier, president of the Aero Club of America, had a great faith in the future of aviation not only as a technical achievement but also as an economic opportunity. He awarded the first Collier Trophy in 1911 to aviation pioneer Glenn H. Curtiss for the development of the sea plane called the "Hydro," a pusher configuration on a single main float.

There were hardly any military or commercial opportunities in 1911 that opened the door to the promise of high-volume production. Now we have the advantage of hindsight. After more than 90 years of Collier Awards, we can see the great history of aviation, made possible by the unbelievably great reductions in engine specific fuel consumption and in engine specific weight. The Collier Awards relating to airplanes were usually made directly to the airframers, and in only about 20 percent of the awards was there any mention of the engine manufacturer. I include the engine manufacturers because "the engine is the heart of the airplane," as Pratt & Whitney's public relations man, George Wheat, stated in the early days of the company.

PISTON ENGINE ERA

Donald Douglas built the U.S. Army Air Service planes that set out to encircle the globe in 1924 (Fig. 2). Two Douglas single-engine World Cruisers circled the globe in 360 hr. Engine reliability was a concern because each aircraft used about nine engines to make the trip. The 12-cylinder, 400-hp liquid-cooled Liberty engine, built in 1918, represented the state of the art in powerplant technology. The Liberty engine, which had been based upon the most recent automotive engine technology from Jesse G. Vincent of Packard and E. J. Hall of Hall-Scott, was created during WWI and put into mass production quickly. The total production run was about 20,000 engines.

The Collier Award Committee in 1927 gave the award to Charles Lawrance, who deserves the credit for getting the U.S. Navy interested in air-cooled engines at the beginning of the 1920s. He believed, based upon his assessment of the French air-cooled engines, that an air-cooled engine would be lighter than the liquid-cooled engines of his day. The Navy suggested to Frederick Rentschler (president of Wright Aeronautical at the time) that he bring Lawrance and his engine into Wright and develop the engine. George Mead and Andrew Willgoos took over the development of the nine-cylinder air-cooled J-1. After Mead and Willgoos left Wright in 1925 to join Rentschler's Pratt & Whitney Aircraft Company, Lawrance continued with the able assistance of Sam Heron at Wright with the development of the J series into the J-5 (the Wright Whirlwind) that powered Charles Lindbergh's epic flight to Paris in 1927.

Aircraft propulsion moved toward commercial use with the invention of the variable-pitch propeller in the early 1930s. When aircraft speeds were low, the take-off propeller pitch and the cruise pitch were roughly compatible. However, as the aircraft cruise speeds increased, there became a big difference between take-off and cruise propeller pitch. Frank Caldwell, of Hamilton Standard, solved this problem with his variable-pitch propeller.

By the mid-1930s, Douglas was creating a blossoming commercial transport market with his DC-2 and principally with his DC-3 twin-engine aircraft, powered by air-cooled engines. Sanford Morse, of General Electric, found a way of improving the altitude performance of piston engines by using the engine exhaust into a turbine to run a supercharging compressor. In retrospect, we can see that Moss came within one step of inventing the aircraft gas turbine in the United States. His supercharger development made a significant improvement in altitude performance for military engines in WWII.

Howard Hughes in 1938 flew a Lockheed-14, twin-engine aircraft powered by Wright R-1820 engines around the world in about 91 hr, a new record. This achievement, of course, was not a demonstration of commercial feasibility but it did show future possibilities when larger aircraft and engines would be available.

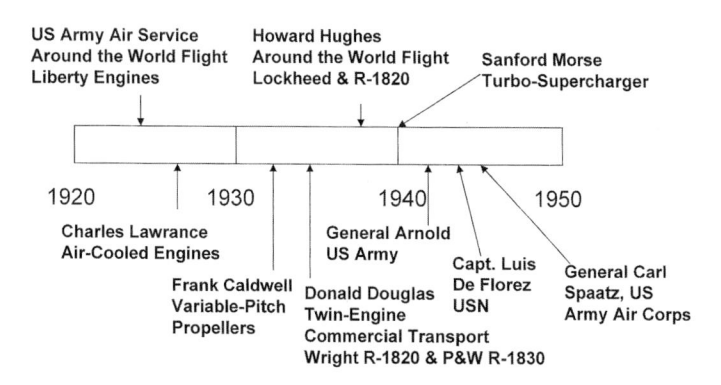

Fig. 2 Time line of Collier Awards that used the piston engine.

The Collier awards from 1942 through 1944 to Army Air Corps Generals Arnold and Spaatz, and to Navy Captain Luis DeFlorez, were for their production, training, and effective use of aviation in WWII. Air-cooled engines provided about 85 percent of the total American engine horsepower in WWII (Pratt & Whitney at 50 percent and Wright Aeronautical at 35 percent). Fifteen percent of the power came from liquid-cooled engines, Allison 1710s and Rolls-Royce Merlins built by Packard.

What one might conclude from the piston engine time line shown in Fig. 2 is that Frederich Rentschler, George Mead, and Charles Lawrance were correct in their vision for air-cooled engines. And since the late 1920s, American aviation progress in the piston engine era demonstrated world leadership in military and commercial aircraft.

Gas Turbine Engine Era

The United States was behind the Europeans in gas turbine technology at the end of WWII. Pratt & Whitney in 1952 was betting the farm on the J57 in the Air Force B-52. That effort paid off, launching other Collier Awardees such as the F-100, the F4D, and the F8U aircraft programs. Gas turbines used by Collier winners shown in Fig. 3 highlights six Pratt & Whitney gas turbine engines, the J57, the J58, the JT9D, the PW4000, the F100, and the F135. In the early 1950s, the state of the art in compressor pressure ratio was about 6:1. The J57 opened the technology path to high-pressure ratio gas turbines in the Boeing B-52. The commercial version of the J57 (JT3C) launched luxury jet travel in the Boeing 707 and the Douglas DC-8. The J58 was in a class by itself as a Mach 3+ cruise engine in the A-11, the YF-12, and principally in the SR-71.

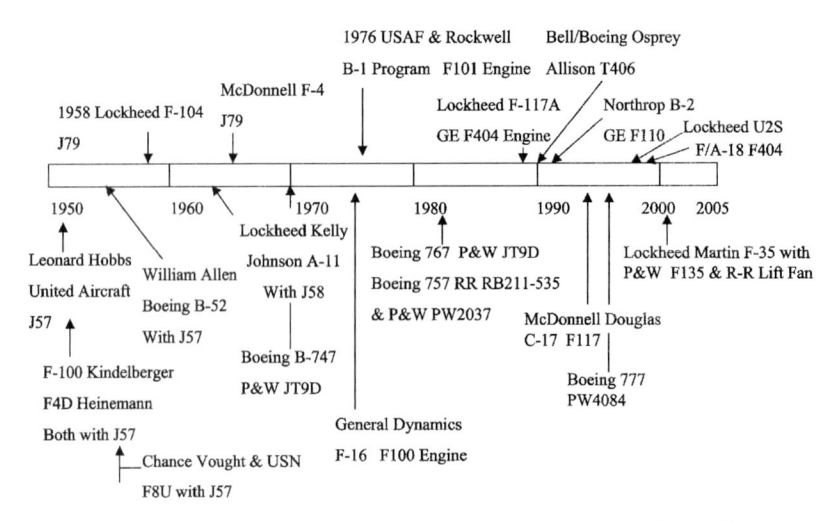

Fig. 3 Time line of Collier Awards that used the gas turbine engine.

The JT9D high-bypass ratio commercial turbofan launched wide-body commercial aircraft, starting with the Boeing 747, and later became the engine for the Boeing 767 and the Airbus A310. The PW4000, which was a further development of the JT9D concept, launched the Boeing 777. The F100 engine technology in the F-15 and F-16 aircraft improved to the F119 configuration, with the added feature of thrust vectoring by the exhaust nozzle in the advanced tactical fighter, the Raptor F-22A. This engine provided a high spool for the advanced F135 engine powering the Lockheed Martin F-35 aircraft. What is striking about this engine–airframe combination is its complete departure from what we normally think of an airframe and engine. For example, in aeronautics we were taught that an airplane produces its own lift when the engine pulls it to a certain air speed where its wing surfaces provide the lift. Traditionally the role of the engine is to pull the airplane to its required speed so that the lifting surfaces (the wings) have enough lift to overcome the weight of the aircraft. What is different in the F-35 is the engine produces the normal thrust as usual, but also provides vectored thrust (up, down, or sideways) from its multi-directional jet nozzle and vertical thrust from a Rolls-Royce forward lift fan, powered by a shaft from the front of the F135 engine. The propulsion system is also providing lift, particularly in those sections of the flight envelope where the aircraft's aerodynamic lift is not always possible because of an unfortunate combination of aircraft altitude and speed.

A number of achievements by General Electric (the J79, the F101, the F404, and the F118) and Allison (the T406) in their respective aircraft are also shown in Fig. 3. While the J57 advanced the state of the art in gas turbines with its 12:1 compressor pressure ratio and dual-spool compressor configuration, General Electric, with a different technical approach to a 12:1 compressor pressure ratio, followed quickly with its J79, using a single spool with variable stators in the compressor for the Lockheed F-104 and the McDonnell F-4 applications.

Then the next generation of military engines, the P&W F100 and the GE F101 combined the dual-spool concept and the variable-stator technologies to produce twice the compressor pressure ratio of the J57 and J79 engines. In the commercial engine technology, General Electric and Pratt & Whitney turbofans use dual spools with variable stators that achieve overall pressure ratios in the 40s.

In the later half of the 1990s, Lockheed was recognized for its U-2 reconnaissance aircraft, which had been operational in the 1950s, first with the J57, then with the J75. After about a half century, this innovative aircraft is now flying with the GE F118 engine, which is also in the Air Force B-2 bomber.

The technology growth in gas turbines has been more impressive than that of piston engines. The thrust has grown from the 3000-lb thrust level just after WWII to about 100,000 lb. The specific fuel consumption has decreased by at least 65 percent for the commercial engines. The advancement in military engine technology has created a degree of aircraft maneuverability never dreamed possible.

The significant increase in engine size and reduction in specific fuel consumption has made huge commercial aircraft possible. Engine reliability has improved so much that commercial airliners fly the Atlantic with twin-engine aircraft without concern. While the aircraft's lift/drag ratio has not improved very much from the days of the Boeing 707 and Douglas DC-8, there has been such a great reduction in engine specific fuel consumption that one could conclude that the engine has been the greater contributor to current commercial and military aircraft performance than improvements in the aircraft technology.

Progress in Airframe Technology

In the following pages we trace commercial aviation progress, because commercial aviation is more easily documented and measured compared to military aviation. What we see in the commercial aircraft engine story does not have much application to military aviation. A significant difference is that the military engines put more emphasis on higher thrust to weight ratio than commercial aircraft engines, where cost-of-ownership is more important. The development of aviation proceeded in the last century along an "S-curve," where the progress was slow at first, then shot up. In the 1980s, it settled at a slower growth rate. It took about 30 years to settle on the configuration for a commercial airliner (Fig. 4). Wings became swept and airfoils were refined, but these are relatively minor changes compared to the evolution from piston engines to the powerful high-bypass turbofans of today.

The Junkers' all-metal monoplanes, flying just before 1920, and the Boeing 247, flying in the early 1930s, were significant steps toward defining the ultimate money-making commercial airliner. Subsequently, the fabulous DC-3, the first airliner to become profitable for the airlines, came into service in 1936 and is still flying. The configuration requires only two engines. The

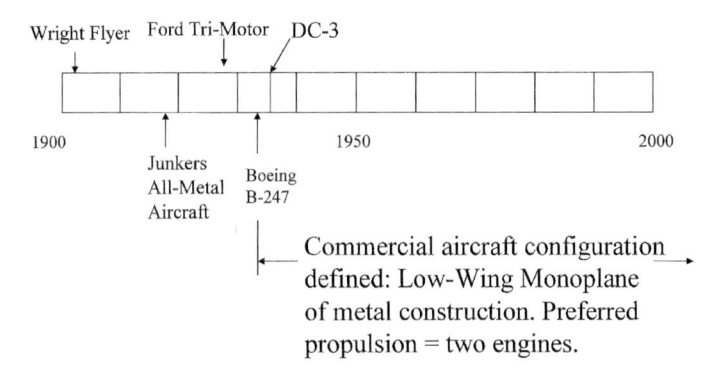

Fig. 4 Time line of the evolution of commercial aircraft configuration.

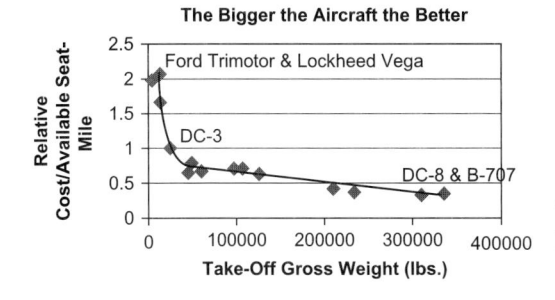

Fig. 5 The bigger the vehicle, the lower the per passenger costs (in then-year dollars) [data from 2].

P&W R-1830 Twin Wasp and the Wright Aeronautical R-1820 were available to power the DC-3. Minimum engine costs (first cost and operating costs) require the minimum number of engines. One engine would be the minimum except that the aircraft is required to climb out after an engine shutdown on take-off. This engine-out requirement defined the practical minimum as two engines. The Ford and Fokker trimotors were popular in their day because they could continue to fly after an engine failure; however, they were costly to operate as shown in Fig. 5. But the figure also shows that the airframers made great strides in the field of structural analysis, materials, and manufacturing technology to build huge structures.

The DC-3 was the first aircraft big enough to be profitable for the airlines. Perceptive readers instantly know that all points subsequent to the DC-3 are not necessarily with only two engines. A further reduction in cost could be achieved in the four-engine examples, when two large engines replaced the four engines. In all the four-engine points, there were no two larger engines available at the time to replace the four engines. The cost per available seat-mile would be lowest if the Boeing 747 975,000-lb gross take-off weight model were added to Fig. 5.

In the first 25 years or so of aviation, it was considered a dangerous, adventurous activity rather than a safe, sensible mode of transportation. Most of the concern in flying was due to the engines, which in those days were not reliable (compared to today's standards). The automobile engines of that time period were not exactly reliable either, but at least an engine failure did not offer the possibility of a suddenly abbreviated life span. In the days of single-engine aircraft, an engine failure in flight left the pilot with an airframe possessing a substantial amount of kinetic and potential energy, which would be reduced to zero in a relatively short time. In the most optimistic scenario, the pilot could make a *dead-stick landing* in a field where the total of the kinetic and potential energies would be dissipated gradually through aerodynamic and mechanical friction drag. The worst-case scenario would be when the total energies were dissipated by structural damage to the aircraft and pilot.

Pratt & Whitney's contribution to aviation was the introduction of dependable engines, achieved through excellence in design and intensive "build 'em & bust 'em" development effort. The engines available for the DC-3 were the

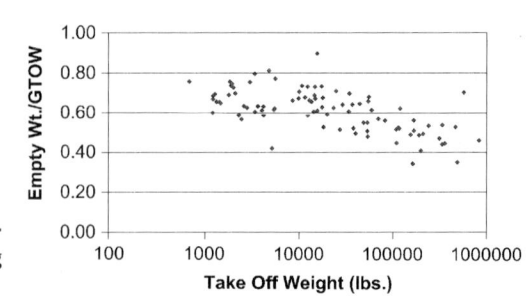

Fig. 6 The ratio of empty to take-off weight decreases with increasing take-off weight [data from 4].

Wright Aeronautical R-1820 and the P&W R-1830, with essentially the same cubic inch displacement. Wright's approach was to wring the maximum power out of nine cylinders in a single row. Pratt & Whitney got the same horsepower out of 14 cylinders in two rows of seven cylinders, a considerably more conservative design. The Pratt & Whitney engine was the most widely used engine in the DC-3 and in the military version, the C-47, because of its reliability [3].

The only way to provide the power as the demand for higher gross weight developed was to go to four engines, because there was no engine big enough at that particular time for a twin-engine installation. The DC-3 was powered by two engines rated at 1000 hp. The DC-4 was about twice the weight of the DC-3 and used four engines (R-1830 at 1100 hp). The DC-6 was about four times the gross weight of the DC-3 and used four engines (R-2800 at 2400 hp).

This same philosophy continued into the jet age, where twin-engine configurations were preferred when the engines were available. Otherwise, three or four engines were used. Engine reliability has improved greatly since the days of the DC-3, where time between overhauls was in the 1600-hr level, to the R-2800 level of 2500–3000 hr in the Convair, Martin, and Douglas aircraft after WWII. In the jet age, the gas turbines had even greater time between overhauls. In some cases, the engines were on the wing for more than three years in commercial service. Time between overhauls (TBO) is not used today as a maintenance requirement on commercial aircraft. It has been replaced by on-condition monitoring, which means the health of the engine is continuously monitored and maintenance action is taken on only those components that require attention.

The reliability of a modern aircraft gas turbine is so high that it is perfectly safe to fly across the Atlantic, for example, in a Boeing 767 or an Airbus A310 because, should one engine require a shutdown, the other engine could power the aircraft to a safe landing 3 hr later after the shutdown. The Federal Aviation Authority (FAA) formally calls this authorization to fly over water routes with twin-engine aircraft ETOPS (extended-range twin-engine operations). It is based on the capability of the twin-engine aircraft to reach a suitable airport after a single-engine shutdown. In fact, commercial aviation

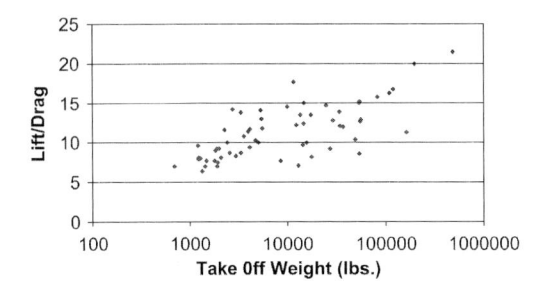

Fig. 7 L/D ratios increased from the early days of biplanes with heavily-braced wings and fixed landing gears [data from 5].

has become so routine as far as the general public is concerned that nobody seems to be concerned about the number of engines.

The ratio of empty/gross weight for the Wright Flyer in 1903 was about 0.76 because the gross weight (of about 700 lb) was equal to the empty weight plus a Wright brother and a small can of gasoline. The DC-3, with a gross weight of 25,000 lb, had an empty weight of 17,720 lb, which gave an empty/gross weight ratio of 0.71. In practical terms, 29 percent of the gross weight was available for payload and fuel. Current large aircraft have almost 60 percent of the gross weight available for payload and fuel. Figure 6 demonstrates the significant progress made in aircraft structural weight.

The gross weight was also a function of time (bigger airplanes being built later as requirements grew) so that the trend would be the same. The two highest lift/drag (L/D) values are those of the B-47 and the B-52 with their very high aspect ratio wings. The commercial aircraft have relatively bigger fuselages with more surface area for drag. Hence the L/D is not as good as those in the B-47 and B-52. The U-2, with its glider-like wingspan, would have an L/D somewhat higher than those shown in Fig. 7.

In summary, we see that there has been great progress in airframes in structural efficiency (to carry more passengers) and in aerodynamic efficiency (higher L/D ratios). As the airframe became more agile, it could handle more power. We explore this in the next section.

AIRFRAME'S DEMAND FOR POWER

A glider is towed to a high altitude, at which condition the glider has both potential energy (from the altitude) and kinetic energy (from the velocity of the tow plane). In the absence of thermal air currents, the glider dives to pick up speed and levels out at a point where the forward velocity provides enough lift to balance the glider's weight. However, this is a transient condition because the drag reduces the glider's forward velocity quickly, which decreases the lift, all of which results in the need for another dive to build up speed to generate enough lift to balance the glider weight for a short time interval. Of course in the meantime the ground is getting closer. Even though soaring is a noble sport, the glider is not a practical means of commercial transportation. One needs an engine for level flight for aircraft. The aircraft engine is a mechanical device that converts the chemical energy in the fuel to propulsion energy to power the aircraft through the air.

Here is a little back-of-the-envelope analysis to illustrate the magnitude of the propulsion requirements for various size aircraft all the way up to a million pounds of gross take-off weight. We are talking about accuracy in the arithmetic within 10 percent. We are not designing airplanes here. We are just trying to get a feel for the magnitude of the engineering challenge in providing engines for the complete spectrum of commercial aircraft. If we take the weight of an aircraft and divide by the aircraft's L/D ratio, we can get an estimate of the aircraft drag, which, of course, is the same level of the thrust at constant flight speed. Look at Fig. 8. If you divide one million pounds of aircraft weight by an L/D ratio of 15, the result is roughly 70,000 lb of cruise thrust. Six hundred mph is slightly over Mach 0.9, and requires about 100,000 hp to keep the aircraft flying. As a rule of thumb this cruise thrust is usually about one-half the take-off thrust. Also, I know that it takes about 30 minutes to climb to the cruise altitude, and by that time the gross weight is lighter than at take-off.

For review, thrust (in pounds) multiplied by aircraft speed (in feet per second), divided by 550 foot pounds per second per horsepower equals thrust horsepower. What kind of engines could meet these requirements? Gas turbines can go all the way down in power and thrust to cover the entire range of aircraft gross weights. However, the piston engines are limited in the maximum power output because of the mechanical complexity of crowding cylinders together. The largest of the piston engines in service were the liquid-cooled Rolls-Royce Eagle (3000 hp) and the P&W R-4360 (3500 hp).

While the commercial aircraft configuration was settled with the DC-3, the aircraft engine configuration had to be reborn around the gross weight of the B-36 bomber, which I believe was the largest piston engine-powered aircraft (with a little help from the J47 jet engines for the 410,000 gross weight aircraft configuration). A couple of huge flying boats used eight and 12 engines but these applications were not very extensive. The largest turboprop powered aircraft would have been the Douglas C-132 at 500,000-lb gross take-off weight, with the PT5 (T57) engine, but the program was canceled. Piston engine companies (Allison, Wright Aeronautical, and Pratt & Whitney) had to start over again in the jet age. Figure 9 illustrates the increasing demand for

Fig. 8 Magnitudes of propulsion requirements.

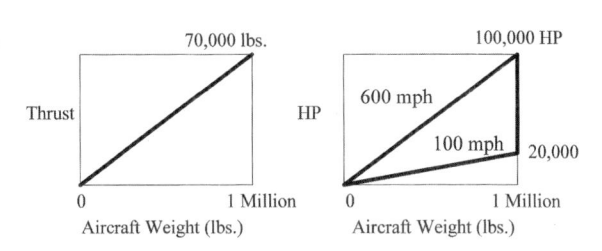

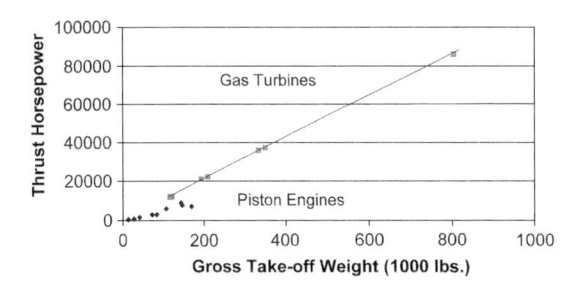

Fig. 9   As weight increases, so does aircraft cruise horsepower.

horsepower as aircraft weight increased. On the other hand, the airframers kept the same aircraft configuration but changed the sweep and the profile of the wings, and used the wing body interaction to maintain high L/D ratios. Those, along with corresponding advancements in avionics, materials, and structures, were not trivial refinements. But they were less significant compared to an all-new engine configuration.

In other words, the airframe was evolving smoothly along a specific track. On the other hand, the aircraft engine was progressing along the piston engine track and then all of a sudden the train of progress had to switch to a completely different track. In the next section, we explore how Pratt & Whitney rose to the challenge of providing more powerful engines to match their customers' evolving needs and technologies.

PROGRESS IN ENGINE TECHNOLOGY

In 1824 a French scientist, Nicholas Leonard Sadi Carnot (1736–1832), became the father of thermodynamics. His monumental work was called *Reflections on the Motive Power of Fire and on Machines Fitted to Develop that Power* [6]. His thermodynamic principles showed that it was possible to convert work into heat. For example, if you mount an outboard motor in a barrel of water, 100 percent of the work done on the water by the motor's propeller blades is converted into heat. However, Carnot explained that the opposite conversion process of heat into work is nowhere near 100 percent. The most efficient conversion of heat energy would be with the use of the Carnot cycle, whose efficiency depended upon the absolute temperatures at which heat was added and rejected. For example, the maximum efficiency of a heat engine (operating in a Carnot cycle) between 60° F and 2500° F would be 82 percent. He would get that number by taking the temperature difference (2500 plus 460) minus (60 plus 460) and dividing this difference by the absolute temperature at which heat was added (2500 plus 460). The practical engineering use of the Carnot cycle is to determine the maximum amount of work that can be obtained between temperatures of heat addition and heat rejection.

The Otto cycle, the most popular form of a piston engine, is named after Nicolaus A. Otto (1832–1891), a German scientist. It is called the four-cycle engine because of the four steps in the process, intake, compression, power, and exhaust. The piston and cylinder produces power only 25 percent of the time because it has a power stroke only for one-quarter of the cycle (two complete revolutions of the crankshaft). This intermittent nature of the power stroke naturally caused vibrations in the mechanical structure and created the need for vibration damping and a flywheel to help smooth out the power impulses.

When Hans von Ohain, in the early 1930s, thought about ways of creating thrust, he favored the steady-flow process of the Brayton cycle—named after U.S. scientist George B. Brayton (1839–1892)—for gas turbines. The steady flow (producing power continuously instead of only 25 percent of the time) sounded attractive. However, the significant difference between the two cycles is the manner in which heat energy is added to the process. The Otto cycle adds the heat energy at constant volume and the Brayton cycle adds the heat energy at constant pressure.

In the Otto cycle the heat is added at essentially constant volume when the piston is near top dead center. Then an abrupt temperature and pressure rise occurs upon ignition. The frequency of the power-producing stroke (about 25 percent of the time) is low enough for the cooling system to keep the metal surfaces at satisfactory temperature levels. The Brayton cycle runs at a constant pressure and temperature at 100 percent of the time, but temperatures have to be lower than those of the Otto cycle to keep the metal structures operating at safe stress levels. The piston engines were more efficient than gas turbines because of their ability to run higher average temperatures. The Brayton cycle was known long before 1900, but the limitation of materials to stand up to the kind of temperatures that made the gas turbine competitive with the piston engine kept the gas turbine out of the market until after WWII.

The sum of engine plus fuel weight can be rewritten as the sum of engine weight per airflow/engine thrust per pound of airflow, plus the specific fuel consumption (pounds per hour per pound of thrust) multiplied by the time duration of the flight (Fig. 10). The engine challenge is to keep the sum of the engine and fuel weight per pound of engine thrust as low as possible. We will take each of these terms and examine what the engine designer can do with each one to minimize weight. The use of the sum of fuel and engine weights for a specific flight is very limited. Different amounts of fuel are burned in the flight segments of taxiing, take-off, climb, cruise, descent, landing, and taxiing. I use the concept here only to illustrate the effects of the major variables, weight/airflow, thrust/airflow, and specific fuel consumption.

Fig. 10   Calculation to minimize weight.

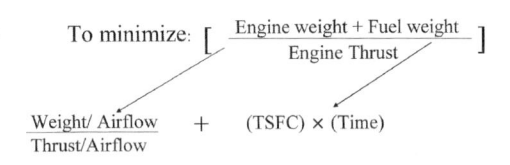

$$\text{To minimize:} \left[\frac{\text{Engine weight} + \text{Fuel weight}}{\text{Engine Thrust}} \right]$$

$$\frac{\text{Weight/ Airflow}}{\text{Thrust/Airflow}} + (\text{TSFC}) \times (\text{Time})$$

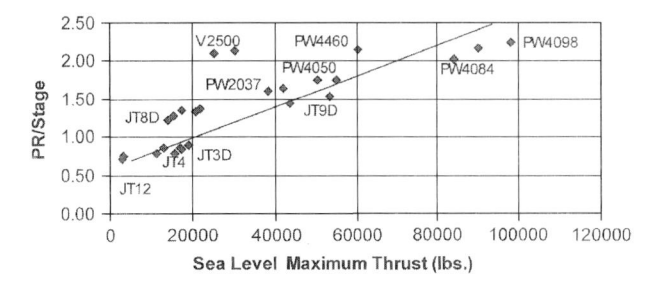

Fig. 11 Improvement in pressure-ratio per each stage of turbomachinery.

ENGINE WEIGHT/AIRFLOW

Research on engine weight per pound of airflow was aimed at minimizing the number of stages of the turbomachinery in the compressors and turbines. There has been steady progress in getting more pressure ratio per each stage of turbomachinery. In Fig. 11 we see improvement in the pressure ratio per compressor stage (the pressure ratio per stage was obtained by dividing the compressor pressure ratio by the number of stages). Another direction was to look at newer materials with improved strength per weight ratio, and even composite materials.

The secret of improving the output per stage lies in managing the development of boundary layers over the airfoils. This means controlling the local velocities on the airfoils. In a conventional airfoil, the local velocities build up quickly into the supersonic range, and the only way for the flow to adjust to the downstream pressure level is with a normal shock, which results in flow separation. On the other hand, by controlling the local velocities (by curvature of the airfoil) into the low supersonic regime, the flow can adjust to slowing down toward the trailing edge without a strong shock wave that would induce boundary layer separation (Fig. 12a). One of the benefits of the controlled-diffusion airfoils has been the thicker leading edges, which have

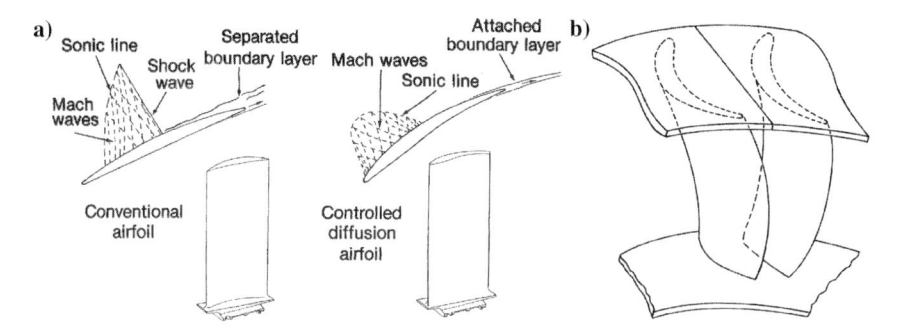

Fig. 12 Improving airfoil efficiency, a) by managing the airflow over the airfoils, and b) by considering three-dimensional flows (courtesy of Pratt & Whitney Archives).

longer operational life because of their greater resistance to erosion compared to the older, conventional airfoils (Fig. 12b).

In the early days of Pratt & Whitney gas turbines, we designed the turbines and compressors according to mean-line design. Then we constructed the root and tip profiles to be compatible with the free-vortex flow conditions. After that, we used straight-line fairing between the root and tip profiles to define the blade and vane shapes. Much later, when the full power of the computer could be used, we took into account radial flows and the behavior of boundary layers at the root and tip of airfoil passages. When all of these factors were taken into account, we ended up with some peculiar shaped passages; for example, see how the airfoil twists at the root in Fig. 13.

As air flows over an airfoil, it creates friction on the airfoil surface, and the velocity of the air next to the airfoil slows down. This part of the airflow is called a *boundary layer*. There are two types of boundary layers, *laminar* (with lower drag) and *turbulent* (with higher drag). The objective of the airfoil designer is to keep the boundary layer laminar as long as possible. A rough surface is known to stimulate the transition from laminar to turbulent. Therefore, the airfoil surface must be perfectly smooth. Compressor and turbine blades (rotating airfoils) and vanes (stationary airfoils) also have boundary layer flows at the airfoil roots and tips. These boundary layer flows go from the pressure side of an airfoil across the root or tip, to the suction surface of an adjacent airfoil. These are problems that the airframe designer does not have, except perhaps at the juncture of the wing and fuselage. Some aircraft use winglets (shrouds in turbine terminology) at the wing tips to cut down the boundary layer flow from the pressure surface to the suction surface. Controlling the diffusion to avoid separation, and the pressure distribution on the airfoil to maintain a laminar boundary layer, are the ways to minimize airfoil profile losses (Fig. 14).

THRUST/AIRFLOW

Higher temperatures tend to produce more thrust per pound of airflow. The objective is not to demonstrate high thrust/airflow for a short time, but to be able to sustain the high performance level over long periods of time.

Fig. 13 Controlled endwall blading (courtesy of Pratt & Whitney Archives).

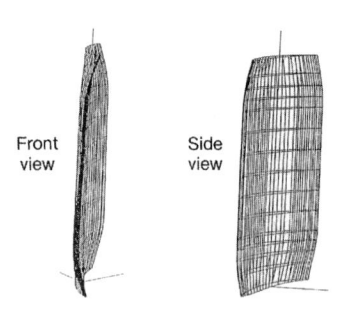

Front view

Side view

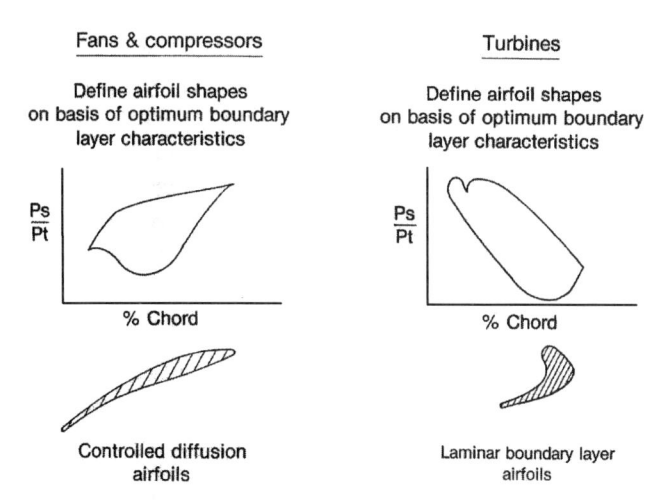

Fig. 14 Two ways of minimizing airfoil profile losses (courtesy of Pratt & Whitney Archives).

The need for high temperatures created a need for protective coatings on metal surfaces to minimize corrosion, plus a need for stronger materials at the higher temperatures.

THRUST-SPECIFIC FUEL CONSUMPTION

The combination of higher turbine inlet temperature, higher overall pressure ratio, and higher bypass ratio offer the potential for reduced specific fuel consumption. The trends have been toward higher and higher values of turbine inlet temperature and overall pressure ratio (Fig. 15). With an eventual goal of 1500° C (about 2700° F) in turbine inlet temperature and an overall pressure ratio in the 40–50 range, more stages of compressors and turbines are necessary, unless even higher output per stage is possible with the turbomachinery. At this stage of the game you can appreciate that the aerodynamics and thermodynamics of the aircraft engine are vastly more complicated than those of the aircraft. In addition, there has been an unbelievable degree of progress made in engine technology compared to aircraft technology.

There is a loss of performance in the engine when there are leaks along the flowpath in the engine. It is essential to run very tight clearances for the rotating parts, and to have small clearances where the seals are located. Figure 16 shows the gas path in an engine. Unfortunately the aircraft engine (in contrast to stationary industrial gas turbines) has to endure substantial gyroscopic forces, and of course the military engines have even greater gyroscopic forces in their operating envelopes.

When the rotors in an engine start to flex the turbomachinery, blades could rub against the outer case. In the past the designer would allow plenty

488 JACK CONNORS

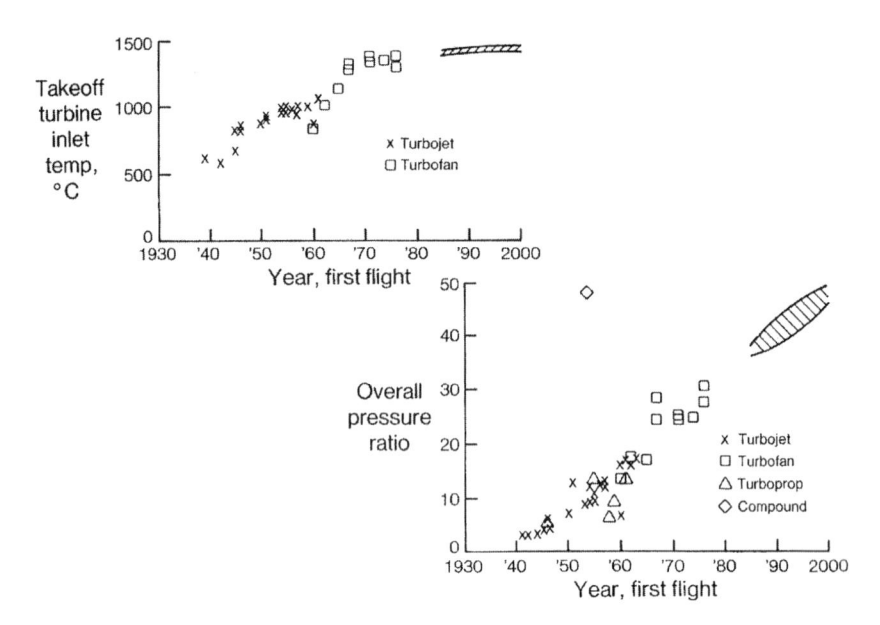

Fig. 15 Progress in turbine temperature and overall pressure ratio (courtesy of Pratt & Whitney Archives).

of room at the blade tips to avoid serious rubs. This came with a performance penalty. The next technique for higher performance was to vary clearance with a process known as *active clearance control* (ACC). This control handles the problem of the relative transient dynamic and thermal growth of the rotor and case. One way to accomplish this was to cool the

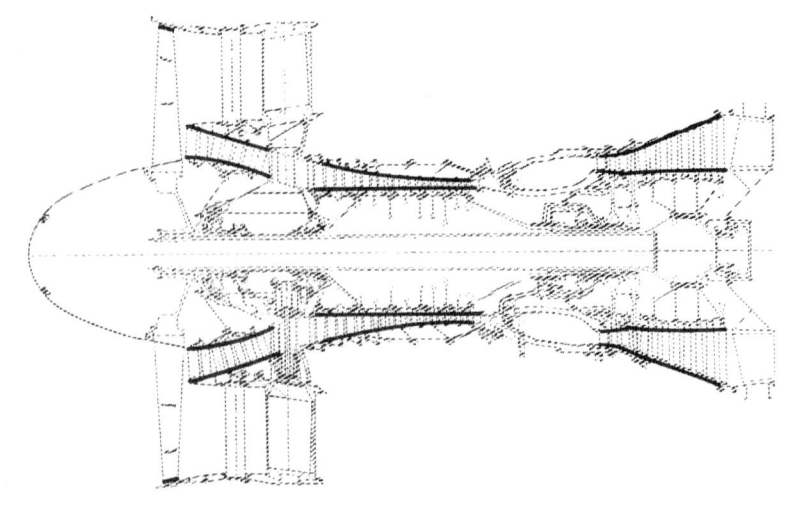

Fig. 16 The engine gas path is shown downstream of the fan (courtesy of Pratt & Whitney Archives).

case when the aircraft was at cruise operating conditions and the engines were not subjected to maneuver loads (as would be the situation during take-off and the subsequent aircraft rotation at the beginning of climb). Abradable rub strips took care of the local rubbing caused by maneuver loads. The ACC took care of the performance requirement when the aircraft was in a steady flight condition.

COMPARISON OF TURBOMACHINERY AIRFOILS WITH AIRCRAFT AIRFOIL

The aerodynamics of the aircraft and the engine have one common nemesis—boundary layer management on airfoil surfaces. The problem is much more difficult in regard to compressor and turbine airfoils for the following reasons:

1) The engine airfoils are tiny (compared to the wing) and are packed together closely. The airflow over the wing is easier to analyze because the airfoil is close to the conditions for an isolated airfoil, which is definitely not the case with turbomachinery.
2) The flow of the boundary layers on the compressor and turbine blades and vanes is much more complex than the aerodynamics of the wing–body interaction because of the secondary flows at the root and tip of airfoils.
3) The turbomachinery airfoils are subject to continuously varying angle of attack from idle speed all the way up to maximum power. In addition, rotating blades encounter rapid variation in angle of attack as the blades sweep through the wakes of the upstream vane cascade.
4) The boundary layers on the rotating blades are subjected to rotational forces, which do not exist for the wing and tail surfaces of the aircraft.
5) The engine has a series of complex problems in delivering the right amount of cooling flow to those stationary and rotating parts that depend upon cooling to keep the metal permissible stress level at its proper value.
6) The airframer can heavily instrument the aircraft wing to observe the flow over its surfaces. The engine test engineer has a much more daunting task to instrument the engine airfoils.

The engine designer, in contrast to the airframe designer, must face government regulations pertaining to the inlet and exhaust of engines. As far as I know there are no restrictions on an airframe wake as there are on engine exhausts in regard to chemical emissions, noise, and visible smoke.

The engine designer has had to come to terms with boundary layer development throughout the engine. The boundary layer behavior is much more complex in a compressor or turbine stage than in an airframe wing. The compressor and turbine blades and vanes are essentially tiny wings inside the engine. I will try to illustrate this complication in engine airfoils in three steps. The first step considers the flow over a wing airfoil. The second shows how the flow is more complicated over the compressor airfoil and the third example will explain why there are fewer turbine stages (compared to compressor stages) in a gas turbine.

Let's start out with a wing airfoil to show why aerodynamics is much more complicated in turbomachinery than on a wing. The pressure downstream of the airfoil is the key factor in determining the lift of an airfoil (Fig. 17a). The lift of an airfoil is the difference in pressure between the pressure and suctions surfaces acting on the airfoil's projected area. In the airplane wing, the pressure downstream of the airfoil is the same as the pressure upstream of the airplane. In the case of a compressor airfoil, the downstream pressure is higher, and a turbine airfoil experiences a lower pressure downstream. In all of these airfoil applications, the limiting pressure rise on the suction surface (upper surface of wing) to the exit pressure limits how much lift the airfoil can provide.

To an observer on the airfoil (the wing) the airfoil is being washed by a stream of air at the velocity of the aircraft. Around the nose of the airfoil, the pressure rises to the full stagnation pressure, and then part of that oncoming flow goes around the suction surface, and the other part goes around the pressure surface. The net difference in pressure between the two surfaces multiplied by the projected area becomes the so-called lift force on the wing. The suction surface has a minimum pressure point after which the pressure on the surface rises back to the value of the ambient pressure. It is very important to make sure this pressure rise on the suction surface is kept below an acceptable limit. An old rule of thumb was to never have the pressure rise exceed half of the dynamic pressure. The dynamic pressure (as a reminder) is one-half the product of density and the velocity squared divided by the acceleration due to gravity (32.2 ft/sec^2). The dynamic pressure, in reference to the pressure rise to ambient pressure level, is calculated at the point of minimum pressure. Figure 17b shows that compressor airfoil has a more challenging goal than that shown in Fig. 17a.

In practical terms, one would redesign the airfoil to increase the minimum pressure on the suction surface so that the subsequent pressure rise to higher than inlet pressure would remain within the acceptable limit. What we learn from these simple examples is that the compressor airfoil cannot take the "full lift" force (as the wing whose static pressures at the inlet and exit are equal) but must have a reduced work input from the shaft power in order to avoid separation. Figure 17c shows that when the exit static pressure of the turbine airfoil is lower than that of the inlet, the pressure rise from the minimum pressure on the suction surface to the exit pressure is less. This means that an airfoil as a turbine blade can provide more power to the shaft by changing the airfoil shape to lower the minimum pressure to a value that is consistent with the pressure rise from minimum to exit static pressure. To review:

1) The airfoil for a wing has the same static pressure ahead of and behind the airfoil.

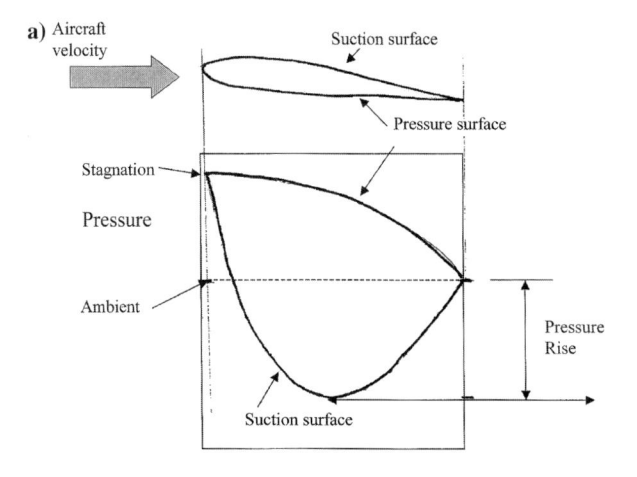

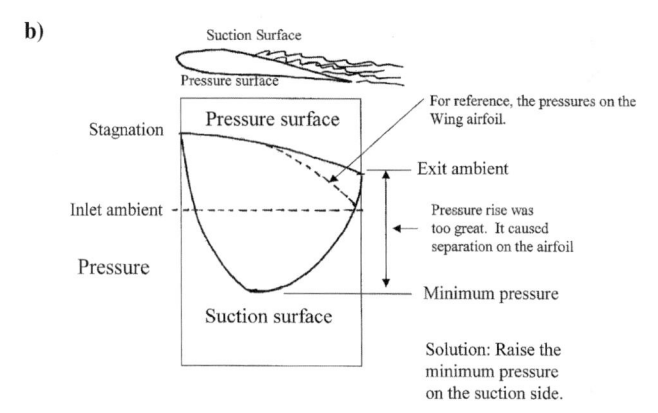

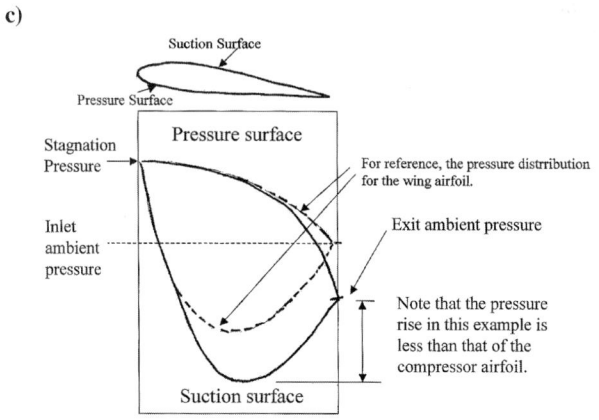

Fig. 17 Pressure distribution on a) a wing airfoil, b) an airfoil used in a compressor, and c) an airfoil used in a turbine.

2) The compressor airfoil has a higher exit static pressure than its inlet static pressure. Therefore, the work input to the airflow from this airfoil (net pressure surface minus suction surface pressures acting over the projected airfoil area) must be less than in the case of the wing airfoil conditions to avoid boundary layer separation on the suction surface.
3) On the other hand, the turbine airfoil can put out more work than the wing airfoil because the exit static pressure is less than that at the inlet to the airfoil.

Now you know why the axial-flow compressor has more stages than the turbine that drives the compressor.

COMPARISON OF AIRFRAME AND ENGINE IMPROVEMENTS

It is interesting to compare the improvement in the airframe and the aircraft engine since the early days of the B-707 and the DC-8, both powered by the JT3C turbojet. How much improvement has there been in airframe drag coefficient since the DC-3? The answer is that there has been hardly any change in the airframe L/D ratio over the past half century, compared to the huge reduction in specific fuel consumptions of engines. For example, the reduction in specific fuel consumption from the early JT3C turbojet to the PW2037 turbofan is about 50 percent (Fig. 18a). The change in aircraft L/D has been essentially zero from the B-707 (L/D = 19–19.5) to the B-747 (L/D = 18) to the B-767 (L/D = 18), using the L/D values from [5]. In other words, the substantial improvement in engine efficiency in the last 50 years plus the airframers' skills in building aircraft of high structural efficiency have made the jet age economically feasible (Fig. 18b).

PROGRESS IN ENGINE CONTROLS

When the Wright Brothers demonstrated their 1910 Wright Flyer, they used the simplest of engine control systems. Their vertical four-cylinder 30-hp engine did not even have a carburetor. Starting the engine by spinning

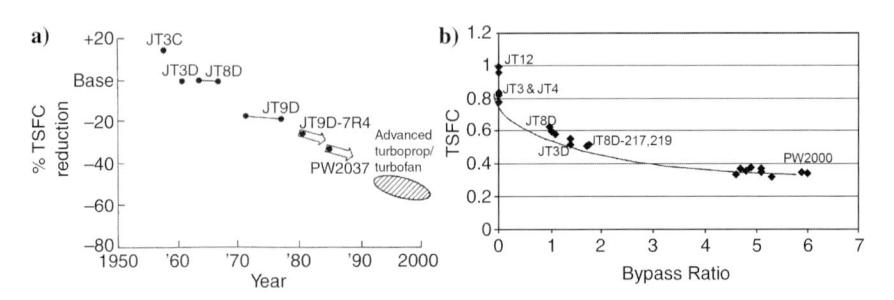

Fig. 18 Thrust specific fuel consumption, a) trend in aircraft engine efficiency (courtesy of Pratt & Whitney Archives), and b) with bypass ratio.

the prop was a tedious task. Thereafter, the pilot just played with the spark advance to vary the engine power. The early gas turbines, like the early piston engines, also had relatively simple controls. The gas turbine in the 1940s was referred to (in contrast to what appeared to be complex piston engines) as a simple rotating element. Then, as more and more was demanded of the engine, the control system became more complex. Complexity grew so quickly that even in the early days of the J57, the lure of an electronic control was irresistible. Its application seemed to make sense. However, the state of the art was not sufficiently advanced for vacuum tubes to operate in a hot, vibratory environment nestled on the periphery of a 10,000-lb thrust engine (Chapter 7). Hydromechanical controls then came to the rescue.

Dick Baseler, J57 development engineer, used the Hamilton Standard JFC3 electronic control in the J57 program. Some of the early flying in the B-52 flight test program was with prototype engines managed by electronic fuel controls. However, the miniature electronic tubes and associated accessories failed to live up to Baseler's requirements, and he made the shift to hydromechanical controls. Hydromechanical controls in the next decade established a great reputation for reliability. Baseler was so burned by electronic controls on the prototype J57s for the B-52 that he refused to consider electronic controls for the rest of his career at Pratt & Whitney. He was famous for his remark, "Show me a good electronic control, and I'll show you a good hydromechanical television set."

The end of the 1950s meant the end of the slide rule as a design tool for gas turbines. The J58 was the last of the slide rule-designed engines. Analog computers were being used to simulate certain transient calculations in heat transfer and stress analysis. In the definition of turbine airfoils in the mid-1950s, Pratt & Whitney was using IBM payroll computers on a part-time basis for straight line fairing of turbine blades from well-defined root and tip airfoil contours (Chapter 8). In the 1960s, company engineers were beginning to use Fortran and mainframe computers to make engine performance estimates and to do stress analysis. Gradually the design process was indeed becoming much more complex and computerized.

After the banishment of the JFC3 at Pratt & Whitney in the early 1950s, I never heard of an electronic control until I got to know Joel Kuhlberg in connection with the early JT10D program before the program became globalized with the search for foreign partners. Figure 19 summarizes the significant events and steps in the evolution of the full authority digital electronic control (FADEC), primarily from a commercial engine viewpoint [7]. Hamilton Standard was working on electronic controls in the mid-1960s for military applications. Pratt & Whitney was also conducting electronic engine control studies, as well as experimental work in the second half of the 1960s for the Advanced Manned Strategic Aircraft and the Tactical Fighter programs. When it came time for the F100 engine competition, the company proposed a

Significant Events

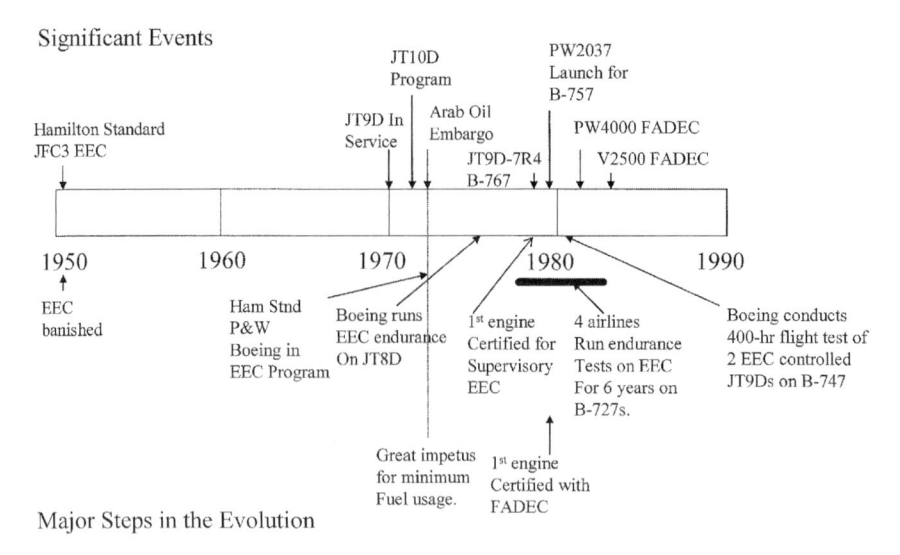

```
                         JT10D              PW2037
                         Program            Launch for
                                            B-757
                         JT9D In   Arab Oil
Hamilton Standard        Service   Embargo           PW4000 FADEC
JFC3 EEC                                    JT9D-7R4
                                            B-767    V2500 FADEC

1950          1960          1970          1980          1990

EEC         Ham Stnd    Boeing runs   1st engine   4 airlines    Boeing conducts
banished    P&W         EEC endurance Certified for Run endurance 400-hr flight test of
            Boeing in   On JT8D       Supervisory  Tests on EEC  2 EEC controlled
            EEC Program                EEC          For 6 years on JT9Ds on B-747
                                                    B-727s.

                              Great impetus   1st engine
                              for minimum     Certified with
                              Fuel usage.     FADEC
Major Steps in the Evolution
```

Fig. 19 Evolution of the FADEC.

hydromechanical engine control with a digital electronic supervisory unit that could make adjustments for exhaust gas temperature and engine rotational speed as a means of improving engine performance.

As time went on, complexity in engine control requirements seemed to grow exponentially (Fig. 20). This complexity started with military engines but the same trend was picked up in the commercial engine area. In the 1970s, Hamilton Standard and Pratt & Whitney developed what became the FADEC. Electronic controls moved into the number one position in steps. The first step was as an

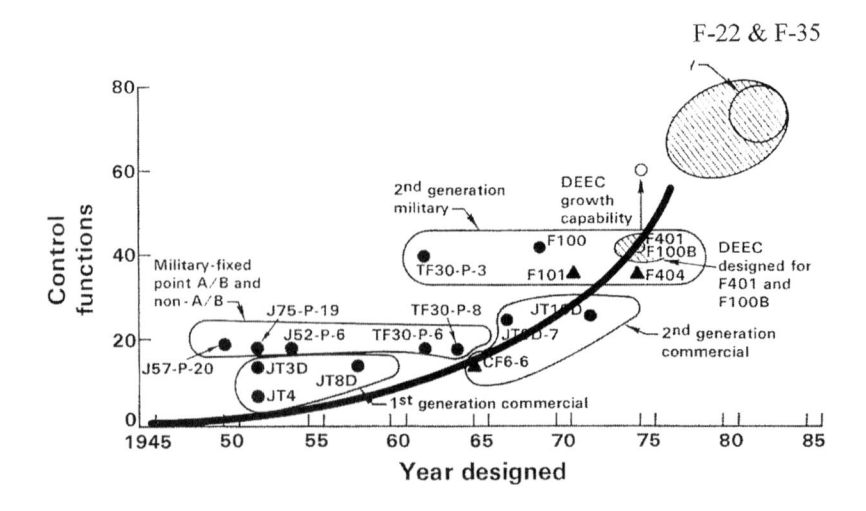

Fig. 20 What was asked of engine controls over time (courtesy of Pratt & Whitney Archives).

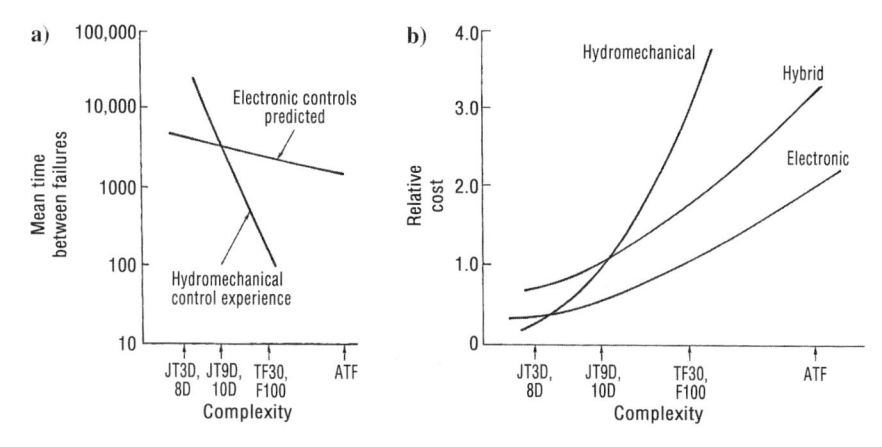

Fig. 21 Assessment of hydromechanical vs electronic controls, by a) cost, and by b) experience (courtesy of Pratt & Whitney Archives).

auxiliary adjunct to the hydromechanical control. If the electronics failed, nothing drastic happened (JT9D-7R4 and F100-PW-100). Then it became a full authority electronic control with hydromechanical backup. Again, if the electronics failed, the engine still put out thrust (F100-PW-200). This trend continued in the training-wheel approach until the electronic control, with built in redundancy, became the whole show. As the industry turned to advanced gas turbines, the cost of hydromechanical controls became prohibitive (Fig. 21).

The FADEC is mounted on the fan case of a commercial engine before the connections are in place. The fan location is the logical position for mounting the FADEC because it is the coolest location on the engine. Figure 22 shows

Fig. 22 The inside view of the FADEC (courtesy of Pratt & Whitney Museum).

what the inside of the FADEC looks like. The FADEC is more reliable compared to hydromechanical controls. In addition, of course, the FADEC handles much more complicated demands than the hydromechanical control. The FADEC's reliability is enhanced by redundancy, which is an easier to do with electronics than it is with hydromechanical redundancy. In the end, the FADEC set new standards of reliability and managed the operation of the engine with a degree of finesse never dreamed of in the hydromechanical days.

THEN AND NOW

About a little more than a half-century ago when the aircraft gas turbine was just coming out of its shell, we referred to it as a simple rotating element. Now it is a sophisticated energy conversion device that converts the chemical energy in hydrocarbon fuel into efficient aircraft propulsion—managed effectively by an electronic fuel control. Military and commercial engines developed in two different directions, as shown in Fig. 23. The military maximum thrust size leveled off around 40,000 lb, while the commercial engine thrust level kept increasing up to 100,000 lb as aircraft grew in size. For a more detailed discussion of the evolution of gas turbine technology I suggest Ben Koff's excellent treatise, "Gas Turbine Technology Evolution: A Designer's Perspective," *Journal of Propulsion and Power*, Vol. 20, No. 4, July–August, 2004.

PRATT & WHITNEY'S FINEST MOMENTS

To conclude this technical history of the engines of Pratt & Whitney, I find it a daunting challenge to select the company's finest moment. Even daring to select Pratt & Whitney's four most defining events is a difficult task. However, the following is my bold selection when I look at Pratt & Whitney's 82-year

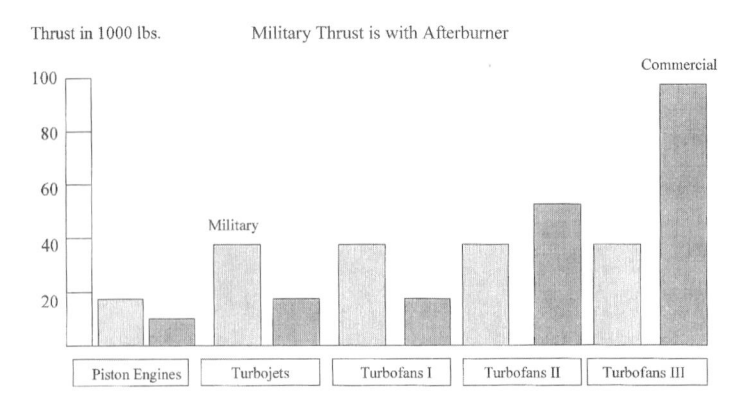

Fig. 23 Pratt & Whitney's progress in propulsion (military engine with afterburner, light gray; commercial engine, dark gray).

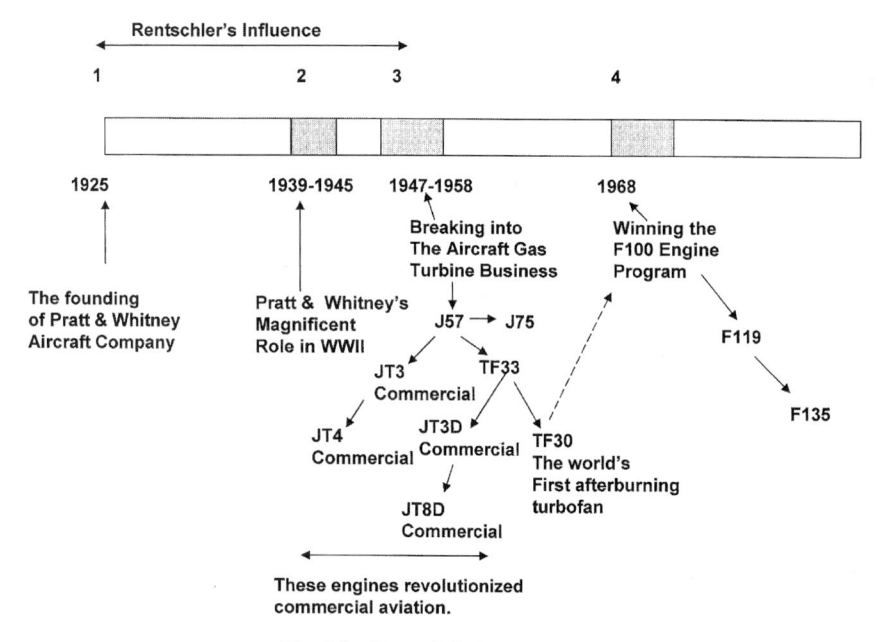

Fig. 24 Four defining moments.

contribution to military and commercial aviation (Fig. 24), three of which Frederick Rentschler participated in before his death in 1956.

The first two events are in the piston engine era of Pratt & Whitney. First, founding the company was a significant event. It was a huge financial risk at that time because the military engine market was dependent upon the U.S. Navy buying engines. Second, Pratt & Whitney's role in WWII is a book in itself. Entering the gas turbine market is third milestone I have selected. Rentschler gave Luke Hobbs the financing, facilities, and personnel he asked for in 1945. In return, Rentschler wanted to dominate the aircraft gas turbine engine business by 1955. Hobbs achieved that goal!

The technology fallout from the J57 created the TF33 turbofan, which was a modification of the J57 front end. The TF33 turbofan powered the B-52H and C-141 transport. This new turbofan technology was the basis for the TF30 turbofan. The experience with the TF30 afterburning turbofan in the F-111 became one factor in the company winning the F100 engine program, which became a must-win campaign for the Florida plant to survive. The fourth defining moment, the F100 program, gave Pratt & Whitney extensive experience in inlet-engine compatibility, an area of expertise that helped greatly in subsequent F-22 and F-35 program competitions.

FOUNDING OF PRATT & WHITNEY

Frederick Rentschler and George Mead founded Pratt & Whitney Aircraft Company in 1925 in Hartford, Connecticut, at a time when aviation was still in the adventurous stage of development. One of the major obstacles to commercial aviation was the unreliability of aircraft engines. The obstacle to military aviation (from Rentschler's viewpoint) was the then-current wisdom that liquid-cooled piston engines were the future of military aviation because they fit into the airplane's streamlined fuselage more neatly than the bulky air-cooled radial engines.

Rentschler and Mead disagreed. They believed the future of aviation would be with air-cooled radial engines, because air-cooled engines were simpler, lighter, and would be more reliable because of their simplicity compared to liquid-cooled engines. The U.S. Navy was the first customer for air-cooled engines. Then the Army Air Corps began to buy them after trying them out, and found out that radial engines could power an airplane to at least the same speed as their favorite liquid-cooled engines. In 1929, William Boeing suggested that he and Rentschler form a new company called the United Aircraft & Transport Company that would provide coast-to-coast airline service; airport services; and manufacture airplanes, propellers, and engines. Boeing learned that the Pratt & Whitney Wasp and Hornet engines were more dependable, and that his mail planes could carry more mail because those engines were lighter than the liquid-cooled Liberty engines in use. Boeing himself gave Rentschler credit for creating commercial aviation.

CONTRIBUTION TO THE ALLIES' WWII EFFORT

Pratt & Whitney organized the automobile companies (Ford, Buick, Chevrolet, and Nash-Kelvinator) and trained them to manufacture Pratt & Whitney engines. It took from 15–18 months to get them into production. They had to duplicate Pratt & Whitney's manufacturing facility. What the automobile companies thought was good precision manufacturing was not good enough for aircraft engine production.

Pratt & Whitney expanded its manufacturing base by lining up about 600 subcontractors and by increasing its manufacturing floor space from 600,000 ft^2 to over 5 million ft^2 to produce with the auto companies (plus Continental and Jacobs) 363,619 engines that powered about 60 different aircraft models, bombers, fighters, reconnaissance, and trainers. No production airplane was ever held up by lack of an engine. In WWII, Pratt & Whitney supplied about half the total horsepower of American engines. Wright Aeronautical, at about 35 percent, and Allison, at about 15 percent, furnished the rest of the power. Pratt & Whitney estimated that the total power output from its engines in WWII was 603,814,723 hp (450,446 megawatts or about half the electrical power generated by public utilities in the United States in 2005).

Rentschler did what was completely unexpected at that time; he gave money back to the government when he found his costs were actually lower than what he negotiated. He created the Renegotiation Act by his actions. In addition, he not only waived the license fee to those building Pratt & Whitney engines, he also helped with technology when other competitive engine companies ran into problems.

BREAKING INTO THE AIRCRAFT GAS TURBINE BUSINESS

In 1945, the company had to cut its piston engine business way back after VJ Day, when $414 million's worth of engine orders were suddenly cancelled. Further bad news struck when Pratt & Whitney found it was not in the new aircraft gas turbine engine business, which was sewed up by Westinghouse and General Electric. There appeared to be no room for the company as both the Navy and Air Force had their preferred engine companies producing turbojet engines.

The only chance was to get into the Air Force B-52, which already had a Wright Aero turboprop (T35) or a Westinghouse turbojet (J40), depending upon which way the Air Force decided to go. If they wanted long range, then they would go with the turboprop. If they wanted speed, they would go with the turbojet, but they could not have range and speed at the same time because the state of the art in gas turbines could not provide such an engine. Pratt & Whitney decided to bet the farm on going with a technology that leapfrogged the competition and gave the B-52 *both* range and speed. The subsequent fallout was a string of fighters, tankers, the U-2 reconnaissance aircraft, and luxury commercial aircraft eventually powered by turbofans (JT3D, JT8D) and the military TF30, the world's first afterburning turbofan. It was the experience solving the F-111's inlet-engine compatibility problems that gave Pratt & Whitney a technology advantage (relative to General Electric) that helped to win the F100 engine program.

F100 ENGINE PROGRAM

Pratt & Whitney's high thrust/weight afterburning turbofan enabled the Air Force to do spectacular maneuvers in the F-15 and F-16 aircraft. These new maneuvers introduced problems that attracted the attention of the press, Congress, the Department of Defense, and its competitor General Electric. The Great Engine War was then highly publicized, with an uncomplimentary, but temporary, tinge to Pratt & Whitney's reputation. Even so, the outcome was a fabulous success with the F100-PW-229 engine model, the top of the line F100 engine. This long, drawn-out struggle put Pratt & Whitney in the position to win the F119 engine program, and subsequently the F135 engine program, because of the company's dedication, integrity, knowledge of the customer's needs, and the ability to satisfy those needs more effectively than any other engine manufacturer.

COMMON ELEMENTS

These four events have certain elements in common:

1) *Desperate situation*—The technical challenges were formidable. The reputations of both the Air Force and Pratt & Whitney were in jeopardy.
2) *Failure was not an option*—Management had considerable financial risk. Failure in this program would have serious business consequences in the future.
3) *Extreme dedication*—Team members did more than their best; they did what was required to win.
3) *Competitive spirit*—Team members were going to do better than any competitor and were not giving up until victory was achieved.

RENTSCHLER'S LEGACY

When Rentschler died in 1956, his company was well established in the jet era. The JT3 and JT4 commercial jet engines launched luxury jet travel with the B-707 and the DC-8. Later the JT8D made travel convenient in the short- and intermediate-range aircraft (B-737, DC-9, and B-727), and in 1970 the JT9D launched luxury wide-body long-range air travel when it powered the B-747. In any review of aviation propulsion history the first witness to be called has to be Frederick Brant Rentschler. Paul Fisher, Director of Public Relations at Pratt & Whitney in 1955, had this to say regarding Rentschler's credentials [8]:

> When I saw Mr. Boeing a couple of years ago in Washington, he made it perfectly clear that Rentschler was the moving spirit that established commercial aviation as we know it today. No one questions that Mr. Rentschler's first rule—that the heart of the airplane was a dependable engine—is the foundation of present commercial flight.

In 1947, when Harvard Professor Robert Schlaifer was writing his book, *The Development of Aircraft Engines*, Rentschler gave him many pages of information on the development of air-cooled engines, and of his own personal experience in aviation [9]:

> I thought I learned one thing out of our small WWI experience, namely, that it seemed that the best airplanes were always equipped with the known best engines and this made me feel that, if we were to get abreast of foreign developments and someday exceed them, it would certainly be necessary to develop American aviation engines which were superior to those of any of our foreign competitors.

Frederick Rentschler looked back at the first 50 years of aviation's progress in 1953, and concluded that 75 percent of the progress was due to the engine. I feel confident that those of us in the aircraft engine business would have no problem in agreeing with his assessment of the power behind aviation's progress. I am sure Pratt & Whitney's competitors agree with me when I remind them of the illuminating experience we have all shared on the defendant's side of the table in the Procurement Department's conference room at an airframer's facility. In that environment we were told that the engine was a relatively minor, insignificant accessory on the superb majestic creation known as the airframe. Also, of course, there was the usual reminder that the airframe weight would not grow and the thrust requirement would stay at its present under-powered airframe level in the specifications. When Rentschler was making his assessment of progress, he was unaware of the three generations of turbofans that would be coming along in the commercial and military transport applications. Had he been aware of the huge reductions in specific fuel consumption from the high-bypass turbofans, he might well have increased his engine credits for aviation progress.

Rentschler is said to have had vision. His vision included not only what might be but what it would take in resources (personnel, facilities, equipment, and financial) to be competitive. In addition, he understood how to manage the resources to design, develop, produce, and market the engine successfully. As he believed, the engine is truly the heart of the airplane. It miraculously converts the chemical energy in the fuel to thrust power for the aircraft. It performs this transformation by controlling the flow of air around hundreds of tiny airfoils (miniscule, compared to the airplane wing) in an environment sometimes hot enough to turn many of our familiar metals into mush. What is truly amazing is that the ambient air flows over the airframe and joins back together at the rear of the aircraft at the speed of the aircraft. Meanwhile, the gas turbine reaches ahead of the aircraft and in a fraction of a second ingests the ambient air and transforms it into high-speed jet(s) for the continuous production of thrust.

Those of us who have spent a significant portion of our lives working at Pratt & Whitney, and are now retired (as are those in the wonderful group who have helped me put this history together), look back with pride at being members of the team that did so much to advance aviation (Fig. 25). My old boss and friend, Dick Mulready, summed up the experience in working at Pratt & Whitney in those great times [10]: "They [the engineers] were so dedicated to making this equipment work that they would spend all kinds of extra hours at it.... Everybody was just happy with the job."

Speaking of teams, it is only appropriate that we give Mr. Rentschler the last word on the essence of Pratt & Whitney's success in the early days with Mead, Willgoos, Brown, Borrup, and Marks [11]:

> These few men were bound together first by strong personal friendships and second because each had complete and mutual respect for the particular abilities of each other. These were the ideal qualities, which made for a team, and only a team can accomplish important things. Naturally

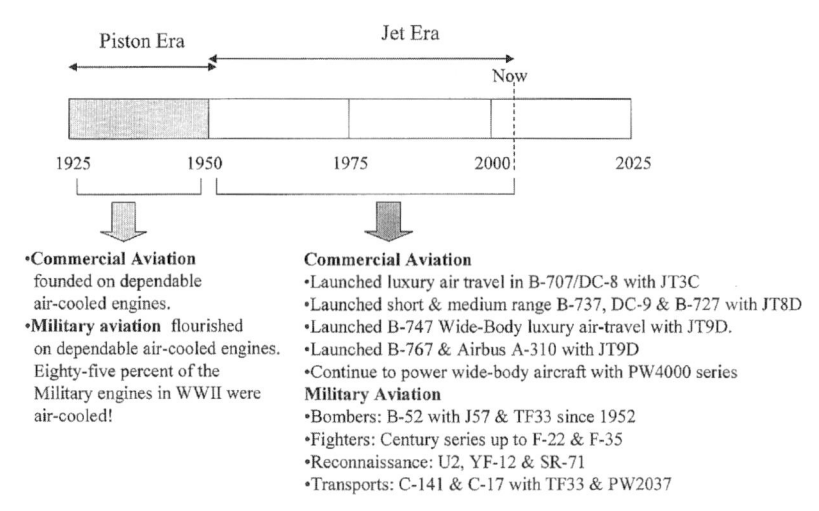

•**Commercial Aviation**
founded on dependable air-cooled engines.
•**Military aviation** flourished on dependable air-cooled engines. Eighty-five percent of the Military engines in WWII were air-cooled!

Commercial Aviation
•Launched luxury air travel in B-707/DC-8 with JT3C
•Launched short & medium range B-737, DC-9 & B-727 with JT8D
•Launched B-747 Wide-Body luxury air-travel with JT9D.
•Launched B-767 & Airbus A-310 with JT9D
•Continue to power wide-body aircraft with PW4000 series
Military Aviation
•Bombers: B-52 with J57 & TF33 since 1952
•Fighters: Century series up to F-22 & F-35
•Reconnaissance: U2, YF-12 & SR-71
•Transports: C-141 & C-17 with TF33 & PW2037

Fig. 25 **Pratt & Whitney's role in aviation.**

we added to the organization as we went along, picking and choosing with the greatest of care and always with complete teamwork uppermost in or minds. I believe this basic team quality has been solidly built into Pratt & Whitney and later into United, and I believe more than anything else this accounts for whatever success that has been experienced.

REFERENCES

[1] Ambrose, S. E., *Undaunted Courage: Meriwether Lewis, Thomas Jefferson and the Opening of the American West*, Simon & Schuster, New York, 1997, p. 52.
[2] Miller, R., and Sawyers, D., *The Technical Development of Modern Aviation*, Praeger Publishers, New York, 1970, pp. 117 and 207.
[3] Holden, H. M., *The Legacy of the DC-3*, Wind Canyon Publishing, Inc., Niceville, Florida, 1997, p. 53.
[4] *Civil Aircraft*, Harper Collins Publisher, Glasgow, Scotland, 1996.
[5] Loftin, Jr., L. K., *Quest for Performance: The Evolution of Modern Aircraft*, NASA, 1985, Appendix A [retrieved Aug. 2009: http://www.hq.nasa.gov/pao/History/SP-468/contents.htm].
[6] Cummins, Jr., C. L., *Internal Fire*, Society of Automotive Engineers, Warrendale, PA, 1989, p. 31.
[7] Molohoskey, A., *The Electronic Engine Control System*, Harbridge House, Inc., Boston, MA, 1984, p. 7 (I am indebted to Joel Kuhlberg for bringing this excellent summary of FADEC to my attention).
[8] Fisher, P., "Letter to C. J. McCarthy, President of Chance Vought," Pratt & Whitney Archives, April 20, 1955.
[9] Rentschler, F. B., "Letter to Robert Schlaifer," Pratt & Whitney Archives, August 25, 1947.
[10] Mulready, D., *Advanced Engine Development at Pratt & Whitney: The Inside Story of Eight Special Projects, 1946–1971*, SAE International, Warrendale, PA, Feb. 2001.
[11] Rentschler, F. B., "An Account of Pratt & Whitney Aircraft Company 1925–1950," p. 23; Rentschler wrote his reflections in 1950 and distributed copies—under the auspices of Pratt & Whitney President Arthur E. Wegner, it was published by the company in Oct. 1986 for internal distribution.

Appendix

PRATT & WHITNEY MEDALLION

1926 to 1945

1945 to 1981

1981 to 1987

1987 to present

The two-inch medallion or logo stamped on every engine produced by Pratt & Whitney has changed over the last eight decades, but three significant features grab our attention: the eagle, the name Pratt & Whitney, and the slogan "dependable engines."

There is no record in the company archives to indicate who selected the eagle, but thanks are due to Jack Rosenthal for locating the original engineering drawing of the medallion (Fig. 1). The original medallion was designed by Harry Gunberg in 1926, checked by Larry Castonguay, and approved by Andy Willgoos.

The eagle has been used as a symbol of imperial power for governments over the many centuries of recorded history. The eagle on the medallion against what could be a blue sky is a natural symbol of aviation.

The name of Pratt & Whitney comes from two gentlemen who ran a fabulous company establishing world standards of precision machinery from 1860 to around 1900. Niles-Bement-Pond took over the Pratt & Whitney Company at the turn of the century and continued as a leader in the manufacture of precision machinery. The name of Pratt & Whitney was known throughout the world in 1925 when Rentschler and Mead founded the Pratt & Whitney Aircraft Company. Accuracy was a word that described the machine tool company.

Rentschler was quick to build on that ready-made image with the message of dependable engines, which of course was the sine qua non of commercial aviation. The assurance of dependability was something the public needed to feel comfortable with air travel.

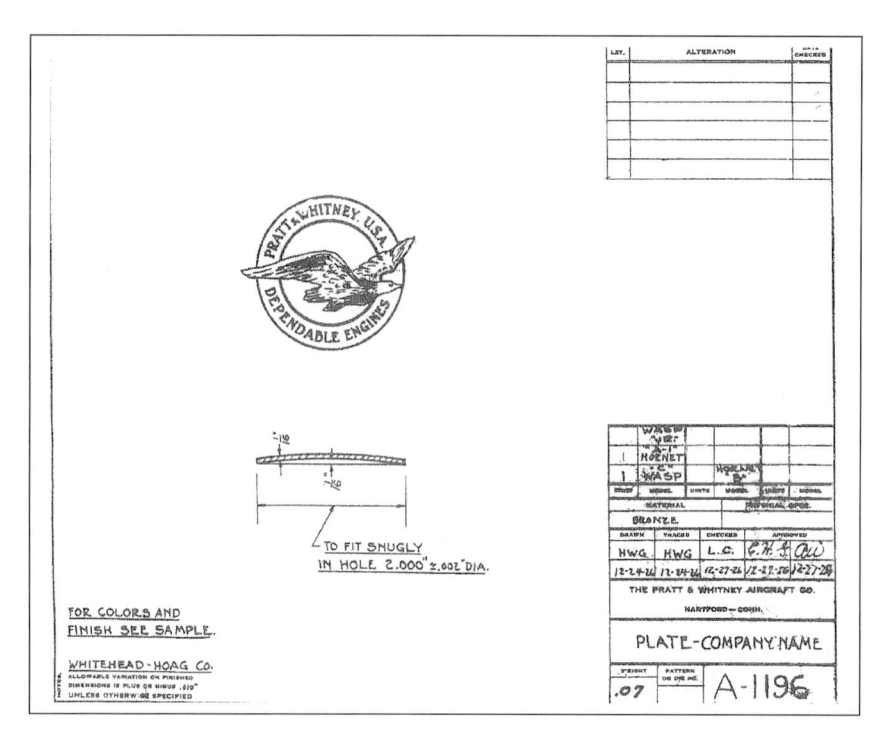

Fig.1 Original engineering drawing (courtesy of Pratt & Whitney).

Afterword

Early Influences

It was in the early years of the jazz age that Louis Armstrong was making history with King Oliver. The first nonstop coast-to-coast airplane flight took place when pilots Kelly and Macready took off from Roosevelt Field in a Fokker T-2 aircraft. Flight time from New York to San Diego was about 27 hours for the 2500 mile trip. It was in this environment on May 8, 1923, the same day and month of Harry Truman's birthday, that my mother had her first child—me. As a screaming little infant, I was unaware of the above events but it turned out that jazz, aviation, and Harry Truman would be weaved seamlessly into the focus of my adult life.

I first saw Louis Armstrong and his band at the Metropolitan Theater in Boston when I was 12 years old, whereupon I decided to become a jazz trumpet player. Since that time I have had the good fortune to talk to Louis on two occasions. I still play regularly with retiree jazz musicians who get a kick out of playing the old standards.

Aviation became my passion at age nine when I made my first solid model airplane—the Granville Brothers Gee Bee Super Sportster R-1 in which Jimmy Doolittle set a speed record in the 1932 Thompson Trophy Race, a record that held for four years. There was an added stimulus when I saw General Italo Balbo's fleet of 25 Savoia-Marchetti S-55A flying boats soar over Boston on their way to the Chicago World's Fair in 1933. One weekend, Clarence Chamberlain dropped by the airport in nearby Squantum offering short rides in his Curtiss Condor aircraft. My father, who was the adventurous type, was quick to take me up. To a little kid like me the Condor seemed like a giant aircraft. As it lumbered down the bumpy runway I wondered if it would get into the air before it entered the Boston Harbor, which was getting alarmingly closer each second. It did.

My affinity with President Truman developed out of sharing the same birthday and later reading about his life. For a time he was my Commander in Chief during WWII while I was serving in the Army Air Corps as Weather Officer. One of the duties of a weather officer was to brief heavily armed, skeptical flight crews on what weather they could expect during their missions.

Such training served me well later on in my career at Pratt & Whitney briefing upper management on my programs in an environment euphemistically called the quarterly review process. A highlight of the Army Air Corps duty was strolling through the maintenance shops looking at those magnificent, powerful engines that helped to win WWII. Maintenance officers convinced me that Pratt & Whitney engines were the best, thus moving my career preferences in a definite direction.

My father was a truck mechanic. Once in a while he would take me to the shop to see the huge vehicles. What impressed me was the loud, powerful roar of the truck engines. One Sunday morning he took me to the shop while he and a couple of other mechanics were working on a truck rear wheel. I was just a little kid, probably not old enough for school, as best as I can remember. What I do recall though was the burst of profanity that ionized the atmosphere when a greasy nut fell from my father's fingers into the mechanism on which he was working. Apparently, the previous hour's work would be for naught because now he would have to take the assembly apart again. Then he got a brilliant idea: Would it be possible for someone with tiny fingers to retrieve the nut? That is when I successfully completed my first task as a mechanic's helper.

My engine preference was further directed toward Pratt & Whitney when I took graduate courses in gas turbines at MIT from Professor Soderberg. Little did I know that he would play such a significant role in Pratt & Whitney's future in the J57 program. His influence at Pratt & Whitney was readily recognized when I started my introduction to practical stress analysis of turbine blades for Pratt & Whitney's first turboprop, the T34. We used the Soderberg diagram to take into account the combined effects of steady and vibratory stresses in estimating turbine blade stress levels.

EARLY DAYS AT PRATT & WHITNEY

In my first few days on the job I had a desk next to Gene Odegaard, a stress analyst. My immediate boss, Owen Welles (assistant project engineer in the Turbine Design and Research Group), gave me a pile of literature to learn about the work the Technical & Research Group was doing. Then, what appeared to be a high-ranking engineer came to Gene's desk and asked Gene to estimate the torsional frequency of the rotor in a specific design of the T34 turboprop.

I vaguely recalled in a stress analysis course under Soderberg how one would use the Holtzer analysis method to calculate the natural frequency of a rotor. The textbook problem I worked on took hours and hours of tedious calculation. That problem was simple compared to what Gene was just asked to do. He was casually given as much time as he needed between then (late morning) and early afternoon. After Gene gave his critical speed estimate to the high-ranking engineer (Perry Pratt, I learned later), I humbly asked how he could do such a complicated calculation so quickly. Gene introduced me to the exciting world of back-of-the-envelope analysis. He said he had some previous calculations plus some test data on rotors. What he did this time was to compare the turbine and compressor moments of inertia with similar cases from the past and then he "scaled" the critical speed by "correcting" for the differences in rotor moments of inertia.

My first impression was that Pratt & Whitney had the finest group of engineers that I could ever hope to work with. My 35 years at Pratt & Whitney was one continuous learning experience, which reinforced that first impression many times over.

ENGINE RELIABILITY—THE KEY TO COMMERCIAL SUCCESS

This point was made early in the development of aviation by Orville Wright. In January 1, 1919 he wrote an article titled "The Future of Civil Flying" for *Aviation and Aeronautical Engineering*, in which he said:

> Although it is now fifteen years since the first flight was made with a heavier-than-air machine, the use of the airplane for commerce and sport has developed but little.... In order to create a real sport it will be necessary to provide means for flying cross-country without risk. There are several ways in which this can be accomplished: (1) The perfection of the flying machine and motor to that degree where forced landings will never be necessary; (2) the establishment of distinctly marked and carefully prepared landing places at such frequent intervals that one could always be reached in case of the sudden stoppage of the motor; and (3) the development of airplanes of such design as to permit a landing in any ordinary field encountered in cross-country flying.

He then discussed the time saving in flying from Dayton to Washington in three hours or to New York in a little over four hours. He continued: "such flights would be exceedingly common if it were not for the danger involved in flying over ground where if the motor should stop suddenly, no safe landing place would be within reach."

This is the challenge Frederick Rentschler wanted to overcome in 1925—to introduce a dependable engine into aviation. William Boeing in his mail contract proved that one could make money flying the mail when there were no unscheduled stops due to lack of engine dependability.

PROPULSION PARADIGM SHIFT

In a remote airfield in northern Germany on August 27, 1939 a new form of propulsion was being demonstrated by Ernst Heinkel and his newly hired powerplant engineer, Hans von Ohain. This historic flight was kept a secret from even the German military. It proved to Heinkel that he could design a unique fighter aircraft powered by a jet engine that would put the Luftwaffe in an invincible position. Shortly after this flight he conducted a demonstration flight for the technology people in the Luftwaffe. This event triggered a massive effort resulting in a jet-powered Me262 fighter and the Arado bomber, neither of which was built by Heinkel. Such is frequently the injustice of being a pioneer.

Aviation was changing in a new direction. A similar scene was repeated again in 1941 when a Gloster demonstration aircraft flew with British Air Force (RAF) Officer Frank Whittle's jet engine. The British military thought so little of this event that they did not think it was worth recording. Thanks to an unauthorized filming, however, we do have a record of this historic flight.

These two inventors met for the first time in 1978 at an engineering conference at Wright Field. Later in 1989 they participated in a program sponsored by Wright Field and AIAA celebrating the 50th anniversary of jet-powered flight.

At the time when von Ohain and Whittle were promoting their engine concepts, jet propulsion was not new. Neither was gas turbine technology. The contemporaries of these gentlemen were knowledgeable about gas turbines but they looked at the application of the jet engines to then-current aircraft speed capability. They were correct in concluding that turbojet engines made no sense in the available aircraft. Whittle and von Ohain thought in terms of much faster aircraft that approached the speed of sound. Jet engines under these flight conditions looked quite appealing.

Meeting the Inventors

In 1990, Whittle and von Ohain were at a joint propulsion meeting in Orlando, Florida sponsored by the AIAA, the ASME, and the SAE. The peak of my engineering career was when my wife and I enjoyed an evening with the two innovators and their wives (Fig. 1). Previously, I met Dr. Von Ohain a couple of times at Wright Field through my friend Cliff Simpson, chief of the Air-Breathing Propulsion Division at Wright Field. I was also privileged to have met Sir Frank two times—once when I escorted him around the Experimental Assembly Area at Pratt & Whitney and another time when he spoke to the World Affairs Society in Hartford at Carbone's Restaurant.

Gary Plourde, a fellow Pratt & Whitney engineer, who did remarkable work in the late-1960s in the discipline of inlet–engine compatibility, saw my wife Evie and me at the meeting and kindly invited us to dinner with the aviation pioneers. At the table Evie sat next to Sir Frank and I was opposite Dr. von Ohain. Evie mentioned to Sir Frank that I had quite a collection of photographs of him when he was a young RAF officer.

AFTERWORD

Fig. 1 Evie Connors and Dr. von Ohain, with Gary Plourde and Mrs. von Ohain in the background.

She went on to exclaim, "I must say you were a handsome officer!" Whereupon Sir Frank, sporting his Cranfield tie, leaned over and kissed her, replying, "Now you tell me, my dear! Where were you when I was young?" Dr. Von Ohain was somewhat more reserved in his comportment but was full of life as he demonstrated later in Sir Frank's suite when he crooned a few lines from the famous Bavarian drinking song, *In Muenchen Steht Ein Hofbräu Haus*. When he came to the *"Ein, Zwei, g'suffa,"* I suspected, by his animated drinking hand, he had more than a novice's experience in handling beer mugs.

As the evening came to a close, I asked Dr. von Ohain if he ever noticed that the process of innovation seemed to proceed in three phases: You, We, and I.

Phase 1: *Project without honor.* The innovator gets no respect. He or she is treated as a pariah pursuing a crazy, impractical concept that will never work and is just a drain on current resources that could be more wisely spent on something more productive. The pronoun most frequently used in this phase is *you*—tying a specific person to the wasteful project.

Phase 2: *Initial Success.* In this phase the innovator has conducted some experimental work, which indicates the idea has some merit. The innovator notices the change in pronoun when the skeptics ask about his wasteful endeavor, "It's beginning to look as if *we* might have something here."

Phase 3: *Full-scale production.* In the third phase, the innovator is nowhere in sight. The program is in the hands of perhaps the most extreme skeptical critics who now talk casually, "*My* program is going quite well!"

The eminent Herr Professor thought for a moment and then broke out in a smile when he concurred, "Yes, there's some truth in that."

Both Whittle and von Ohain, after their great innovative efforts, were put off to the side while established engine companies took over the development and production of jet engines. In Germany von Ohain went on to design and demonstrate the HeS 011(2900-lb thrust), which was probably the most advanced jet engine in Germany by the time the war ended in 1945. Then he graciously accepted our government's invitation to become an American.

Sir Frank did not make out as well even though he was knighted. The British government took away his company and engine work. Rolls-Royce carried on further developments of Whittle's work. However, he did have the thrill of flying a Gloster Meteor III aircraft powered by his engine technology, telling us at the 1989 AIAA meeting in Dayton, Ohio, "flat out along the high-speed course at Herne Bay at a height of 50 feet."

Golden Eagles

Bruce Torell (standing, center row) former President of Pratt & Whitney, had Fig. 2 taken at the Hartford Golf Club in 2000. This is the gang that got me started on this book. These *Golden Eagles* represent about 1500 man-years of aircraft engine experience. The senior members of this group worked

Fig. 2 The Golden Eagles of Pratt & Whitney.
Row 1: Dave Motyka, Charlie Steffens, Elton Sceggel, Don Jordan, Gordon Bywaters, Frank Murphy, Bill McGaw, John Gavin, Dick Palatine, Jack Connors.
Row 2: Al Oberg, Grady McRae, Hans Stargardter, Lou Daukas, Dick Smith, Frank Manna, Dick Henry, Bruce Torell, Sid Satar, Roy Beckett, Walt Doll, Pete Thomas, Hugh Crim, Don Brendal.
Row 3: Cliff Horne, Dave Phinney, Nils Carlson, Arne Oberg, Art Brown, Bill Martens, Larry Carlson, Jerry Kester, Jim Bruner, Ted Slaiby, Phil Hopper, Bob Toft, Ed Brown, Pete Talbot, Carl Bristol.

for Pratt & Whitney before WWII and made substantial contributions to piston engine development for the war effort. After the war they continued to make great innovations for gas turbines.

Closing Comments

Fifteen chapters ago you received the preflight briefing and now have completed the flight through more than 80 years of Pratt & Whitney's engine history. I hope you have enjoyed the journey, which has been made as pleasant as possible by three crew members (Fig. 3), whom I want to point out and thank for their great efforts.

- Jesse Hendershot, retired Pratt & Whitney Customer Service Center instructor, spent more than 2000 hours as a volunteer over about an eight-year period to organize the Pratt & Whitney Archives. There would have been no extensive history of the company without Jesse's efforts. He also made valuable suggestions in reviewing my drafts.
- Gene Montany, retired vice president of Pratt & Whitney's Strategic Planning Group. Gene contributed material, helped me recall crucial events, and reviewed my drafts to keep me on the factual path of events.
- Pat DuMoulin, AIAA Senior Editor of Books, who bore the brunt of the Herculean effort to reshape my voluminously worded text, photographs, and graphics into a readable and attractive publication.

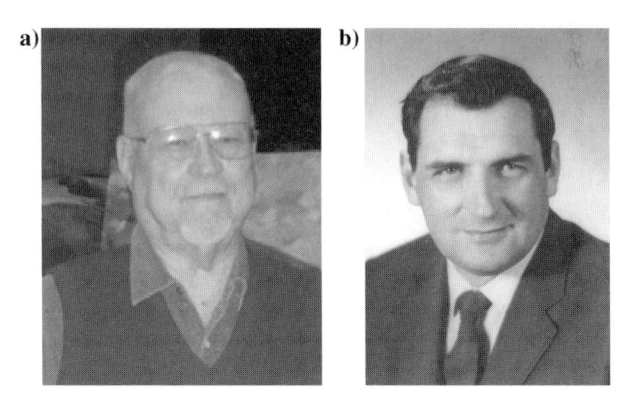

Fig. 3 a) Jesse Hendershot, and b) Gene Montany.

INDEX

Note: Page numbers in *italic* denote figures

AAM. *See* air-to-air missile
Abernethy, Bob
   in bypass bleed cycle, 325–326
ACC. *See* Active clearance control
Active clearance control (ACC), 488
Advanced Medium STOL Transport
      (AMST), 427, 429
   STOL capability of YC-15, 429
Advanced tactical fighter (ATF), 453
   See also F119; F-22
Advanced Technology Engine Gas
      Generator (ATEGG), 453
Advanced Turbine Engine (ATE), 427
Advanced Turbine Engine Gas Generator
      (ATEGG), 398
AEDC. *See* Arnold Engineering
      Development Center
Aeronautical Systems Division (ASD), 388
Affordability, aircraft, 467
Aft fan, 291
   See also thrust augmenter
   aft fan concept of, *292*
Air Commerce Act (1926), 6
Air Mail Act
   of 1925. *See* Kelly Bill
   of 1934. *See* Black-McKellar Bill
Airfoil
   boundary layer, 486
      types of, 486
   complications in engine, 489–492
   controlled diffusion airfoils, 438
   efficiency increase in
      engine, *485*, *486*, 485–486
   loss reduction in engine, *487*
   pressure distribution on, *491*
   turbomachinery vs aircraft, 489

Airframe
   and aircraft engine development
      thumb rule, 404
   technology progress, 478–481
Airline operations
   hub and spoke type, 432
Air-cooled engines, 32, 34
   achievments of Rentschler in, 27
   dominance of, 48
   growth in horsepower, *47*
   horsepower and cylinder displacement, *33*
   vs liquid cool engines, 109–110, *111*
   models of, 145–146
   power increases, *45*
   of 3000-hp, 110–111
   weight/horse power, *32*
   in World War II, *24*
Air-to-air missile (AAM), 334
Air-to-surface missile (ASM), 334
Allison J33, *203*
   See also Nene engine
American Society of Mechanical Engineers
      (ASME), 280, 339
AMST. *See* Advanced Medium STOL
      Transport
Andersen, Bill, 419
Arnold Engineering Development Center
      (AEDC), 378, 387
ASD. *See* Aeronautical Systems Division
ASM. *See* Air-to-surface missile
ASME. *See* American Society of
      Mechanical Engineers
ATE. *See* Advanced Turbine Engine
ATEGG. *See* Advanced Technology Engine
      Gas Generator; Advanced Turbine
      Engine Gas Generator

ATF. *See* Advanced tactical fighter
Aviation
  airframe technology progress in, 478–481
  Collier awards in, 474
  commercial
    DC-3 aircraft in, 479
    engines used in, 478–479
    evolution of, *478*
    Sir Winston Churchill's view on, 6
  engine controls progress in, 492–496
  engine technology progress in, 483–492
  history of, 473–474
  L/D ratios of biplanes in, 481
  military, roles of, 473
  Morrow Committee recommendations for, 6
  pioneers, 48
    R-1340 Wasp, 31, 33
  Pratt & Whitney's contribution to, 479–480
  President Coolidge's views about, 5
  propulsion requirements for, *482*
  "S-curve" development of, 478
  stakeholders, *4*
Axial-flow compressors, 25

B-52 aircraft
  blade length and rpm in, 240
  events of flight of, *239*
  with J57 engine, *243*
  J57 installation problem in, 238
  re-designation of, 241
B-767 twin, 418
"Back of the envelope", 161, 297, 507
Baffles, 35, *36*, *93*
  in cooling mechanisms, 93, *150*
Baseler, Dick, *235*
  Baseler's engine, 235
  in hydromechanical controls, 493
  in Pratt & Whitney, 234–236
Baseler's engine, 235
BBN. *See* Bolt, Beranek, and Newman
Beckwith, Gordon, 365JT9. *See* J91 engine
Black-McKellar Bill, 21
Blade
  of JT9D and JT10D, 434
Blue Two program, 459
BMEP. *See* Brake mean effective pressure

Boeing
  Air Transport Company, 18
  team, 233–234
  367–80 (Dash-80), *253*
Bolt, Beranek, and Newman (BBN), 268
Bootstrap cycle. *See* RL10 cycle
Brake mean effective pressure (BMEP), 37, 39, 149
  for air-cooled engines, 40
  constant-volume combustion, 39
  factors affecting, 39
  as MEP, 40
  range for rotary engines, *41*
Brayton cycle
  and Otto cycle comparison, 284
  steady flow process of, 284
  vs 304 engine, *314*
Brown, Bill, *204*
  in J42 engine, 204
  in J48 afterburner, 210–211
Brown, Walter F., 7
Bursey, Roger, *389*
  in F100, 389
Bypass engine, 291
Bypass ratio, 297

Caldwell, Frank
  in improving Spitfire, 138
  variable pitch propeller by, 137, 475
CANEL. *See* Connecticut Aircraft Nuclear Engine Laboratory
Carnot cycle, 483
Carnot, Nicholas Leonard Sadi
  Carnot cycle, 483
  in conversion of heat into work, 483
CBR. *See* chemical-biological-radiological
Centrifugal
  engine
    progression of turbine blades in, *213*
  flow compressors, 25
  load
    density relationship, 331
Chance Vought F4U-5, 138
  R-2800-32W in, 138
Chemical-biological-radiological (CBR), 459
CJ-805 turbojet, 402
CJ805-23, 402

INDEX 515

Clark, Lionel B.
  along with Doolitle, Bill and school's faculty, 193
  in Pratt & Whitney Aircraft, 189–191
CNS. *See* Common Nacelle System
Commercial Products Division (CPD), 440
Common Nacelle System (CNS), 416
Comparison
  of engine parts in various gas turbines, 433
  of JT9D and JT10D blades, *434*
Compound engine
  compound R-4360, *155, 156*
  conventional supercharged vs simple, *154*
  exhaust gas energy recovery in, 155–158
  turbosupercharged, *155*
Compounding process, 154
Compression ratio, 88
  increased, 88
    X-38 with, 89
    X-39 with, 89
Compressor blade
  improved strength of, 331
Connecticut Aircraft Nuclear Engine Laboratory (CANEL), 283
Connors, Jack, *508, 430*
  in Advanced Turbine Engine (ATE), 427
  as director of marketing–Europe, 421–423
    difference in marketing at Europe, 423
  with Ed Granville, 430
  in IAE, *440*
  in JT10D demonstrator engine, 427
  as Manager of Current Engines, 439
Cost per seat mile, *479*
CPD. *See* Commercial Products Division
Crankcase, *53*, 93
CX-HLS program, 404
Cycle time, 271
Cylinder
  bore improvement, 67
  count, 43, 44
    world record, 44
  displacement comparison, *68*

de Havilland Aircraft Company, 138
de Havilland aircrafts (DHs), 63

Department of Defense (DoD), 4, 428
Designations, engine, 216
Detonation, 42
  conditions of, 42–43
  effect of, 39
  inhibiters, 42
  inlet temperature reduction, 43
    as a safety margin, 43
  intercoolers effect on, 43
  as knock, 42
  octane number, 42
  supercharger effect on, 43
  temperature effect on, 43
Development testing and flight-testing experience, *381*
DHs. *See* de Havilland aircrafts
Direct injection, 143
  water injection, 143–145
Direct Operating Cost (DOC), 371
DOC. *See* Direct Operating Cost
DoD. *See* Department of Defense
Doll, Walt
  in designing compressor for J91, 281
  in JT4 designing, 269–270
  in Pratt & Whitney, 132–134
  in two-spool compressor, 221

EBF. *See* externally blown flaps
Empty to takeoff weight ratio, 480
Engine surge, 176
Engine, aircraft, 482
  airfoil profile of
    aircraft airfoil compared with, 489
    complications in, 489–492
    improving efficiency in, *485*, 485–486
    ways to minimize losses in, *487*
  and airframe improvements comparison, 492
  components of, *378*
  components replacement reasons in, 379
  control requirements complexity, *494*
  controlled endwall blading of, *486*
  controls improvement, 492–496
  costs associated with, 448
  cruise horsepower vs GTOW, *483*
  cycle time, 271
  design goals for
    fighter aircraft, 379
  development
    time line during WWII, *163*

Engine, aircraft (*Continued*)
  development procedure, *377*, 377–378
    altitude test, 378
    flight test, 378
    in flying test bed, 378
    sea-level static testing, 378
  designations, 216
  exhaust application, 429
  FADEC in, 493–496, *495*
  gas path of, *488*
  high-temperature effect on, 486–487
  hydromechanical vs electronic controls of, *495*
  loiter time, 271
  market development, 4–7, 218
    relationship stability, 196
  Nuclear propulsion engine of, 284
    J87 (GE X211), 284
    J91, 284
  possible movement directions, 378–379, *379*
  progress, 483–492
  reached production status, *67*
  structure, 93–94
  technology
    development and progress, 483–492
  thrust-specific fuel consumption in, 487–489, *492*
  ways of achieving design goals, 379
  weight
    comparison, *64*
    reduction calculation, *484*
ETOPS. *See* Extended twin-engine operations
Extended twin-engine operations (ETOPS), 435, 480
Externally blown flaps (EBF), 429

F100 engine, *386, 391, 392*
  demonstrator engine and events, 389–391
  description, 391
  development, 393
  as F100-PW-229, *451*
    specification, 452
    tests in AEDC, 452
  in F-15, *465*
  in F-16, *392*
  with multi-axis thrust vectoring nozzles, *465*
  problem encountered in, 394–395
  product success, 395
  production with NATO nations, 393
  thrust/weight comparison, *393*
  time line, *385*
F119, 452, *455, 457*
  engine description, 455–456
    blisk, 456
    thrust comparison, *456*
    turbomachinery stages comparison, 456
  engine mockup, 462
  events in, 453–455, *458*
  Gillette, Frank in, 453
    2006 Collier Award, 454
  IPD benefits in, 462–463
  stealth airplane, 454
  supersonic persistence, 454
  time line, *453*
  trade-off studies, *453*
F135 engine
  applications for, 469
  Collier award (2001), 469
  description, 467–468
  multi-tasking engine, *468*
  performance summary, 469
  propulsion system, *468*
  time line, *467*
F15 Eagle, 385
F-22, 457, *458*
FAA. *See* Federal Aviation Authority
FADEC. *See* Full authority digital electronic control
Federal Aviation Authority (FAA), 480
Flight test
  aircrafts of Pratt & Whitney, *249*
  J57-powered aircraft for, 248
  pilot and engineers, *249*
Florida Research & Development Center (FRDC), 315, 329, *336*
Flying boats, 80, 82
Flying machine, early, *62*
FOD. *See* foreign object damage
Ford Motor Company, 162
Foreign object damage (FOD), 206
Four-cycle engine. *See* Otto cycle
FRDC. *See* Florida Research & Development Center
Free-piston gas generator, 164–166
  free-piston compressor, *165, 168*
Free-piston gas turbine
  concept of, 163
  contract with MIT, 163–164

INDEX 517

experimental experience, 169
features of, 166
first gas turbine engine program, 166
in four-engine bomber, 164
single-engine pursuit plane, 164
supercharged diesel and turbine concept, 167
French engine company, 427–428
Front-fan turbofan concept of, *292*
Full authority digital electronic control (FADEC), 433, 493–496
evolution of, *494*

Gas turbine engine era, 1, 476–478
Collier awards time line in, *476*
General Electric achievements in, 477
JT9D turbofan engine in, 477
piston engine era vs, 2–3
Pratt & Whitney gas turbine engines, 476
technology growth in, 477–478
Gas turbine engine, 61
blade, improved strength of, 331
business of, 9, 182–183, 213–214, 227, 476, 499
etymology of, 161
parts comparison in various, *433*
potential, 172
propeller engine
early research facilities in, 174, 175, 178
Hobbs in, 173
Pratt in, 174
PT1, 166, 167
PT2 engine, 176, *181*
Technical & Research Group for, 174
temperature and pressure ratio, *488*
Willgoos, Andrew in, 61
General Electric company
nuclear propulsion engine of, 284
progress in gas turbines, 197–198
seizing U.S. Air Force market, 25
turbofan production in, 300
German company BMW, 69
Gillette, Frank, *454*
in F119, 453
2006 Collier Award, 454
Granville, Ed, *236*
contribution in J57, 237
Great Engine War, 393–395

Gross take-off weight (GTOW), 172
vs cruise horsepower, 172
Grumman, 200
in founding panther, 201
GTOW. *See* gross take-off weight
Guggenheim, Daniel, 6
in developing aviation, 8
in financing Lindbergh's trip, 6, 8

Hartford Graduate Center, *372*
Heat engine
baffles, 35, *36*
efficiency of, 34–35
heat transfer coefficient of, 35
heat transfer rate in, 35
surface to volume ratio in, 38, *39*
Hendershot, Jesse, 141, *511*
High-speed aircraft propulsion, 369
Hobbs, Leonard "Luke", *23, 61, 237, 242*
in aircraft gas turbines, 123
Army Corps of Engineers, 61
brainchild engine, 99
criterion, 258
engine projects proposed by, 215
philosophy in T45 program, 219
reflections on jet engines, 254–255
into turbine–propeller engine, 173
Hollow exhaust valve, 95
Hooker, Stanley
analysis on twin-spool, 222
Hornet engine, 19, 51, 114, 171, 498,
R-1690, 68–69, 75, 76, 77–78, 147
R-1860, *79*, 80–81, 180
R-2180, *96*, 98, 99, 115, *119*, 119–120
Horsepower, 37
equation for, 37
octane number and, *42*
octane rating effect on, 42
reasons for increased, 45

IAE. *See* International Aero Engines
ID. *See* Inner diameter
Ideal cylinder. *See* square cylinder
IED. *See* Initial Engine Development
IEDP. *See* Initial Engine Development Program
ILFPS. *See* LiftFan™ Propulsion System
Initial Engine Development (IED), 389
Initial Engine Development Program (IEDP), 385
Inlet temperature reduction, 43

Inner diameter (ID), 228
Integrated Product Development (IPD), 458
    benefits, 462–463
    engine mockup, 461, *462*
    Reliability and Maintenance
            (R&M), 459
        goals, 459
        to increase, 459
    support tool involvement, 460–461
        lifting adapter tool, *462*
    supportability awareness, 459
    supportability reviews, 460
        display board, *460*
        supportable design, *461*
Intercoolers, 43
    benefits, 43
International Aero Engines (IAE), 439
    members of collaboration, *440*
    V2500 engine, 441
        applications for, 442
        description, 441
        division of responsibilities, 441
        events associated with, *442*
        specification of, 441
        time line, 441–442
International Fuel Cells, 374
    fuel cell powerplants, 374
IPD. See Integrated Product Development

J42 engine
    See also Nene engine
    data at sea level, 207
    efficiency of, 206–207
    F9F-2 Panthers' efficiency with, *208*
    overhauls and foreign object damage in, 205–206
    significance of, 206
    time line, *205*
J48 engine
    with afterburner, *209*
    without afterburner, *210*
    applications of, 210
    configurations of, 209
    in Cougar, *210*
    data at sea level, 212
    in F-93A, *211*
    in F-94C, *211*
J52 engine
    with afterburner, *279*
    airfoil efficiency improvement, 274–275

applications of, 272
    military, 275–278
burner assembly of, *274*
dash 408 version of, 278–279
description of, 272
general configuration of, *273*
in Grumman A-6A Intruder, *278*
in Hound Dog missile, *276*
low turbine of, 275
non-afterburning, *273*
plug nozzle of J52-P-3, *277*
in Skyhawk, *277*
specifications of, 272
time line, *271*
two-spool engine design of, *273*
J57 engine, *233*
    applications of, 242–245, 246–247
    in Convair F-102, *245*
    description of, 232–233
    in Douglas A3D Skywarrior, *245*
    in F-100, *245*
    in F-101 with speed record, *250*, 251
    flight testing, 248
    genesis of, 216–218
    as J57-P-23 with afterburner, *244*
    in KC-135, 243
    in Lockheed U-2, *244*
    principal engineers of, *223*
    problems in installation of, 238
    roller bearing problem in, 230–231 (*See also* Materials research and development department)
    sequence of events in, 231–232
    specifications of, 246–247
    success of wasp waist, 237–238
    test run of, 233–234
J58 engine, *322, 327*
    applications for, 332
    in Blackbird, *333*
    with bypass bleed cycle, *327*
    J58 compressor map, *325*
    J58-P-2 engine, *324*
    at Mach (3.2), 324
    McDermott, Jack in, *323*
    military version of, 324
    speed of, 332
    time line, *322*
J60. *See* JT12 engine
J75 (JT4) engine, 257, *259*
    applications of, 261, 262–263, 264–268

comparison, 259, *260*
description of, 258–259
in F-105, *266*
in F-106, *267*
J75 program progress, *261*
J75-P-13 in U-2B and U-2C, *264*
redesigning of, 259–260
Rickenbacker, Eddie for, 257
specifications of, 262–263
J91 engine, *279*
compressor configuration of, 281
description of, 281–283
nine-stage transonic compressor of, *282*
as nuclear propulsion engine, 284–285
performance, 282
time line, *280*
X-291 as, *283*
Jet engine
incorporating split compressors, 222
Jet fuel (JP), 312
Jet-powered aircrafts
noise suppressors in, 269
permissible noise levels for initial regulations in, 268
unit of noise measurement PNdb, 268
Jet propulsion
European developments in, 199–200, *200*
Jet turbine (JT), 215
Johnson, Kelly
in reconnaissance aircraft configurations, 312
U-2 development program, 312
Joint strike fighter (JSF), 466
JP. *See* Jet fuel
JSF. *See* joint strike fighter
JT. *See* Jet turbine
JT10 turbofan, *294*
See also PT5 turboprop engine; J75 engine
with afterburner, *296*
designation of, 295
experimental engine of, *297*
tailpipe of, *297*
JT10D
See also PW2000 engine
Douglas, McDonnell in, 429–430
Fiat in, 440

MTU in, 440
need of foreign partners, 431
Rolls Royce in, 432
study data, 432
time line, *428*
JT11-20, 324, 329
JT12 engine, *286*
applications of, 286, 287
bird-proof, 286–289
description of, 285
in Jetstar, *288*
modified as, 289
JFTD12 turboshaft engine, *289*, *290*
in Sabreliner, *288*
specifications of, 287
time line, *285*
JT3 turbojet
See also JT3D
comparison with JT3 turbofan, *304*
conversion into JT3D, 301–303
JT3-10A engine model
configuration of, 231
JT3A engine
See also J57
evolution of, 217
JT3C in a Boeing 707 nacelle, *254*
JT3D, *305*
applications of, 306, 307
commercial, 309
military, 307–308
in C-141 Starlifter, *307*
in Douglas DC-8-50, *309*
in RB-57F, *308*
serendipitous outcome of, 307
specifications of, 306
time line, *301*
X-248 as, *302*
JT4. *See* J75 engine
JT8 (J52), *341*
in B-727, 348–352
role of Jordan, Don, *351*, 351–352
role of Taylor, Stan, *350*, 350–351
JT8. *See* J52 engine
JT8D engine, *355*
achievements, *364*
as center engine, *357*
in B-727-200, *357*
in B-737, 358
commercial applications, 355, 356
in Super Caravelle, *355*

JT8D engine (*Continued*)
  comparison with JT3D, *354*
  in Douglas AMST, 363
  engine description of, 353–354
  exhaust mixer, *361*
  inflight shutdown rate, 362
  as JT8D-200, *358, 359*
    applications, 360
    specifications, 360
  as JT8D-209, *359*
  JT8D-17 and JT8D-209, 360
  origin of, *354*
  specification of, 356
  time line, *353*
JT9D engine, *410, 411, 412*
  accumulated engine flight
    hours, *414*
  in B-747, 412
  Collier Award (1970), *423*
  compared with JT3D, 410–411
  complications in development
    increase in GTOW, 406–407, *407*
    financial predicament at Boing, 407
  description, 409–412
  development events in, 415
  flight test, 412
    in B-52, 412
  JT9D family, 415–416
    models with increasing thrust, *416*
    rerate engines, *416*
  in passenger service, 415
  safety record of, 415
  specification and applications of, 413
  start of concept, 405–406
  time line, *398*
  Toft, Bob, *408*, 408–409
JT9D-7R4
  in A310, 421, 422
  in B-767, 421
  campaign at United Airlines, 418–419
    Andersen, Bill, *418*
    Dolores, Pinwar, 419
    Latimer, Howie, 419
    Rudolph, Don, *418*
    Wagner, Terry, 419
  guaranteed maintenance plan for
    airlines, 420
  international marketing, 421
  marketing plan, 420
  provision for non-stop transcontinental,
    420

Rosati, Bob, *418*
  presentation at United Airlines, 421
JTF10A (TF30), *342, 346*
  See also TF30
  applications of, 344–345, 346
  in F-111A, *347*
  JTF10A-1 (TF30-P-2), *343*, 346
  specifications of, 344–345
  TF30-P-412A in F-14, *348*
  time line, *342*
JTF14
  -E for C-5, 402
  engine description, 403
  demonstrator engine, *402, 403*
  specification of, 402
JTF16 lift cruise engine, *399*
  and TF30, *399*
JTF17 duct-burning turbofan, *380, 381*
  engine description, 380–381

Kelly Bill, 6
Kennedy, Jim, *422*

L/D. *See* Lift/Drag
LACE. *See* Liquid air condensing
  engine (LACE)
Lawrance, Charles, *12*
  contribution in air cooled
    engine, 474–475
Lethality, aircraft, 467
Liberty engine, 474
Lift/Drag (L/D), *481*
LiftFan™ Propulsion System (ILFPS), 469
Light Weight Gas Generator (LWGG), 397
  demonstrator engine, *398*
Light-weight fighter (LWF), 392
Lindbergh, Charles
  Atlantic crossing by, 6, 8
  "Lucky Lindy" travel by, 8
  in Pratt & Whitney, 195
Line replaceable units (LRUs), 448
Lippincott, Harvey, *195*
  in Pratt & Whitney, 194
Liquid air condensing engine
  (LACE), 280
Liquid-cooled engines
  comparison with air-cooled engines,
    147–149
  cylinder sizes of, *149*
  Fedden's report on, 157–158
  growth in horsepower, *47*

INDEX   521

H-2600 Sleeve Valve Engine, *147*
H-3130/H-3730 Sleeve Valve Engine,
    146–147, *147, 148*
Packard 2500 engine, 33
R-2060 Yellow Jacket, *146*
weight/horse power, *32*
Wright E-4 engine, 32
Liquid-cooling system, 32
Lockheed Martin Corporation, 25, 190,
    452, 466, 468, 469
Loiter time, 271
LRUs. *See* line replaceable units
LWF. *See* light-weight fighter
LWGG. *See* Light Weight Gas Generator

Mach speed
  afterburning J58, 324
  airflow and pressure ratio correction,
    324–325
  bypass bleed cycle, 325–326, 326, 332
  first by J58, 324
  jet nozzle, 327–328
    blow-in door ejector, *328, 329*
    functions, 328
    primary flow, 328
    secondary air flow, 328
    tertiary [external] air flow, 328
  Mach 3 cruise capability, 324
  non-afterburning J58, 324
  pulsetrap, 329
  pure ramjet, 325
  take-off and Mach 3 operating
    points, *325*
  thermal efficiency, 325
Mack Truck Company, 10
MacRobertson air race, 5
Master rod, 51, *52*
Materials research and development
    department, 337–339
  directionally solidified metal, 337, *338*
  Gatorizing, 338
  powder metallurgy, 338–339
  single-crystal, 337, *338*
  Titanium, 337 (*See also* J57 engine)
McDermott, Jack, 323
  in J58 program, 324
  in R-2800 program, 323
McNary-Watres Act of 1930, 7
Mead, George J., 8, *11*
  article about, 58
  epilogue for, 58–60

flying experience with Wasp, 57,
    57–58
ideal cylinder of, 31–32, 38
initiated twin-row radial air-cooled
    engine, 66
in internal combustion engine, 31
and Hobbs, 23
Mead-Wright 12-cylinder, *50*
in R-1340 WASP, 31, 33–34
with Rentschler, 31
set standard, 63–64
and T (Tornado) engines, 49
  T-3 engine, 50
in Wasp, 50–57
at Wright Aeronautical, 12
in WWII, 159
Mean effective pressure (MEP), 37
MEP. *See* mean effective pressure
MHI. *See* Mitsubishi Heavy Industries
Mitsubishi Heavy Industries (MHI), 449
Montany, Gene, 235, 275, 333, 349, 367,
    409, 431, *511*
Motoren & Turbinen Union (MTU), 332
MTU. *See* Motoren & Turbinen Union

NACA. *See* National Advisory Committee
    for Aeronautics
NAMs. *See* nautical air miles
National Advisory Committee for
    Aeronautics (NACA), 59
Nautical air miles (NAMs), 417
Nene engine, 26, *203, 204*
  description of, 203–204
  flow path, 203
  Grumman and, 200–201
  as J42, 204–205
NEPA. *See* Nuclear Energy for Propulsion
    of Aircraft
Niles-Bement-Pond Company, 14, 15
Nine-cylinder Lawrance J-1, 41
Noise level
  definition, 269
  perceived (PNdb), 268
Nuclear Energy for Propulsion of Aircraft
    (NEPA), 283
Nuclear Propulsion engine, 284
  J87 (GE X211), 284
  J91, 284

Octane numbers, 42
OD. *See* Outer diameter

Otto cycle, 34, *35*, 36, *37*, 483
  four steps in, 484
  schematic representation of, *35*
Outer diameter (OD), 228

P&WA. *See* Pratt & Whitney Aircraft
Parkins Ring Cowl, 66
Parkins, Wright, *65*
  in designing wasp waist J57s, 229–231
  in engine drag reduction, 66
    Parkins Ring Cowl, 66
  in Pratt & Whitney, 65
  in R-2000 engine's crankcase endurance, 101, 107
Patrol bomber, 129
Perceived noise level in decibels (PNdb), 268
Performance numbers (PN), 42
  See also octane numbers
PFRT. *See* Preliminary Flight Rating Test
Piston
  etymology of, 34
  working principle of, 34
Piston engine, 34, *37*, 61
  compounding process in, 154
  constant-volume combustion, 39
  cylinder development progress in, *150*
  cylinder size limitation, 38
  cylinders quantity in, 43, 44
  detonation effect on, 39
  detonation inhibiters in, 42
  development in, 150–154
  displacement per cylinder, 43, *44*
  energy distribution in, *153*
  evolution of, *152*
  fuel consumption of, *157*
  fuels improvement in, 150, *151*
  horsepower of, 37
    growth in, *149*
  inlet temperature reduction in, 43
  lead plated and indium flashed, 140–141
  octane number and horsepower of, 42
  Otto cycle in, 34, 35, *37*
  piston diameter effect on, 38
  piston speed, 44
  power stroke and flame fronts in, *36*, 38
  rotational speed, 44
  stroke and bore, *38*
  surface/volume effect on, 38, 39
  test hours required to develop, *153*
  time between overhauls (TBOs) for, *152*
  Willgoos, Andrew in, 61
Piston engine era, 474–476
  Caldwell, Frank in, role of, 475
  Lawrance's, Charles, contribution in, 474–475
  Morse, Sanford in, role of, 475
  time line of Collier awards in, *475*
Pitch-up, 251
PN. *See* Performance numbers
PNdb. *See* perceived noise level in decibels
Podolny, Bill, *373*, 373–374
  and fuel cell, *374*
Pratt & Whitney
  achievements on 40th anniversary, 397
  at Aero Propulsion Laboratory, 382–388
  Aircraft (P&WA), 62–64, 66, 80, 82, 98, 109–110,
  aircraft gas turbine business of, 9, 182–183, 213–214, 227, 476, 499
  Aircraft Service School, 191–192
  army and navy engine training school by, 192–194, *193*
  with automobile companies, 498
  in aviation, *502*
  BMW in, 3
  business strategies, 15, 26–27, *311*
  in Canada, 285
  CANEL facility, 283–284
  for competitors, 198–199
  dependable engine at, 507
  engine designations at, 215–216
  engine projects of, 17, 26, 140–141, 221, 284, 293, 310, 496, 499
  facility expansion, 23, 334–337
  FADEC of, 493–494
  Fiat in, 440
  finest moments of, 8–9, 496–497, *497*
  flight operations engineering and, 195–196
  flight test organization, 125, *126*, 127–130
  founding of, 8–11, 15, 497–498
  French company and, 342
  vs GE, 393–395
  materials research and development department, 337–339
  medallion, *503, 504*

INDEX 523

missile airframe design work of, 334
Mitsubishi in, 3
MTU in, 440
name origin, 16
Rolls-Royce and, 202, 432, 435
service department of, 189, 194
slips, 172, 382, 394
Tool Company, 9, 99, 503
UA&TC by, 9, 18–22
Pratt, Perry W., 121, *122*
  in aviation industry, 280
  developer of R-2800C engine, 121
  in jet engine business, 121
  with Hobbs, 123
  note on JT3-10A, 228
  with Parkins, Wright, 123
  in Pratt & Whitney aircraft gas turbines, 123
Preliminary Flight Rating Test (PFRT), 330
Prop turbine (PT), 215
Propeller mechanism
  principle of, 161
Propeller reduction gear, *94*
Propulsion
  by Heinkel, Ernst, 507–508, 510
    and von Ohain, Hans, 508, 510
  progress, *496*
  by Whittle, Frank, 508, 510
Propulsive efficiency, 220
PT. *See* Prop turbine
PT1, 166, *167*, *170*
  estimated performance of, *171*
  not "abomination", 171
  vs piston engine performance, 170
  power plant under test, 169
  Soderberg's contribution in, 169
  turbine experimental experience, 169
PT2 engine, *176*, *181*
  applications for, 178–179, 182
  in B-17, *181*
  competitive specific weight of, *180*
  engine description of, 176–177
  engine surge, 176
  equivalent shaft hp vs year, *179*
  equivalent shaft specific fuel consumption, *180*
  flight test of, 181–182
  negative torque sensor in, 182
  power output of, 177
  take-off power in, 179
  test run events, 178
  time line of, *177*
PT3 engine
  See also J52 engine
  description of, 270
  military designation of, 270
  progress in developing, 270
PT5 turboprop engine, *293*
  Coar, Dick in, 294, *296*
  comparison with turbojet JT3, *293*
  development program of, 294
  flight test, *294*
  mockup of, *220*
  Mulready, Dick designer of, *296*
  reduction gear of, *295*
  time line, *294*
Pulitzer Race, *48*
PW1000G
  design principle, 470–471
  features of, 470
  noise levels, 472
  PurePower™, 469
  reduction gear system, *470*, 472
  turbomachinery stages, 471
PW2000 engine, 434
PW2037, *437*, *438*
  applications for, 439
  in C-17, 436, *439*
  as F117-PW-100, 436
  in Delta 757-200, 436
  engine description
    clearance reduction, 438
    with controlled diffusion airfoils, 438
    with FADEC, 437
    gyroscopic deflection reduction, 438–439
  engine models, 437
  in fuel economy, 435
  in Ilyushin-96, 439
  sequence of events in, *436*
PW4000 engines, *443*, *444*, *445*
  in A300-600, *447*
  in A310-300, *447*
  applications for, 445–447
  in B-747, *447*
  in B-777, *446*
  description, 443–445
  in MD-11, *447*
  PW4052 engine, 442, *443*
  specification of, 447
  turbomachinery stages, 450

PW6000, 448, *450*, *451*
  in A318, *451*
  applications for, 451
  features of, 449–450
  with MHI and MTU, 449
  specifications of, 448
  time line, *449*
  turbomachinery stages, *449*, 450

R&D. *See* research & development
R&M. *See* Reliability and Maintenance
R-1340 Wasp, 16–18, 68, *69*
  in air mail service, 18
  air-cooled radial, 33
    hp vs displacement, 33
    goal of, 33
  applications for, 70–74
  birthplace of, 17
  in Boeing Air Transport
      Company, 18
  etymology of, 17
  with fuel injection, *144*
  in military, 18
  specifications of, 69
  Vought, Chance in, 17
R-1535 engine
  performance curves to Howard
      Hughes, 141
R-1535 Twin Wasp Jr. engine, 95–96
  applications for, 97
  in Howard Hughes' H-1 Racer, 96
  specifications of, 96
R-1690 Hornet A, 68–69
  applications for, 77–78
  in BMW, 69
  features of, 69, 75
  in Junkers W34, *75*
  late-model Hornet, *76*
  in Sikorsky S-42, *76*
  specifications of, 75
R-1830 Twin Wasp engine, 83,
      89, *111*
  applications for, 90–92
  casing of, 93
  efficiency of, 130–132
  engine structure, 93–94
  hollow exhaust valve in, *95*
  in Luftwaffe, 138–139
  master rod bearing problem
      in, 139
  propeller reduction gear of, 94

specifications of, 89
and supercharger, 133–134
valve lubrication in, automatic, 94
R-1860 Hornet B, *79*
  applications for, 81
  bore size limiting factor, *79*
  specifications of, 80
R-2000 Twin Wasp engine
  applications for, 108
  difference in features of, 99, 101
  specifications of, 107
R-2180A Twin Hornet A engine, *96*, 98
  applications for, 99
  specifications of, 98
R-2180E engine, *119*
  applications for, 120
  description, 115, 119–120
  specifications of, 119
R-2270 twin-row engine, experimental,
      83, *88*
  specifications of, 88
R-2800 Double Wasp engine, 99, *100*
  applications for, 102–106
  features of, 99, 101
  in Grumman Hellcat, *100*
  in Republic P-47, *100*
  specifications of, 101
  in Vought Corsair, *100*
R-2800C program
  Perry Pratt experience in, 136–137
R-4360 engine, *107*, *111*, *113*
  cooling system improvement, 112,
  as "corncob engine," 107
  exhaust gas energy recovery,
      155–158
  features of, 111–112
  fuel injection in, 112–113
  'rocker boxes' orientation, *112*
  specifications of, 115
  test flight, 127–130, *128*
  under test, *114*
  Wasp Major applications, 116–118
  X-108 in first flight, *115*
  as X-Wasp, 111
R-985 Wasp Jr., *82*
  applications for, 84–87
  specifications of, 83
Radial engines, 39, *41*
  BMEP for, *41*
  improvement in air-cooled, 66
  weight per horsepower for, *46*

INDEX 525

Rapid Solidification Rate (RSR), 338–339
Regression Simulation of Turbine Engine Performance (RSTEP), 372
Reliability and Maintenance (R&M), 459
   goals, 459
   to increase, 459
   supportability awareness, 459
Rentschler, Frederick B., 1, *10*
   accomplishments of, 27–28
   "defining moments" in the period of, 8–9
   final words of, 28
   founding Pratt & Whitney, 8–9, 14
   legacy of, 10, 500–502
   1945 business plan, 26–27
   Renegotiation Act of, 498–499
   UA&TC, 9, 19, 22
   United Aircraft Corporation and, 21
   in Wright Aeronautical, 12–14
   in Wright-Martin Company, 9–11
Request for Proposal (RFP), 386
Required Operation Capability (ROC) document, 427
Rerate engines, 416
   thrust specific fuel consumption (TSFC), *417*
   variation based on thrust size, *417*
Research & development (R&D), 428
Return on Investment (ROI), 371
RFP. *See* Request for Proposal
Rigs, 134
RL-10
   cycle, 320 (*See also* 304 heat exchanger)
   as National Historic Mechanical Engineering Landmark, 339
   rocket engine, *320*, 321
ROC document. *See* Required Operation Capability (ROC) document
Rocket engine, *320, 321*
   See also Suntan project
   achievements of, 321
   test run of, 321
   working principle of, 319–320
ROI. *See* Return on Investment
Rolls-Royce
   Conway bypass engine, 297, 300
   progress, *201*
Rosati, Bob, 418
Rotary engine, 39, *40*
   BMEP for, 41
   features of, 40
   weight/horsepower for, *46*

RSR. *See* Rapid Solidification Rate
RSTEP. *See* Regression Simulation of Turbine Engine Performance
Rudolph, Don, 419

SAA. *See* South African Airways
SAE. *See* Society of Automotive Engineers
Second generation turbofan, *421*
Sens, Bill
   at Pratt & Whitney, *300*, 300–301
   view on liquid hydrogen fuel, 312–313
Short take-off and vertical landing (STOVL), 466
Sleeve, 146
Smith, Arthur, *186*
   in Pratt & Whitney, 185–186
Society of Automotive Engineers (SAE), 34
Soderberg, C. R., *255*
   in compressor design, 227
   and Pratt on JT3-10A, 228
South African Airways (SAA), 138
Space Power Opportunity Planning Study (SPOPS II), 368
Space propulsion
   study conclusions, 369
Space Propulsion Opportunity Planning Study (SPOPS), 368
Space technology, 372
   Apollo fuel cell, *374*, 374–376
Space Transportation Booster Engine (STBE), 372
Specific fuel consumption, 298
Specific weight, 45
SPOPS II. *See* Space Power Opportunity Planning Study
SPOPS. *See* Space Propulsion Opportunity Planning Study
Square cylinder, 31, 38
SST. *See* supersonic transport aircraft
STBE. *See* Space Transportation Booster Engine
Stearman Aircraft Company, 19
STF200
   compared to the JT3D, *401*
   higher bypass turbofan in, 401
   advantage of, 401–402
STF200C, *400*
   variations, 401
STOVL. *See* short take-off and vertical landing

Suntan program
  See also 304 project
  morphed into the RL10 program, 335
  time line, *322*
Superchargers, 43, 133, *56*
  higher efficiency in, 150
  improving, *151*
Supersonic afterburner, 251
Supersonic transport aircraft (SST), 370
Supersonic transport engine program, 379
  conclusions of, 371
  objectives of, 370
  parameters in, 370–371
  phases of, 370
  regression analysis, 371–372
    in engine cycles optimization, 371–372
    in rocket engine optimization, 372
Supportability, aircraft, 467
Survivability, aircraft, 467

T (Tornado) engines, 49
T57. *See* PT5 turboprop engine
Tactical Fighter Experimental (TFX), 346, 385
TALON X. *See* Technology for Advanced Low Nitrogen Oxide
Tay, 202
  See also Nene engine
  design of, 208–209
Taylor, Stan, *350*
  in J52 for B-727, 350–351
TBOs. *See* time between overhauls
TEA. *See* Tri-ethyl-aluminum
Technology for Advanced Low Nitrogen Oxide (TALON X), 472
Temperature and pressure ratio, *488*
Test pilot, 126, 145, 248
TF30 low-pressure fan/compressor, *343*, 346
TF30-P-8 nonafterburning engine, *346*
TF33. *See* JT3D
TFX. *See* Tactical Fighter Experimental
Thermal efficiency, 220
304 engine
  See also RL10
  with afterburner, *313*
  vs Brayton cycle, *314*
  concerns and properties of fuel in, 314
  18-stage hydrogen turbine in, *315*
  engine cycle for, *313*
  heat exchanger in, 314, *315*
  on test in FRDC, 315–316, *316*
304 project, 310
  etymology of, 311
  failure in, 316–317
  time line, *311*
Thrust augmenter, 402
  See also aft fan
Thrust specific fuel consumption (TSFC), 301, *417*
Thrust vectoring, 463, *464*, 466
  Burse, Roger, 463
"Tillie". *See* Theose E. Tillinghast
Tillinghast, Theose E., 120, *121*
  contributions of, 120
  General William E. Mitchell Memorial Award to, 121
  in Pratt & Whitney, 120
Time between overhauls (TBOs), *152*, 397
  of Pratt & Whitney engines, 397
Transportation milestone, 474
Tri-ethyl-aluminum (TEA), 280
TSFC. *See* Thrust specific fuel consumption
Turbine. *See* gas turbine engine
Turbofan, 291
  applications of, 307
  birth of, 291–292
  comparison with turbojet and turboprop, 297–298
  first-generation, 401
    compared with second generation, 411
  front-fan concept, 292
  JT10 after burning, 294–296
  Parkins's approach in, 298–299
  second generation
    time line, *421*
Turbojet. *See* Jet turbine
  time line, *258*
Turboprop. *See* Prop turbine
Twin-row radial air-cooled engine, 66
Twin-spool turbojet
  JT3-8 engine, 224
    description of, 224–225
    high rotor from, *225*
  JT3-10 engine
    description of, 226
    experimental engine X-184, *226*

UA&TC. *See* United Aircraft & Transport Corporation

UARL. *See* United Aircraft Research Laboratory
United Aircraft & Transport Corporation (UA&TC), 9
  breakup of, 19–22
  coast-to-coast service in, 21
  formation of, 18–19
United Aircraft Corporation, 21, 172
United Aircraft Research Laboratory (UARL), 298
Upper surface blowing (USB), 429
USB. *See* upper surface blowing

V2500 engine, 441
  applications for, 442
  description, 441
  division of responsibilities, 441
  events associated with, *442*
  specification of, 441
  time line, 441–442
Valve lubrication, automatic, 94
von Ohain engine, 199

Warden, Henry E., *240*
  efforts in launching B-52, 239–242
Waring, Dana, 134–136
Wasp engine, 16–18, 26, 32, 50–58, 66–68, 79, 142, 143, 171, 180, 216, 237, 464, 498
  accessory and blower section of, 54–55, *55*
  cylinder head of, *54*
  forged aluminum crankcase of, *53*
  installation in Corsair O2U, *56*, 56–57
  major parts of, 51
  master rod design in, 51
  nose section in, *53*
  one-piece master rod in, *52*
  power section of, 54, *55*
  quality of, 56
  R-1340, 33, 36, 68, 69, 70–74, *144*, 147,
  R-1535, 84–87, 95–96, 97, 141
  R-1830, 83, *89*, 90–92, 93–95, *111*, 130–134, 138–139, 148
  R-2000, 99, 101, 107, 108
  R-2800, 99, *100*, *101*, 102–106, 107, 108, 136–137, 138
  R-4360, *107*, *111*, 112, *113*, *114*, *115*, 116–118, 127–130, 149, 155–158
  R-985, *82*, 83, 84–87, 133, 190
  source of whine in, 142–143

  supercharger of, *56*
  two-piece crankshaft in, *52*
Wasp Jr. *See* R-1535
  See also Wasp engine
Wasp Major. *See* R-4360
  See also Wasp engine
Wasp R-1340, 16–18
Wasp waist, 216, 228, 229, 233
  J57s roller bearing problem in, 230–231
  Brook, Niles insight on, 230
Water injection, 143
Water-cooled engines, 32–33
  See also liquid-cooled engines
  growth in horsepower, 47
WBS system. *See* Work Breakdown Structure (WBS) system
Weaver, Bill, *383*
Wells, Ed, 434–435
Westinghouse
  in axial-flow turbojets, *198*
  in developing gas turbines, 197, 198
  in Navy turbojets, 198
  seizing U.S. Navy market, 25
Whittle engine, 199
  See also centrifugal engine
Whittle, Frank
  aft fan concept of, *292*
  front-fan turbofan concept of, *292*
Willgoos, Andrew Van Dean, *17*, *183*
  honoured posthumously, *183*, 184–185
  in Wasp R-1340, 16–17
William Boeing, 9
  in forming UA&TC, 9, 19
Work Breakdown Structure (WBS) system, 409
World War II (WWII), 1, 22, 23, 27, 182, 185, 196
  engine production, 60
Wright Aeronautical Corporation, 10, 12, 15, 18, 19, 24, 27, 48, 68, 154, 157, 218, 223
  air-cooled engines, 33, 40
  liquid-cooled engines, 31–32, 44
  in World War II, 27, 476
Wright-Martin Company, 10–11, 31, 32, 49, 184
WWII. *See* World War II

X-176, *225*

YC-15 demonstrating STOL capability, *429*

SUPPORTING MATERIALS

Many of the topics introduced in this book are discussed in more detail in other AIAA publications. For a complete listing of titles in the Library of Flight Series, as well as other AIAA publications, please visit www.aiaa.org.

AIAA is committed to devoting resources to the education of both practicing and future aerospace professionals. In 1996, the AIAA Foundation was founded. Its programs enhance scientific literacy and advance the arts and sciences of aerospace. For more information, please visit www.aiaafoundation.org.